Strum

1972

Basic Linear Networks

for Electrical and Electronics Engineers

Holt, Rinehart and Winston Series
in Electrical Engineering, Electronics, and Systems

Other Books in the Series:

George R. Cooper and Clare D. McGillem, Methods of Signal and System Analysis
Samuel Seely, Electronic Circuits
Mohammed S. Ghausi and John J. Kelly, Introduction to Distributed-Parameter Network: With Application to Integrated Circuits
Benjamin J. Leon, Lumped Systems
Shu-Park Chan, Introductory Topological Analysis of Electrical Networks
C. T. Chen, Introduction to Linear System Theory
Roger A. Holmes, Physical Principles of Solid State Devices

Basic Linear Networks

for Electrical and Electronics Engineers

BENJAMIN J. LEON

Professor of Electrical Engineering

Purdue University

PAUL A. WINTZ

Associate Professor of Electrical Engineering

Purdue University

HOLT, RINEHART AND WINSTON, INC.

New York Chicago San Francisco Atlanta
Dallas Montreal Toronto London Sydney

Library of Congress Catalog Card Number: 71–125461
SBN: 03–078325–9
Printed in the United States of America
0 1 2 3 22 9 8 7 6 5 4 3 2 1

Preface

This book was written for a first course in circuit analysis. With present-day integrated circuits, both monolithic and hybrid, RLC circuits and linear active circuits that are the models of transistors must be considered together; the authors believe that this togetherness of active and passive elements must be taught from the start. Thus, this text considers the basic building blocks for electronic networks to be R, L, C; controlled sources; and independent sources. Even in the chapter on filters (Chapter 9), the passive filters are presented with a view toward their use as coupling networks between amplifying devices. In all chapters, networks with controlled sources are used extensively.

Throughout the text, general methods are emphasized. Special techniques that apply to restricted classes of networks are given only to illustrate points in the general theory and its applications. The examples are precisely that—examples to illustrate concepts. The student should not look for a catalog of tricks. Instead, he should develop special techniques from the general theory, although it is, of course, often best at the sophomore level to work with special simple cases before going to the general case. This text tries to balance generality with learning in simple steps.

The text was written for a one-semester course to be taken during the second semester of the sophomore year. The students will have completed one and a half years of calculus and will concurrently be taking a mathematics course in differential equations. They will have also completed a one-semester physics course in electricity and magnetism and a one-semester course in electrical engineering with a textbook by Hayt and Hughes (Reference [2] in the Bibliography). For students without this background, this text can serve for a two-quarter or even a two-semester course.

The specific mathematics requisites for the course include elementary calculus, the solution of linear algebraic equations, and complex number arithmetic. The students are eased into matrix notation very gradually in the text. It is introduced in Chapter 5 as a bookkeeping method, and matrix manipulations are added gradually, as needed, through the remaining seven chapters. The symbolism is defined in Appendix I. Often students complain when they see matrices introduced in electrical engineering before the subject has been introduced in their mathematics courses; although the gradual introduction of matrices in this text does not eliminate the initial complaints, it does quell them quickly. The authors' experience is that by the end of the term all students handle matrix notation with ease. For students wishing more instruction on matrices, the inexpensive paperback book by Tropper (Reference [38] in the Bibliography) is a good reference.

Differential equations are not presented until Chapter 10 so that the students may get maximum benefit from their concurrent mathematics course in differential equations. If the students have not seen differential equations before, the instructor may have difficulty completing the text in one semester. For students who have completed a course in differential equations prior to this course, Chapters 10, 11, and 12 can be presented immediately after Chapter 3.

With regard to preparation in physics, the text was written with the assumption that the student has a basic knowledge of electricity and magnetism. He should be familiar with Coulomb's law, Ampere's law, Faraday's law, and Ohm's law. A very elementary knowledge of the operation of transistors is assumed in the discussion of models in Chapter 1. This knowledge is good for motivational reasons, but is not really required for the remainder of the text.

Chapters 1 and 2 introduce the notation used and connect this part of the text with material that students have had in their physics course. For students in schools that have a first electrical engineering course that introduces devices and some basic analytic techniques, Chapters 1 and 2 are reviews. Such students will be already aware of the need for the systematic methods of network analysis presented in the subsequent chapters of the text. For others, the examples in Chapter 2 begin to demonstrate the need for systematic methods. Accompanying laboratory work can also be used to motivate the students for the more mathematical aspects of network theory.

In Chapter 3 the most fundamental problem of network theory, the formulation of network equilibrium equations, is presented in detail. Basic principles based on the concepts of network graphs are stressed, starting from simple ideas and building to general methods. Specialized techniques, such as mesh and node-to-datum equations, are presented but not emphasized.

Chapter 4 presents the solution of network equations in the important special case of sinusoidal steady-state excitation. Here the student becomes familiar with complex amplitudes for sinusoidal voltages and currents. Complex impedances and admittances are introduced in this chapter. By considering this special case first, the student can understand some meaningful engineering problems early in the course and can do interesting experiments in an accompanying laboratory.

In Chapter 5, port parameters are presented first for 1-port networks, then for 2-port networks, and finally for n-port networks. In Chapter 6, a systematic method of network analysis suitable for handling a very large class of linear, time-invariant networks with a general program on a digital computer is presented. This method is based on the indefinite admittance matrix, a generalization of the classical node-to-datum analysis method, and was chosen because of its generality and simplicity. The authors believe that, with the availability of the digital computer, the students should learn how to formulate general problem-solving algorithms; special methods of analysis are becoming less important. Furthermore, if the student learns to formulate an algorithm, he can formulate his own special ones in his later engineering career as the need arises.

Some of the important engineering properties of amplifiers are presented in Chapter 7. Such important considerations as gain, input and output impedances, and isolation are defined; then a series of examples is presented to show how the quantities are computed. The important points are the concepts of amplifiers, rather than the particular properties of the particular devices used in the examples.

The behavior of networks as a function of frequency is discussed in Chapters 8 and 9. The basic concepts, including transfer functions, resonance, poles and zeros, and magnitude and phase plots, are the topics of Chapter 8. In Chapter 9, the first real design techniques of the course are introduced. Filter or coupled-amplifier design is an elementary design technique that shows the student the practical use of network models. The concepts of magnitude and frequency scaling are put to practical use in Chapter 9.

Chapter 10 returns to the equilibrium equations of Chapter 3 to present a general analysis method for arbitrary excitations for simple networks. In Chapter 11, state variables and matrix methods are used to show the general solution of linear network equations. In Chapter 12, the general methods are related to the special case of sinusoidal steady-state analysis. The student is shown the general approach, but he is reminded that simpler special methods for less general problems can also be formulated and executed. The basis for these three chapters is time-domain analysis of differential equations by elementary mathematical techniques. The symbolism of Laplace transforms is introduced, but the subtleties of Laplace transforms are not discussed, as they are not needed.

BENJAMIN J. LEON
PAUL A. WINTZ

Lafayette, Indiana
March 1970

Contents

Signal Quantities

Time functions: lower case italic and special font Greek

v	voltage
i	current
q	charge
p	power
f,g,x,u,w	generic signals
ϕ	magnetic flux
λ	flux linkage

Complex amplitudes of steady-state sinusoidal signals: the same letter as the associated time function but in italic capitals; for example

V	voltage complex amplitude
I	current complex amplitude

Vectors of signal quantities: the same letter in boldface italics; for example

$\boldsymbol{v}$	vector of voltages
$\boldsymbol{i}$	vector of currents

V	vector of voltage complex amplitudes
I	vector of current complex amplitudes

Magnitudes and phases of steady-state sinusoidal signals: Greek letters

ν	voltage magnitude
ϕ	voltage phase
ι	current magnitude
θ	current phase

Effective and average values of steady-state sinusoidal quantities: the following symbols

$\mathcal{V}$	effective voltage
$\mathcal{I}$	effective current
$\langle p \rangle$	average power

Transfer Functions

Complex valued scalar transfer functions: special font symbols as follows

$\mathfrak{z}$	impedance
y	admittance
$\mathfrak{h}$	element of the hybrid parameter matrix
$\mathfrak{a},\mathfrak{b},\mathfrak{c},\mathfrak{d}$	elements of the transfer parameter matrix
y	element of the indefinite admittance matrix
$\mathcal{H}$	generic transfer function

Transfer function matrices: boldface roman capitals

$\mathbf{Z}$	impedance matrix
$\mathbf{Y}$	admittance matrix
$\mathbf{H}$	hybrid parameter matrix
$\mathbf{T}$	transfer parameter matrix
$\mathbf{I}$	indefinite admittance matrix

Special mathematical quantities

$U(t)$	unit step defined in Equation (12-60)
$\delta(t)$	unit impulse defined on page 410
$\mathbf{1}$	unit (or identity) matrix

Roman and Greek letters of the regular font are used for miscellaneous constants, independent variables, and network parameters.

Boldface roman lower case letters are used for various real parameter matrices.

CHAPTER

1

Physical Circuits and Network Models

1-1 Introduction

We live in a world of radio, television, airplanes, spacecraft, computers, automated factories, and nuclear power plants. Electronic circuits make them work.

The significant physical quantities associated with electronic circuits are voltages and currents. Our primary interest is in those voltages and currents that can be measured outside the various devices at the junctions where these devices are connected. These junctions (connection points) are called *terminals*. The basic experimental procedure for determining terminal voltages and terminal currents is to measure them with appropriate instruments. A voltage (potential difference) between a pair of terminals can be measured by placing a voltmeter across the pair of terminals; terminal currents can be measured by placing an ammeter in series with the terminal. These measurement procedures are shown schematically in Figure 1-1, where the devices are symbolically shown as boxes with wires connecting them to their terminals. The ammeter indicates the current flowing through terminal a of device #3, while the voltmeter indicates the voltage between terminals b and c of device #3.

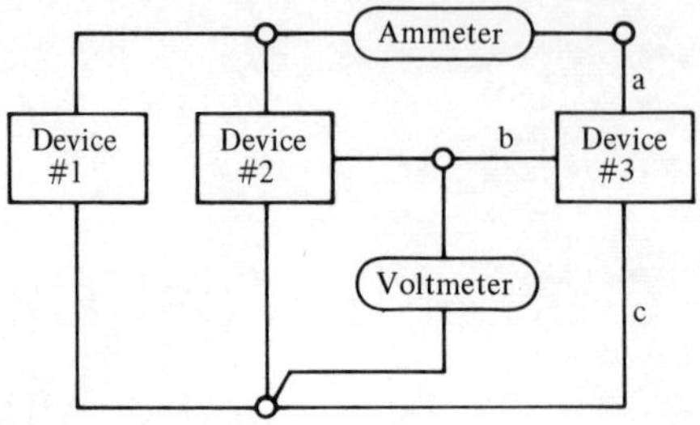

Figure 1-1 A schematic diagram.

The job of an engineer is to build and operate machines that perform useful functions. The machines of electrical and electronic engineers operate on voltages and currents. These electrical quantities are processed by the machines so that the overall operation has some purpose. An electric-power distribution system is an example of such a machine. The transistor radio described by Figure 1-2 is another. Let us briefly discuss some of the engineering problems associated with the AM broadcast receiver of Figure 1-2.

This radio receiver was designed to sense certain electromagnetic radi-

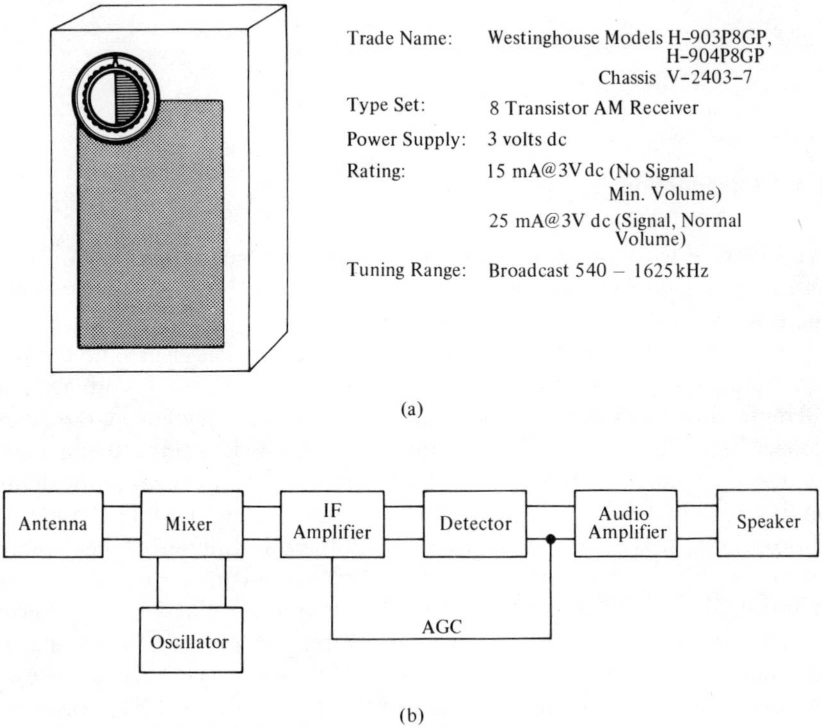

Figure 1-2 (a) Broadcast receiver. (b) Its block diagram.

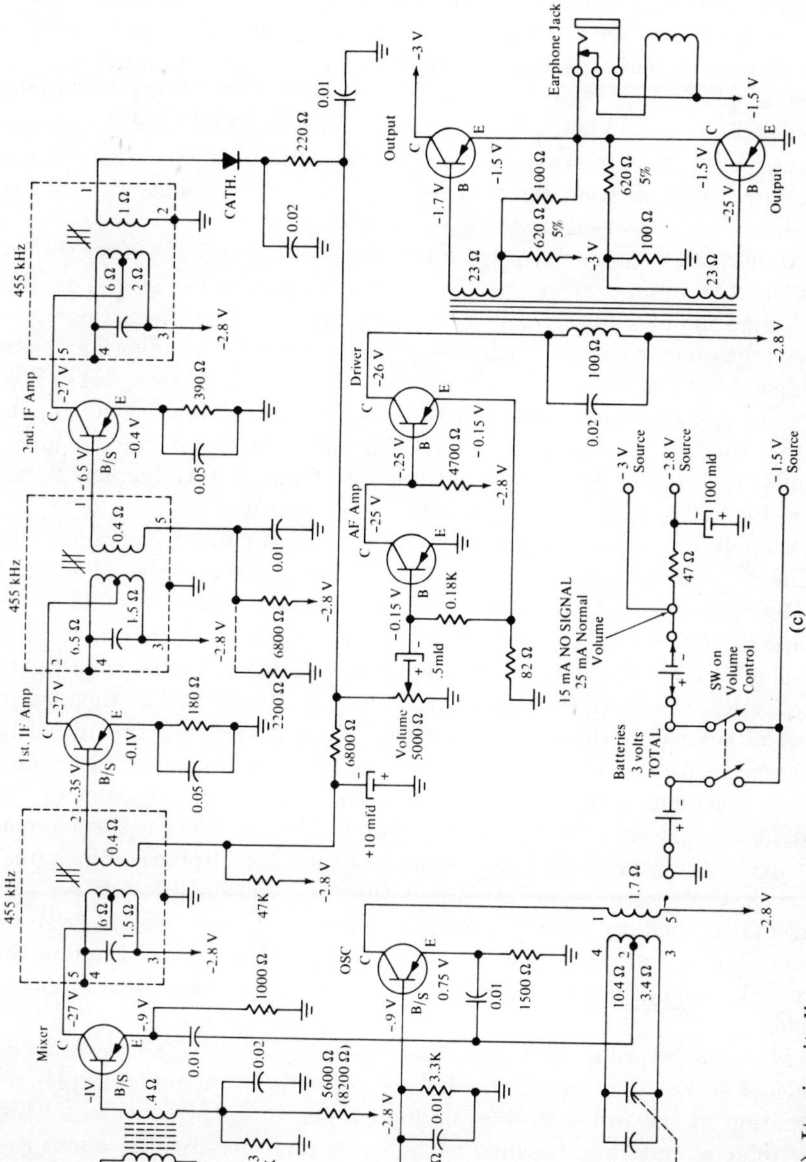

Figure 1-2 (c) Its circuit diagram.

ations produced by AM broadcast stations and to generate sound waves related in a very particular way to these transmissions. A generic term that includes sound waves, electromagnetic radiation, and other similar physical phenomena that vary with time is *signal*. Thus the radio receiver, like most machines of interest to electrical and electronic engineers, is a signal processor. In this case the input signal is the electromagnetic radiation from the broadcast station, and the output signal is the sound wave emitted from the speaker.

An invention many years ago and subsequent experience have shown that an AM radio receiver can be built with the blocks shown in the block diagram of Figure 1-2(b). The antenna converts the radio waves to voltages and currents. All the pairs of wires in the diagram have voltages between them and currents flowing through them. Each box performs a specific operation on the voltage and current at its input, causing the voltage and current at its output to have a particular form. Finally, the voltage and current at the left of the speaker box are transformed into the sound waves emitted by the speaker.

Each box in Figure 1-2(b) represents an assortment of transistors, resistors, capacitors, transformers, and so forth, connected as illustrated in Figure 1-2(c). If one looked inside the radio case, he would see a device corresponding to each element in this schematic diagram. To completely understand the operation of the radio, one must investigate the physical processes that go on inside each device as well as the overall effect of connecting many such devices. In this text, we are primarily concerned with the latter, that is, the overall effect of connecting a number of electronic devices, given that we know the characteristics of the individual devices. Device manufacturers usually specify the characteristics of the devices they sell in terms of their signal-processing characteristics, that is, the transformations they perform on the voltages and currents applied to their terminals. Hence, one can study, analyze, design, build, and operate electronic circuits without understanding the physical phenomena that cause the transformations inside the individual devices.[1]

Most engineering advances take place by small improvements on previous work. The understanding of previous work, for example, the operation of the radio receiver of Figure 1-2, is called analysis. Most electronic circuits are designed by doing an analysis, making minor corrections that should improve the operation, repeating the analysis, making additional corrections, and so forth, until an acceptable design evolves. Circuit analysis is the primary topic of this text. A few basic

[1] For another example, we point out that one can successfully study, analyze, design, build and operate a ferris wheel constructed from a motorized erector set without understanding how motors transform electrical energy into mechanical energy and without understanding the chemical processes going on inside the batteries that drive the motor.

circuits with known properties are introduced in several chapters to show how such designs are generated.

Our purpose is to analyze electronic circuits, which consist of a number of electronic devices connected in various configurations. Some devices have two terminals, some have three terminals, some have four, and some have more. Hundreds of devices—including resistors, capacitors, transistors, coils, tunnel diodes, operational amplifiers, thermistors, gyrators, potentiometers, switches, transformers, rotators—are presently available, and more are invented each year. In order to handle all of these devices as well as those yet to be invented, we must bring some order into the picture. The ordering of this chaos is brought about by modeling.

In Section 1-2, we introduce the concept of modeling and present the basic building blocks for constructing models of electronic devices. In Section 1-3, we show how to use these building blocks to construct models for some popular electronic devices.

1-2 The Nine Network Elements

A *model* is a symbol drawn on a piece of paper and characterized by one or more equations. To distinguish the model from the physical circuit, we refer to the model as a *network* and its components as *network elements.* Hence, for any physical circuit (an interconnection of circuit elements), there is a corresponding network (an interconnection of network elements). The network elements and the interconnection rules for networks are precisely defined. The analysis and synthesis of electronic networks (models) is the basic problem of this text. The relation between the models and the physical circuits is alluded to from time to time. In the laboratory, the engineer learns to relate his knowledge of networks (mathematical models) to the construction of electronic circuits (hardware). The correlation between the formulas derived from the model and the voltages and currents measured on the corresponding circuit is good, provided the modeling of the individual elements is good.

A very large class of linear electronic circuit elements can be modeled by one or more of nine linear network elements—five network elements with one pair of terminals (two terminals) and four network elements with two pairs of terminals (four terminals).

As illustrated in Figure 1-3(a), a 2-terminal network element has one voltage and one current associated with its terminal pair. The voltage v is the voltage across its terminal pair, and the current i is the current into one of the terminals (and out the other); that is, the voltage v is the electric potential at the plus terminal relative to the electric potential at the minus terminal (the potential difference), and i is the current in the direction of the arrow. Each two-terminal network element is defined by one equation relating the element voltage to the element current. Note that such an equation does not fix the values of the voltage and current; rather, it describes the relationship between them. Hence, each two-terminal network element constrains its voltage and current to be related in a particular way.

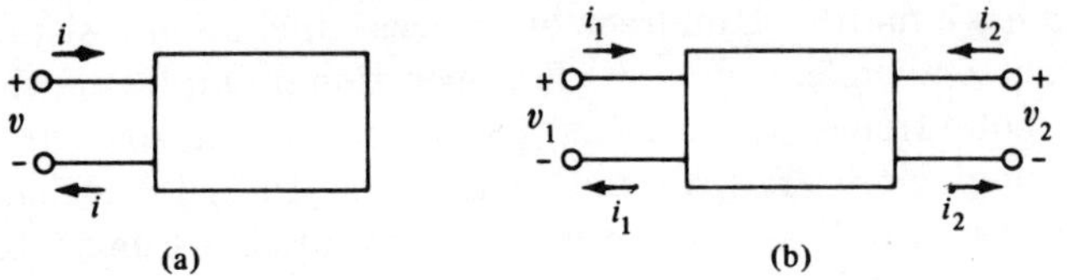

Figure 1-3 (a) Two-terminal network element. (b) Four-terminal network element with paired terminals.

As illustrated in Figure 1-3(b), a network element with two pairs of terminals has two voltages and two currents associated with it. The voltage v_1 and the current i_1 are associated with one terminal pair, and the voltage v_2 and the current i_2 are associated with the other terminal pair. Hence, each such network element can be defined by two equations relating the four quantities v_1, i_1, v_2, i_2. Again, these constraint equations do not determine the values of the four voltages and currents, but they do fix the relationships between them. Before proceeding, we point out that the currents and voltages associated with electronic circuits (and their models) are generally functions of time. We choose to denote time-varying voltage by v rather than $v(\mathrm{t})$ and time-varying currents by i rather than $i(\mathrm{t})$ in order to keep the notation as simple as possible. Hence, v is the value of the voltage (in volts) at time t, and i is the value of the current (in amperes) at time t.

The resistance. The two-terminal network element whose voltage v is some constant multiplied by its current i at each instant of time is called a *resistance.* The symbol for the resistance is presented in Figure 1-4.

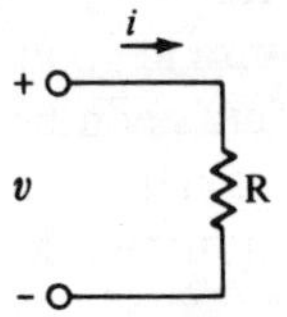

Figure 1-4 The symbol for a resistance.

The voltage at the plus terminal relative to the minus terminal is v, and the current into the plus terminal is i. The defining equation for a resistance is given by

$$v = \mathrm{R}i \qquad \textbf{(1-1)}$$

where the constant of proportionality R is called the value of the resistance, or, in short, the *resistance.* Note that the resistance defines only the relationship between the element voltage and the element current; it does not specify the values of the voltage and current. We also point out that R may be positive, negative, or zero. If $R = 0$, the voltage is zero for any current, and the terminals are said to be *short-circuited,* or *shorted.*

Sometimes it is more convenient to define the resistance by a formula in which the current equals a constant multiplied by the voltage. In this case the constant is called the *conductance,* and the symbol G is used. That is, Equation (1-1) can be written in the form

$$i = Gv \tag{1-2}$$

where $G = 1/R$. G too may be positive, negative, or zero. If $G = 0$, the current is zero for any voltage, and the terminals are said to be *open-circuited,* or *open.*

Since current and voltage are physical quantities that can be related to the basic physical quantities (length, mass, time, and charge), the resistance and conductance also have physical dimensions. Voltage is given in volts, current in amperes, resistance in ohms, and conductance in mhos. In terms of length, mass, time, and charge, these dimensions are as follows:

$$\text{one volt is one } \frac{\text{kilogram(meter)}^2}{\text{(second)}^2\text{ coulomb}}$$

$$\text{one ampere is one } \frac{\text{coulomb}}{\text{second}}$$

$$\text{one ohm is one } \frac{\text{kilogram(meter)}^2}{\text{second (coulomb)}^2}$$

$$\text{one mho is one } \frac{\text{second (coulomb)}^2}{\text{kilogram(meter)}^2}$$

Equation (1-1) defines a linear, time-invariant resistance. It is also possible to define a nonlinear resistance, a time-varying resistance, and a resistance that is both nonlinear and time-varying. A time-varying resistance is simply that; the resistance R changes with time. For a nonlinear resistance, the relationship between the element voltage v and the element current i are related at every instant in time by graph of i versus v, as illustrated by the example presented in Figure 1-5. Nonlinear and time-varying resistances are beyond the scope of this text. Hence, in this text the term "resistance" implies the relationship given in Equation (1-1), with R constant with time and the i-v graph a straight line with slope $G = 1/R$.

Some physical devices with two wire leads are called *resistors.* Under

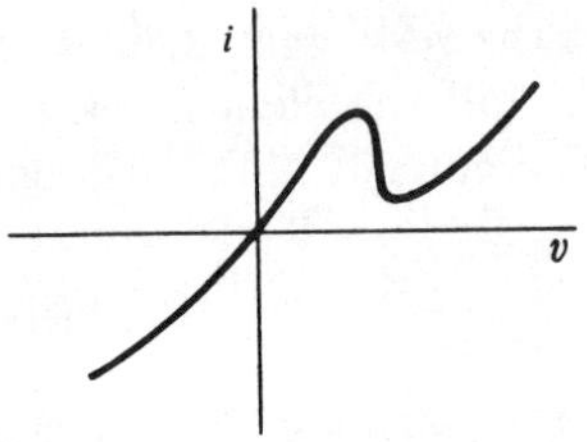

Figure 1-5 $i = v$ curve for a nonlinear resistance.

a wide range of conditions the measured resistor voltage and resistor current are related by a constant so that it is reasonable to model a resistor (circuit element) with a resistance (network element) under these conditions. We point out, however, that outside these conditions the resistance is no longer a reasonable model for a resistor. For example, for a voltage that is too large, the resistor material may get hot enough so that the electrical properties of the material change; for an even larger voltage, the device may fail (melt, burn, and so forth), producing a drastic change in electrical properties. The engineer learns the range of applicability of the model by experience and by reading the manufacturer's specifications.

The capacitance. There are devices that store electric charge proportioned to the applied voltage. Such devices are called *condensers* or *capacitors.* The network element used to model such devices is called the *capacitance.* The symbol for the capacitance is presented in Figure 1-6. As with the resistance, the voltage polarity and current direction

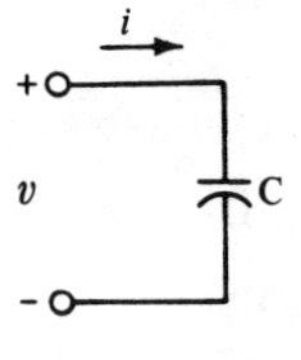

Figure 1-6 The symbol for a capacitance.

are chosen so that the current enters the plus terminal. The capacitance is defined in terms of the voltage v across its terminals and the charge q stored inside the network element. The equation relating q to v is given by

$$q = Cv \tag{1-3}$$

where C is a constant called the value of the capacitance or, in short, the *capacitance.* Hence, at each instant in time, the charge q is C times the voltage v. Since current is the time rate of change of charge, the voltage-current relationship for the capacitance is given by

$$i = \frac{dq}{dt} = C\frac{dv}{dt} \tag{1-4}$$

Sometimes it is more convenient to write the voltage as a constant multiplied by the charge. In this case the symbol $S = 1/C$, called the *elastance,* is used, and we write

$$v = Sq \tag{1-5}$$

or

$$v = S\int i\,dt \tag{1-6}$$

The units of the capacitance C are farads with dimension coulombs per volt, while the elastance has dimension volts per coulomb. The integral in Equation (1-6) is left indefinite. In a specific network, conditions other than the presence of the capacitance as an element determine the limits of integration.

Time-varying and nonlinear capacitances can also be defined. A nonlinear capacitance has a more complicated q-v graph (perhaps something like the i-v graph of Figure 1-5), while for time-varying capacitances, C is a function of time, so that Equations (1-4) and (1-6) must be written $i = d(Cv)/dt = C\,dv/dt + v\,dC/dt$. For the time-varying elastance, Equation (1-6) still applies. In this text we consider only the linear, time-invariant capacitance.

Probably the best correlation between a simple device and a single network element is the air or vacuum dielectric capacitor and the linear capacitance. Consider two parallel plates in air connected by wires to terminals, as shown in Figure 1-7. When a voltage is placed across the

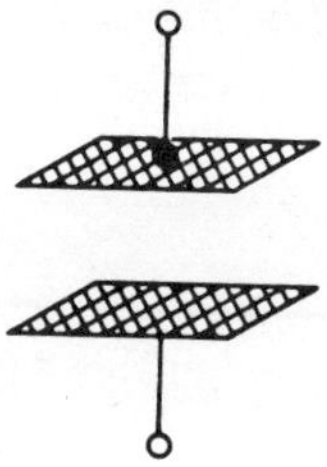

Figure 1-7 A parallel-plate capacitor.

terminals, plus on top and minus on the bottom, a charge is induced on the plates. A positive charge collects on the top plate, and a negative charge on the bottom plate. The net charge in the capacitor is zero, since the positive charge on the top plate is exactly equal to the negative charge on the bottom plate. The charge is directly proportional to the voltage. The constant of proportionality of charge to voltage is called the capacitance. Thus, for the device at fixed voltage, the measured charge and measured voltage are related by $q = Cv$.

If the impressed voltage on the terminals of the capacitor of Figure 1-7 varies with time, then the charge will also vary with time. In this case the current, which is the time derivative of the charge, will not be zero. This current must flow through the wires from the terminals to the plates. But when current flows through a wire, a voltage is required to overcome the resistance of the wire. If the current is not too large, the voltage is proportional to the current. Such a voltage-current relation is exactly the definition of the linear resistance. Thus a more refined model for the capacitor (physical device) of Figure 1-7 is the network

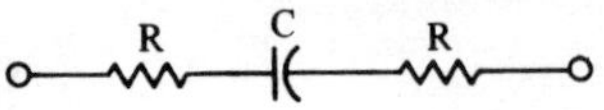

Figure 1-8 A more precise model for a capacitor.

(model) of Figure 1-8. In most applications the wire leads are very short, and the resulting R is small enough to be neglected.

The inductance. From basic physics (Ampere's law), we recall that a current has a magnetic field associated with it. Hence, by winding a length of wire into a coil, one can construct a simple device that stores magnetic flux proportional to the applied current. Such devices are called *inductors* or *coils.* The network element used to model these devices is called the *inductance.* The symbol for the inductance is presented in Figure 1-9. If we let ϕ be the magnetic flux and n the number

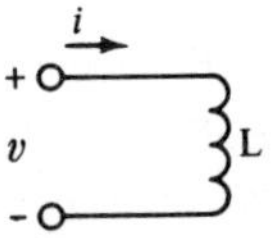

Figure 1-9 The symbol for an inductance.

of turns, then $\lambda = n\phi$ is called the *flux linkage.* The linear, time-invariant inductance is defined as having a current–flux-linkage relationship given by

$$\lambda = Li \tag{1-7}$$

where the constant of proportionality L is called the value of the inductance or, in short, the *inductance.* The units of L are henries with dimensions webers per ampere.

Furthermore, according to Faraday's law, a changing magnetic field induces a voltage equal to the rate of change of flux linkage. Hence,

$$v = \frac{d\lambda}{dt} = L\frac{di}{dt} \tag{1-8}$$

gives the voltage-current relationship for the L-henry inductance. Equations (1-7) and (1-8) can also be written in the form

$$i = \Gamma\lambda \tag{1-9}$$

$$i = \Gamma \int v \, dt \tag{1-10}$$

where $\Gamma = 1/L$ is called the *reciprocal inductance.*

Time-varying inductances and nonlinear inductances can also be defined, but these more complicated models are beyond the scope of this text.

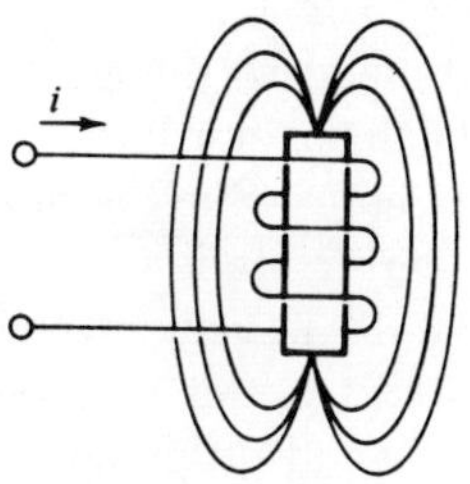

Figure 1-10 A coil.

Consider the coil of wire (inductor) shown in Figure 1-10. Let us assume that a current i is flowing down through the coil. Then a magnetic field with flux lines, as shown, is also present. If there is no magnetic material such as iron in the magnetic-field region, then the measured magnetic flux is proportional to the measured current. Even if iron is present, the flux and current may be proportional if the current is not too large. We restrict our discussion to devices where the measured flux-current relation is linear.

When the current and flux vary with time, a voltage is generated in each turn of the coil. This voltage is proportional to the derivative of the total flux through that turn. Since the turns are in series, the total voltage from the top of the coil to the bottom is the sum of the voltages in the separate turns. To account for all turns and their fluxes, a quantity λ called the flux linkage is used. For a fixed coil with a linear current-flux relation, the current–flux-linkage relation is also linear. Using the symbol L as the constant of proportionality, we find $\lambda = \mathrm{L}i$ and $v = \mathrm{d}\lambda/\mathrm{dt} = \mathrm{L}\,\mathrm{d}i/\mathrm{dt}$. Hence, the relationship between these (measured) physical quantities is identical to the equation used to define the inductance.

In the preceding argument, we neglected the fact that the coil is made of wire and that a voltage is required to push a current through it. Thus, for a more precise model of an inductor, we should add a resistance in series with an inductance, as illustrated in Figure 1-11. The induc-

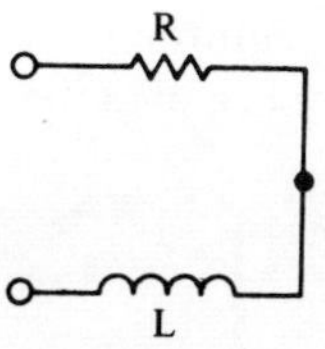

Figure 1-11 Model for an inductor.

tance accounts for the ability of the inductor to store magnetic flux, while the resistance accounts for the resistive effect of the length of wire used to construct the coil. A still more precise model for the coil could be obtained by noting that two turns, one above the other, with a voltage between them, is like a little capacitor. Thus, there should be a small capacitance in parallel with the inductance to give a good model of the physical device. Such arguments lead to a continuum of effects and a statement that continuous fields that take both time and space into account, are necessary to describe a coil. Such a statement is true for a complete description. For engineering design purposes, the first approximation of a simple inductance model is very often valid. Then, to establish behavior of the designed circuit under more stringent operating requirements, the first- and the second-order corrections can be included if necessary.

The independent voltage source. The *independent voltage source* or, in short, *voltage source*, is a two-terminal network element with the symbol presented in Figure 1-12. The voltage source always maintains the volt-

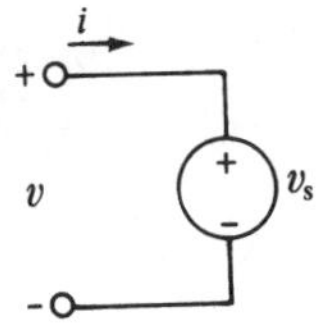

Figure 1-12 The independent voltage source.

age across its terminals at v_s volts (with the polarity indicated by the plus and minus signs inside the circle), independent of the current drawn from it or forced into it by other network elements connected to its terminals. Hence, this network element is defined by the equation

$$v = v_s \tag{1-11}$$

where v is the element voltage, and v_s is the value of the voltage source in volts. v_s may be a constant or a function of time. Equation (1-11) is sometimes called the *source equation* for the independent voltage source.

The voltage source derives its name from the fact that it is the basic model for certain physical devices that supply electrical energy. The 60-cycle alternating current generators that supply our domestic electricity can be modeled by a voltage source, as illustrated in Figure 1-13(a). A 12-volt car battery can be modeled by a voltage source in series with a resistance, as illustrated in Figure 1-13(b). There are other situations in

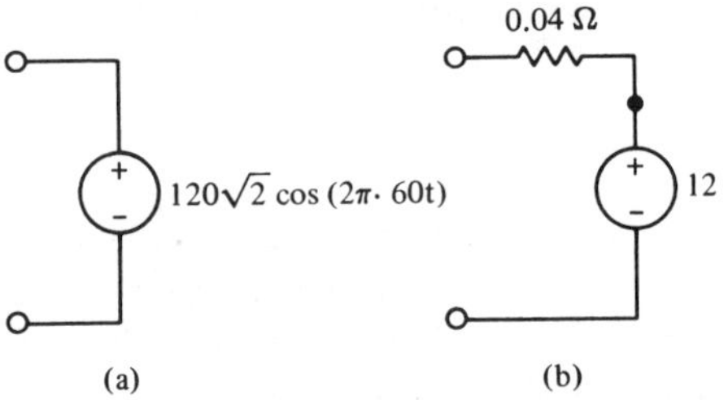

Figure 1-13 (a) Model for a 120-volt household wall socket. (b) Model for a car battery.

which the voltage-source model is useful, even though the corresponding physical device is not supplying electrical energy. Any device for which the voltage is known can be modeled by a voltage source.

The independent current source. The symbol for the *independent current source* or, in short, *current source,* is presented in Figure 1-14. The current source maintains a current of i_s amperes (in the direction of the

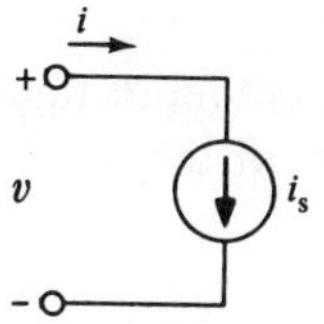

Figure 1-14 The independent current source.

arrow inside the circle), independent of its voltage, that is, no matter what is connected across its terminals. Hence, the equation for this network element is given by

$$i = i_s \tag{1-12}$$

where i is the element current, and i_s is the value of the current source in amperes. In general, i_s is a time function. Equation (1-12) is sometimes called the *source equation* for the independent current source.

An example of a physical device that can be modeled by a current source is the Van de Graaff electrostatic generator. Other physical devices can be modeled by current sources. Also, any device for which the current is known can be modeled by a current source.

The independent sources (both voltage and current) are called independent because their voltage (or current) is independent of other considerations. On the other hand, many electronic devices (such as transistors) have more than one pair of terminals, where the voltage (or current) at one pair of terminals depends on the voltage (or current) applied at a different pair of terminals. To model these devices, network elements called *controlled sources* are required. We next introduce four such network elements, each of which has four terminals (two terminal pairs), as in Figure 1-3(b). These elements are defined by two constraint equations governing the four quantities (two currents and two voltages) of interest.

Voltage-controlled voltage source. The symbol for a *voltage-controlled voltage source* is presented in Figure 1-15. The voltage and current at the left-hand pair of terminals are v_1 and i_1, and the voltage and current at the right-hand terminal pair are v_2 and i_2, as shown. The defining equations for this network element are given by the pair of equations

$$\begin{aligned} i_1 &= 0 \\ v_2 &= \mu v_1 \end{aligned} \tag{1-13}$$

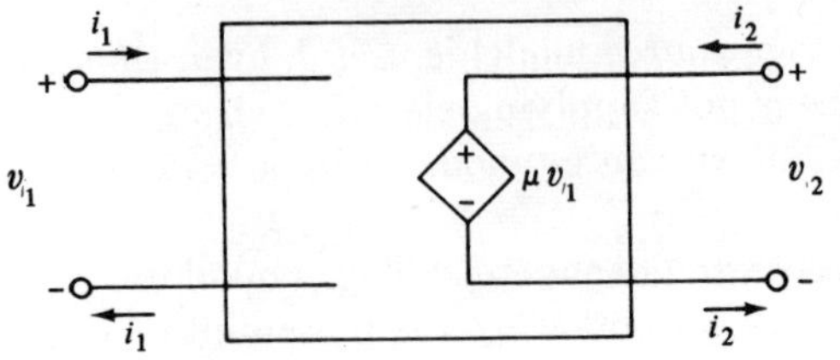

Figure 1-15 Voltage-controlled voltage source.

where μ is a dimensionless constant. Hence, the left-hand terminals are an open circuit, and the voltage across the right-hand terminals is μ multiplied by the voltage across the left-hand terminals. The voltage v_1 is called the *control voltage* or, in short form, the *control*, because it "controls" the voltage v_2 (hence the name "controlled source"). The right-hand side of the voltage-controlled voltage source is similar to the independent voltage source in the sense that the voltage at the terminal corresponding to the plus end of the diamond-shaped symbol is maintained at a specified voltage relative to the terminal corresponding to the minus end. However, the independent voltage-source voltage v_s is independent of all other network considerations, whereas the controlled voltage-source voltage is completely specified by another network parameter (the voltage v_1). The second equation of Equations (1-13) is often called the *control equation* because it relates the source voltage to the control voltage.

The voltage-controlled current source. The symbol for a *voltage-controlled current source* is presented in Figure 1-16. The defining equations for

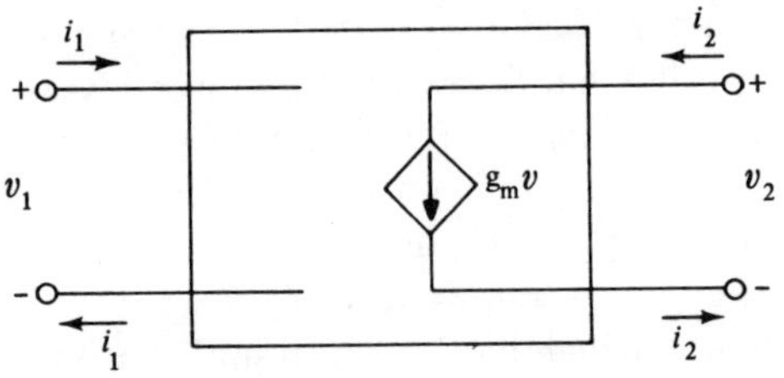

Figure 1-16 Voltage-controlled current source.

this network element are given by

$$\begin{aligned} i_1 &= 0 \\ i_2 &= g_m v_1 \end{aligned} \tag{1-14}$$

where the constant g_m has the units of amperes per volt and is called the *transconductance* or the *mutual conductance*. Hence, the left-hand terminals are open, while the right terminals are connected to a current source. The current-source current is g_m multiplied by the voltage v_1.

Since the voltage v_1 determines the current-source current, it is called the *control voltage*, or the *control*. The diamond-shaped symbol with the arrow inside indicates that the current through it is maintained at $g_m v_1$ amperes for any v_2. Hence, the current-source current is independent of the voltage v_2, but it is completely determined by the voltage v_1. The second equation of Equations (1-14) is called the *control equation* because it specifies how the current-source current depends on the control voltage.

The current-controlled voltage source. The symbol for the *current-controlled voltage source* is presented in Figure 1-17. The defining equations

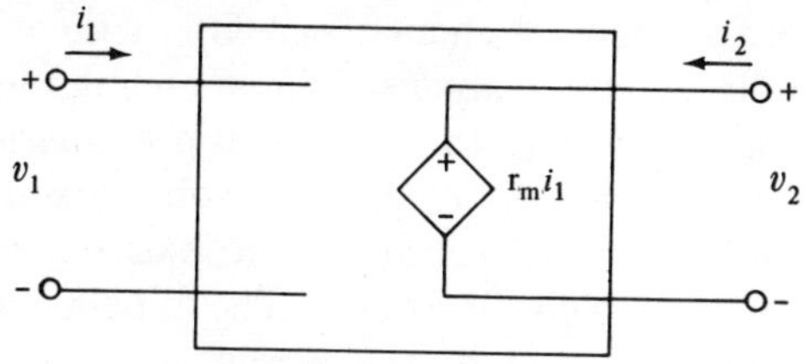

Figure 1-17 Current-controlled voltage source.

for this network element are given by

$$\begin{aligned} v_1 &= 0 \\ v_2 &= r_m i_1 \end{aligned} \qquad \textbf{(1-15)}$$

Hence, the left-hand terminals are a short circuit, while the right-hand terminals are connected to a voltage source of value $r_m i_1$ volts, with the polarity indicated by the plus and minus signs inside the diamond. The constant of proportionality r_m has the dimensions of volts per ampere (ohms) and is called the *mutual resistance*. Because the value of the voltage source is maintained at $r_m i_1$ volts, i_1 is called the *control current*, or the *control*. The second equation of Equations (1-15) is called the *control equation* for the current-controlled voltage source.

The current-controlled current source. The symbol for the remaining combination, the *current-controlled current source* is presented in Figure 1-18. The defining equations are given by

$$\begin{aligned} v_1 &= 0 \\ i_2 &= \alpha i_1 \end{aligned} \qquad \textbf{(1-16)}$$

where α is a dimensionless constant. Note that the left-hand terminals are shorted, while the right-hand terminal-pair current is maintained at αi_1 amperes in the direction of the arrow inside the diamond. i_1 is called the *control current*, or the *control*. The equation $i_2 = \alpha i_1$ is called the *control equation*.

These four controlled sources are linear and time-invariant; they are

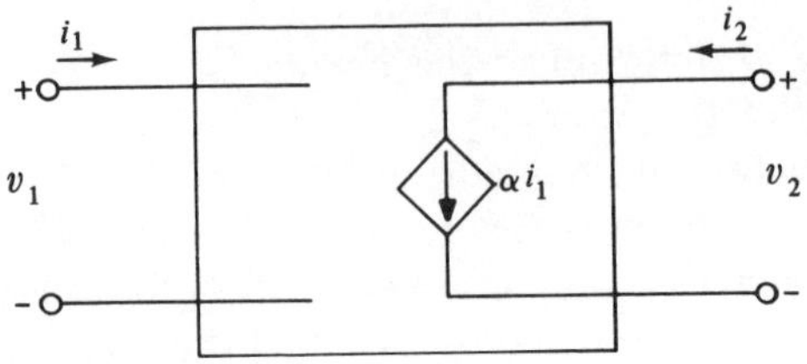

Figure 1-18 Current-controlled voltage source.

the controlled sources used in this text. Some networks require nonlinear and time-varying controlled sources, but these more sophisticated models are beyond the scope of elementary circuit theory.

All four controlled sources are used in modeling electronic devices. The operational amplifier and (to a crude approximation) some triode vacuum tubes have the voltage-controlled voltage source as a model. The voltage-controlled current source is the basic model for the field-effect transistor and the pentode tube. The bipolar transistor can be modeled to a first approximation by the current-controlled current source. For all these devices, the simple controlled-source network element is only a first approximation. Better models require that one or more resistances, capacitances, and so forth, be added to the model. More elaborate models for some of these physical devices are discussed in Section 1-3.

1-3 Models of Physical Devices

In Section 1-2, we introduced the nine network elements and indicated how they can be used to model some simple circuit elements such as resistors, capacitors, coils, and automobile batteries. In this section, we present some more elaborate models of some more sophisticated electronic devices, such as transistors and transformers.

The FET. A common device that illustrates the use of controlled sources in modeling is the field-effect transistor (FET). This device consists of a piece of semiconductor with two terminals, an insulating layer, and a conducting section connected to a third terminal, as shown in Figure 1-19.

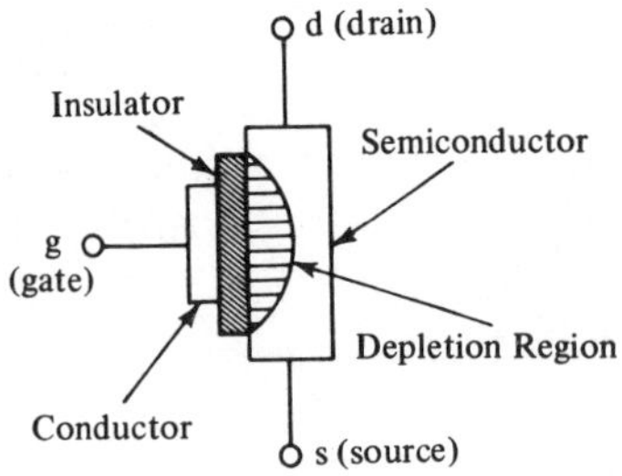

Figure 1-19 Physical construction of an FET.

The important property of the semiconductor, so far as the operation of the FET is concerned, is that there is only a finite number of charge carriers to provide for current flow. Furthermore, these carriers have a finite maximum mobility under reasonable operating voltages. Thus, for the slab of semiconductor alone, with an applied voltage positive at terminal s relative to terminal d, the current is proportional to the voltage for low voltages. At higher voltages, the current saturates because there is a finite number of charges, and the speed of their movement is limited. The resulting current-voltage characteristic of a slab of semiconductor is that of Figure 1-20(a). The charge carriers may be negative (electrons),

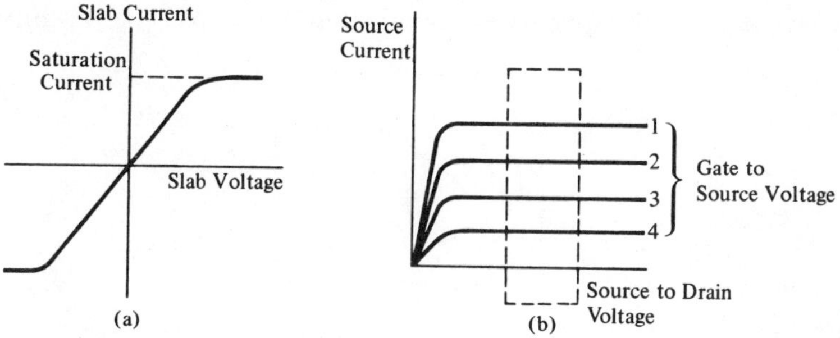

Figure 1-20 (a) Voltage-current characteristics for a slab of semiconductor. (b) FET characteristics.

positive (holes), or some of both, depending on the semiconductor material used. When the majority of the carriers is negative, the semiconductor is called n-type, and when positive, p-type.

To understand the FET, let us suppose the semiconductor slab in Figure 1-19 is p-type. If a positive voltage with respect to the slab is applied at terminal g, then the holes near the gate will be repelled. The area from which the holes move is depleted of carriers. This region, called the depletion region, is shown in Figure 1-19. Current cannot flow in this region, since there are no carriers. Thus, in the center of the slab, the current-carrying cross-sectional area is reduced. The result is an increase in the slab resistance (at voltages below saturation) and also a decrease in the saturation current. The higher the gate voltage, the larger the depletion region, the higher the resistance, and the lower the saturation current. Hence, by changing the gate voltage (relative to the slab), we change both the slope and the saturation current of the slab current-voltage characteristic. The upper right-hand quadrant of the slab current-voltage relationship for several different values of gate voltage is presented in Figure 1-20(b). When the FET is operated in this region, terminal s is called the *source*, terminal d is called the *drain*, and terminal g is called the *gate*.

The complete FET characteristic is nonlinear and requires a nonlinear-circuit model. If, by controlling the other elements that are connected to the FET, we can restrict the terminal voltages and currents to values corresponding to the region inside the dotted box in Figure 1-20(b), then the characteristic of the FET is a family of constant current lines. The value of the current into terminal s (and out terminal d) depends on the voltage at terminal g relative to terminal s. The voltage across terminals s and d does not affect the current, because operation is in the current saturation region. This is exactly the definition of a voltage-controlled current source. For operation inside the dotted box in Figure 1-20(b), the FET of Figure 1-19 can be modeled as shown in Figure 1-21. The constant g_m depends on the physical construction of the device, that is, on the spacing of the horizontal lines in Figure 1-20(b).

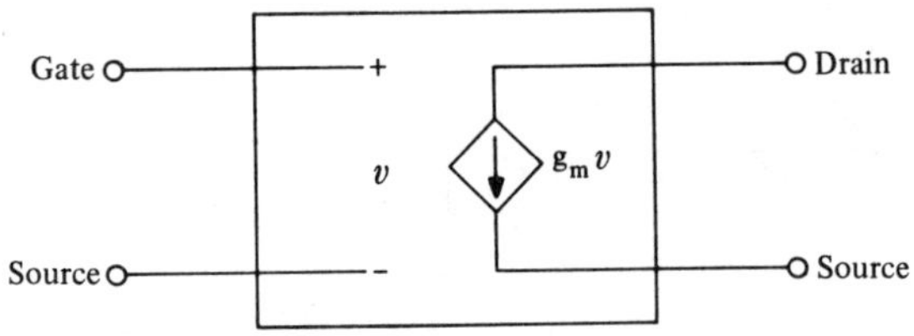

Figure 1-21 FET model.

The model of Figure 1-21 is only a first approximation to the operation of the FET, even when the terminal voltages and currents are confined to the box in Figure 1-20(b). Note that the conductor, insulator, and semiconductor-slab layers are of the same construction as a capacitor. Hence, there is a capacitive effect between terminals g and s and between terminals g and d. Furthermore, from measurements on the real device, we find that even for a constant gate voltage, the slab current is not entirely independent of the slab voltage. A slight increase of slab current with slab voltage is observed, so that the parallel lines of Figure 1-20(b) are not exactly horizontal but slope upward to the right. This can be accounted for in the model by adding a large resistance (a conductance equal to the slope) in parallel with the current source. We thus arrive at the more reasonable model of the FET presented in Figure 1-22(a). This model contains two capacitances, one resistance, and one voltage-controlled current source—a total of four network elements. Further refinements are also possible by adding more network elements, but the model of Figure 1-22 is reasonable for most applications. For convenience, we usually use the FET model of Figure 1-22(b), which is identical to the model of Figure 1-22(a). Finally, the symbol for the FET circuit element is presented in Figure 1-22(c).

The bipolar transistor. Another three-terminal device in common engineering use is the bipolar transistor. The physics of this device is more

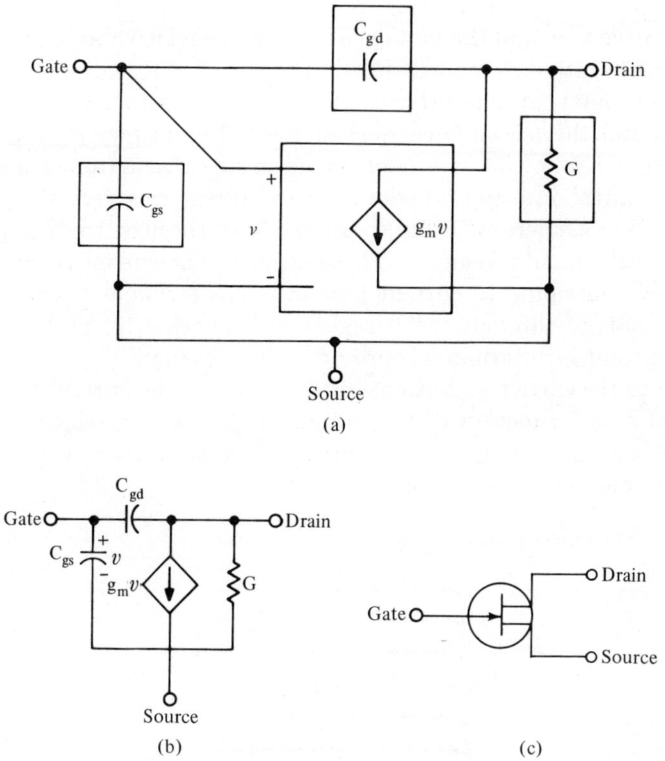

Figure 1-22 (a) A complete FET model consisting of two capacitances, one conductance, and one voltage-controlled current source. (b) An equivalent but more esthetically pleasing model. (c) Symbol for the FET circuit element.

subtle than that of the FET, but an elementary argument that justifies a reasonable model can be formulated. The bipolar transistor consists of three layers of semiconductor material of alternating type—either n-p-n or p-n-p. For our discussion, let us consider the n-p-n arrangement of Figure 1-23. In normal linear operation, the voltage at termi-

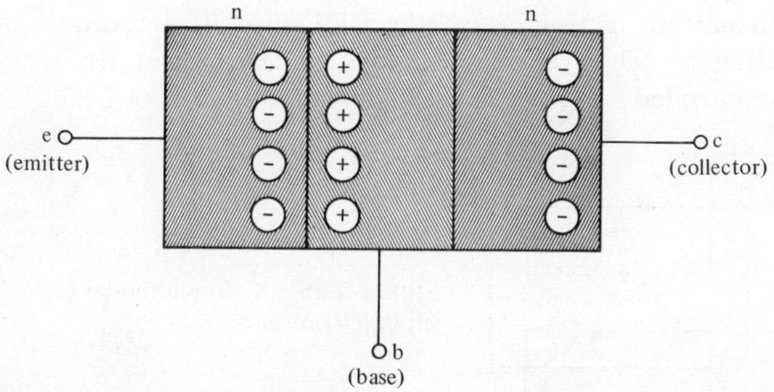

Figure 1-23 Physical construction of a bipolar transistor.

nal c relative to e, and the voltage at terminal e relative to b, are positive. Thus, the negative charge carriers in the right-hand region are attracted toward terminal c, the positive carriers in the central region move toward the left, and the negative carriers in the left-hand region move toward the p region as shown. Many of the negative carriers from the left-hand n region diffuse into the p region. If the p region is very thin, most of the negative carriers will diffuse all the way through the p region and into the right-hand n region. These negative carriers move on to terminal c and contribute to current flow into this terminal. Those carriers that do not go through the p region are canceled by positive carriers there; current into terminal b provides these carriers.

Due to the carrier migrations outlined above, the current flowing into terminal c as a function of the voltage at terminal c relative to terminal b (with the current out of terminal e as a parameter), is given by the family of curves in Figure 1-24.

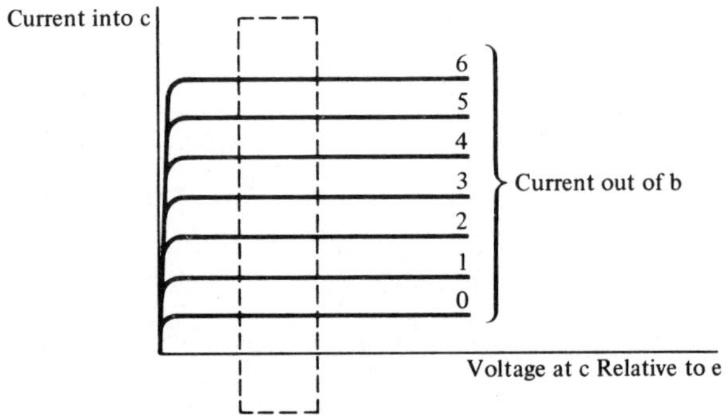

Figure 1-24 Transistor characteristics.

In the region of the dotted box in Figure 1-24, the characteristic of the bipolar transistor, like that of the field-effect transistor, is a family of parallel lines of virtually constant current over a large range of voltages. However, the parameter that controls the current into terminal c is the current out of terminal e (the FET current source was controlled by a voltage). Thus, a simple model for the bipolar transistor is the current-controlled current source, arranged as in Figure 1-25.

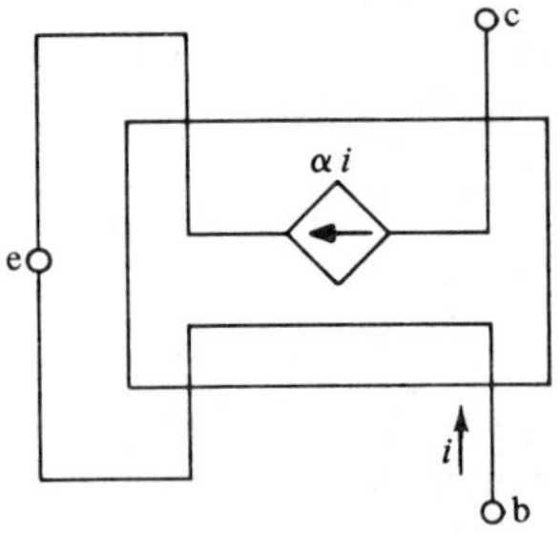

Figure 1-25 A simple model for a bipolar transistor.

A more refined model that also accounts for the resistance of the semiconductor material as well as the capacitance associated with the junction between the right-hand n region and the center p region (see Figure 1-23) is presented in Figure 1-26(a).[2] Note that this network utilizes

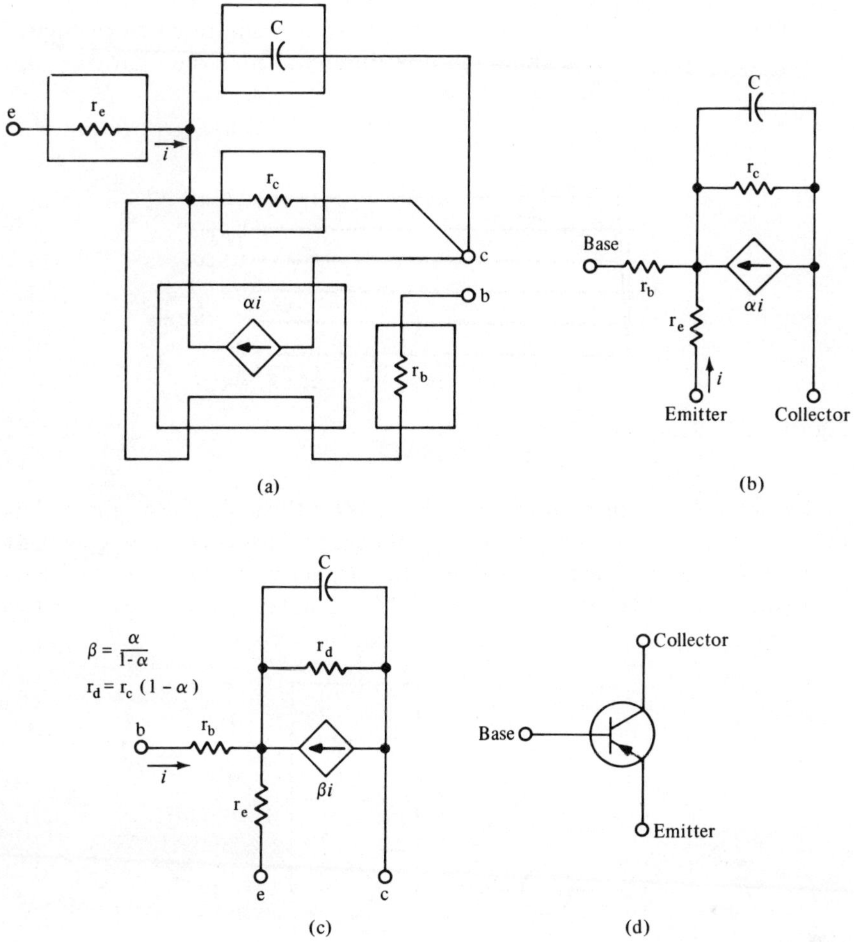

Figure 1-26 Bipolar transistor model.

five network elements. The c terminal is called the *collector*, the b terminal the *base*, and the e terminal the *emitter*. The network model for the transistor is usually drawn as in Figure 1-26(b), where r_b is called the base resistance, r_e the emitter resistance, and r_c the collector resistance. Note that the model of Figure 1-26(b) is equivalent to the model of Figure 1-26(a). Another transistor model is presented in Figure

[2] In keeping with common usage, lower case r is used for resistance internal to a transistor.

1-26(c). With $\beta = \alpha/(1 - \alpha)$ and $r_d = r_c(1 - \alpha)$, this model is equivalent to the model of Figure 1-26(b). (The concept of an equivalent network is precisely defined in Section 2-3. See also Problem 2-6.) The standard symbol for the bipolar transistor circuit element is presented in Figure 1-26(d).

Some transistors have characteristics like the one presented in Figure 1-27. This transistor characteristic is different from the characteristic

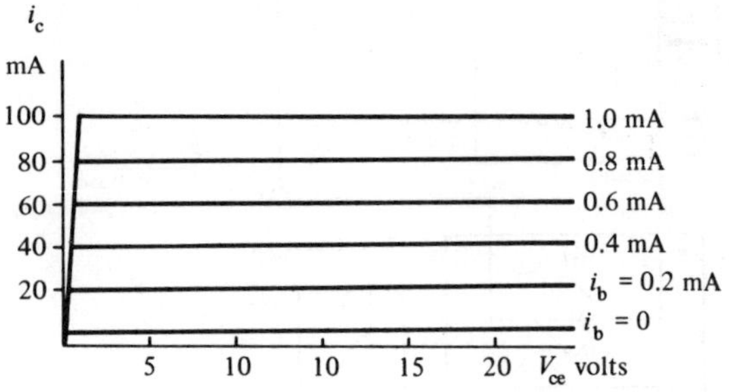

Figure 1-27 Transistor characteristics.

of Figure 1-24, in that the $i_b = 0$ line passes through the origin. This can be accounted for in the transistor model by inserting an independent current source of (constant) value i_0 in parallel with the controlled current source, as illustrated in Figure 1-28. The transistor characteristics

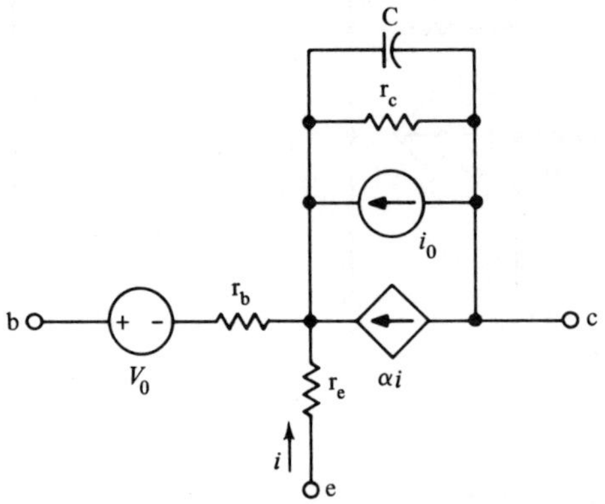

Figure 1-28 Transistor model for characteristics of Figure 1-27.

of Figures 1-24 and 1-27 say nothing of the voltage-current relationship at the base. The model of Figure 1-26(b) contains only a resistance r_b in the base leg. Laboratory measurements on transistors indicate that a better model results if we insert an independent voltage source of value v_0 (a small constant) in series with r_b, as illustrated in Figure 1-28.

Coupled coils. In Section 1-2, we discussed a physical device called the inductor and its model, the inductance. Recall that an inductor is a coil of wire wound on a suitable frame, such as a hollow cardboard cylinder, as illustrated in Figure 1-29(a). The current i generates a magnetic

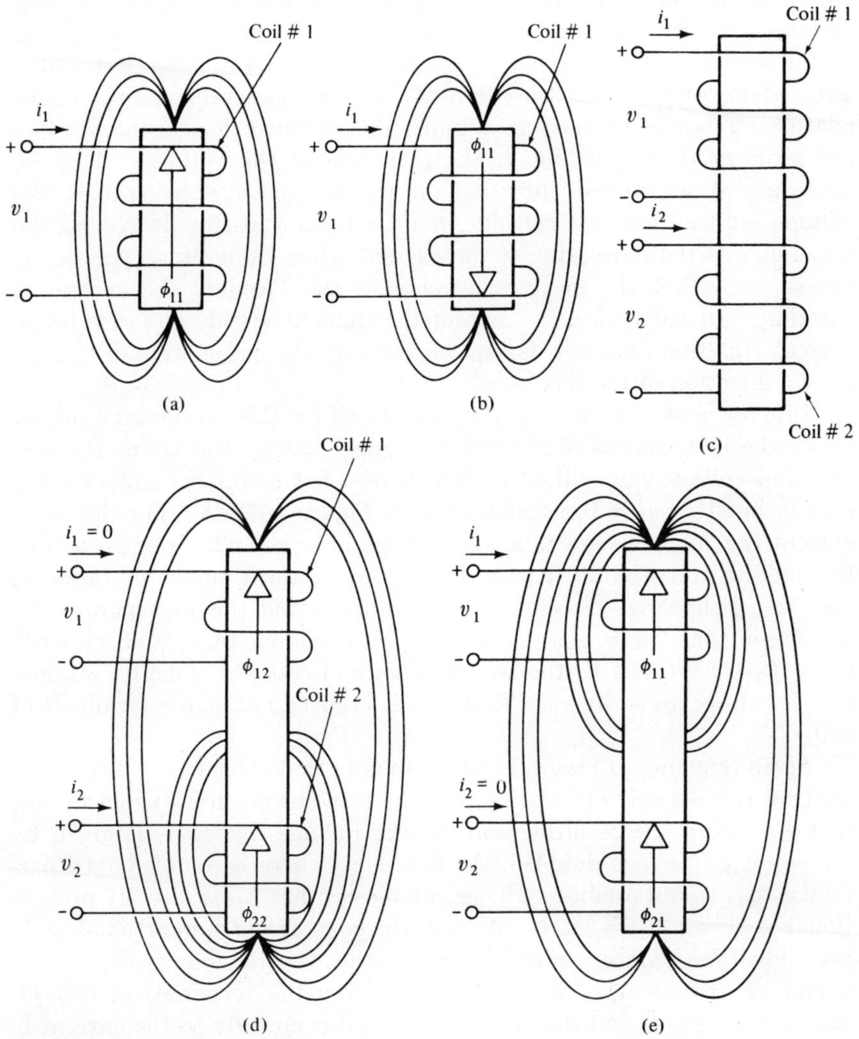

Figure 1-29 Flux in coupled coils.

field, which, according to the right-hand rule, results in a magnetic flux up through the cylinder and down around the outside, as illustrated in Figure 1-29(a). Let ϕ_{11} be the flux-linking coil #1 due to current i_1 and n_1 the number of turns. Then, the flux linkage $\lambda_{11} = n_1\phi_{11}$ is directly proportional to the current; that is, $\lambda_{11} = L_1 i_1$, and the constant of proportionality is called the *self-inductance* (or *inductance*) of coil #1. Finally, the voltage induced across the coil terminals is the time rate of change of the flux linkage; that is, $v_1 = d\lambda_{11}/dt = L_1\, di_1/dt$.

Suppose that we had wound coil #1 on the cylinder in the opposite direction, as illustrated in Figure 1-29(b). Then the right-hand rule dictates that the direction of the flux is down through the cylinder and up the outside. The flux linkage is still given by $\lambda_{11} = n_1\phi_{11}$, but is now directed downward inside the cylinder. For this case the voltage-current relationship is still given by $v_1 = L_1\, di_1/dt$, provided the current i_1 enters the terminal corresponding to the plus side of the voltage. That is, according to our current direction and voltage polarity conventions, the voltage induced across a coil due to a curent in that coil always has the same sign as the derivative of the current when the voltage polarity is chosen such that the positive terminal is the terminal the current is entering. In particular, the sign of the right-hand side of the voltage-current equation does not depend on the way the coil is wound, that is, on the direction of the flux.

Suppose now that we wind a second coil on the cardboard cylinder below the first coil, as illustrated in Figure 1-29(c), and adopt the current and voltage conventions shown there. Let us further suppose that $i_1 = 0$, but that $i_2 \neq 0$, as suggested by Figure 1-29(d). For this case, current i_2 generates some flux, all of which links coil #2. Let ϕ_{22} be this flux-linking coil #2 due to current i_2, and n_2 the number of turns on coil #2. Following the same argument as in the previous paragraph, we reason that $\lambda_{22} = n_2\phi_{22}$ is directly proportional to i_2, and we write $\lambda_{22} = L_2 i_2$. We call L_2 the *self-inductance* of coil #2. Finally, we note that a voltage given by $v_2 = L_2\, di_2/dt$ is induced across the terminals of coil #2.

Again referring to Figure 1-29(d), we note that some of the flux generated by i_2 links coil #1. Let ϕ_{12} be the flux linking coil #1 due to current i_2. Then the resulting coil #1 flux linkage due to i_2 is given by $\lambda_{12} = n_1\phi_{12}$. If i_2 is doubled, the flux everywhere is doubled. In particular, ϕ_{12} would double. Hence, we reason that λ_{12} is directly proportional to i_2, say $\lambda_{12} = Mi_2$. We call the constant of proportionality M the *mutual inductance*. Finally, we note that the flux linkage λ_{12} induces a voltage of $v_1 = d\lambda_{12}/dt = M\, di_2/dt$ across the terminals of coil #1. Recall that $i_1 = 0$ and that this voltage is due entirely to the current in coil #2.

We next consider the situation illustrated by Figure 1-29(e). We have the same coils wound on the same cylinder, but this time we set $i_2 = 0$ and $i_1 \neq 0$. As discussed in reference to Figure 1-29(a), current i_1 generates a magnetic field, and the flux-linking coil #1 due to i_1 is ϕ_{11}; the flux linkage is $\lambda_{11} = n_1\phi_{11} = L_1 i_1$; and the induced voltage is $v_1 = L_1\, di_1/dt$. The presence of coil #2 does not change these quantities.

Meanwhile, at coil #2, we observe a flux ϕ_{21} caused by current i_1, with the associated flux linkage $\lambda_{21} = n_2\phi_{21}$. Furthermore, the flux linkage λ_{21} is directly proportional to the current i_1, that is, $\lambda_{21} = Mi_1$, and $v_2 = d\lambda_{21}/dt = M\, di_1/dt$. The constant of proportionality M is the same constant discussed in reference to Figure 1-29(d). The fact that the constant turns out to be the same in both cases regardless of the geometry of the two coils is not obvious; nevertheless, it is true.[3]

We are now prepared to discuss the case illustrated in Figure 1-30, where both currents are nonzero. Both currents generate flux so that

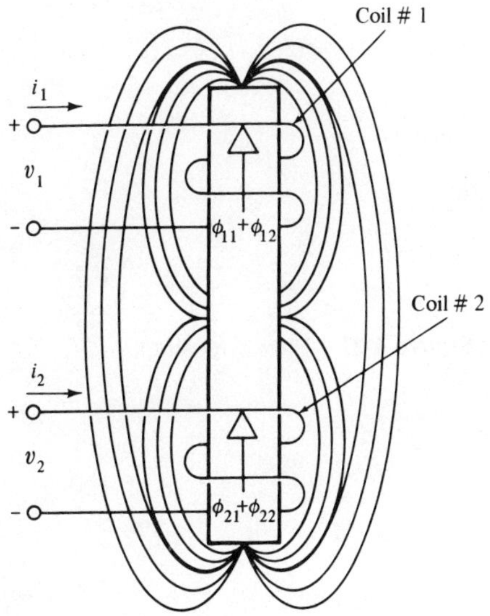

Figure 1-30 Flux aiding.

the total flux is the sum of the fluxes produced by the currents individually. Hence, the total flux-linking coil #1 is $\phi_{11} + \phi_{12}$ (ϕ_{11} is due to i_1, and ϕ_{12} is due to i_2). It follows that the total coil #1 flux linkage is

$$n_1(\phi_{11} + \phi_{12}) = n_1\phi_{11} + n_1\phi_{12} = \lambda_{11} + \lambda_{12} = L_1 i_1 + M i_2$$

Consequently, the voltage induced across the terminals of coil #1 is

$$v_1 = \frac{d}{dt}(\lambda_{11} + \lambda_{12}) = \frac{d\lambda_{11}}{dt} + \frac{d\lambda_{12}}{dt} = L_1\frac{di_1}{dt} + M\frac{di_2}{dt}$$

The voltage v_1 is due in part to the flux generated by i_1 and in part to the flux generated by i_2. Note that if $i_2 = 0$, then $v_1 = L_1\, di_1/dt$; it is due only to the self-inductance of coil #1, as discussed in reference to Figure 1-29(a). On the other hand, if $i_1 = 0$, then $v_1 = M\, di_2/dt$; it is due only to the magnetic coupling between the coils.

[3] For further elaboration see W. R. Smythe, *Static and Dynamic Electricity* (New York: McGraw-Hill, 1950), 2nd ed., p. 311.

By a similar argument, we reason that the total flux-linking coil #2 is $\phi_{21} + \phi_{22}$, the flux linkage is $n_2(\phi_{21} + \phi_{22}) = n_2\phi_{21} + n_2\phi_{22} = \lambda_{21} + \lambda_{22} = Mi_1 + L_2i_2$, and the induced voltage is given by

$$v_2 = \frac{d}{dt}(\lambda_{21} + \lambda_{22}) = \frac{d\lambda_{21}}{dt} + \frac{d\lambda_{22}}{dt} = M\frac{di_1}{dt} + L_2\frac{di_2}{dt}$$

We now consider the slightly different arrangement illustrated in Figure 1-31. The top coil is the same as before, but the bottom coil is wound on the cylinder in the opposite direction.

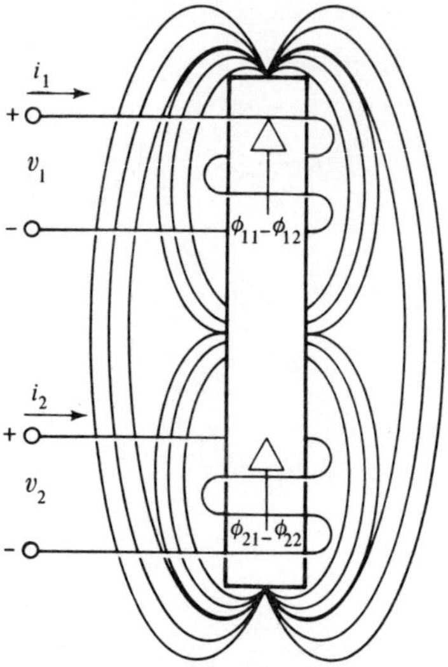

Figure 1-31 Flux opposing.

Consider the flux in coil #1. The flux ϕ_{11} due to current i_1 is the same as determined in reference to Figure 1-30; that is, its direction according to the right-hand rule is up through the cylinder and down the outside. However, applying the right-hand rule to coil #2, we find that it now generates a flux down through the cylinder and up the outside. Consequently, the flux ϕ_{12} in coil #1 due to current i_2 is directed downward, as illustrated in Figure 1-31. Hence, the total flux linking coil #1 in the upward direction is $\phi_{11} - \phi_{12}$. It follows that the coil #1 flux linkage is $n_1(\phi_{11} - \phi_{12}) = \lambda_{11} - \lambda_{12}$ and the voltage induced is $v_1 = d(\lambda_{11} - \lambda_{12})/dt = d\lambda_{11}/dt - d\lambda_{12}/dt = L_1\,di_1/dt - M\,di_2/dt$. This term may be either positive or negative, depending on which of the two terms is larger.

Next, consider the flux in coil #2. The coil #2 current i_2 generates flux ϕ_{22} in the downward direction, while i_1 generates flux ϕ_{21} in the upward direction. The total flux in the downward direction is $\phi_{22} - \phi_{21}$ so that the resulting coil #2 flux linkage is $n_2(-\phi_{21} + \phi_{22}) = -\lambda_{21} + \lambda_{22}$, and the induced voltage is

$$v_2 = \frac{d}{dt}(-\lambda_{21} + \lambda_{22}) = -\frac{d\lambda_{21}}{dt} + \frac{d\lambda_{22}}{dt} = -M\frac{di_1}{dt} + L_2\frac{di_2}{dt}$$

It is important to note that the self-inductance term is always positive. That is, according to our current direction and voltage polarity conventions, the voltage-current relationship for an inductance is $v = +L\, di/dt$, provided the current enters the terminal corresponding to the positive side of the voltage. Hence, the contributions to both v_1 and v_2 in Figure 1-31 due to the self-inductances are positive. The sign of the mutual-inductance term can, in general, be either positive or negative. If the fluxes due to the two currents are additive, the mutual-inductance terms are positive; if the two currents produce fluxes in opposite directions, the mutual-inductance terms are negative.

Finally, we define the *coupling coefficient*

$$k = \frac{M}{\sqrt{L_1L_2}} \qquad \textbf{(1-17)}$$

It can be shown that due to physical considerations $M \leq \sqrt{L_1L_2}$. Hence, the magnitude of the coupling coefficient is confined to values between zero and one, that is,

$$0 \leq k \leq 1 \qquad \textbf{(1-18)}$$

Note that $k = 0$ corresponds to $M = 0$ (no magnetic coupling between the coils), while $k = 1$ corresponds to the maximum possible coupling $M = \sqrt{L_1L_2}$ (all the flux generated in one coil also links the other coil and vice versa). Hence, the coupling coefficient is a measure of the amount of magnetic coupling between the coils, with $k = 0$ indicating no magnetic coupling and $k = 1$ indicating perfect magnetic coupling.

The preceding discussion was concerned with a type of physical device—two coils of wire positioned such that they were magnetically coupled. We now present a two-terminal pair network element that can be used to model this circuit element.[4] This four-terminal network model is presented in Figure 1-32(a); it has four electrical quantities associated with it—v_1, v_2, i_1, i_2. The equations relating these four quan-

[4] A network using only the elements defined in Section 1-2 can be constructed to model a pair of coupled coils (see Problem 2-5). Because of the complexity of that model and the frequent use of coupled coils in engineering practice, the simpler symbol of Figure 1-32 is used.

tities are

$$v_1 = L_1 \frac{di_1}{dt} + M \frac{di_2}{dt}$$
$$v_2 = M \frac{di_1}{dt} + L_2 \frac{di_2}{dt} \quad \text{(1-19)}$$

The constants L_1 and L_2 are called *self-inductances,* and M is called the *mutual inductance.* The dots indicate the signs of the mutual-inductance terms. If the two currents both enter (or leave) dotted terminals, the sign is positive; if only one of the two currents enters a dotted terminal, the sign is negative. Hence, the equations associated with Figure

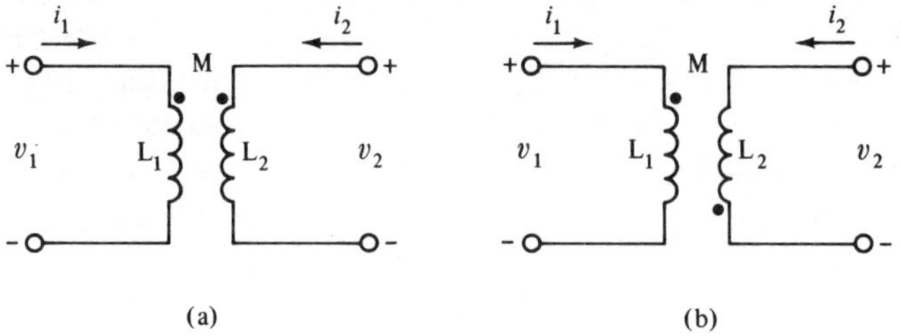

Figure 1-32 Coupled coil model.

1-32(a) are as given in Equation (1-19), while the equations for the network of Figure 1-32(b) are given by

$$v_1 = L_1 \frac{di_1}{dt} - M \frac{di_2}{dt}$$
$$v_2 = -M \frac{di_1}{dt} + L_2 \frac{di_2}{dt}$$

1-4 Power

Consider the two-terminal network element illustrated in Figure 1-33. We define the *power* (sometimes called *instantaneous power*) absorbed by

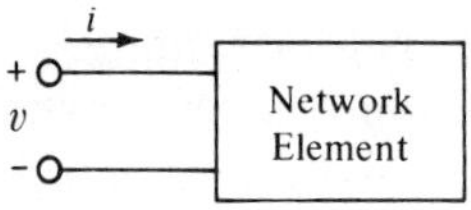

Figure 1-33 Two terminal element.

the network element to be the product of the element voltage multiplied by the current into the positive terminal, that is,

$$p = vi \quad \text{(1-20)}$$

The units of p are watts. One watt is one volt-ampere. This definition is valid for all two-terminal network elements. If the power is positive, the element is absorbing or dissipating power. If the power is negative, the element is giving up or delivering power. If the voltage or current is time-varying, then the power is generally varying as a function of time, that is, $p = p(\mathrm{t})$.

For a resistance the power formula can be related to the resistance (or conductance) in a variety of ways. Indeed, substituting Equations (1-1) and (1-2) into Equation (1-20), we obtain

$$p_{\mathrm{R}} = vi = i^2\mathrm{R} = v^2/\mathrm{R} = v^2\mathrm{G} = i^2/\mathrm{G} \tag{1-21}$$

Thus, a positive resistance always dissipates power, while a negative resistance always delivers power.

The capacitance and the inductance store energy. We now show that the energy relationships for the capacitance and the inductance are consistent with the power definition given by Equation (1-20).

The energy stored in a capacitance is one-half the product of the charge and the voltage, that is,

$$\mathcal{E}_{\mathrm{C}} = \tfrac{1}{2}qv \tag{1-22}$$

Now, power is the rate of change of energy with time. Hence, the power dissipated in the capacitance is given by

$$p_{\mathrm{C}} = \frac{\mathrm{d}}{\mathrm{dt}}\left[\tfrac{1}{2}qv\right] = \tfrac{1}{2}v\frac{\mathrm{d}q}{\mathrm{dt}} + \tfrac{1}{2}q\frac{\mathrm{d}v}{\mathrm{dt}} \tag{1-23}$$

But from Equation (1-14) we note that $\mathrm{d}q/\mathrm{dt} = \mathrm{C}\,\mathrm{d}v/\mathrm{dt} = i$. Hence, substituting these quantities into Equation (1-23) yields

$$p_{\mathrm{C}} = \tfrac{1}{2}vi + \tfrac{1}{2}q\frac{i}{\mathrm{C}} \tag{1-24}$$

Finally, according to Equation (1-3), $q/\mathrm{C} = v$ so that we have

$$p_{\mathrm{C}} = \tfrac{1}{2}vi + \tfrac{1}{2}vi = vi \tag{1-25}$$

Note that since $q = \mathrm{C}v$, Equation (1-22) can be written in the form

$$\mathcal{E}_{\mathrm{C}} = \tfrac{1}{2}\mathrm{C}v^2 \tag{1-26}$$

Hence, the energy stored in the capacitance is positive for positive C and negative for negative capacitances. Furthermore, $p = \mathrm{d}\mathcal{E}_{\mathrm{C}}/\mathrm{dt}$ so that the capacitance absorbs power when the energy is increasing and delivers power when the energy is decreasing. If the stored energy is not changing, the power absorbed is zero. This is true even if the energy stored is not zero.

An analogous argument for the inductance is left as an exercise. (See Problem 1-6.)

The definition given by Equation (1-20) is also valid for independent sources. Hence, independent sources can either absorb power or deliver power.

For the network elements having two pairs of terminals, as in Figure 1-34, the total power absorbed is the sum of the powers absorbed by

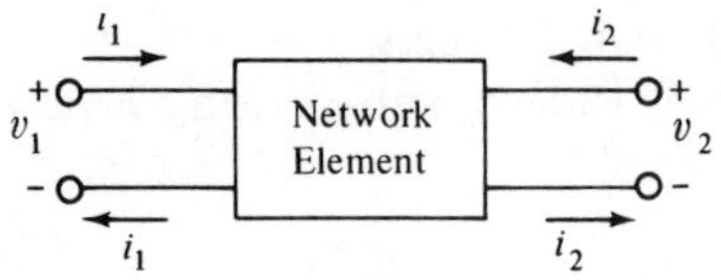

Figure 1-34 Two terminal pair element.

each of the two terminal pairs, that is,

$$p = v_1 i_1 + v_2 i_2 \tag{1-27}$$

Recall that all of the controlled sources had either $i_1 = 0$ or $v_1 = 0$. Hence, $v_1 i_1 = 0$ for all four controlled sources, so that the power entering the control terminal pair is zero. On the other hand, the power entering the terminal pair connected to the (controlled) current or voltage source can be either positive or negative.

Finally, we introduce the concept of average power. By *average power* we mean the time average of the instantaneous power, that is,

$$\langle p \rangle = \lim_{T \to \infty} \frac{1}{2T} \int_{-T}^{+T} p(t)\, dt \tag{1-28}$$

The average power $\langle p \rangle$ is a number and is a more "gross" measure of power than the instantaneous power p. This definition is convenient when the instantaneous power is varying periodically. For the case of periodically-varying power (p is periodic), Equation (1-28) reduces to

$$\langle p \rangle = \frac{1}{\text{period duration}} \int_{\text{one period}} p(t)\, dt \tag{1-29}$$

PROBLEMS

1-1 What are the dimensions of capacitance in terms of length, mass, time, and charge?

1-2 What are the dimensions of inductance in terms of length, mass, time, and charge?

1-3 Show that the formula in Equation (1-8), $v = L\, di/dt$, has correct dimensions. That is, show that voltage has the same dimension as (henries) (amperes per second).

1-4 For the network shown [Figure P1-4(a)], characterize the cascade of two controlled sources as a single controlled source. That is, replace Figure P1-4(a) by Figure P1-4(b). The question mark in front of v_1 in the box is a number, and the question mark in the diamond is either $\pm$ or an arrow.

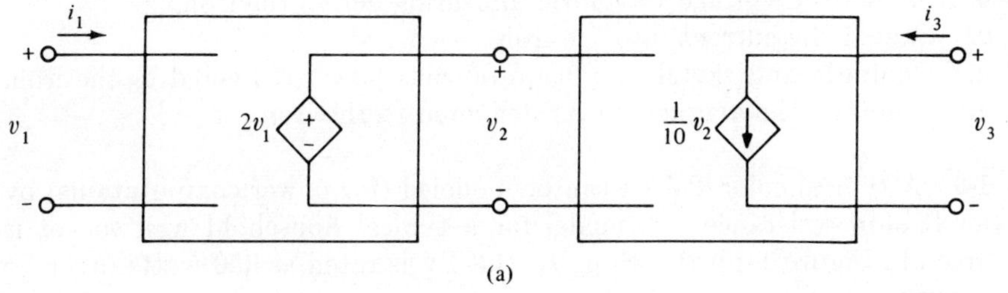

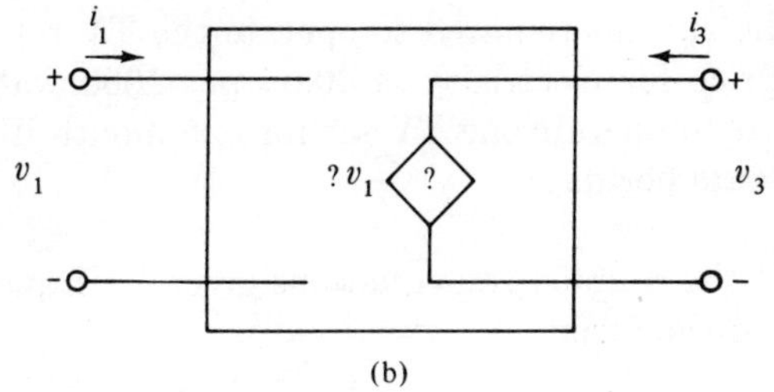

Figure P1-4

1-5 The operational amplifier is a voltage-controlled voltage source with extremely high value of μ. Often these devices are constructed with two inputs and one output such that the output voltage is proportional to the difference between the two input voltages. Construct a model of the circuit described in Figure P1-5, using network elements defined in this chapter.

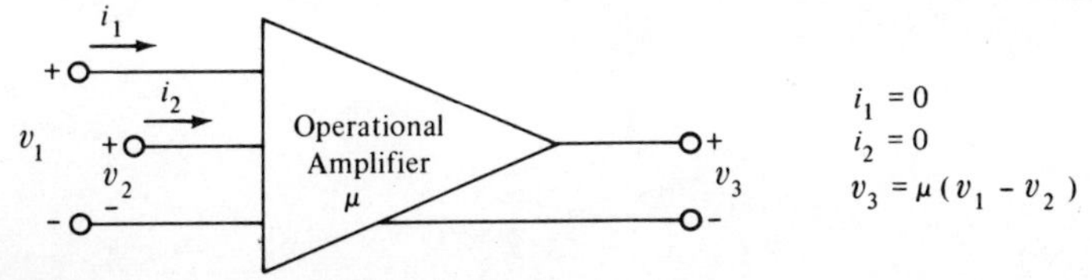

Figure P1-5

1-6 Show that the energy stored in an inductance is $\frac{1}{2}Li^2$.

1-7 Using the power equation for a two-terminal pair network, Equation (1-27), and the definition of the coupled coil model, Equations (1-19), show that if k [Equation (1-17)] is less than one, then the total energy stored is always positive.

1-8 A model for a standard household wall socket is given in Figure 1-13(a). A typical electric iron is nothing more than a big resistor for converting electrical energy into heat and can be modeled by a 20 Ω resistance.

a. Sketch a model of this iron plugged into the wall socket.
b. Sketch the voltage waveform appearing across the iron.
c. Sketch the current into the iron.
d. Compute and sketch the instantaneous power delivered to the iron.
e. Compute the average power delivered to the iron.

1-9 A typical color TV set can be modeled (for power computations) by an R-ohm resistance. A model for a typical household wall socket is given in Figure 1-13(a). Suppose the TV is rated as 350 watts (average power).

a. Compute the current required to operate the TV set.
b. A typical price for electricity is $0.05 per 1000 watt-hours. How much does it cost to operate the TV set for one month if it is used at an average of 10 hours per day.

1-10 Show that the coupled coil equations given by Equation (1-19) can also be written in the form

$$i_1 = \frac{L_2}{L_1L_2 - M^2}\int v_1\,dt - \frac{M}{L_1L_2 - M^2}\int v_2\,dt$$

$$i_2 = -\frac{M}{L_1L_2 - M^2}\int v_1\,dt + \frac{L_1}{L_1L_2 - M^2}\int v_2\,dt$$

CHAPTER

2

Network Laws and Their Application

2-1 Introduction

In Section 1-2 we defined the nine network elements in terms of the constraints they place on the voltages and currents at their terminals. These network element laws do not determine the values of the voltages and currents, but do fix certain relationships between them. Recall also that the relationships imposed by the network element equations depend only on the network elements; they do not depend on how the network element is connected to other network elements.

In this chapter we present two physical laws that place further constraints on the network element voltages and currents. Whereas the network element laws depend only on the network elements and not on the way they are connected, the laws we now introduce depend only on the way the network elements are connected, but not on the particular network elements involved. These laws are known as *Kirchhoff's current law* and *Kirchhoff's voltage law.*

2-2 Kirchhoff's Laws

Suppose that we take some of the network elements defined in Chapter 1 and connect them, as illustrated in Figure 2-1. Each box in Figure 2-1 represents one of the network elements. Note that each of these network elements is a two-terminal element, except element #5, which has

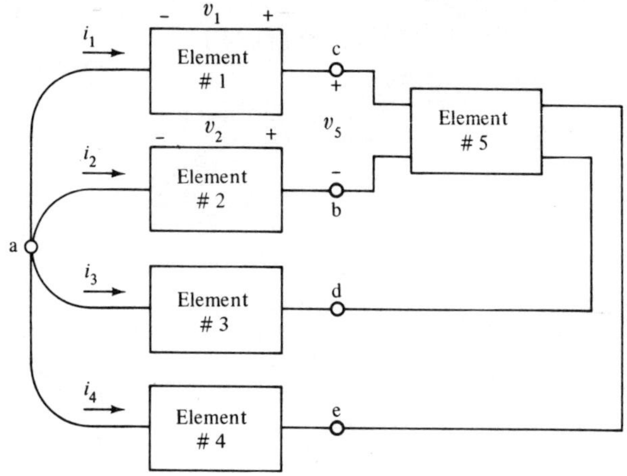

Figure 2-1 A network.

four terminals. These network elements are connected by a fusion of their terminals. The common terminals (labeled a, b, c, d, e in Figure 2-1) are called *nodes*, and the resulting interconnection of network elements is called a *network*. The corresponding interconnection of circuit elements is called a *circuit*. In the circuit the junctions corresponding to the nodes are the places where the wire leads are soldered together.

Kirchhoff's current law (KCL). Recall that none of the nine network elements introduced in Chapter 1 is capable of storing current. The current into one terminal of a two-terminal network element (say, for example, the capacitance) is equal to the current out the other terminal. The same is true for each terminal pair of each of a the two terminal pair network elements. Furthermore, we do not allow current to collect at the nodes. Hence, the total net current into (or out of) any node is zero. For example, for node a in the network of Figure 2-1, we could write $i_1 + i_2 + i_3 + i_4 = 0$. A similar equation could be written at each of the other nodes. Note that this equation does not specify any of the network currents; rather, it places a constraint on the relationship between the currents. Also note that this constraint depends only on the way the network elements are connected and not on what they are. For example, the equation holds whether element #1 is a resistance, a capacitance, or a voltage source.

Kirchhoff's current law (KCL) states that the sum of currents into (or out of) any node of any network is zero. This rule can be expressed in the form of the equation

$$\sum_{j} i_j = 0 \tag{2-1}$$

where the i_j are the currents into (or out of) a node. One such equation can be written for each node of any network.

Kirchhoff's voltage law (KVL). Refer to Figure 2-1 and note that the voltage at node a relative to node c is the element #1 voltage v_1; furthermore, the voltage at node b relative to node a is the element #2 voltage v_2; finally, the voltage at node c relative to node b is v_5 (the voltage across the left-hand terminal pair of element #5). The potential difference encountered going from node a to node c is independent of the route of travel. For example, the potential difference encountered going from node a to node b (v_2) and then from node b to node c (v_5) is the sum of these two ($v_2 + v_5$). But this must equal the potential difference encountered going from node a to node c via element #1 (v_1). Hence, we conclude that $v_1 = v_2 + v_5$. As a consequence, the sum of voltages around any closed loop must be zero. That is, in Figure 2-1, the voltage at node a relative to node a is zero. But this voltage is also the sum of the voltages going from node a to node b to node c, and then back to node a via element #1. Therefore, $0 = v_2 + v_5 - v_1$. This is the same as our previous equation $v_1 = v_2 + v_5$. Note that this equation does not specify the voltages v_1, v_2, and v_5. Rather, it places a constraint on their values. This constraint equation depends only on the way the network elements are connected, and not on what the particular network elements are.

Kirchhoff's voltage law (KVL) is the general statement of this rule. KVL states that the sum of voltages going from node to node around any closed path is zero. In equation form we write

$$\sum_{j} v_j = 0 \tag{2-2}$$

where the v_j are the voltages around any closed path connecting one or more nodes. One such equation can be written for every closed path for any network.

Since the electronic network models are mathematical entities, Kirchhoff's laws can be looked on as definitions for the networks. These laws also have physical significance for the corresponding circuits. The current law (KCL) is a statement of conservation of charge. If the current flowing into a node were not zero, then the charge, which is the integral of the current, would accumulate at that node. By charge conservation, accumulation of positive charge at one node requires accumulation of negative charge at another. The capacitance allows for such accumulation of charge pairs *inside* an element. We cannot allow the nodes to have such properties if the network rules are to be consistent.

The voltage law (KVL) can be stated as the uniqueness of the voltage between two nodes and the transitive nature of the voltage between nodes. That is, the voltage between node a and node b can be measured directly, or it can be measured as the sum of the voltage between node a and an intermediate node c plus the voltage between nodes c and b. The voltage between two nodes is independent of the path of measurement. This property of voltage is identical to the physical property of electric potential. Thus, the network models are consistent with the physical concepts of electric fields, the fields being confined inside the elements.

2-3 Some Network Theorems

There are several network theorems that are a consequence of the linear nature of the network elements and the laws that govern them. These theorems reduce the effort involved in applying Kirchhoff's laws to networks. In this section, we state the theorems and give examples of their application to networks whose equations are algebraic equations. In Appendix II, the theorems are proved for more general networks where the equations are differential equations.

Voltage division and current division. Consider the network presented in Figure 2-2, in which the voltage v_s is applied across the series combination

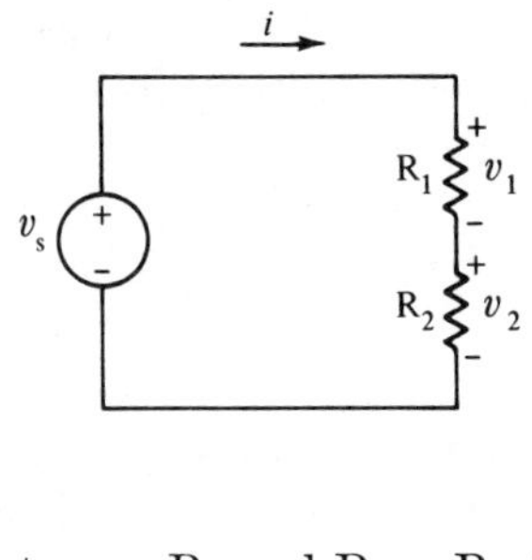

Figure 2-2 Series resistances.

of resistances R_1 and R_2. By KVL we know that

$$v_s = v_1 + v_2 \tag{2-3}$$

The network divides the voltage v_s into two components, v_1 and v_2; hence, it is called a *voltage-divider network,* or a *voltage divider.* Let us compute the two voltages v_1 and v_2 to see how they depend on R_1 and R_2.

The easiest way to compute v_1 and v_2 is first to compute the current i.

From Figure 2-2 we note that

$$\begin{aligned} v_1 &= R_1 i \\ v_2 &= R_2 i \end{aligned} \tag{2-4}$$

Substituting Equations (2-4) into Equation (2-3), we obtain

$$v_s = R_1 i + R_2 i = (R_1 + R_2) i \tag{2-5}$$

Solving for i we have

$$i = \frac{v_s}{R_1 + R_2} \tag{2-6}$$

Finally, substituting i back into Equations (2-4), we obtain

$$\begin{aligned} v_1 &= \frac{R_1}{R_1 + R_2} v_s \\ v_2 &= \frac{R_2}{R_1 + R_2} v_s \end{aligned} \tag{2-7}$$

We conclude that the voltage v_s divides proportionally to the resistances, that is, $v_1 \sim R_1$, and $v_2 \sim R_2$. If the resistances are equal ($R_1 = R_2$), then the voltage divides into two equal parts ($v_1 = v_2 = v_s/2$). If $R_1 > R_2$, then $v_1 > v_2$. For example, if $R_1 = 1$, and $R_2 = 3$, then $v_1 = \frac{1}{4}$, and $v_2 = \frac{3}{4}$.

We next consider the network presented in Figure 2-3, in which the

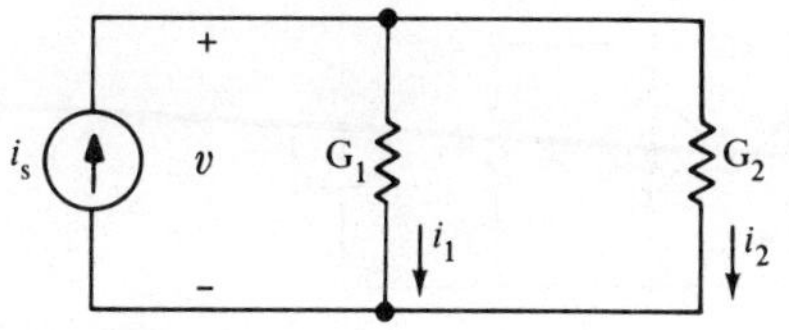

Figure 2-3 Parallel resistances.

current source i_s is applied to a parallel combination of conductances. From KCL we know that

$$i_s = i_1 + i_2 \tag{2-8}$$

so that the network has the effect of dividing the current i_s into two components, i_1 and i_2, and is called a *current-divider network*, or a *current divider*. The easiest way to compute i_1 and i_2 is first to compute the voltage v. Now,

$$\begin{aligned} i_1 &= G_1 v \\ i_2 &= G_2 v \end{aligned} \tag{2-9}$$

and substituting Equation (2-8) into Equation (2-9) yields

$$i_s = G_1v + G_2v = (G_1 + G_2)v \tag{2-10}$$

or

$$v = \frac{i_s}{G_1 + G_2} \tag{2-11}$$

Therefore, we find from using Equation (2-11) in Equations (2-9) that

$$\begin{aligned} i_1 &= \frac{G_1}{G_1 + G_2} i_s \\ i_2 &= \frac{G_2}{G_1 + G_2} i_s \end{aligned} \tag{2-12}$$

Hence, the current divider divides the currents proportional to the conductances. For $G_1 = G_2$, $i_1 = i_2$; for $G_1 > G_2$, $i_1 > i_2$, and so forth.

Superposition. The superposition theorem states that the voltage across or the current through any element in a network containing several independent sources can be found by considering these sources one at a time, computing the voltage or current due to each and adding the results. When one independent source is being considered, the other independent sources are set to zero. An independent voltage source is set to zero by setting its voltage to zero, say $v_s = 0$; an independent current source is set to zero by setting its current to zero, say $i_s = 0$.

As an example of the application of the superposition theorem, consider the network presented in Figure 2-4. This network represents a

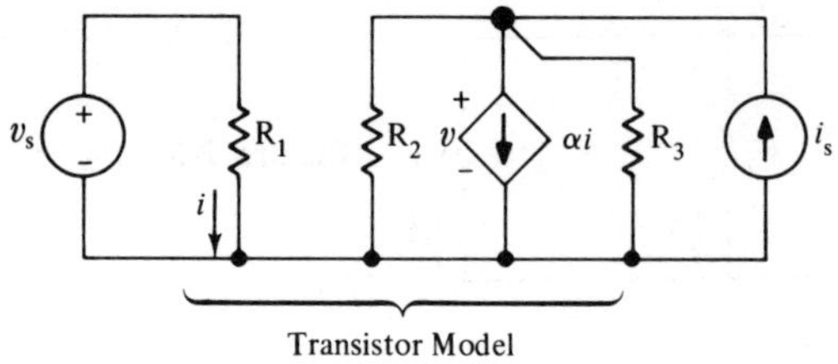

Figure 2-4 Example for superposition.

simple model of a bipolar transistor driven by a voltage source on the left and a current source with a parallel resistance on the right. Let us compute the voltage v across the controlled source. We first compute v with both independent sources active and then use the superposition theorem to obtain the same result.

Applying KVL to the left-most loop of the network of Figure 2-4, we write

$$v_s - R_1 i = 0 \tag{2-13}$$

Applying KCL at the upper right-hand node, we write

$$i_s - \frac{v}{R_3} - \alpha i - \frac{v}{R_2} = 0 \tag{2-14}$$

We now solve Equation (2-13) for i ($i = v_s/R_1$) and substitute this into Equation (2-14) to obtain the answer

$$v = \frac{R_2R_3}{R_2 + R_3} i_s - \frac{R_2R_3\alpha}{R_1(R_2 + R_3)} v_s \tag{2-15}$$

We now show that we get the same answer by using the superposition theorem, which states that we can use the network of Figure 2-5(a) to find the contribution to v due to v_s; then use the network of Figure 2-5(b) to compute the contribution to v due to i_s; then add the results to obtain v.

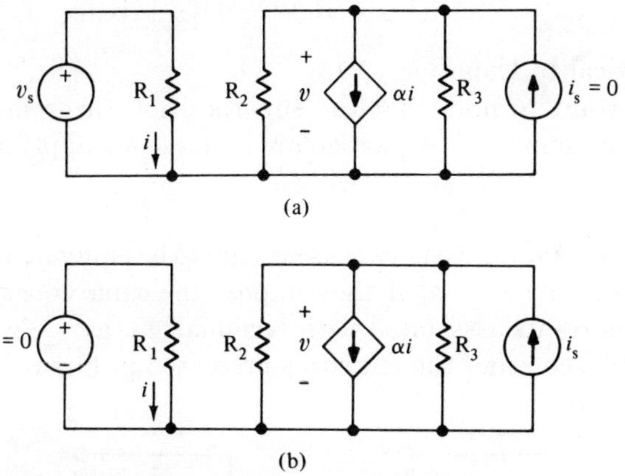

Figure 2-5 Application of superposition.

For the network of Figure 2-5(a) we observe that

$$i = \frac{v_s}{R_1} \tag{2-16}$$

and write a KCL equation at the upper right-hand node

$$\frac{v}{R_2} + \frac{v}{R_3} + \alpha i = 0 \tag{2-17}$$

Substituting Equation (2-16) into Equation (2-17), we obtain

$$v = \frac{-R_2R_3\alpha}{R_1(R_2 + R_3)} v_s \tag{2-18}$$

For the network of Figure 2-5(b) we observe that $i = 0$, so that the controlled current source current is zero; that is, $\alpha i = 0$, since $i = 0$. Hence, the KCL equation at the upper right-hand node is

$$\frac{v}{R_2} + \frac{v}{R_3} - i_s = 0 \tag{2-19}$$

or, solving for v, we obtain

$$v = \frac{R_2R_3}{R_2 + R_3} i_s \tag{2-20}$$

The superposition theorem states that, with both independent sources active, the voltage v is the sum of the two contributions due to v_s and i_s. Adding Equations (2-18) and (2-20) we obtain

$$v = -\frac{R_2R_3\alpha}{R_1(R_2 + R_3)} v_s + \frac{R_2R_3}{R_2 + R_3} i_s \tag{2-21}$$

which is identical to Equation (2-15).

It is important to note that the superposition theorem applies only to independent sources. In particular, it does not apply to controlled sources.

Equivalent networks. Two networks are said to be equivalent with respect to terminals $a_1, a_2, \cdots, a_n$ if they impose the same constraints on the voltages and currents associated with terminals $a_1, a_2, \cdots, a_n$.

For example, consider the networks presented in Figure 2-6. For the

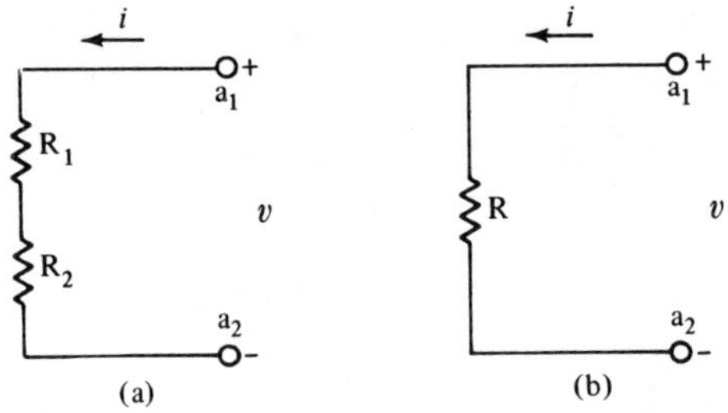

Figure 2-6 Equivalent networks for $R_1 + R_2 = R$.

network of Figure 2-6(a) the voltage across the resistance R_1 is R_1i volts; likewise, the voltage across R_2 is R_2i volts. Hence, according to KVL, the voltage v is the sum of these two voltages:

$$v = R_1i + R_2i = (R_1 + R_2)i \tag{2-22}$$

For the network of Figure 2-6(b), we can write

$$v = Ri \tag{2-23}$$

We conclude that for $R = R_1 + R_2$, the two networks are equivalent with respect to terminals a_1 and a_2, since the voltage-current equations [Equations (2-22) and (2-23)] are identical for this case. Hence, a network consisting of two resistances in series is equivalent to a network consisting of a single resistance, provided $R = R_1 + R_2$.

As a second example of equivalent networks, we consider the two networks presented in Figure 2-7. For the network of Figure 2-7(a), we

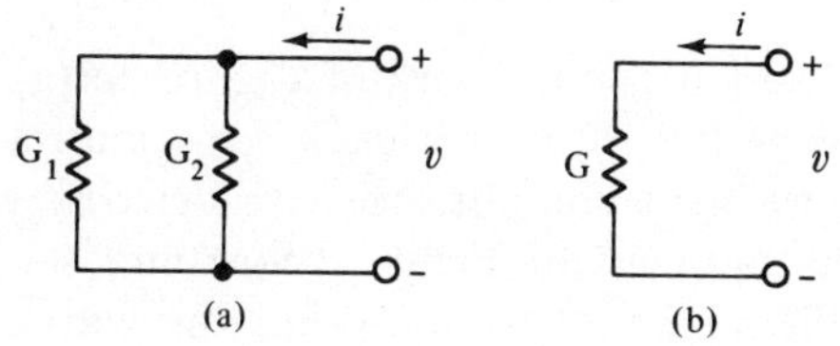

Figure 2-7 Equivalent networks for $G_1 + G_2 = G$.

note that the current through the conductance G_1 is given by G_1v and the current through G_2 is G_2v. Hence, by KCL, the current i into the upper terminal is the sum of the currents through G_1 and G_2:

$$i = G_1v + G_2v = (G_1 + G_2)v \tag{2-24}$$

For the network of Figure 2-7(b), we write

$$i = Gv \tag{2-25}$$

Hence, the two networks are equivalent, provided $G = G_1 + G_2$. This situation is often described by the statement that "conductances in parallel add."

Thevenin's theorem. We now consider the somewhat more complicated situation presented in Figure 2-8 and pose the following question: Can

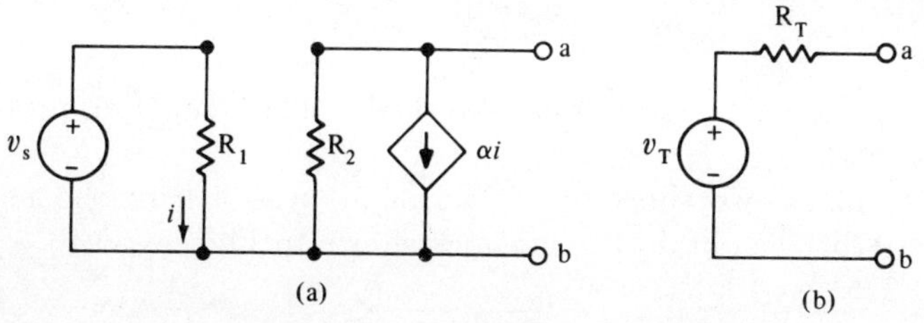

Figure 2-8 An example of Thevenin's theorem.

we find a v_T and an R_T such that the two networks of Figure 2-8 are equivalent with respect to terminals a and b? Rather than ponder this question, let us proceed as if the answer were yes, and attempt to compute a v_T and an R_T such that the networks are equivalent.

We first consider the network of Figure 2-8(b) and reason that if terminals a and b are left open-circuited, then the current into terminal a is zero. It follows that for this situation the voltage across the resistance R_T is zero. Consequently, the open-circuit voltage v_{oc} across terminals a and b is, by KVL, equal to v_T, that is,

$$v_{oc} = v_T \qquad (v_{oc} = \text{open-circuit voltage at terminal a relative to terminal b}) \tag{2-26}$$

We further reason that if the network of Figure 2-8(a) is equivalent to the network of Figure 2-8(b), then their open-circuit voltages must be identical. Hence, we next compute the open-circuit voltage at terminals a and b of the network of Figure 2-8(a) and set v_T equal to this quantity. From Figure 2-8(a) we note that the voltage across R_1 is v_s, so that the current i is given by

$$i = \frac{v_s}{R_1} \tag{2-27}$$

We next note that the current-controlled current source current is

$$\alpha i = \frac{\alpha}{R_1} v_s \tag{2-28}$$

The same current flows through the resistance R_2, and we conclude that the open-circuit voltage at terminal a with respect to terminal b is given by

$$v_{oc} = -\frac{\alpha R_2}{R_1} v_s \tag{2-29}$$

Hence, equating the open-circuit voltages yields

$$v_T = -\frac{\alpha R_2}{R_1} v_s \tag{2-30}$$

Having computed v_T, we are half-finished. The remaining problem is to compute R_T.

Note that if we short-circuit terminals a and b of the network of Figure 2-8(b), we obtain the network shown in Figure 2-9(b). The cur-

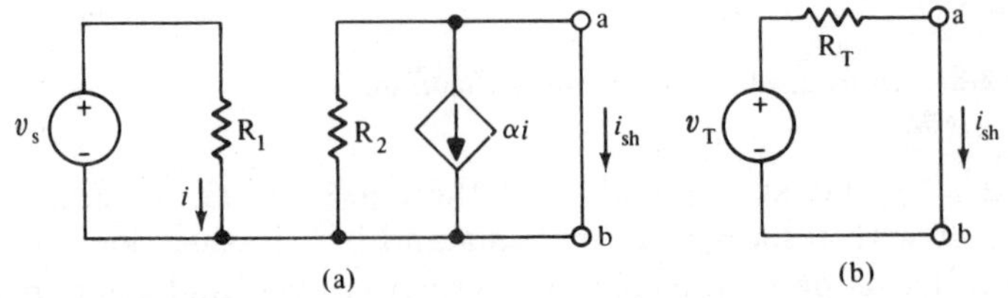

Figure 2-9 Equivalence of short circuit currents.

rent down through the short circuit is given by

$$i_{sh} = \frac{v_T}{R_T} \tag{2-31}$$

or

$$R_T = \frac{v_T}{i_{sh}} \tag{2-32}$$

or, since $v_T = v_{oc}$, we can write

$$R_T = \frac{v_{oc}}{i_{sh}} \tag{2-33}$$

Hence, for the network of Figure 2-8(b), the resistance R_T is given by the ratio of the open-circuit voltage to the short-circuit current. Furthermore, if the two networks of Figure 2-8 are equivalent, their short-circuit currents must be identical. Hence, we can compute i_{sh} by shorting terminals a and b of the network of Figure 2-8(a) and computing the current flowing down in this short circuit. Shorting terminals a and b results in the network presented in Figure 2-9(a). We note that the current through R_1 is still given by $i = v_s/R_1$, and the controlled source current is $\alpha i = (\alpha/R_1)v_s$. However, since the voltage across R_2 is zero, the current through R_2 is zero. Hence, by KCL,

$$i_{sh} = -\frac{\alpha}{R_1} v_s \tag{2-34}$$

Combining Equations (2-29, 2-32, 2-34) we have

$$R_T = \frac{v_{oc}}{i_{sh}} = \frac{-(\alpha R_2/R_1)v_s}{-(\alpha/R_1)v_s} = R_2 \tag{2-35}$$

It is also possible to compute R_T by a somewhat different approach. This procedure becomes apparent by referring to Figure 2-8(b) and noting that if we set $v_T = 0$, then looking into terminals a and b, we see only the resistance R_T. Hence, if we set $v_T = 0$ and connect an independent current source to terminals a and b, as illustrated in Figure 2-10(a), then the

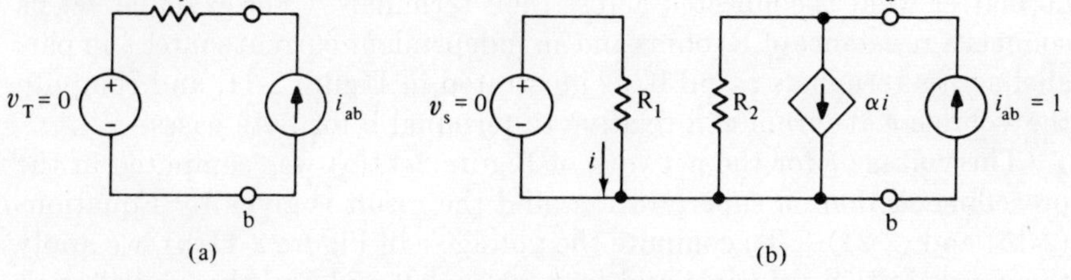

Figure 2-10 An alternative method for computing R_T.

voltage at terminal a relative to terminal b is given by

$$v_{ab} = R_T i_{ab} \tag{2-36}$$

or

$$R_T = \frac{v_{ab}}{i_{ab}} \tag{2-37}$$

Hence, an alternative method for computing R_T for any network is to set all independent sources to zero, to connect an independent current source to terminals a and b, and to compute the voltage v_{ab}. R_T is given by the ratio $R_T = v_{ab}/i_{ab}$. Equation (2-37) is valid for any choice of i_{ab}. A particularly convenient choice is $i_{ab} = 1$, for then

$$R_T = v_{ab}, \qquad i_{ab} = 1 \tag{2-38}$$

Note that we could also have connected an independent voltage source of value v_{ab} to terminals a and b, computed the resulting current, and then used Equation (2-37).

As an example of this procedure, we compute R_T for the network of Figure 2-8(a). As illustrated in Figure 2-10(b), we set $v_s = 0$ and connect a 1-ampere current source to terminals a and b. $v_s = 0$ causes $i = 0$, and so the current-controlled current source current $\alpha i = 0$. Hence,

$$v_{ab} = i_{ab} \cdot R_2 = 1 \cdot R_2 = R_2 \tag{2-39}$$

which is identical to the result obtained previously [Equation (2-35)].

It is important to note that when applying this method, only independent sources are set to zero. In particular, controlled sources are not set to zero. In the example illustrated by Figure 2-10(b), the current-controlled current source was not arbitrarily set to zero as part of the procedure; it simply happened in this particular example that the control current went to zero when we set $v_s = 0$.

We have thus far shown that if we set $v_T = (-\alpha R_2/R_1)v_s$ and $R_T = R_2$, the networks of Figure 2-8 are equivalent, relative to terminals a and b if we leave the terminals open-circuited (they both have the same open-circuit voltage) or short-circuited (they both have the same short-circuit current). In fact, the two networks of Figure 2-8 are equivalent no matter what is connected across their terminals. For example, let us connect a resistance of R_3 ohms and an independent current source i_s in parallel across terminals a and b, as illustrated in Figure 2-11, and compute the voltage v at terminal a relative to terminal b for both cases.

The voltage v for the network of Figure 2-11(a) was computed in the preceding section on superposition, and the result is given by Equations (2-15) and (2-21). To compute the voltage v in Figure 2-11(b), we apply the superposition principle and first set $i_s = 0$ and find the contribution to v due to v_s. Using the voltage divider principle defined by Equation

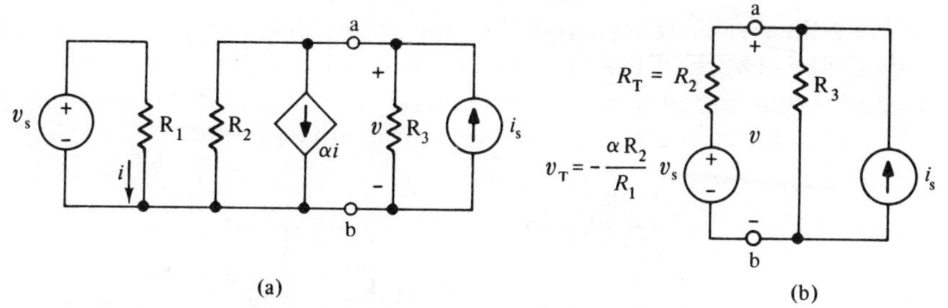

Figure 2-11 Application of Thevenin's theorem.

(2-9), we find

$$v = \frac{R_3}{R_T + R_3} v_T = \frac{R_3}{R_2 + R_3}\left(\frac{-\alpha R_2}{R_1} v_s\right) = \frac{-\alpha R_2 R_3}{R_1(R_2 + R_3)} v_s \tag{2-40}$$

In order to compute the contribution to v due to the independent current source, we set $v_s = 0$ and use the result illustrated in Figure 2-5(b). With the independent voltage source equal to zero, $i = \alpha i = 0$, and the independent current source sees two conductances ($1/R_2$ and $1/R_3$) in parallel. But this is equivalent to a single conductance of value

$$G = \frac{1}{R_2} + \frac{1}{R_3} = \frac{R_2 + R_3}{R_2 R_3} \tag{2-41}$$

Hence,

$$v = \frac{i_s}{G} = \frac{R_2 R_3}{R_2 + R_3} i_s \tag{2-42}$$

Therefore, the voltage v of the network of Figure 2-11(b) is given by the sum of Equations (2-40) and (2-42), that is,

$$v = \frac{-\alpha R_2 R_3}{R_1(R_2 + R_3)} v_s + \frac{R_2 R_3}{R_2 + R_3} i_s \tag{2-43}$$

Comparing Equation (2-43) to Equation (2-15) or Equation (2-21), we note that this is the same result as obtained by solving the network of Figure 2-11(a).

The existence of an equivalent network consisting of an independent voltage source v_T and a resistance R_T for any network consisting of the network elements defined in Chapter 1 (excluding the frequency-dependent elements-capacitances, inductances, and coupled coils) can be proved. The statement of this rule is called *Thevenin's theorem*, v_T is sometimes called the *Thevenin equivalent voltage* and R_T the *Thevenin equivalent resistance* or, in short, the *Thevenin resistance.* A proof of the theorem is given in Appendix II. In Chapter 4 we introduce the concept of impedance and generalize the statement to include networks containing capacitances, inductances, and coupled coils.

Norton's theorem. Thevenin's theorem states that any network has an equivalent network of the form shown in Figure 2-12(a). *Norton's theorem* states that any network has an equivalent network of the form shown in Figure 2-12(b). A proof is given in Appendix II.

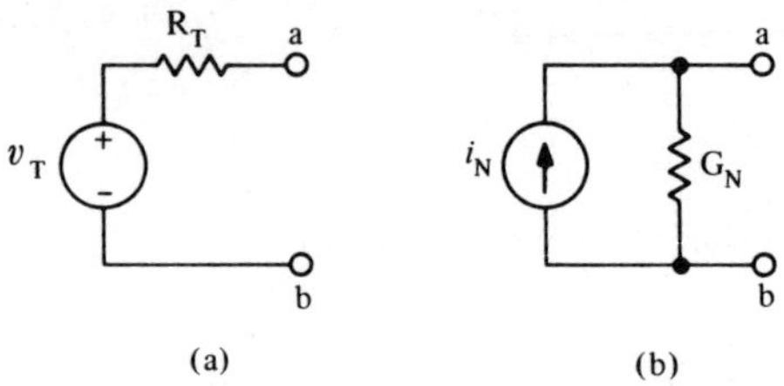

Figure 2-12 Equivalent networks for $i_N = v_T/R_T$ and $G_N = 1/R_T$.

We now determine the relationship between the parameters of the Norton equivalent network (i_N and G_N) and the parameters of the Thevenin equivalent (v_T and R_T).

If we connect a short circuit across the terminals a and b of the Norton equivalent network, as illustrated in Figure 2-13(a), the short-circuit current i_{sh} is given by $i_{sh} = i_N$. For the Thevenin equivalent

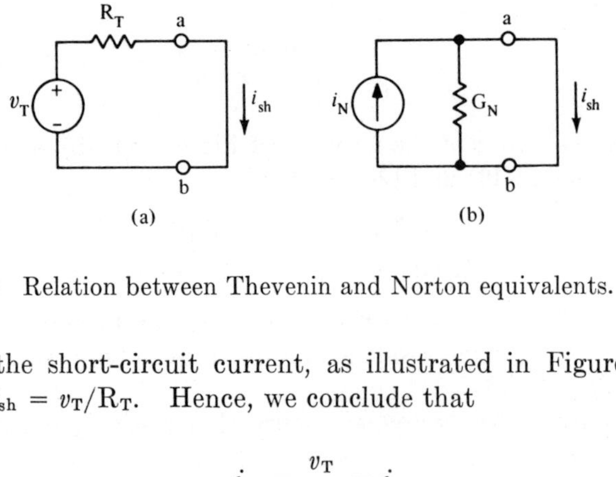

Figure 2-13 Relation between Thevenin and Norton equivalents.

network, the short-circuit current, as illustrated in Figure 2-13(b), is given by $i_{sh} = v_T/R_T$. Hence, we conclude that

$$i_N = \frac{v_T}{R_T} = i_{sh} \tag{2-44}$$

Referring to Figure 2-12(a), we note that the open-circuit voltage for the Thevenin equivalent network is given by v_T. For the Norton equivalent network, the open-circuit voltage is i_N/G_N. Hence,

$$v_T = \frac{i_N}{G_N} \tag{2-45}$$

Finally, substituting Equation (2-44) into Equation (2-45) and solving for G_N, we find

$$G_N = \frac{1}{R_T} = \frac{i_{sh}}{v_{oc}} \tag{2-46}$$

To give an example, we now compute the Norton equivalent network relative to terminals a and b for the network of Figure 2-14(a). We

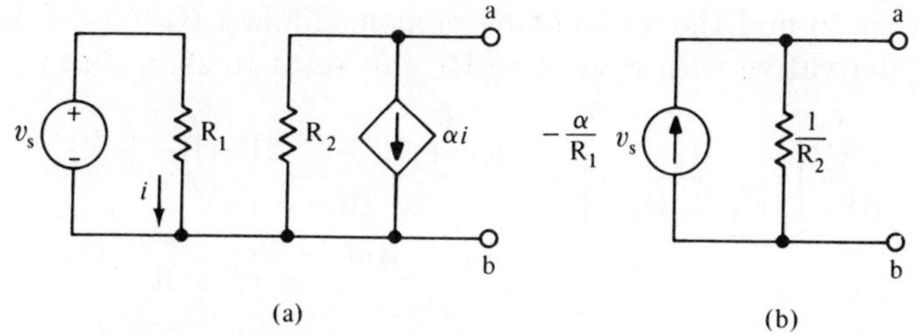

Figure 2-14 Equivalent networks.

have already found the open-circuit voltage and the short-circuit current for this network. [See Equations (2-29) and (2-34).] Hence,

$$\begin{aligned} i_N &= i_{sh} = -\frac{\alpha}{R_1} v_s \\ G_N &= \frac{i_{sh}}{v_{oc}} = \frac{1}{R_2} \end{aligned} \tag{2-47}$$

The Norton equivalent network is presented in Figure 2-14(b). i_N is often called the *Norton-equivalent current source* and G_N, the *Norton-equivalent conductance.*

Power transfer. In the previous section we stated that any two-terminal network containing no frequency-dependent elements has a Thevenin equivalent network, as illustrated in Figure 2-15. Suppose that we now

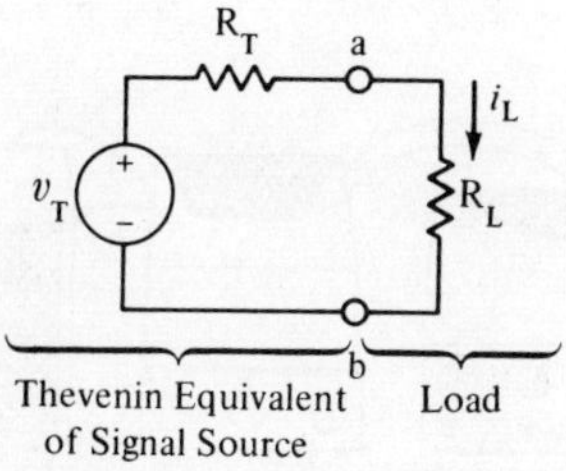

Figure 2-15 The power transfer problem.

connect a resistance R_L across terminals a and b and pose the following question: For a given R_T, what value of R_L will maximize the power delivered to the load from the signal source? This question is easy to answer, and we proceed as follows.

The current through R_L is given by $i_L = v_T/(R_T + R_L)$. Hence, the power dissipated in R_L is given by

$$p_L = i_L^2 R_L = \left(\frac{v_L}{R_T + R_L}\right)^2 R_L = v_T^2 \left[\frac{R_L}{(R_T + R_L)^2}\right] \tag{2-48}$$

In order to find the value of R_L that maximizes $R_L/(R_T + R_L)^2$, we take the derivative with respect to R_L and set it to zero, that is,

$$\begin{aligned} \frac{d}{dR_L}\left[\frac{R_L}{(R_T + R_L)^2}\right] &= \frac{(R_T + R_L)^2 - 2R_L(R_T + R_L)}{(R_T + R_L)^4} \\ &= \frac{(R_T + R_L) - 2R_L}{(R_T + R_L)^3} = 0 \end{aligned} \tag{2-49}$$

Equation (2-49) is satisfied if $(R_T + R_L) - 2R_L = 0$ or

$$R_L = R_T \tag{2-50}$$

We conclude that the power delivered to the load is a maximum if the load resistance R_L is "matched" to the source resistance R_T. In Chapter 4 we generalize this result to include both sources and loads containing frequency-dependent elements.

As an application of this result, consider the signal source modeled by the network of Figure 2-14(a). Then, if we were asked to select the load resistance (to be connected across terminals a and b) that maximizes the power transfer from the network to the load, we should choose $R_L = R_2$, since the Thevenin equivalent resistance relative to terminals a and b is R_2.

As a second example, we consider the hi-fi system presented in Figure 2-16. A typical speaker can be modeled by a resistance of 8 ohms,

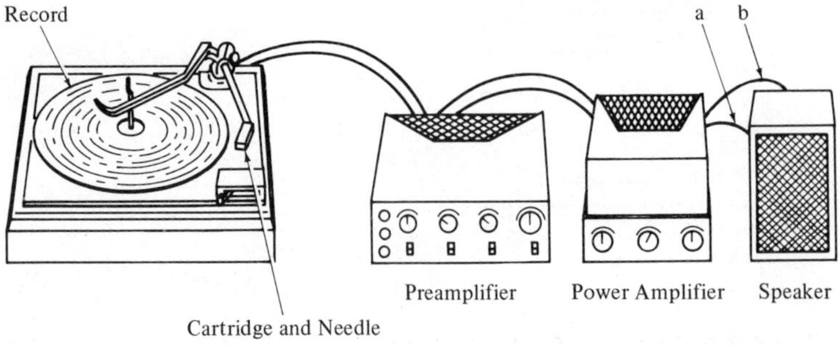

Figure 2-16 A hi-fi system.

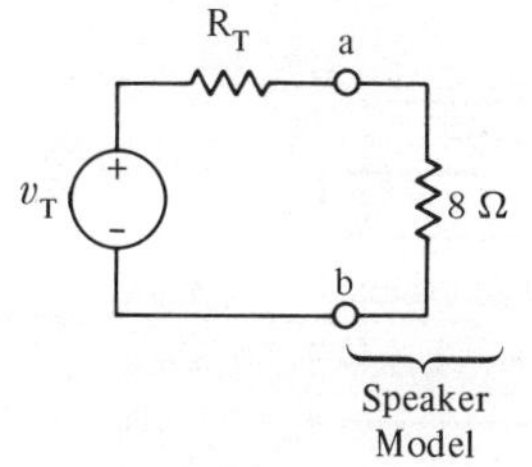

Figure 2-17 Thevenin equivalent of hi-fi system with respect to speaker terminals.

as in Figure 2-17. Since most hi-fi enthusiasts want as much power as possible out of the speaker, the power amplifier should be designed such that its Thevenin equivalent resistance with respect to the speaker terminals is 8 ohms.

Consistency of the power relations—Tellegen's theorem. As pointed out above, electronic networks are mathematical entities with a significant relationship to physical circuits. In the network context, power and energy have been defined. Such definitions are useful only if this defined power has the same properties as physical power. The basic property is conservation of energy. In terms of a network, conservation of energy means that the algebraic sum of all power flow into the elements of a network must be zero. In other words, all energy dissipated or stored in network elements must be supplied by other elements. In physical circuits, this statement is certainly true. The truth of the statement for network models is a special case of a well-established theorem — Tellegen's theorem.

The statement of the theorem is as follows: Consider a network of n two-terminal elements. (For a controlled source, the source part is a two-terminal element.) Let v_i be the voltages across these elements and i_i the currents through them. So long as the v_i satisfy Kirchhoff's voltage law and the i_i satisfy Kirchhoff's current law, then

$$\sum_{i=1}^{n} v_i i_i = 0$$

The proof of the theorem requires only that the v_i and i_i satisfy KVL and KCL, respectively. The element definitions that relate the v's to the i's do not enter the proof. Thus energy is conserved for any network.

2-4 Biasing of Transistors and the Transistor Model

Consider the 2N3705 transistor with characteristic curves as given in Figure 2-18(a). The low-frequency model for this transistor, when the transistor is operating inside the boxed region of the characteristic curves,

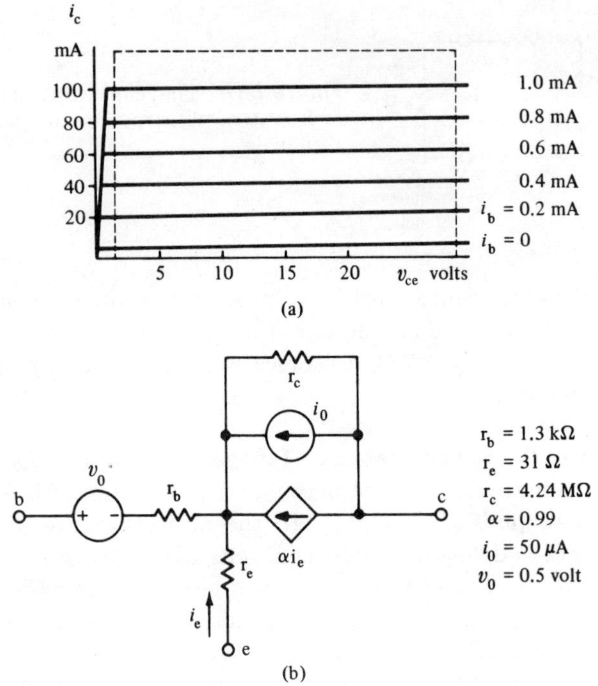

Figure 2-18 2N3705 transistor characteristics.

is given in Figure 2-18(b). The models of Figures 2-18(a) and 2-18(b) are equivalent in the sense of Section 2-3, provided operation is restricted to the boxed region. That is, from Figure 2-18(a) we note that if v_{ce} = 10 volts and i_b = 0.20 mA, then i_c = 12 mA. Hence, if we go to Figure 2-18(b) and connect a 10-V independent voltage source between the collector and emitter terminals and a 0.20-mA independent current source between the base and emitter terminals then we would also find that i_c = 12 mA.

We next note that, since the values of the independent sources (i_0 and v_0) in the transistor model are small, all of the transistor voltages and currents are small. Hence, the state of the transistor, with no external independent sources connected to it, would be very close to the origin of Figure 2-18(a). In particular, it would be outside the boxed region.

One of the most commonly used transistor circuits is shown in Figure 2-19(a). Here we have connected a "signal source" (modeled by an independent voltage source in series with a resistance) between the base and emitter terminals and a "load resistor" (modeled by a resistance) between the collector and emitter terminals. When properly designed, this circuit exhibits "voltage gain"; that is, the output voltage across the load resistance R_L is larger than the input signal voltage v_s. In order to use the transistor model of Figure 2-18(b) for the transistor in

Figure 2-19(a), as illustrated in Figure 2-19(b), we must somehow guarantee that the transistor operation be restricted to the boxed region of the characteristic curves.

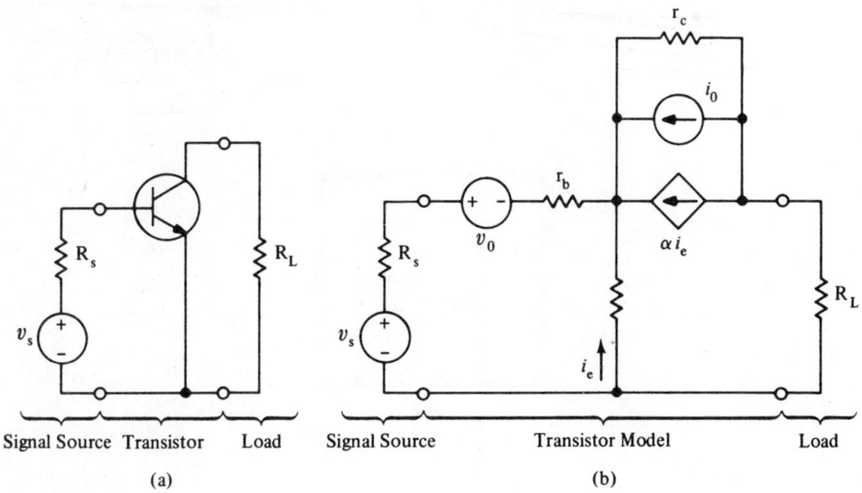

Figure 2-19 Transistor with source and load.

One way to keep the transistor operation within the allowed region is to add an auxiliary circuit to the transistor circuit in order to move the quiescent state of the transistor from near the origin to some point within the boxed region. This point (which gives the state of the transistor with no signal applied, that is, $v_s = 0$) is called the *operating point.* Suppose that we wish to establish the operating point shown in Figure 2-20(a). This can be accomplished by connecting an independent (dc) voltage source v_{bs} and two resistances, R_1 and R_2, to the transistor, as shown in Figure 2-20(b). In order to choose values for v_{bs}, R_1, and R_2 that give the desired operating point, we use the transistor model, as illustrated in Figure 2-20(c). Since the transistor-model parameters specified by the manufacturer are only crude approximations (some of the actual parameter values may vary by 100 percent or more from transistor to transistor), certain approximations that are only more or less valid can be made.

We start by choosing the bias voltage supply v_{bs} to be some convenient value. The only constraint is that the value be small enough so that the transistor will not be damaged in the event that R_2 gets shorted and the full bias voltage v_{bs} appears across the transistor terminals. To choose R_2 we note that R_2 is normally much smaller than R_1 and r_c, so that the current through R_2 is approximately $\alpha i_e + i_0$. Therefore, to a reasonable approximation, R_2 is given by the voltage across R_2, $(v_{bs} - v_{ce_0})$,

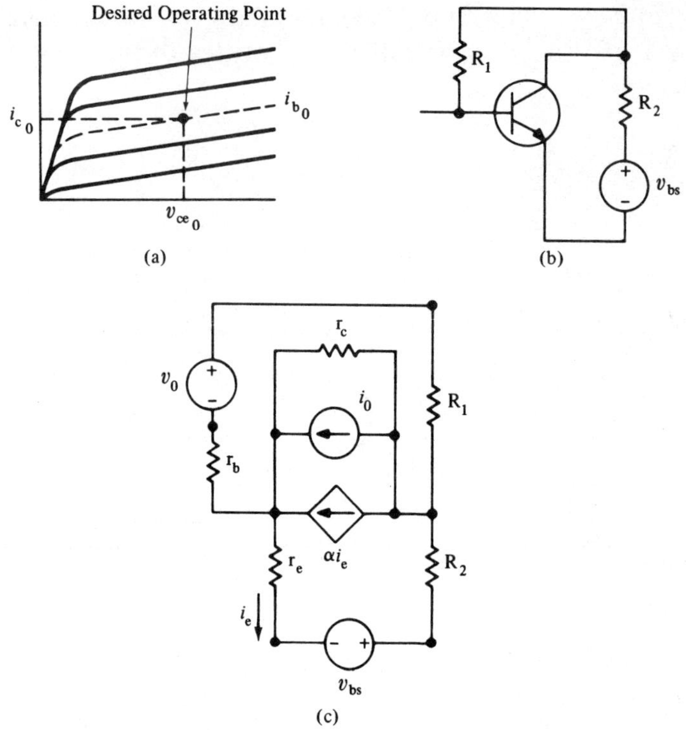

Figure 2-20 Transister biasing.

divided by the current through R_2, $(\alpha i_{b_0} + i_0)$; that is,

$$R_2 \approx \frac{v_{bs} - v_{ce_0}}{\alpha i_{b_0} + i_0} \tag{2-51}$$

In order to select a value for R_1 we first note that the internal resistance of the transistor looking into the base and emitter terminals is very small. That is, if we were to compute the Thevenin equivalent resistance with respect to the base and emitter terminals we would find $R_T \approx 0$. Hence, the voltage at the connection point between the voltage source v_0 and the resistance r_b relative to the emitter is almost zero. Hence, by KVL, the voltage across R_1 is essentially $v_{ce_0} - v_0$. The current through R_1 is the required base current i_{b_0}. Therefore

$$R_1 \approx \frac{v_{ce_0} - v_0}{i_{b_0}} \tag{2-52}$$

The auxilliary network (consisting of v_{bs}, R_1, and R_2), which we added to obtain the desired operating point, is variously called the *bias power supply*, the *bias supply*, or the *biasing circuit*.

We next note that if we reconnect the signal source and load resistance to the network of Figure 2-20(c), as shown in Figure 2-21(a), the operating point is affected. For example, adding R_L across the collector and emitter terminals results in a (Thevenin) equivalent resistance of $R_2R_L/(R_2 + R_L)$ across the collector and emitter terminals. Hence, it appears that we have two choices. One is to incorporate R_L into the

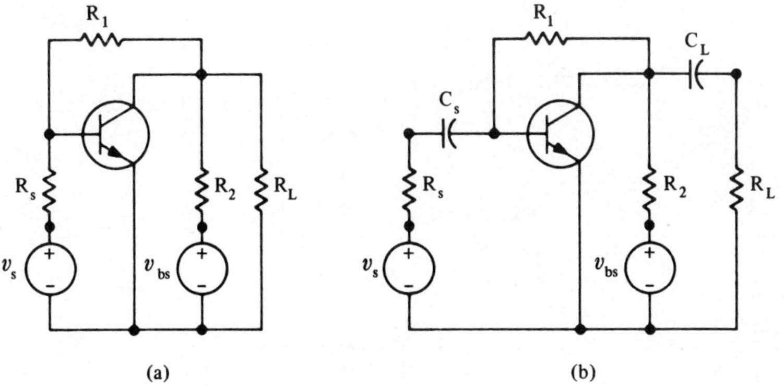

Figure 2-21 Use of blocking capacitors.

bias network by choosing R_2 such that the parallel combination of R_2 and R_L satisfies Equation (2-51); that is,

$$\frac{R_2R_L}{R_2 + R_L} = \frac{v_{bs} - v_{ce_0}}{i_{b_0} - i_0} \tag{2-53}$$

A better choice is to isolate R_L from the transistor and its bias supply by adding a capacitance C_L, as illustrated in Figure 2-21(b). Since the capacitance is an open circuit to dc currents, it "blocks" any dc biasing current from going through R_L and is, therefore, called a *blocking capacitor*. The signal source can also be isolated from the dc bias supply if we insert another blocking capacitor C_s between them. In most applications it is possible to choose values for C_s and C_L such that they are essentially short circuits to the signal currents, while being open circuits to the bias currents. Therefore, the transistor in Figure 2-21(b) is biased at the operating point for any values of R_s and R_L so long as no signal is applied. That is, with $v_s = 0$, the transistor base current is i_{b_0}, the collector current is i_{c_0}, and the collector voltage relative to the emitter is v_{ce_0}. When a signal is applied, the transistor voltages and currents change. Indeed, according to the principle of superposition, the values of these voltages and currents are the sum of the values due to the bias supply v_{bs} and the signal v_s. Since the bias supply voltage v_{bs} is a constant, its contribution to i_b, i_c, and v_{ce} stays constant at i_{b_0}, i_{c_0}, and v_{ce_0}, and so on. Hence, the effect of the signal is to cause the transistor voltages and currents to deviate from the operating point. The transistor model remains valid as long as the signal source does not drive the transistor voltages and currents outside the boxed region.

The biasing arrangement discussed in the preceding paragraphs is only one of a number of ways to bias a transistor in order to obtain a

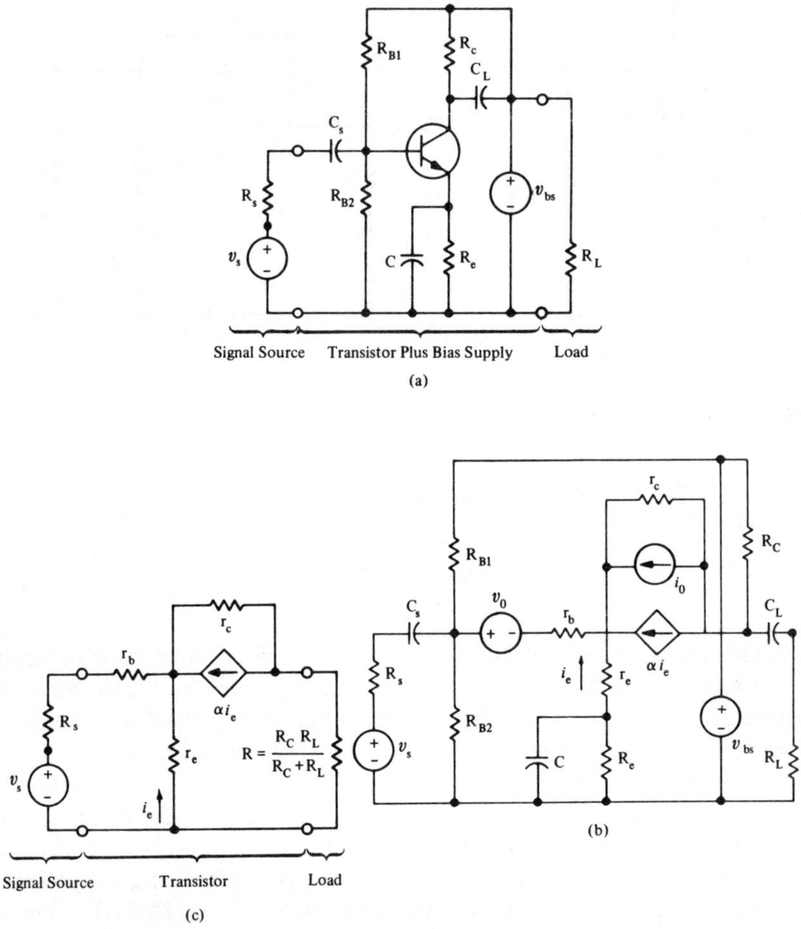

Figure 2-22 Separation of bias and signal.

suitable operating point. Another very common transistor circuit, consisting of a transistor, a signal source, a load, and a bias supply, is presented in Figure 2-22(a). Here the bias supply consists of the (dc) independent voltage source v_{bs}; the four resistances R_c, R_e, R_{B1}, and R_{B2}; the two blocking capacitances C_s and C_L; and the *bypass capacitance* C. The bypass capacitance C is required because the resistance R_e is a necessary part of the biasing system (a dc bias current goes through R_e, causing a dc voltage drop across R_e that helps establish the operating point), but is detrimental to signal operation. C is chosen so that it is an open circuit to the dc bias current, which goes through R_e while the signal current bypasses R_e via the bypass capacitance.

If we now insert the transistor model of Figure 2-18(b) into the network of Figure 2-22(a), we obtain the network of Figure 2-22(b). Although this complete model is required in order to compute all the network voltages and currents, we point out that once the biasing network is designed, we are usually only interested in the output voltage, that is, the voltage across the load resistance R_L. The easiest way to compute the output voltage is to use the principle of superposition: Set all the independent sources to zero except one, and compute the output voltage due to it. Do this for each independent source and sum the results.

Note that the network of Figure 2-22(b) contains four independent sources. Note further that three of these, v_0, i_0, and v_{bs}, are dc sources; hence, they cause only dc voltages and currents. However, the blocking capacitance C_L blocks any dc current from going through R_L. Since no dc currents due to v_0, i_0, or v_{bs} go through R_L, there is no voltage across R_L due to these sources. Hence, the voltage across R_L is due to only the independent source v_s and can be computed (by superposition) with the other three independent sources set to zero. Furthermore, the two blocking capacitances and the bypass capacitance can be approximated with short circuits when we are computing the output due to the signal source. Finally, we point out that R_{B1} and R_{B2} are almost always chosen large enough so that the currents through them are negligibly small, and so that it is reasonable to approximate them with open circuits. After making these approximations, we end up with the network of Figure 2-22(c). We conclude that, except for R_c, the entire bias supply can be neglected. Therefore, in the following chapters we often model transistor circuits as in Figure 2-19(a) and Figure 2-22(c). The bias resistance R_c is understood to be accounted for in load resistance, and it is understood that when building the circuit the bias supply must be added in order to establish an operating point for which the transistor model is valid.

PROBLEMS

2-1 In Figure 2-2, let $v_s = 10$, $R_1 = 1$. Find R_2 such that $v_2 = 9$.

2-2 Let $i_s = 1$ and $G_1 = 1$ in Figure 2-3. Find G_2 such that $i_2 = 0.6$.

2-3 A 9-volt transistor radio is to be run off a 12-volt auto battery. (See Figure P2-3.) The radio requires 5 mA of current. A resistance voltage

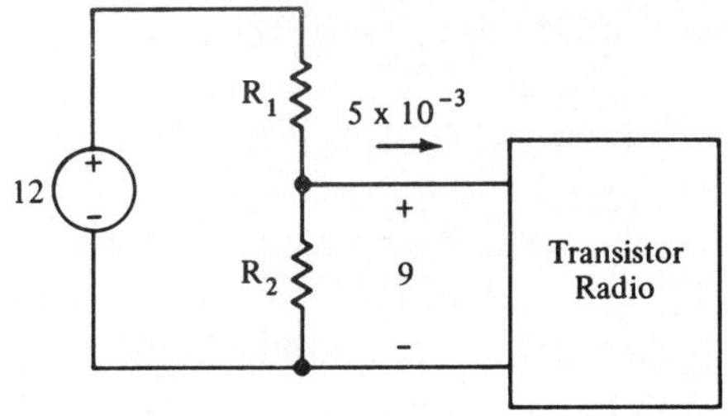

Figure P2-3

divider can be used to make the transistion. Find appropriate values of R_1 and R_2 for the network of the figure.

2-4 Derive a formula for the equivalent resistance of two resistances in parallel (see Figure P2-4).

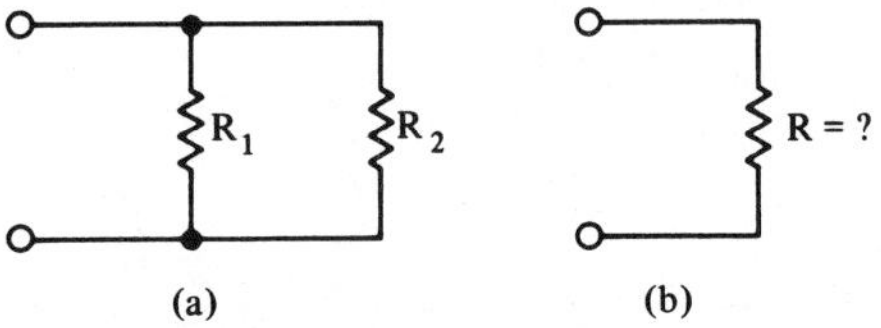

Figure P2-4

2-5 Show that the network shown in Figure P2-5 is equivalent to the coupled coil model of Figure 1-32. Hint: Write a pair of equations in v_1, i_1, v_2, i_2 for this network and show that they are identical to Equations (1-19).

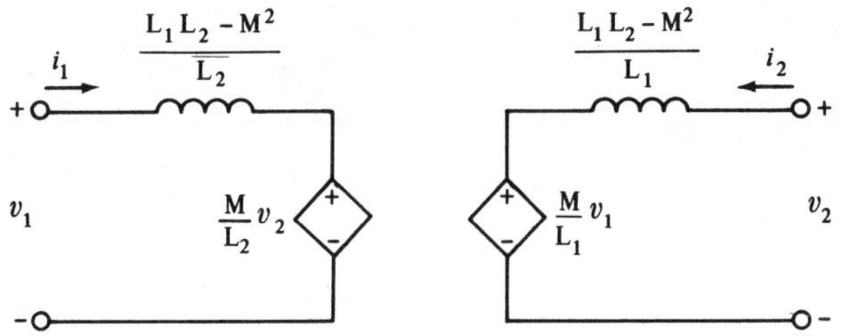

Figure P2-5

2-6 Show that the network of Figure 1-26(c) is equivalent to the network of Figure 1-26(b).

2-7 Find the equivalent resistance of the circuit shown in Figure P2-7. That is, find v_1/i_1.

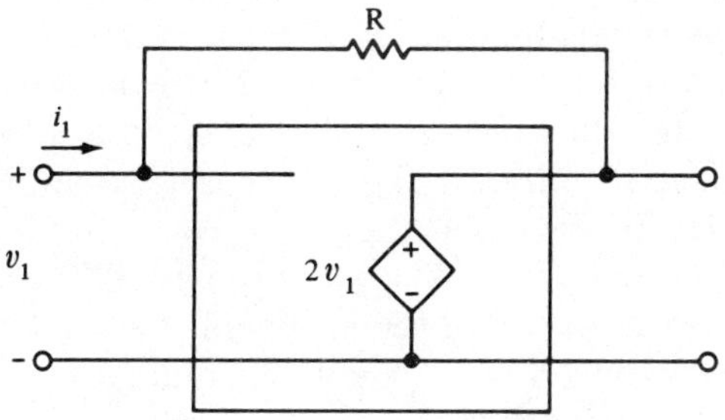

Figure P2-7

2-8 What is the equivalent capacitance of two capacitances in series? Repeat for two in parallel.

2-9 What is the equivalent inductance of two inductances in series? Repeat for two in parallel.

2-10 Consider the pair of coupled coils terminated in a resistance, as shown in Figure P2-10. Let the coupling coefficient be one. Find the input resistance. Show that this resistance is well defined as $M \to \infty$ so long as L_1/L_2 remains constant.

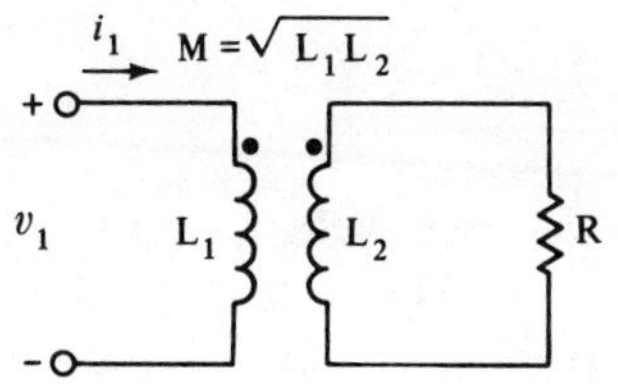

Figure P2-10

2-11 For Figure P2-11, sketch the power delivered to R_2 as a function of R_1.

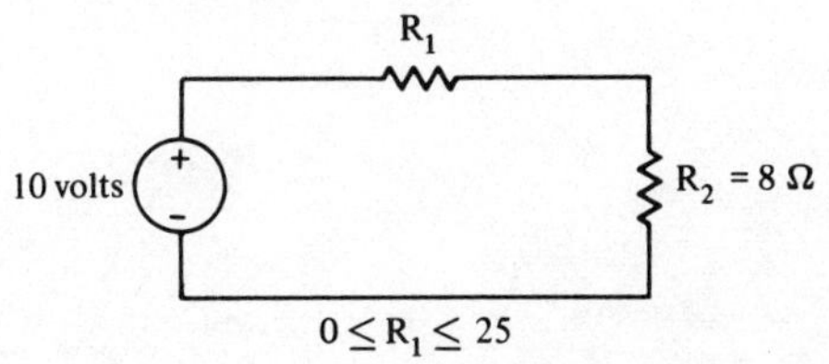

Figure P2-11

2-12 In many circuits, power delivered to resistors is dissipated as heat. The power delivered to resistances in the network model is a measure of this heat dissipation. Power is not linear, so the theorems of this chapter do not apply to power dissipation. For the network of Figure 2-10, with $v_s = 1$, $R_1 = 1$, $R_2 = 100$, $\alpha = 0.9$, $R_3 = 10$, $i_s = 0$, show that the power dissipated in R_T is not the same as the sum of the powers dissipated in R_1, R_2, and the controlled source. The power dissipated in R_3 will be the same in both cases, since v is the same.

2-13 For the network shown in Figure P2-13

a. Use the voltage-divider technique to compute v_1.

b. Use the voltage-divider technique to compute v_2.

c. Use KVL to compute v_0.

d. For $v_s = 10$, $R = 10^3$, compute and sketch v_0 versus R_x for R_x between 900 and 1100.

e. State an application of this network. Hint: Suppose the resistance R_x is the model for a thermistor, which is a device whose resistance is directly proportional to the ambient temperature.

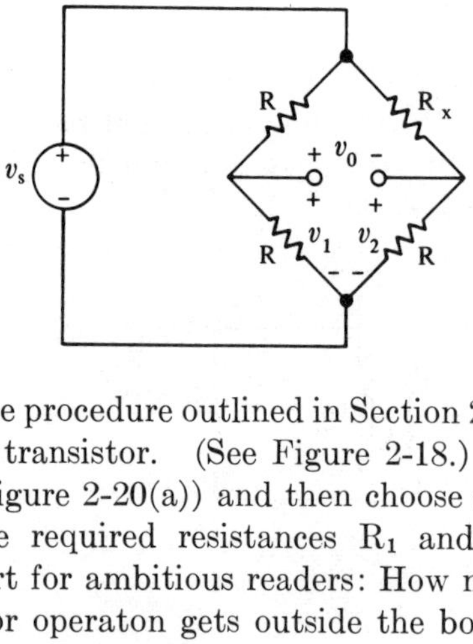

Figure P2-13

2-14 Use the procedure outlined in Section 2-4 to design a bias supply for the 2N3705 transistor. (See Figure 2-18.) First, choose an operating point (see Figure 2-20(a)) and then choose v_{bs} to be about twice v_{ce_0} and compute the required resistances R_1 and R_2. (See Figure 2-20(b).) Optional part for ambitious readers: How negative can v_s become before the transistor operaton gets outside the boxed region (i_b goes to zero)? Assume $R_s = R_L = 0$.

CHAPTER

3 Network Equations

3-1 Introduction

Electronic circuits consist of a number of electrical and electronic devices such as resistors, capacitors, inductors, transformers, transistors, connected in various configurations. Each of these devices can be modeled by one or more network elements, as described in Chapter 1. Therefore, an electronic circuit consisting of a number of these devices connected in any particular way can be modeled by a number of network elements connected in the appropriate arrangement. Any such interconnection of a number of linear network elements is called a *linear network.* (See Figure 3-1.)

Any linear network can be described by a set of linear differential equations. The purpose of this chapter is to present a method for writing such a set of equations. We could, of course, simply write down a number of equations for any linear network by making use of the basic physical laws discussed in Chapter 2. For example, for the network of Figure 3-1(c) we could use Ohm's law to write one equation for each resistance, and we could use KCL and KVL to write a number of additional equa-

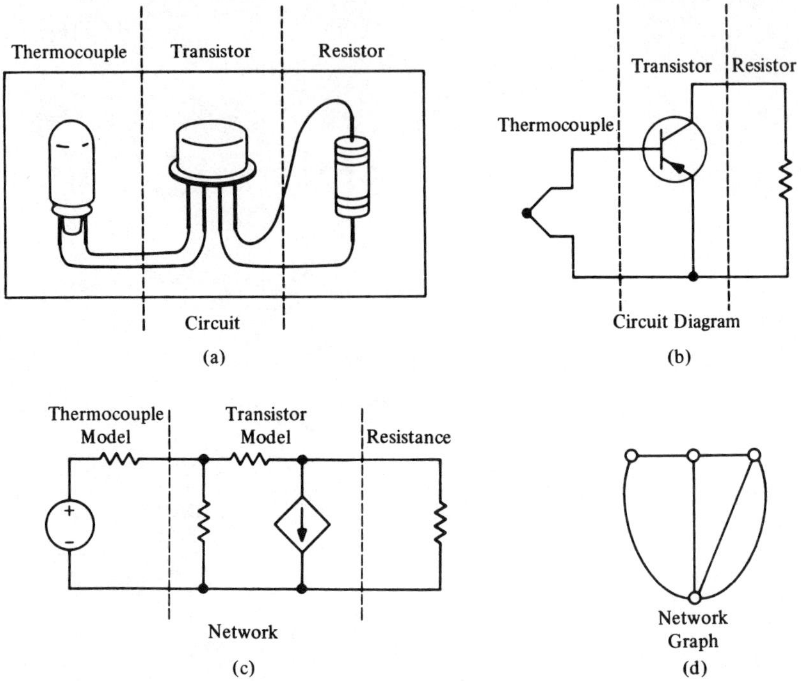

Figure 3-1 (a) An electronic circuit. (b) Its circuit diagram. (c) Its network model. (d) Its network graph.

tions. But how would we know when we had enough equations? How many equations are required to describe a network? Which equations should we write? We need a systematic method that not only tells us how many equations we need, but also which equations to write and how to write them. A very efficient method for achieving these desiderata is based on the rather abstract mathematical theory of topology. Hence, we introduce the concept of network topology in Section 3-2. In Section 3-3, we use these concepts to write a set of network equations, and in Section 3-4, we show how to reduce these equations to a smaller number of equations with a smaller number of unknowns. Sections 3-2, 3-3, and 3-4 lead up to the main result of this chapter, which is presented in Section 3-5, where we present two algorithms (sets of rules) for writing network equations. An alternative method for writing network equations is presented in Section 3-6.

The important material of this chapter is concentrated in Sections 3-2 and 3-5. Sections 3-3 and 3-4 give a justification for the method presented in Section 3-5. The examples of Section 3-3 and the whole of Section 3-4 could be omitted on first reading. On the other hand, the examples of Section 3-3 and the material and examples of Section 3-4 allow a gradual buildup to the important material in Section 3-5. In any event, the reader should not attempt to memorize any of the material of Section 3-4 or to become bogged down in the examples of Sections 3-3 or 3-4.

3-2 Network Topology

In Figure 3-1 we present a picture of an electronic circuit, its circuit diagram, and its network model, or network. A *network graph* is constructed by replacing all network elements with lines and placing a small circle at each point where two or more elements are connected. The network graph for the network of Figure 3-1(c) is shown in Figure 3-1(d). The circles are called *nodes,* and the lines between nodes are called *branches.* We use b to denote the number of branches, and n the number of nodes in the network graph. Hence, for the network graph of Figure 3-1(d), we have $b = 6$ and $n = 4$.

Observe that each branch has both a voltage and a current associated with it. The branch voltage is the voltage across the corresponding element in the network, and the branch current is the current through the corresponding element in the network. It follows that a network consisting of b branches can be completely described by these b voltages and b currents, a total of 2b quantities. Normally most of these quantities are not known. However, since a set of 2b linearly independent equations in 2b unknowns has a unique solution, a b-branch network can be uniquely described by a set of 2b linearly independent equations in these 2b voltages and currents. (A set of N equations is said to be linearly independent if none of the equations can be written as a linear combination of the $N - 1$ remaining equations.) Thus, we have answered the question of how many equations are needed to describe a network—2b.

To assist us in determining which equations to write, we first separate the network graph into two parts—a tree and a cotree. Any set of branches that connects all the nodes of a network graph without forming any closed paths is called a *tree.* It follows that all trees consist of $n - 1$ branches. The set of branches not in the tree is called the complement of the tree or the *cotree.* Hence, all cotrees consist of $b - (n - 1) = b - n + 1$ branches. A number of trees for the network graph of Figure 3-1 are presented in Figure 3-2. Note that each tree (solid lines) defines a unique cotree (broken lines). The branches in a tree are called *tree branches;* cotree branches are called *links.*

The definitions of a tree and a cotree contain no reference to the types of network elements that the branches represent. For example, all four trees of Figure 3-2 are valid trees for the network shown. Later, when we write and solve network equations, we will find that some trees are more convenient than others. Hence, whenever possible, we

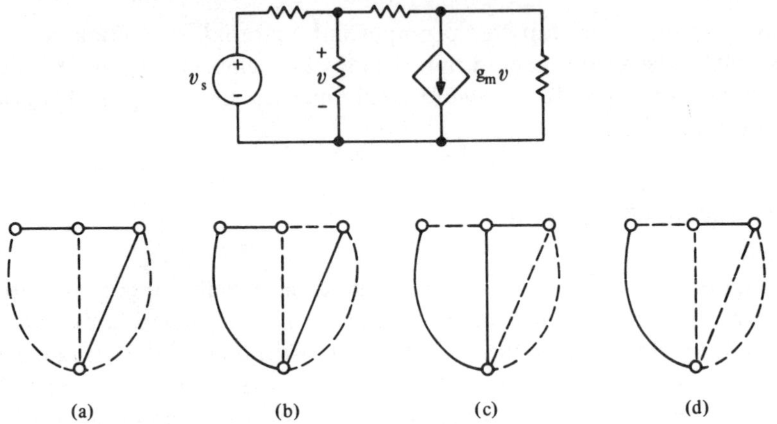

Figure 3-2 A network and its network graph divided into various trees (solid lines) and cotrees (broken lines).

should choose a tree that is convenient in the sense that it simplifies the task of writing network equations.

In Section 3-3 we show that *tree branch voltages* and *link currents* have particular significance, so that it is convenient to associate voltages with tree branches and currents with links. Therefore, voltage sources (both independent and controlled) should be put in the tree; current sources (both independent and controlled) should be put in the cotree. The same holds for the control quantities for controlled sources. Control voltages (for both voltage-controlled voltage sources and voltage-controlled current sources) should, whenever possible, be made tree branch voltages. Similarly, the control currents for current-controlled current sources and current-controlled voltage sources should, whenever possible, be made link currents.

In Chapter 10 we introduce other special tree selections that result in particularly convenient network equations. For the present, we select a tree according to the following constraints:

1. Always put voltage sources in the tree.
2. Always put current sources in the cotree.
3. Whenever possible, put control voltages in the tree.
4. Whenever possible, put control currents in the cotree.

Hence, for the network presented in Figure 3-2, the tree indicated by (c) is preferred to the other three trees shown.

3-3 Systematic Formulation of Network Equations

In this section we present a systematic method for writing a set of 2b linearly independent equations for any b-branch network. This method is based on the topology concepts introduced in Section 3-2.

A systematic method for obtaining one equation for each branch is simply to write the element equation for that branch. For example, if the branch corresponds to an independent source, we write the source equation; if it corresponds to a controlled source, we write the control equation; if it corresponds to an R, L, or C, we write the element equation. The resulting b equations are referred to as the b source, control, and element equations.

In order to write a second equation for each branch, we first separate the network graph into a tree and a cotree and then handle the tree branches and links in different ways. For each tree branch, we use KCL to write the tree branch current as a sum of link currents. This results in n − 1 KCL equations. For each link, we use KVL to write each link voltage as a sum of tree branch voltages. This results in b − n + 1 KVL equations. Since every branch is either a tree branch or a link, the total number of KCL and KVL equations is b, that is, (n − 1) + (b − n + 1) = b.

In summary, 2b linearly independent equations can be obtained for any b branch network by writing the following:

1. b source, control, and element equations (one for each branch).
2. n − 1 KCL equations (one for each tree branch—set the tree branch voltage equal to a sum of link voltages).
3. b − n + 1 KVL equations (one for each link—set the link current equal to a sum of tree branch currents).

A proof that the resulting 2b equations are linearly independent is given in Appendix II.

It is a relatively simple matter to write the b source, control, and element equations. It is also easy to write the b − n + 1 KVL equations by noting that, for each link, there is a closed path of tree branches connected across the link. On the other hand, the n − 1 KCL equations are not always obvious. Therefore, as an aid in writing the KCL equations, we introduce a new concept—the *cut set*.

We reason as follows: None of the network elements is capable of storing current. (All of the network elements defined in Chapter 1 consisted of either one or two terminal pairs and, for each terminal pair, the current into one terminal is the same as the current out the other terminal.) Hence, no group of network elements can store current. Consequently, if we were to blow up a balloon such that part of a network were inside the balloon and part outside, then the total current into the balloon would be zero. This is illustrated in Figure 3-3(a), where we show four wires poking through the surface of the balloon. Since current cannot pile up inside the balloon, the sum of the currents into (or out of) the balloon must be zero, that is, $i_a + i_b + i_c + i_d = 0$. Next, recall that each of the n − 1 KCL equations must relate a tree branch current to one or more link currents. Therefore, to obtain the KCL equation for the *i*th tree branch, we blow up a balloon such that the *i*th tree branch is the only tree branch piercing the surface of the balloon. (Any number of links can penetrate the balloon.) It follows that the current into the balloon through the tree branch must equal the sum of the currents out of the balloon through the links. The set of branches that pierces the balloon is called a *cut set*.

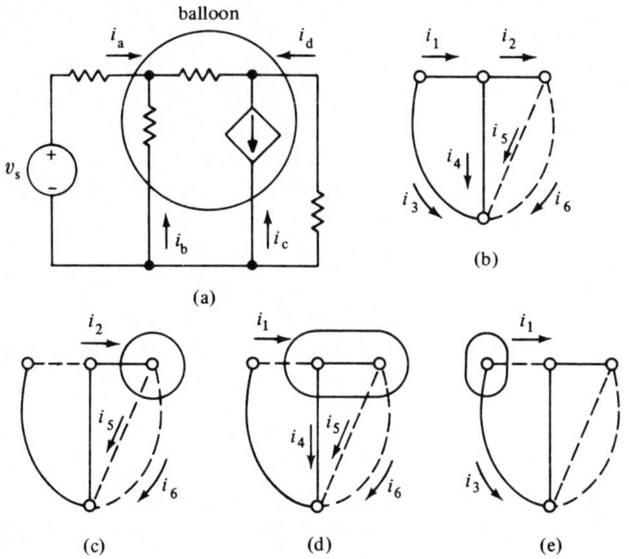

Figure 3-3 Cut sets.

To give an example, we consider again the network of Figure 3-3(a). First we construct a network graph and separate it into a tree and a cotree, as illustrated in Figure 3-3(b). For the tree branch with current i_2, we construct the balloon (cut set) shown in Figure 3-3(c) and set the current into the balloon via the tree branch equal to the sum of the currents out through the links, that is, $i_2 = i_5 + i_6$. For the tree branch with current i_4, we construct the cut set shown in Figure 3-3(d) and again set the current out through the tree branch equal to the sum of the current in through the links, that is, $i_4 = i_1 - i_5 - i_6$. Finally, for the third tree branch, we form the cut set shown in Figure 3-3(e) and write $i_3 = -i_1$.

Note that it is all right to have one or more tree branches (or links) entirely inside the balloon, as in Figure 3-3(d). The branches of consequence are not the ones inside the balloon or those outside, but rather those that penetrate it, that is, the cut set.

We now present some examples of writing sets of linearly independent network equations. The first three examples are quite simple, and the equations are obvious once a tree and cotree are constructed and labeled. Example 3-4 is more complicated, and we construct some cut sets as an aid in writing the equations.

Example 3-1

In this example we write a set of 2b = 12 linearly independent equations for the b = 6 branch network of Figure 3-4(a).

We start by constructing the network graph illustrated in Figure 3-4(b). Then we separate the network graph into a tree and a cotree, as illustrated in Figure 3-4(c). Note that we have placed both the voltage source and the control voltage for the voltage-controlled current source in the tree, and the current source in the cotree. Next, we assign a voltage and current symbol (including polarity) to each branch, as illustrated in Figure 3-4(d).

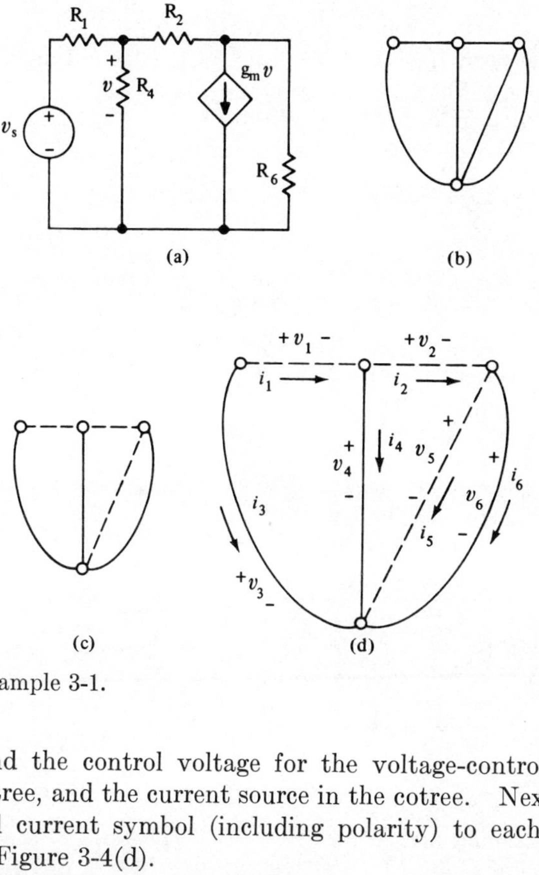

Figure 3-4 Example 3-1.

We can now write a set of twelve linearly independent equations in the twelve variables defined in Figure 3-4(d). The b = 6 source, control, and element equations are

$$\begin{aligned} v_1 &= R_1 i_1 \\ v_2 &= R_2 i_2 \\ v_3 &= v_s \\ v_4 &= R_4 i_4 \\ i_5 &= g_m v_4 \\ v_6 &= R_6 i_6 \end{aligned} \tag{3-1}$$

Note that for branches 1, 2, 4, and 6, we wrote element equations; for branch 3, which corresponds to the independent voltage source, we wrote a source equation; for branch 5, which corresponds to the voltage-controlled current source, we wrote a control equation.

The n − 1 = 3 KCL equations are

$$\begin{aligned} i_3 &= -i_1 \\ i_4 &= i_1 - i_2 \\ i_6 &= i_2 - i_5 \end{aligned} \tag{3-2}$$

Here we have used KCL to write each of the tree branch currents i_3, i_4, and i_6 in terms of the link currents i_1, i_2, and i_5.

The b − n + 1 = 3 KVL equations are

$$\begin{aligned} v_1 &= v_3 - v_4 \\ v_2 &= v_4 - v_6 \\ v_5 &= v_6 \end{aligned} \tag{3-3}$$

Here we have expressed each of the link voltages v_1, v_2, and v_5 in terms of tree branch voltages via KVL.

Equations (3-1), (3-2), and (3-3) are twelve linearly independent equations in the twelve variables $v_1, i_1, v_2, i_2 \cdots, v_6, i_6$.

Example 3-2

In this example we write a set of linearly independent equations for the network of Figure 3-5(a).

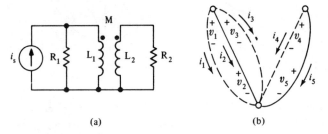

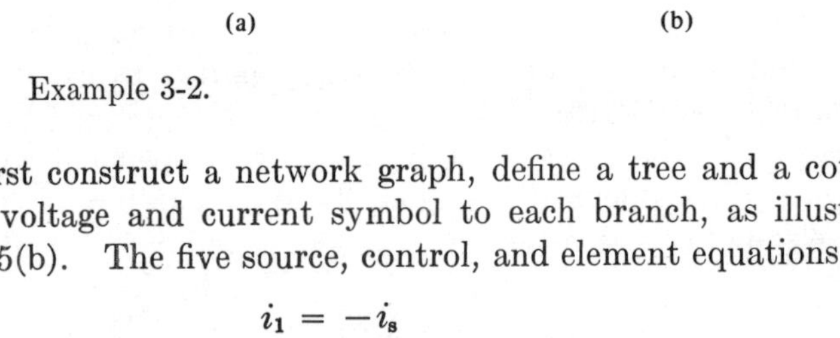

Figure 3-5 Example 3-2.

We first construct a network graph, define a tree and a cotree, and assign a voltage and current symbol to each branch, as illustrated in Figure 3-5(b). The five source, control, and element equations are

$$\begin{aligned} i_1 &= -i_s \\ v_2 &= R_1 i_2 \\ v_3 &= L_1 \frac{di_3}{dt} + M \frac{di_4}{dt} \\ v_4 &= M \frac{di_3}{dt} + L_2 \frac{di_4}{dt} \\ v_5 &= R_2 i_5 \end{aligned} \tag{3-4}$$

Note that the coupled coils require a pair of coupled element equations. (See Section 1-3.)

The KCL equations are given by

$$\begin{aligned} i_2 &= -i_1 - i_3 \\ i_5 &= -i_4 \end{aligned} \tag{3-5}$$

and the KVL equations are given by

$$\begin{aligned} v_1 &= v_2 \\ v_3 &= v_2 \\ v_4 &= v_5 \end{aligned} \tag{3-6}$$

Equations (3-4), (3-5), and (3-6) are ten linearly independent equations in the ten unknown branch voltages and branch currents.

Example 3-3

In this example we write a set of network equations for the network presented in Figure 3-6(a).

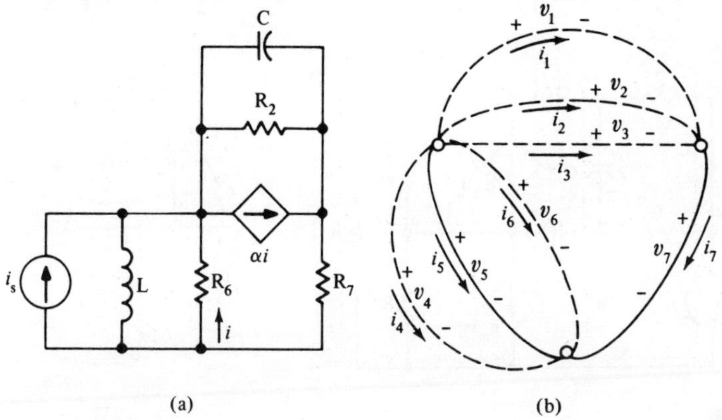

Figure 3-6 Example 3-3.

We start by constructing a tree and defining a voltage and current symbol for each branch, as shown in Figure 3-6(b). The fourteen linearly independent equations are

$$\left.\begin{aligned} i_1 &= C\frac{dv_1}{dt} \\ v_2 &= R_2 i_2 \\ i_3 &= -\alpha i_6 \\ i_4 &= -i_s \\ v_5 &= L\frac{di_5}{dt} \\ v_6 &= R_6 i_6 \\ v_7 &= R_7 i_7 \end{aligned}\right\} \text{ seven source, control, and element equations} \tag{3-7}$$

$$\left.\begin{aligned} i_5 &= -i_1 - i_2 - i_3 - i_4 - i_6 \\ i_7 &= i_1 + i_2 + i_3 \end{aligned}\right\} \text{two KCL equations} \tag{3-8}$$

$$\left.\begin{aligned} v_1 &= v_5 - v_7 \\ v_2 &= v_5 - v_7 \\ v_3 &= v_5 - v_7 \\ v_4 &= v_5 \\ v_6 &= v_5 \end{aligned}\right\} \text{five KVL equations} \tag{3-9}$$

Example 3-4

As a final example of the procedure for writing the 2b network equations, let us consider a more complicated case—a case where the control variable cannot be conveniently located and where it takes some thought to get the correct KVL and KCL equations for the chosen tree. For this purpose consider the network of Figure 3-7. To keep the notation as simple as possible, we assume that all resistances are R ohms, all capacitances are C farads, and both controlled sources have parameter g_m.

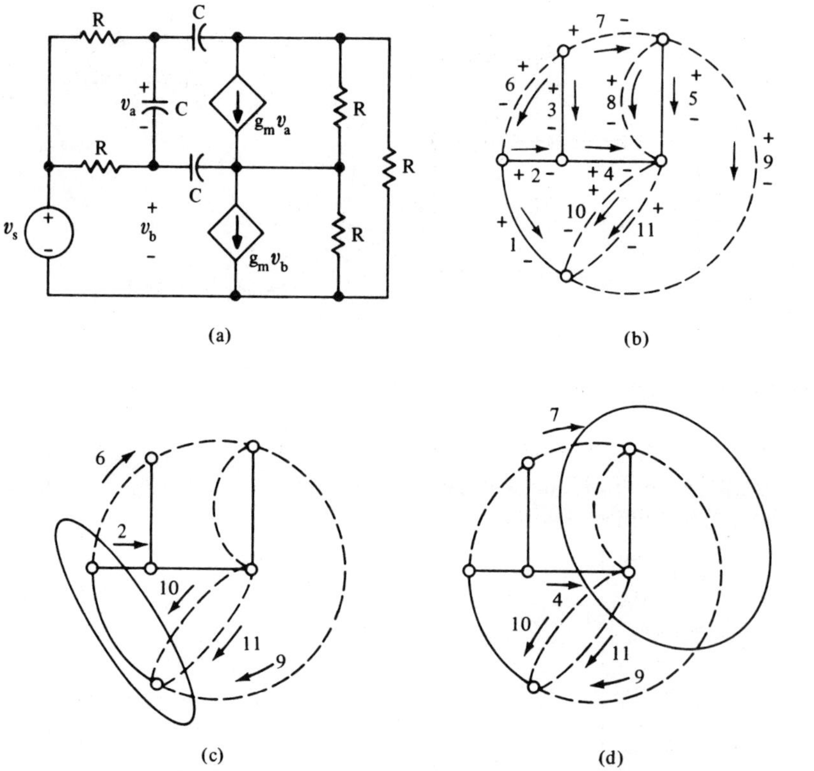

Figure 3-7 Example 3-4.

Since the control voltage v_b is not a branch voltage, we cannot choose a tree such that v_b is a tree branch voltage.[1] We can, however, choose a tree such that v_b is a sum of tree branch voltages. Since there are six nodes, all trees have five branches, and the tree of Figure 3-7(b) is as convenient as any. The remaining six branches are in the cotree.

The source equation is given by

$$v_1 = v_s \tag{3-10}$$

The element equations are given by

$$\begin{aligned} v_2 &= i_2 \mathrm{R} \\ v_3 &= \frac{1}{\mathrm{C}} \int i_3 \, \mathrm{dt} \\ v_4 &= \frac{1}{\mathrm{C}} \int i_4 \, \mathrm{dt} \\ v_5 &= i_5 \mathrm{R} \\ v_6 &= i_6 \mathrm{R} \\ v_7 &= \frac{1}{\mathrm{C}} \int i_7 \, \mathrm{dt} \\ v_9 &= i_9 \mathrm{R} \\ v_{11} &= i_{11} \mathrm{R} \end{aligned} \tag{3-11}$$

The control equations can be written in terms of the network variables by noting that the control voltage v_b is $v_1 - v_2$. Hence, the control equations are given by

$$\begin{aligned} i_8 &= \mathrm{g_m} v_3 \\ i_{10} &= \mathrm{g_m}(v_1 - v_2) \end{aligned} \tag{3-12}$$

The KVL equations that express link voltages in terms of tree branch voltages are easily written if we consider the links one at a time. A tree contains no loops (closed paths). When one link is added to a tree, one loop is formed. The KVL equation around this loop expresses the link voltage in terms of tree branch voltages. Thus we have

$$\begin{aligned} v_6 &= v_2 - v_3 \\ v_7 &= v_3 + v_4 - v_5 \\ v_8 &= v_5 \\ v_9 &= v_5 - v_4 - v_2 + v_1 \\ v_{10} &= -v_4 - v_2 + v_1 \\ v_{11} &= -v_4 - v_2 + v_1 \end{aligned} \tag{3-13}$$

The KCL equations that express the tree branch currents in terms of link currents require more thought. For branches such as branch 5

[1] We could assume a fictitious branch across the terminals that define v_b as a branch with infinite resistance or zero conductance. Such added complexity is unnecessary, as the example demonstrates.

that have one node connected only to links, the equation is obvious. It is given by

$$i_5 = i_7 - i_8 - i_9 \tag{3-14}$$

Similarly

$$\begin{aligned} i_3 &= i_6 - i_7 \\ i_1 &= -i_9 - i_{10} - i_{11} \end{aligned} \tag{3-15}$$

For the other two tree branches (2 and 4), the KCL equations are not so obvious, and so we construct cut sets. For branch 2, we construct the cut set shown in Figure 3-7(c) and write

$$i_2 = -i_6 + i_{10} + i_{11} + i_9 \tag{3-16}$$

For branch 4, we construct the cut set of Figure 3-7(d) and write

$$i_4 = -i_7 + i_{10} + i_{11} + i_9 \tag{3-17}$$

We could also think of the equation for the tree branch current i_3 in terms of link currents as coming from a cut set. Here the surface enclosed only the node joining branches 3, 6, and 7. The three branches form the cut set. Since a tree has no loops, it is easy to see that each tree branch defines one cut set, consisting of itself plus a number of links. Cut sets are to tree branches as loops are to links.

3-4 Reduction of the Number of Network Equations

By applying the strategy outlined in Section 3-3 to any b-branch network, we obtain 2b linearly independent equations in the 2b branch voltages and branch currents. This set of 2b equations in 2b unknowns can easily be reduced to either n − 1 equations in the n − 1 tree branch voltages or to b − n + 1 equations in the b − n + 1 link currents by making either of two sequences of substitutions.

Reduction to equations with the link currents as unknowns. The 2b network equations can be reduced to b − n + 1 equations in the b − n + 1 link currents by first substituting the source, control, and element equations in the KVL equations, and then substituting the KCL equations in these equations. This can be accomplished in three steps:

1. Discard those KVL equations corresponding to links with current sources. In their place, write source equations (for those links containing independent current sources) and control equations (for those links containing controlled sources). The remaining KVL equations are left unaltered. Hence, we still have b − n + 1 equations (one for each link).
2. Eliminate all voltage symbols in the equations obtained from step 1 by substituting the source, control, and element equations.

This results in b − n + 1 equations, in which the unknowns are the tree branch currents and link currents.

3. Eliminate all tree branch currents from the equations resulting from step 2 by substituting the appropriate KCL equations. Since the KCL equations give each tree branch current in terms of link currents, all tree branch currents can be eliminated. The result is b − n + 1 equations (in the b − n + 1 link currents).

Example 3-5

In Example 3-1 we wrote twelve linearly independent equations for the network of Figure 3-8. In this example we reduce these 2b = 12 equations in the twelve branch voltages and currents to b − n + 1 = 3 equations in the three link currents i_1, i_2, and i_5.

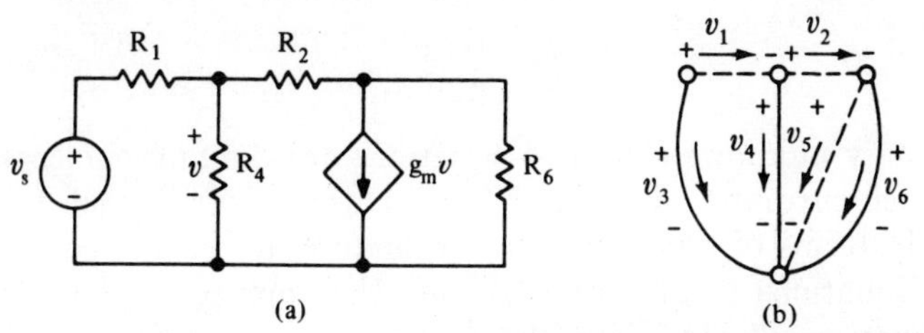

Figure 3-8 Example 3-5.

The original twelve equations are given by

$$v_3 = v_s \} \text{ source equation} \tag{3-18}$$

$$i_5 = g_m v_4 \} \text{ control equation} \tag{3-19}$$

$$\left.\begin{aligned} v_1 &= R_1 i_1 \\ v_2 &= R_2 i_2 \\ v_4 &= R_4 i_4 \\ v_6 &= R_6 i_6 \end{aligned}\right\} \text{ element equations} \tag{3-20}$$

$$\begin{array}{ccc} \text{tree branch currents} & & \text{link currents} \\ \hline i_3 & = & -i_1 \\ i_4 & = & i_1 - i_2 \\ i_6 & = & i_2 - i_5 \end{array} \left.\vphantom{\begin{array}{c}a\\a\\a\\a\end{array}}\right\} \text{ KCL equations} \tag{3-21}$$

$$\begin{array}{ccc} \text{link voltages} & & \text{tree branch voltages} \\ \hline v_1 & = & v_3 - v_4 \\ v_2 & = & v_4 - v_6 \\ v_5 & = & v_6 \end{array} \left.\vphantom{\begin{array}{c}a\\a\\a\\a\end{array}}\right\} \text{ KVL equations} \tag{3-22}$$

Recall that the 3 KVL equations were obtained by writing each link voltage as a sum of tree branch voltages. From Figure 3-8(b) we note that links 1 and 2 correspond to resistances, but that link 5 corresponds to a voltage-controlled current source. Therefore, according to step 1, we discard the KVL equation corresponding to this source ($v_5 = v_6$) and replace it by the control equation ($i_5 = g_m v_4$). Therefore, at the completion of step 1, we have

$$\begin{aligned} v_1 &= v_3 - v_4 \\ v_2 &= v_4 - v_6 \\ i_5 &= g_m v_4 \end{aligned} \tag{3-23}$$

The next step is to substitute the source, control, and element equations into Equations (3-23) in order to eliminate all voltage symbols. Using Equations (3-18) to (3-20) in Equations (3-23) yields

$$\begin{aligned} R_1 i_1 &= v_s - R_4 i_4 \\ R_2 i_2 &= R_4 i_4 - R_6 i_6 \\ i_5 &= g_m R_4 i_4 \end{aligned} \tag{3-24}$$

The only unknowns in these equations are the link currents and the tree branch currents. But the KCL equations express each tree branch current in terms of link currents. Therefore, for each tree branch current in Equations (3-24), we substitute the appropriate KCL equations from Equations (3-21). The result is

$$\begin{aligned} R_1 i_1 &= v_s - R_4(i_1 - i_2) \\ R_2 i_2 &= R_4(i_1 - i_2) - R_6(i_2 - i_5) \\ i_5 &= g_m R_4(i_1 - i_2) \end{aligned} \tag{3-25}$$

We now have three equations in the three link currents. They can be solved by any of the methods of linear algebra.

Reduction to equations with the n − 1 tree branch voltages as unknowns. The 2b network equations for any b-branch network can be reduced to n − 1 equations in the n − 1 tree branch voltages by starting with the n − 1 KCL equations and substituting the source, control, and element equations and the KVL equations into these equations. This is easily accomplished in three steps:

1. Discard those KCL equations corresponding to tree branches with voltage sources. In their place, write the source equations (for independent sources) and the control equations (for controlled sources). The remaining KCL equations are left unaltered. Hence, we still have n − 1 equations (one for each tree branch).
2. Eliminate all current symbols in the equations resulting from step 1 by substituting the source, control, and element equations. This results in n − 1 equations, in which the unknowns are the tree branch voltages and link voltages.

3. Eliminate all link voltages in the equations resulting from step 2 by substituting the appropriate KVL equations. Since the KVL equations give each link voltage in terms of tree branch voltages, all link voltages can be eliminated. The result is n − 1 equations in the n − 1 tree branch voltages.

Example 3-6

In this example we reduce the twelve equations for the six-branch network of Figure 3-9(a) to three equations in the three tree branch voltages v_3, v_4, and v_6.

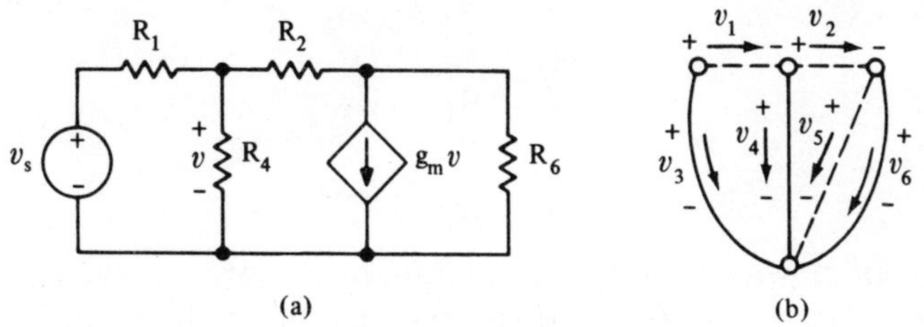

Figure 3-9 Example 3-6.

The twelve equations for the tree and voltage and current symbols defined by Figure 3-9(b) were derived in Example 3-1 and are given by

$$v_3 = v_s\} \text{ source equation} \tag{3-26}$$

$$i_5 = g_m v_4\} \text{ control equation} \tag{3-27}$$

$$\left.\begin{aligned} v_1 &= R_1 i_1 \\ v_2 &= R_2 i_2 \\ v_4 &= R_4 i_4 \\ v_6 &= R_6 i_6 \end{aligned}\right\} \text{element equations} \tag{3-28}$$

$$\left.\begin{array}{ccc} \overbrace{\text{tree branch currents}} & & \overbrace{\text{link currents}} \\ i_3 & = & -i_1 \\ i_4 & = & i_1 - i_2 \\ i_6 & = & i_2 - i_5 \end{array}\right\} \text{KCL equations} \tag{3-29}$$

$$\left.\begin{array}{ccc} \overbrace{\text{link voltages}} & & \overbrace{\text{tree branch voltages}} \\ v_1 & = & v_3 - v_4 \\ v_2 & = & v_4 - v_6 \\ v_5 & = & v_6 \end{array}\right\} \text{KVL equations} \tag{3-30}$$

We start with the three KCL equations that give the tree branch currents in terms of the link currents. From Figure 3-9(b) we note that tree branches 4 and 6 correspond to nonsource elements, but that branch

3 corresponds to an independent voltage source. Therefore, we replace the KCL equation for this tree branch ($i_3 = -i_1$) with the source equation ($v_3 = v_s$). Hence, after completing step 1, we have

$$\begin{aligned} v_3 &= v_s \\ i_4 &= i_1 - i_2 \\ i_6 &= i_2 - i_5 \end{aligned} \tag{3-31}$$

The second step is to use the source, control, and element equations to eliminate all current symbols from Equations (3-31). Substituting Equations (3-26) to (3-28) into Equations (3-31) yield

$$\begin{aligned} v_3 &= v_s \\ \frac{v_4}{R_4} &= \frac{v_1}{R_1} - \frac{v_2}{R_2} \\ \frac{v_6}{R_6} &= \frac{v_2}{R_2} - g_m v_4 \end{aligned} \tag{3-32}$$

The only unknowns in Equations (3-32) are tree branch voltages and link voltages. But the KVL equations give each link voltage in terms of tree branch voltages. Hence, substituting one of Equations (3-30) for each link voltage in Equations (3-32) yield

$$\begin{aligned} v_3 &= v_s \\ \frac{v_4}{R_4} &= \frac{v_3 - v_4}{R_1} - \frac{v_4 - v_6}{R_2} \\ \frac{v_6}{R_6} &= \frac{v_4 - v_6}{R_2} - g_m v_4 \end{aligned} \tag{3-33}$$

Example 3-7

In this example we reduce the fourteen equations for the seven-branch network of Figure 3-10(a) to five equations in the five link currents. We shall also reduce these fourteen equations to two equations in the two tree branch voltages.

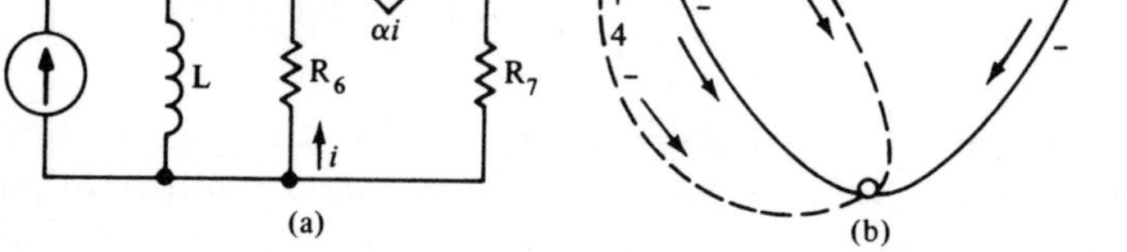

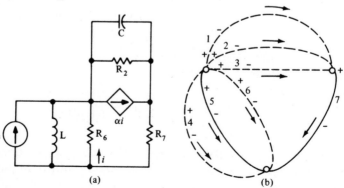

Figure 3-10 Example 3-7.

The network equations for the network and tree in Figure 3-10 were derived in Example 3-3, where we found

$$
\begin{aligned}
i_1 &= C\frac{dv_1}{dt} \\
v_2 &= R_2 i_2 \\
i_3 &= -\alpha i_6 \\
i_4 &= -i_s \\
v_5 &= L\frac{di_5}{dt} \\
v_6 &= R_6 i_6 \\
v_7 &= R_7 i_7
\end{aligned}
\tag{3-34}
$$

$$
\begin{aligned}
i_5 &= -i_1 - i_2 - i_3 - i_4 - i_6 \\
i_7 &= i_1 + i_2 + i_3
\end{aligned}
\tag{3-35}
$$

$$
\begin{aligned}
v_1 &= v_5 - v_7 \\
v_2 &= v_5 - v_7 \\
v_3 &= v_5 - v_7 \\
v_4 &= v_5 \\
v_6 &= v_5
\end{aligned}
\tag{3-36}
$$

To reduce these fourteen equations to five equations in the five link currents, we work first with the five KVL equations given by Equations (3-36). Two of the links correspond to current sources; therefore, we replace the link 3 equation ($v_3 = v_5 - v_7$) with the control equation ($i_3 = -\alpha i_6$) and the link 4 equation ($v_4 = v_5$) with the source equation ($i_4 = -i_s$). We now have

$$
\begin{aligned}
v_1 &= v_5 - v_7 \\
v_2 &= v_5 - v_7 \\
i_3 &= -\alpha i_6 \\
i_4 &= -i_s \\
v_6 &= v_5
\end{aligned}
\tag{3-37a}
$$

We next eliminate all voltage symbols from these equations by using Equations (3-34). This results in

$$
\begin{aligned}
\frac{1}{C}\int i_1\,dt &= L\frac{di_5}{dt} - R_7 i_7 \\
R_2 i_2 &= L\frac{di_5}{dt} - R_7 i_7 \\
i_3 &= -\alpha i_6 \\
i_4 &= -i_s \\
R_6 i_6 &= L\frac{di_5}{dt}
\end{aligned}
\tag{3-37b}
$$

Equations (3-37) contain link currents on the left-hand sides and tree branch currents on the right-hand sides. As a final step, we eliminate the tree branch currents (i_5 and i_7) by using Equations (3-35). The result is five equations in the five link currents:

$$\begin{aligned}
\frac{1}{C}\int i_1\,dt &= L\frac{d}{dt}(-i_1 - i_2 - i_3 - i_4 - i_6) - R_7(i_1 + i_2 + i_3) \\
R_2 i_2 &= L\frac{d}{dt}(-i_1 - i_2 - i_3 - i_4 - i_6) - R_7(i_1 + i_2 + i_3) \\
i_3 &= -\alpha i_6 \\
i_4 &= -i_s \\
R_6 i_6 &= L\frac{d}{dt}(-i_1 - i_2 - i_3 - i_4 - i_6)
\end{aligned} \tag{3-38}$$

To reduce the fourteen equations given by Equations (3-34), (3-35), and (3-36) to two equations in the two tree branch voltages, we work first with the two KCL equations. Since none of the tree branches corresponds to voltage sources, step 1 can be neglected. Next, we eliminate all currents from Equations (3-35) by using the source, control, and element equations. Note that in order to eliminate i_3, we must make a sequence of two substitutions: First, we use the branch 3 (control) equation to replace i_3 with $-\alpha i_6$, then we use the branch 6 element equation to eliminate i_6; that is, $i_3 = -\alpha i_6 = -\alpha(v_6/R_6)$. Hence, we have

$$\begin{aligned}
\frac{1}{L}\int v_5\,dt &= -C\frac{dv_1}{dt} - \frac{v_2}{R_2} + \frac{\alpha v_6}{R_6} + i_s - \frac{v_6}{R_6} \\
\frac{v_7}{R_7} &= C\frac{dv_1}{dt} + \frac{v_2}{R_2} - \frac{\alpha v_6}{R_6}
\end{aligned} \tag{3-39}$$

Equations (3-39) are two equations in terms of the two branch voltages and the five link voltages. But each link voltage is expressed in terms of the tree branch voltages by Equations (3-36); substituting Equations (3-36) into Equations (3-39) yields two equations in the two tree branch voltages:

$$\begin{aligned}
\frac{1}{L}\int v_5\,dt &= -C\frac{d}{dt}(v_5 - v_7) - \frac{(v_5 - v_7)}{R_2} + \frac{\alpha v_5}{R_6} + i_s - \frac{v_5}{R_6} \\
\frac{v_7}{R_7} &= C\frac{d}{dt}(v_5 - v_7) + \frac{(v_5 - v_7)}{R_2} - \frac{\alpha v_5}{R_6}
\end{aligned} \tag{3-40}$$

3-5 Direct Formulation of the Reduced Equations

In Section 3-3 a strategy for formulating a linearly independent set of 2b network equations for any b-branch, n-node network was presented. Then, in Section 3-4, a systematic method for reducing these 2b equations in 2b unknowns to either b − n + 1 equations in the b − n + 1 link currents or to n − 1 equations in the n − 1 tree branch voltages was demonstrated. In this section we combine the results of the previous sections into two algorithms for writing the reduced equations directly from the network graph.

The link current algorithm. The following algorithm is a set of rules for writing b − n + 1 equations in the b − n + 1 link currents directly from the network graph.

1. Construct a tree according to the procedure of Section 3-2:
a. Put all voltage sources in the tree.
b. Put all current sources in the cotree.
c. If possible, put all control voltages in the tree.
d. If possible, put all control currents in the cotree.
2. Assign a current symbol and direction to each link.
3. For each link corresponding to an independent current source, write a source equation relating the link current to the source current.
4. For each link corresponding to a controlled current source, write a control equation relating the link current to other link currents.
5. For each remaining link write a KVL equation of the form {link voltage = Σ tree branch voltages} in terms of link currents and known quantities.

Example 3-8

To illustrate this algorithm we shall write b − n + 1 = 3 equations in the three link currents for the network of Figure 3-11(a).

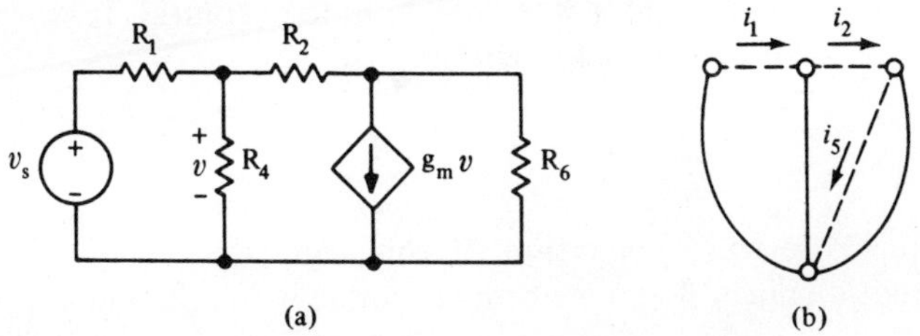

Figure 3-11 Example 3-8.

First, we construct an appropriate tree, say the tree of Figure 3-11(b), and assign a current symbol and direction to each link, say i_1, i_2, and i_5, as in Figure 3-11(b). Since the network contains no independent current sources, we continue to step 4. According to step 4, we must write a control equation for the controlled current source current in terms of other link currents. The control equation $i_5 = g_m v$ can be written in terms of link currents by noting that $v = (i_1 - i_2)R_4$, so that we write

$$i_5 = g_m R_4(i_1 - i_2) \tag{3-41}$$

Finally, according to step 5, we write

$$\begin{aligned} R_1 i_1 &= R_4(i_2 - i_1) + v_s \\ R_2 i_2 &= R_4(i_1 - i_2) + R_6(i_5 - i_2) \end{aligned} \tag{3-42}$$

Equations (3-41) and (3-42) are the same as Equations (3-25) obtained in Example 3-5 by first writing a set of $2b = 12$ equations and then reducing them to three equations in the three link currents via the procedure of Section 3-4.

The tree branch voltage algorithm. The following algorithm is a set of rules for writing $n - 1$ equations in $n - 1$ tree branch voltages directly from the network graph.

1. Construct a tree according to the procedure of Section 3-2:
a. Put all voltage sources in the tree.
b. Put all current sources in the cotree.
c. If possible, put all control voltages in the tree.
d. If possible, put all control currents in the cotree.
2. Assign a voltage symbol and polarity to each tree branch.
3. For each tree branch corresponding to an independent voltage source, write a source equation relating the tree branch voltage to the source voltage.
4. For each tree branch corresponding to a controlled voltage source, write a control equation relating the tree branch voltage to other tree branch voltages.
5. For each remaining tree branch, write a KCL equation of the form {tree branch current = Σ link currents} in terms of tree branch voltages and known quantities.

Example 3-9

Let us illustrate the application of this algorithm by writing a set of $n - 1$ equations in $n - 1$ tree branch voltages for the network of Figure 3-12(a).

The first two steps direct us to construct a tree and to assign voltage symbols and polarities to each tree branch. One valid tree and set of symbols is shown in Figure 3-12(b). As directed by step 3, we write

$$v_3 = v_s \tag{3-43}$$

for the tree branch corresponding to the independent voltage source. Because the network does not contain any controlled voltage sources,

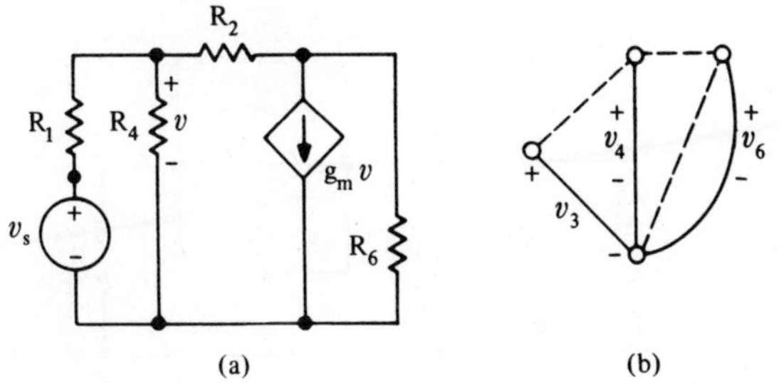

Figure 3-12 Example 3-9.

step 4 can be neglected. Following step 5, we use KCL at the upper center node to set the current down through branch 4 equal to the sum of the two link currents into this node; that is,

$$\frac{v_4}{R_4} = \frac{v_3 - v_4}{R_1} + \frac{v_6 - v_4}{R_2} \tag{3-44}$$

and we apply KCL at the upper right-hand node to set the current down through branch 6 equal to the sum of the link currents entering the node; that is,

$$\frac{v_6}{R_6} = \frac{v_4 - v_6}{R_2} - g_m v_4 \tag{3-45}$$

Equations (3-43), (3-44), and (3-45) are the same as Equations (3-33), which were obtained by first writing twelve equations in the twelve branch voltages and currents and then systematically reducing them to three equations in the three tree branch voltages.

Example 3-10

To illustrate the application of both algorithms, let us consider the network presented in Figure 3-13 along with its network graph.

To write $b - n + 1 = 5$ equations in the five link currents, we first construct a tree and assign current symbols and directions to each link, as indicated by Figure 3-13(c). Link 6 corresponds to an independent current source, so we use the source equation to write

$$i_6 = -i_s \tag{3-46}$$

Link 2 corresponds to a controlled current source, so we use the control equation to write

$$i_2 = -\alpha i_4 \tag{3-47}$$

For the remaining three links, we use KVL to set the link voltage equal

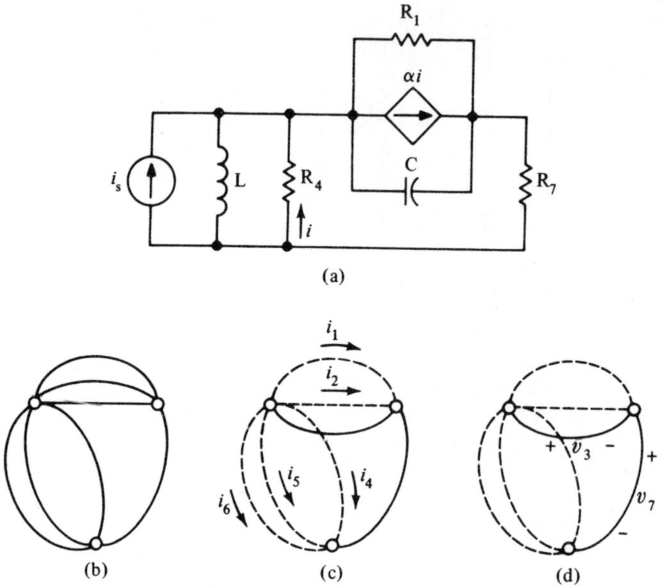

Figure 3-13 Example 3-10.

to a sum of tree branch voltages; that is,

$$
\begin{aligned}
i_1 R_1 &= \frac{1}{C}\int(-i_1 - i_2 - i_4 - i_5 - i_6)\,dt \\
i_4 R_4 &= \frac{1}{C}\int(-i_1 - i_2 - i_4 - i_5 - i_6)\,dt + R_7(-i_4 - i_5 - i_6) \\
L\frac{di_5}{dt} &= \frac{1}{C}\int(-i_1 - i_2 - i_4 - i_5 - i_6)\,dt + R_7(-i_4 - i_5 - i_6)
\end{aligned}
\quad \textbf{(3-48)}
$$

Hence, we have written five linearly independent equations in the five link currents i_1, i_2, i_4, i_5, and i_6.

To write $n - 1 = 2$ equations in the two tree branch voltages, we first construct a tree and then assign voltage symbols and polarities to the two tree branches, as indicated in Figure 3-13(d). Since none of the tree branches corresponds to voltage sources, we use KCL to write each tree branch current as a sum of link currents; that is,

$$
\begin{aligned}
C\frac{dv_3}{dt} &= -\frac{v_3}{R_1} + \alpha\frac{v_3 + v_7}{R_4} - \frac{v_3 + v_7}{R_4} - \frac{1}{L}\int(v_3 + v_7)\,dt + i_s \\
\frac{v_7}{R_7} &= -\frac{v_3 + v_7}{R_4} - \frac{1}{L}\int(v_3 + v_7)\,dt + i_s
\end{aligned}
\quad \textbf{(3-49)}
$$

Networks with coupled coils. Recall that the $b - n + 1$ link current algorithm dictates that we assign link current symbols and then write

KVL equations in terms of these link currents. On the other hand, the n − 1 tree branch voltage algorithm dictates that we assign tree branch voltage symbols and then write KCL equations in terms of these tree branch voltages. Networks containing coupled coils can be analyzed by applying either of these algorithms. However, since the element equations for a pair of coupled coils express the coil voltages in terms of the

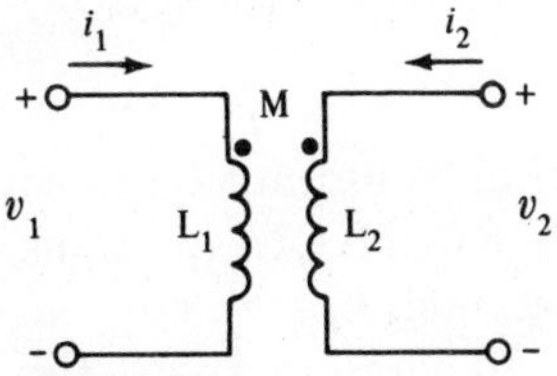

Figure 3-14 Coupled coils.

coil currents, that is (see also Figure 3-14),

$$\begin{aligned} v_1 &= \frac{d}{dt}[L_1 i_1 + M i_2] \\ v_2 &= \frac{d}{dt}[M i_1 + L_2 i_2] \end{aligned} \qquad \textbf{(3-50)}$$

it is apparent that the most convenient algorithm for analyzing networks containing coupled coils is the b − n + 1 link current algorithm, provided that the coil currents are link currents.

Hence, for networks containing coupled coils, we choose a network graph and a tree such that the branches corresponding to the coils are in the cotree, and we then use the b − n + 1 link current algorithm.

Example 3-11. In this example we use the b − n + 1 link current algorithm to write a set of linearly independent network equations for the network of Figure 3-15. We start by constructing a tree such that the branches corresponding to both coils are in the cotree, as illustrated in Figure 3-15. We assign link current symbols, and, since there are no

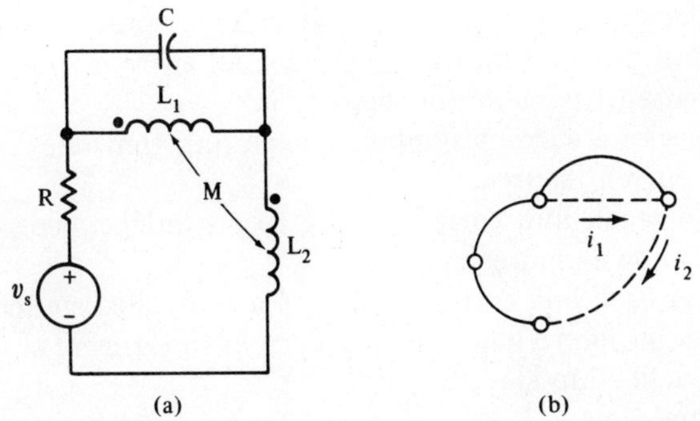

Figure 3-15 Example 3-11.

current sources, we use KVL to express each link voltage as a sum of tree branch voltages:

$$\begin{aligned}\frac{d}{dt}(L_1 i_1 + M i_2) &= \frac{1}{C}\int (i_2 - i_1)\,dt \\ \frac{d}{dt}(M i_1 + L_2 i_2) &= \frac{1}{C}\int (i_1 - i_2)\,dt - R i_2 + v_s\end{aligned} \tag{3-51}$$

The $n - 1$ tree branch voltage algorithm would not be as convenient for our problem because it requires that we write KCL equations in terms of tree branch voltages. This implies that each branch current must be expressed in terms of the branch voltages. Therefore, to use this method we would have to first solve Equations (3-50) for i_1 and i_2 as functions of v_1 and v_2 before writing the KCL equations. No particular problem arises in carrying this out, but it is more involved than the $b - n + 1$ link current algorithm.

Further shortcuts. The experienced network analyst can write an even smaller number of equations directly from the network graph, provided the network contains one or more sources. His strategy is to substitute the source and control equations obtained according to steps 3 and 4 directly into the step 5 equations without first writing them down. For a network containing s_v voltage sources (independent and controlled) and s_c current sources (independent and controlled), he can, therefore, directly write either $b - n + 1 - s_c$ equations in the $b - n + 1 - s_c$ link currents that do not correspond to current sources, or he can directly write $n - 1 - s_v$ equations in the $n - 1 - s_v$ tree branch voltages that do not correspond to voltage sources. The two strategies are summarized in the following algorithms.

An algorithm for writing $b - n + 1 - s_c$ equations in the $b - n + 1 - s_c$ nonsource link currents is as follows:

1. Construct a tree. Place all voltage sources and control voltages (if possible) in the tree; place all current sources and control currents (if possible) in the cotree.
2. Assign a current symbol to each link that does not correspond to a current source.
3. For each link corresponding to an independent source, assign the (known) source symbol.
4. For each link corresponding to a controlled source, use the control equation to label the link current in terms of the current symbols defined in steps 2 and 3.
5. For each link that does not correspond to a current source, write a KVL equation of the form {link voltage = Σ tree branch voltages} in terms of the symbols defined in steps 2, 3, and 4.

Example 3-12

In this example we illustrate the technique for writing directly $b - n + 1 - s_c = 2$ equations in two link currents for the network of Figure 3-16(a).

To write two equations in two link currents, we first construct a tree, as in Figure 3-16(b), and assign current symbols and directions to only those links that do not correspond to current sources. For the link corresponding to the controlled current source, we use the control equation

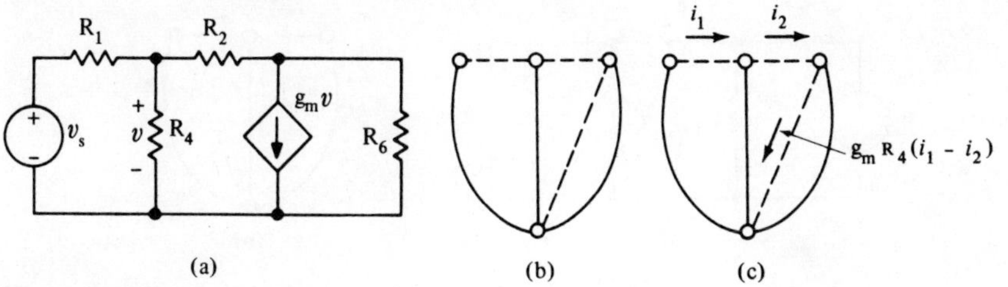

Figure 3-16 Example 3-12.

to write this link current in terms of the link currents already defined, as indicated in Figure 3-16(c). Finally, for each link not corresponding to a current source, we use KVL to express the link voltage as a sum of tree branch voltages in terms of the symbols defined in Figure 3-16(c). From Figure 3-16(c) we write

$$\begin{aligned} R_1 i_1 &= v_s - R_4(i_1 - i_2) \\ R_2 i_2 &= R_4(i_1 - i_2) - R_6[i_2 - g_m R_4(i_1 - i_2)] \end{aligned} \tag{3-52}$$

The same two equations can be obtained by substituting Equation (3-41) into Equations (3-42).

An algorithm for writing $n - 1 - s_v$ equations in the $n - 1 - s_v$ nonsource tree branch voltages is as follows:

1. Construct a tree. Place all voltage sources and control voltages (if possible) in the tree; place all current sources and control currents (if possible) in the cotree.
2. Assign a voltage symbol to each tree branch that does not correspond to a voltage source.
3. For each tree branch corresponding to an independent voltage source, assign the (known) source symbol.
4. For each tree branch corresponding to a controlled voltage source, use the control equation to label the tree branch voltage in terms of the symbols assigned in steps 2 and 3.
5. For each tree branch that does not correspond to a voltage source write a KCL equation of the form {tree branch current = Σ link currents} in terms of the symbols defined in steps 2, 3, and 4.

Example 3-13

In this example we illustrate the technique for writing directly $n - 1 - s_v = 2$ equations in two tree branch voltages for the network of Figure 3-17(a).

We first construct a tree and assign voltage symbols and polarities to only those tree branches that do not correspond to voltage sources. For the tree branch corresponding to the independent voltage source, we assign the known voltage symbol, as shown in Figure 3-17(b). Now,

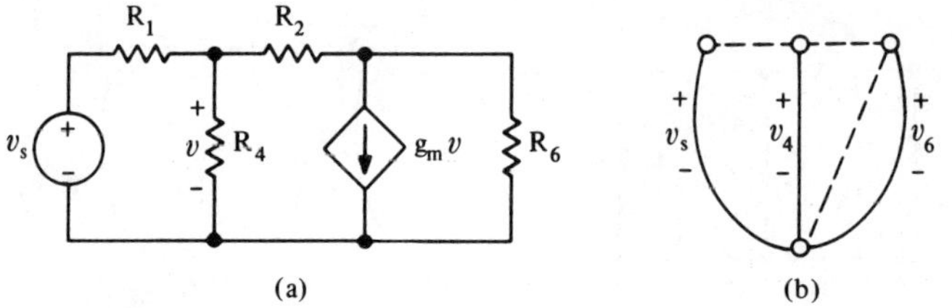

Figure 3-17 Example 3-13.

for the two tree branches not corresponding to voltage sources, we use KCL to express the tree branch current as a sum of link currents, but we use only the assigned symbols. Hence, from Figure 3-17(b), we write

$$\frac{v_4}{R_4} = \frac{v_s - v_4}{R_1} + \frac{v_6 - v_4}{R_2}$$
$$\frac{v_6}{R_6} = \frac{v_4 - v_6}{R_2} - g_m v_4 \tag{3-53}$$

The same equations can be obtained by substituting Equation (3-43) into Equations (3-44) and (3-45).

3-6 Loop and Nodal Analysis

Two network analysis strategies, called loop analysis and nodal analysis, are in widespread use. In this section we present the loop and nodal methods and discuss their relationship to the two strategies defined by the two algorithms of Section 3-5.

Loop analysis. The *loop* method is convenient for analysing networks containing no sources other than independent voltage sources. The method is best illustrated by an example.

Consider the network presented in Figure 3-18. The loop method dictates that we assign a loop current to each closed path that does not contain any other closed paths in its interior, as shown in Figure 3-18.

Then, one KVL equation is written in terms of the assigned loop currents around each of the $b - n + 1$ loops. For example, for the network of Figure 3-18, we have for loops 1, 2, and 3, respectively, the

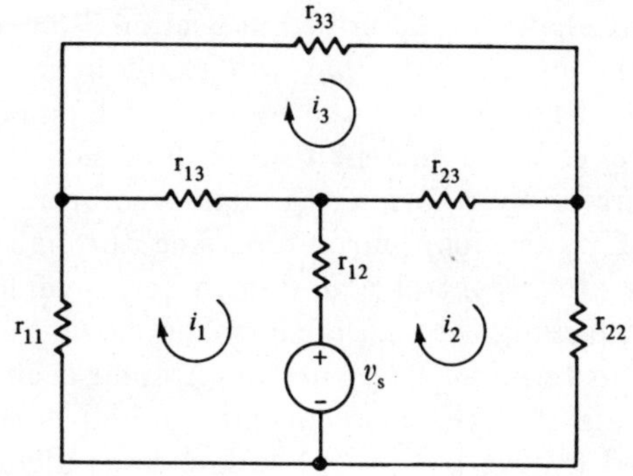

Figure 3-18 Network for loop analysis.

following equations:

$$
\begin{aligned}
0 &= i_1 r_{11} + (i_1 - i_3) r_{13} + (i_1 - i_2) r_{12} + v_s && \text{loop 1} \\
0 &= -v_s + (i_2 - i_1) r_{12} + (i_2 - i_3) r_{23} + i_2 r_{22} && \text{loop 2} \\
0 &= (i_3 - i_1) r_{13} + i_3 r_{33} + (i_3 - i_2) r_{23} && \text{loop 3}
\end{aligned}
\tag{3-54}
$$

These equations can also be rearranged into a standard form in which the voltage source quantities appear on the left-hand side and the loop currents and their coefficients on the right-hand side.

$$
\begin{aligned}
-v_s &= (r_{11} + r_{12} + r_{13}) i_1 + (-r_{12}) i_2 + (-r_{13}) i_3 \\
v_s &= (-r_{12}) i_1 + (r_{12} + r_{22} + r_{23}) i_2 + (-r_{23}) i_3 \\
0 &= (-r_{13}) i_1 + (-r_{23}) i_2 + (r_{13} + r_{23} + r_{33}) i_3
\end{aligned}
\tag{3-55}
$$

We now observe that the standard form of loop equations is of the general form

$$
\begin{aligned}
v_{s1} &= R_{11} i_1 - R_{12} i_2 - R_{13} i_3 \\
v_{s2} &= -R_{21} i_1 + R_{22} i_2 - R_{23} i_3 \\
v_{s3} &= -R_{31} i_1 - R_{32} i_2 + R_{33} i_3
\end{aligned}
\tag{3-56}
$$

where i_ℓ is the ℓth loop current; v_{sj} is the sum of voltage-source voltages around loop j with polarity that would tend to make i_j positive; $R_{j\ell}$ is the sum of resistances common to loop j and loop ℓ if currents i_j and i_ℓ flow through $R_{j\ell}$ in the same direction, or, the negative of the sum of resistances common to loop j and loop ℓ if currents i_j and i_ℓ flow through $R_{j\ell}$ in opposite directions.

For the more general case of a b-branch, n-node network consisting of resistances, inductances, capacitances, and independent voltage sources, the generalization is straightforward.

The loop method presented here is identical to the method dictated by the link-current algorithm presented in Section 3-5. The loop method can also be modified to handle networks containing current sources and controlled voltage sources. Again, the modified loop method is identical to the link current algorithm described in Section 3-5. That is, in order to define a set of loop currents that will result in a set of b − n + 1 linearly independent equations in the b − n + 1 loop currents, we must first choose a tree, assign a current to each link, convert each link current to a loop current by returning through tree branches only, write a source equation for each loop current passing through an independent current source, write a control equation in terms of loop currents for each loop current passing through a controlled current source, and write a KVL equation in terms of loop currents around each remaining loop. Hence, we conclude that the loop-current method is equivalent to the method prescribed by the b − n + 1 link-current algorithm of Section 3-5. Note, however, that even though the b − n + 1 equations can always be written in the form of Equation (3-56), the coefficients $R_{j\ell}$ can be interpreted as (±) the sum of resistances common to loops j and ℓ only for networks containing no controlled voltage sources or independent or controlled current sources. The loop method is not stressed in this text because, for networks containing controlled sources, it has no significant advantage over the link-current algorithm.

Nodal analysis. The *nodal* or *node-to-datum* method is convenient for networks containing no voltage sources or any controlled current sources. We illustrate this method by considering the network of Figure 3-19.

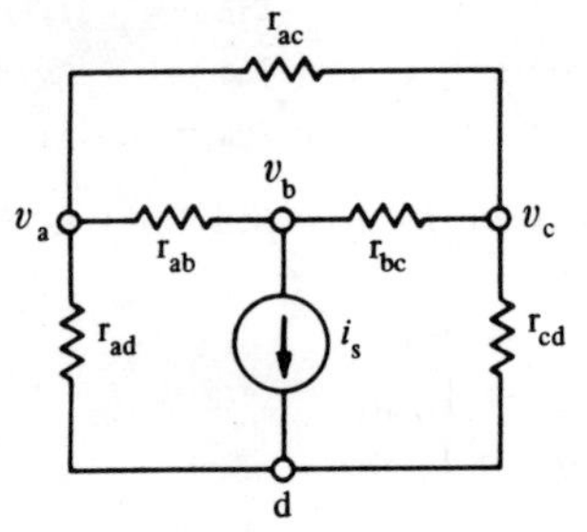

Figure 3-19 Network for nodal analysis.

After arbitrarily selecting the bottom node to be a *reference* or *datum* node, we assign voltage symbols to each of the remaining n − 1 = 3 nodes. v_j is the voltage at the jth node relative to the datum node, as shown in Figure 3-19. Hence, the v_j are called nodal voltages. The nodal method prescribes that we write a KCL equation in terms of the nodal voltages at each of the n − 1 nodes other than the reference node. For example, for the network of Figure 3-19, we write at nodes a, b, and c, respectively,

$$
\begin{aligned}
0 &= \frac{v_a - v_c}{r_{ac}} + \frac{v_a - v_b}{r_{ab}} + \frac{v_a}{r_{ad}} \\
0 &= \frac{v_b - v_a}{r_{ab}} + \frac{v_b - v_c}{r_{bc}} + i_s \\
0 &= \frac{v_c - v_a}{r_{ac}} + \frac{v_c - v_b}{r_{bc}} + \frac{v_c}{r_{cd}}
\end{aligned}
\tag{3-57}
$$

Or, by a simple rearrangement of terms, these equations can be written in the standard form for nodal equations:

$$
\begin{aligned}
0 &= \left(\frac{1}{r_{ad}} + \frac{1}{r_{ab}} + \frac{1}{r_{ac}}\right) v_a + \left(-\frac{1}{r_{ab}}\right) v_b + \left(-\frac{1}{r_{ac}}\right) v_c \\
-i_s &= \left(-\frac{1}{r_{ab}}\right) v_a + \left(\frac{1}{r_{ab}} + \frac{1}{r_{bc}}\right) v_b + \left(-\frac{1}{r_{bc}}\right) v_c \\
0 &= \left(-\frac{1}{r_{ac}}\right) v_a + \left(-\frac{1}{r_{bc}}\right) v_b + \left(\frac{1}{r_{ac}} + \frac{1}{r_{bc}} + \frac{1}{r_{cd}}\right) v_c
\end{aligned}
\tag{3-58}
$$

Equations (3-58) are of the form

$$
\begin{aligned}
i_{sa} &= G_{aa}v_a + G_{ab}v_b + G_{ac}v_c \\
i_{sb} &= G_{ba}v_a + G_{bb}v_b + G_{bc}v_c \\
i_{sc} &= G_{ca}v_a + G_{bc}v_b + G_{cc}v_c
\end{aligned}
\tag{3-59}
$$

where v_ℓ is the voltage at node j relative to the reference node; $G_{j\ell}$ is the sum of admittances common to node j for $j = \ell$ and the negative sum of admittances common to nodes j and ℓ for $j \neq \ell$; i_{sj} is the sum of independent current-source currents common to node j with current direction into node j.

For the more general case of an n-node network containing only independent current sources, resistances, capacitances, and inductances, the normal forms for the n − 1 KCL equations are similar, except that they now contain integral and derivative terms.

The utility of the nodal method is now clear. For networks containing no voltage sources or controlled current sources, the n − 1 KCL equations can be written down by simple inspection of the network.

The nodal method can be modified to handle networks containing voltage sources and controlled current sources. For these cases, however, a general form for equations such as Equations (3-59) cannot be established. The modified nodal method is essentially the same as the n − 1 tree branch voltage algorithm method when a particular tree is chosen. The particular tree required is one with a tree branch leading from the reference node to each of the n − 1 nonreference nodes. For nodes not connected directly to the datum node via a branch, a zero conductance branch must be added between the node and the reference node. Hence, each nodal voltage corresponds to a tree branch voltage. This choice of a node-to-datum tree is a very convenient one for the tree branch voltage algorithm. A detailed discussion of nodal analysis for this choice of tree is the central topic of Chapter 6. This method leads to network analysis algorithms readily implemented on a digital computer.

PROBLEMS

3-1 Find all distinct trees and their corresponding cotrees for each of the network graphs shown. (See Figure P3-1.)

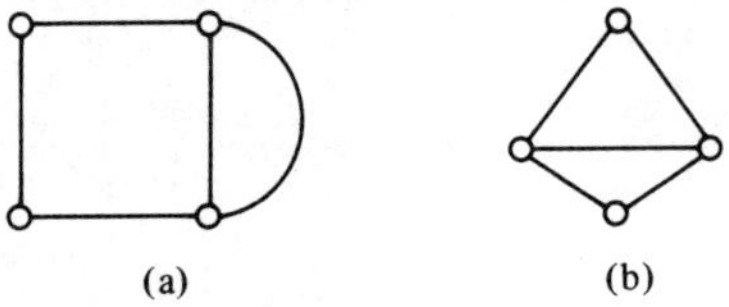

Figure P3-1

3-2 For the transistor amplifier circuit shown in Figure P3-2

a. Draw a network model.

b. Make a network graph, assign a voltage and current symbol to each branch, and define a tree and cotree.

c. Write b source, element, and control equations.

d. Write n − 1 KCL equations.

e. Write b − n + 1 KVL equations.

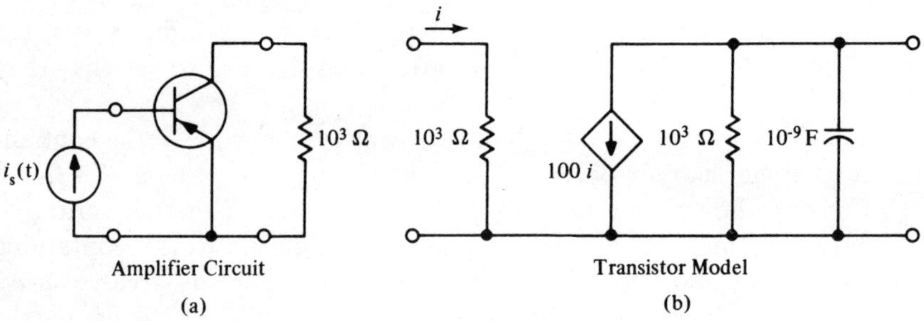

Figure P3-2

3-3 Write a set of 2b linearly independent equations for each of the networks shown in Figure P3-3.

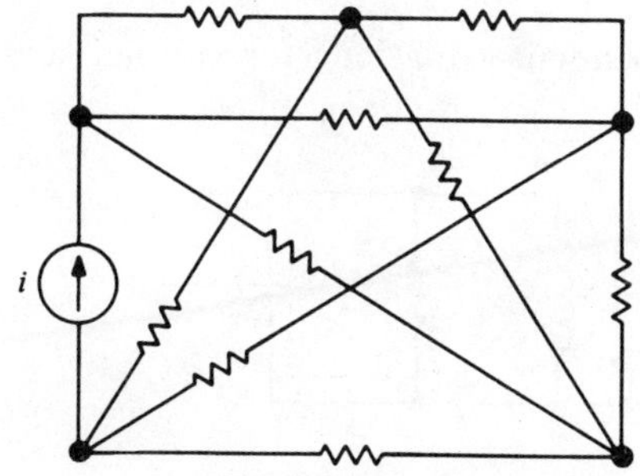

(a)

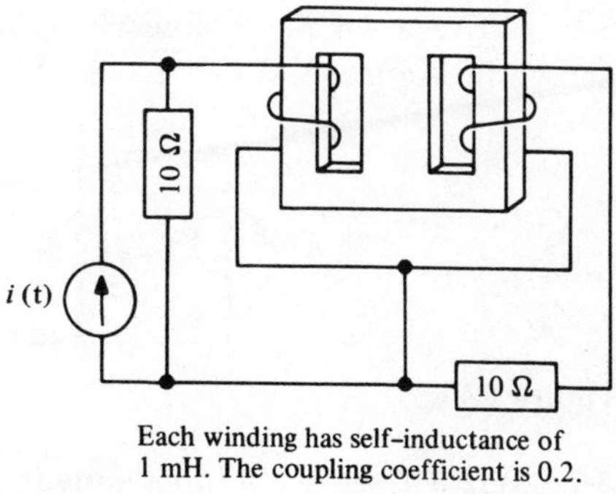

(b)

Figure P3-3

3-4 For the FET amplifier circuit shown in Figure P3-4

a. Draw a network model.

b. Make a network graph and define a tree and cotree.

c. Write the b source, element, and control equations.

d. Write the n − 1 KCL equations.

e. Write the b − n + 1 KVL equations.

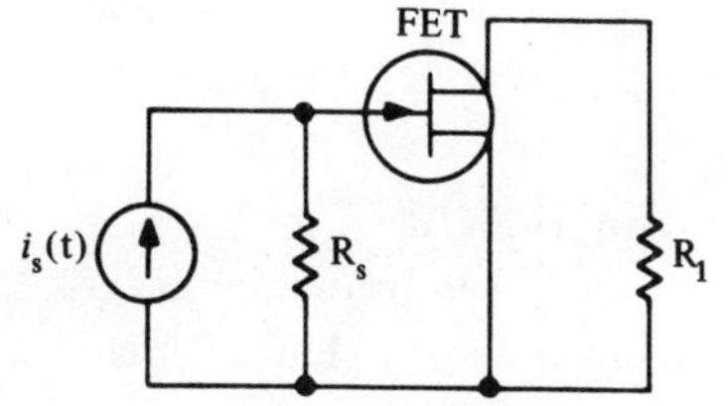

(a)

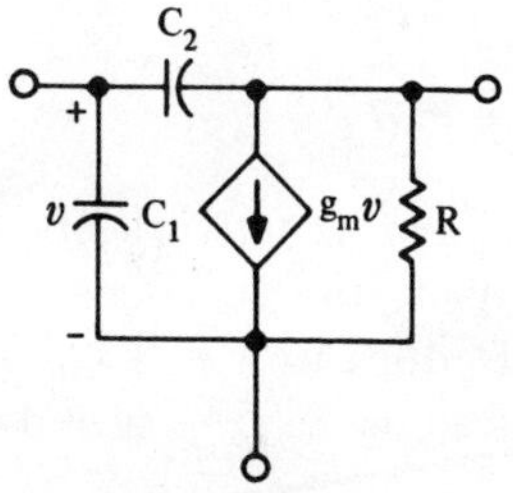

(b)

Figure P3-4

3-5 Repeat Problem 3-4 for the transistor amplifier network shown in Figure P3-5.

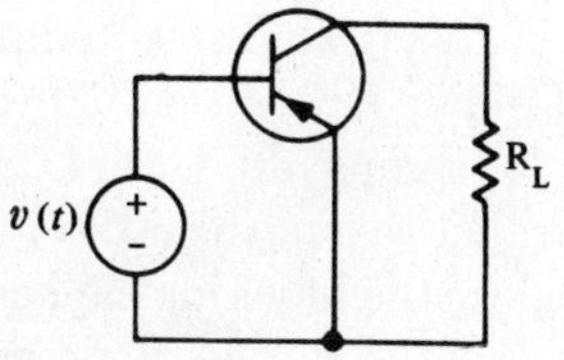

(a)

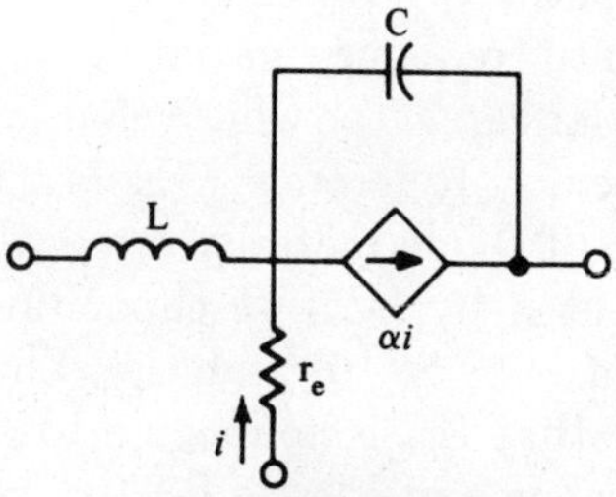

(b)

Figure P3-5

3-6 Write a set of 2b linearly independent equations for the network shown in Figure P3-6.

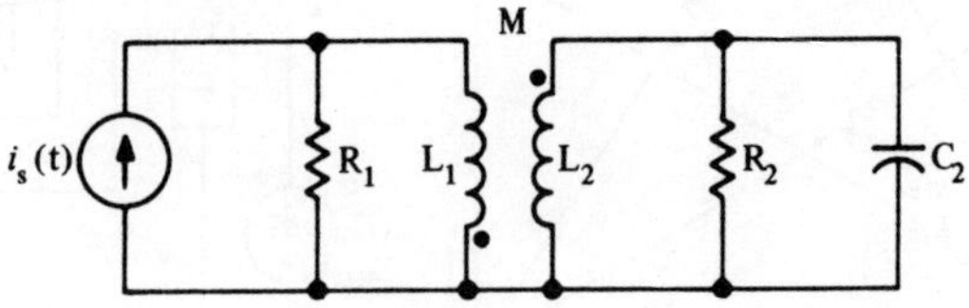

Figure P3-6

3-7 Write a set of 2b independent equations for each network shown in Figure P3-7.

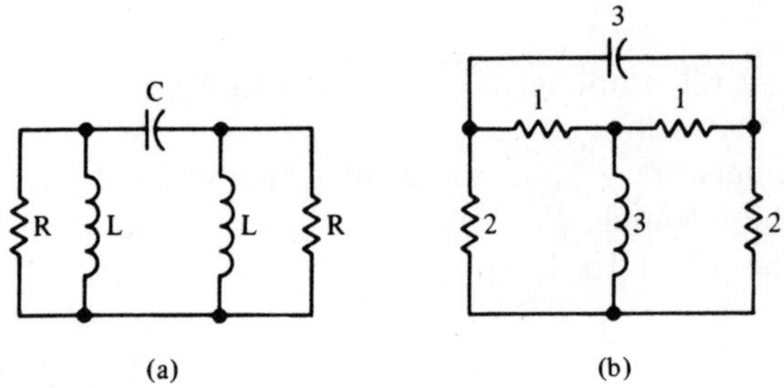

Figure P3-7

3-8 For the network of Problem 3-4

a. Write the 2b linearly independent equations.

b. Reduce to n − 1 equations in the n − 1 tree branch voltages.

c. Reduce to b − n + 1 equations in the b − n + 1 link currents.

3-9 For the network of Problem 3-5

a. Write the 2b equations.

b. Reduce to n − 1 equations in the n − 1 tree branch voltages.

c. Reduce to b − n + 1 equatons in the b − n + 1 link currents.

3-10 The 1.1-MHz oscillator senses temperature changes and translates them into frequency variations. The main temperature-sensing element is a mylar capacitor whose characteristics yield a 0.5 kHz/°C temperature coefficient. Reference: GE Transistor Manual, 7th ed., p. 350 (circuit), p. 571 (TD characteristics). The IN3714 tunnel diode (TD) can be represented by a 25-pF capacitance in parallel with a (negative) conductance of -18×10^{-3} mhos. You can neglect the blocking capacitor and assume that E_{bb} is connected to ground. For the tunnel diode oscillator circuit shown in Figure P3-10

a. Construct a tree and a cotree.

b. Write a set of 2b linearly independent equations.

c. Reduce to n − 1 equations in the n − 1 tree branch voltages.

d. Reduce to b − n + 1 equations in the b − n + 1 link currents.

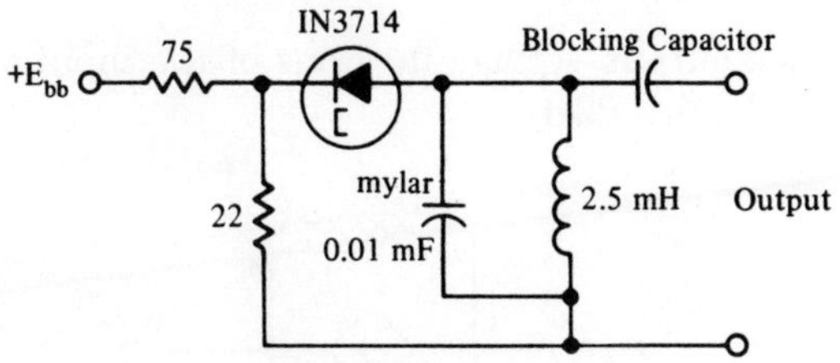

Figure P3-10

3-11 For the network of Problem 3-2

a. Write a set of 2b linearly independent equations.

b. Reduce to b − n + 1 equations in the link currents.

c. Reduce to n − 1 equations in the tree branch voltages.

3-12 Repeat Problem 3-11 for the network shown in Figure P3-12.

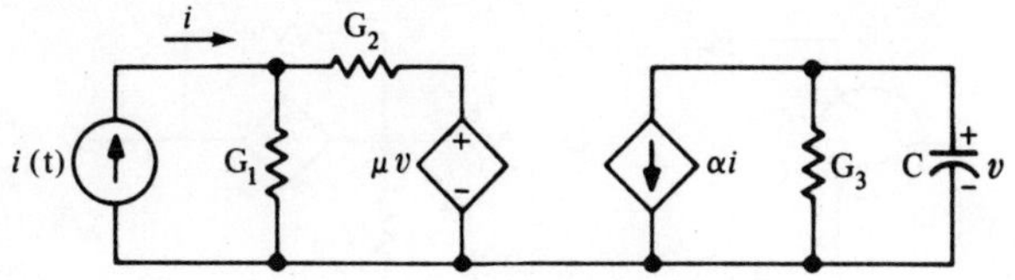

Figure P3-12

3-13 The network shown in Figure P3-13 is often used as a coupling network in radio receivers. Write a set of 2b linearly independent equations for this network and reduce to four equations.

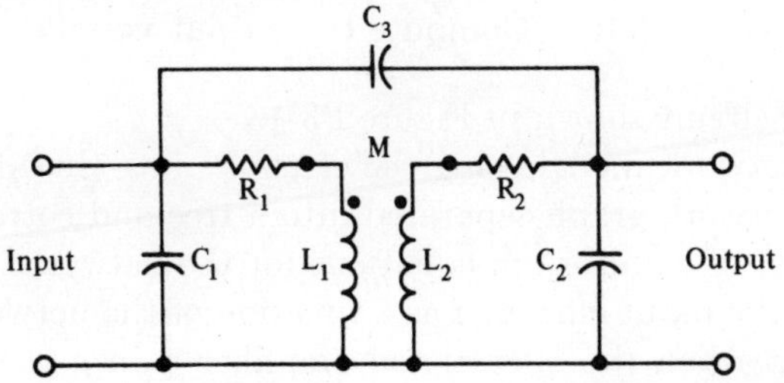

Figure P3-13

3-14 Write a set of differential equations that characterize the network shown in Figure P3-14.

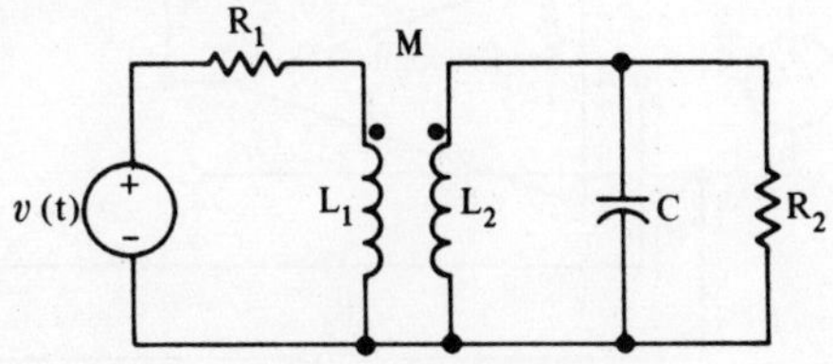

Figure P3-14

3-15 Compute the output voltage in terms of the input voltage for the network shown in Figure P3-15.

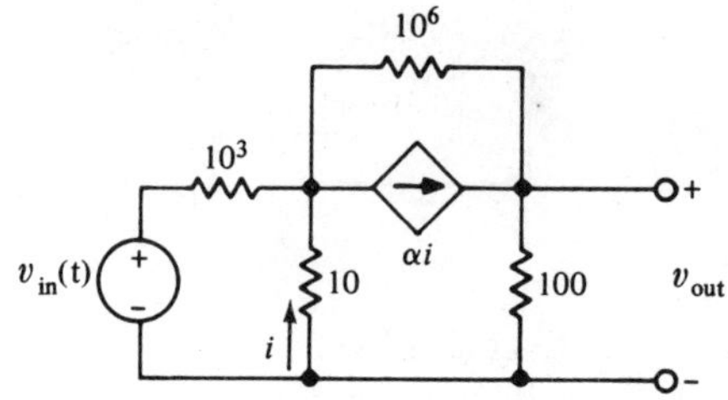

Figure P3-15

3-16 Compute the Thevinen and Norton equivalent networks for the network shown in Figure P3-16.

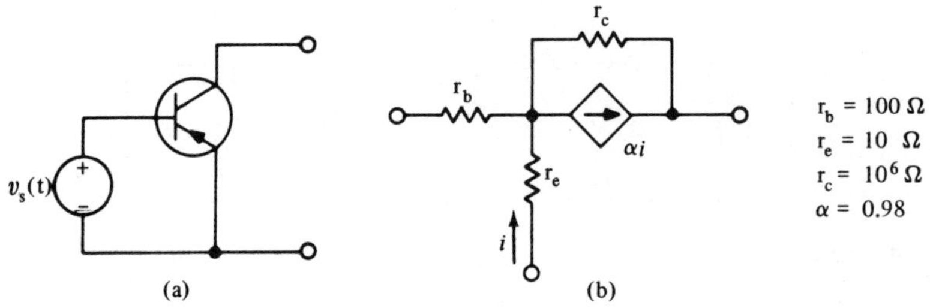

Figure P3-16

3-17 A 10-kΩ load resistor is connected across the output terminals of the network of Problem 3-16. Compute the output voltage.

3-18 For the circuit shown in Figure P3-18

a. Draw a network model using the elements introduced in Chapter 1.
b. Draw a network graph separated into a tree and cotree.
c. Write a set of equilibrium equations for the network.
d. Solve for the input current i as a function of the network parameter.
e. Show that when the operational amplifier gain μ is very large, the input current is independent of μ.

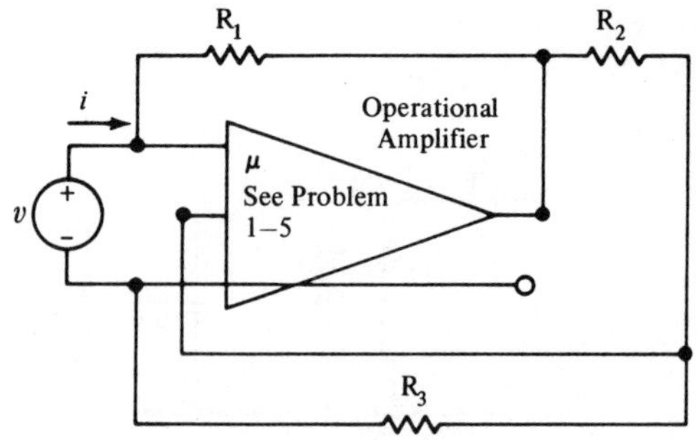

Figure P3-18

3-19 The bridge circuit shown in Figure P3-19 can be used to measure various quantities that produce a change in the resistor R, for example, temperature. Assume the meter has zero resistance and compute the meter current i in terms of R. Sketch i versus R for $0 < R < 2$. Hint: Assume the meter has resistance r. Make a network graph and choose a tree. Use one of the shortcut algorithms to write a set of equations. Solve the equations for i. Take the limit as $r \rightarrow 0$.

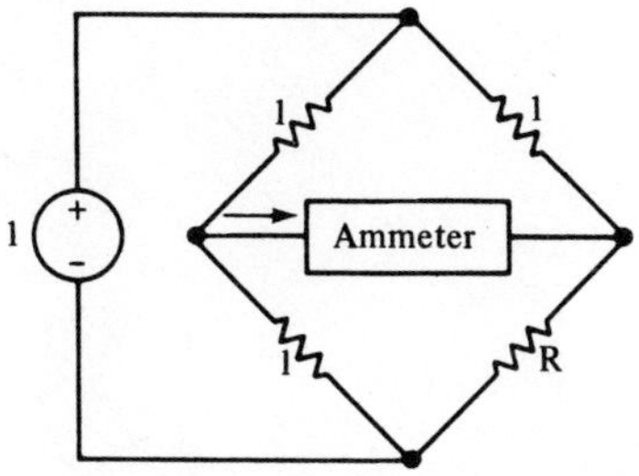

Figure P3-19

3-20 Figure P3-20(a) shows a pair of coupled coils and the equations defining them. Part (b) shows a network that can be made equivalent to the coupled coil network by the proper choice of L_A, L_B, and L_C. Choose L_A, L_B, L_C (in terms L_1, L_2, and M) so that the equations governing the two networks are identical.

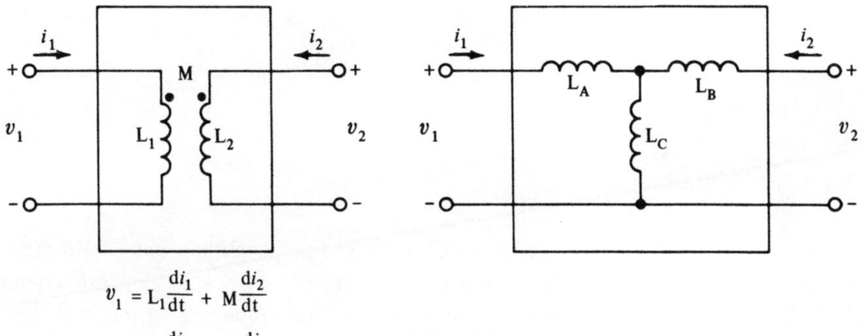

$$v_1 = L_1\frac{di_1}{dt} + M\frac{di_2}{dt}$$

$$v_2 = M\frac{di_1}{dt} + L_2\frac{di_2}{dt}$$

(a) (b)

Figure P3-20

CHAPTER

4

Sinusoidal Steady-State Analysis

4-1 Introduction

In Section 3-6 we presented two algorithms for writing a set of linearly independent network equations for any linear network. For networks containing no inductances or capacitances, the network equations are linear algebraic equations and can be solved by any of the methods of linear algebra. For networks containing one or more inductances or capacitances, the equations are linear differential equations. A general method for solving any set of linear differential equations is presented in Chapter 10.

In this chapter we show that for the special case of sinusoidal sources, the network differential equations can be reduced to linear algebraic equations and can, therefore, be solved by any of the methods of linear algebra. By a sinusoidal source we mean an independent voltage or current source whose voltage or current is a sine wave that exists for all time; that is, it is of the form $A \cos (\omega t + \theta)$ for $-\infty < t < +\infty$. Sinusoidal sources are idealizations of sources that have been in operation for a long time; what constitutes a long time depends on the transient properties of the network containing the source. The transient properties of networks are the subject of Chapter 10.

4-2 Network Equations for Sinusoidal Forcing Functions

One very important property of the linear network elements defined in Chapter 1 is that, if the element voltage is a sinusoid, then the element current is a sinusoid of the same frequency. Since two sinusoids of the same frequency can differ only in amplitude and phase, we can express

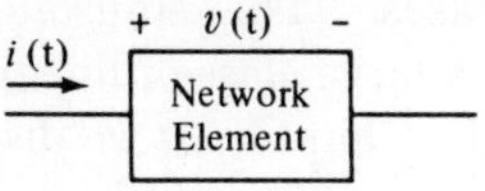

Figure 4-1 A general element.

the element voltage and element current in the following form (see Figure 4-1):

$$\begin{aligned} v(t) &= \nu \cos(\omega t + \phi) \\ i(t) &= \iota \cos(\omega t + \theta) \end{aligned} \tag{4-1}$$

where ν and ϕ represent the amplitude and phase of the voltage $v(t)$, and ι and θ are amplitude and phase of the current $i(t)$. We shall assume that we know the radian frequency ω;[1] then knowledge of the amplitude and phase of each waveform is equivalent to knowledge of the waveform.

Any linear combination of sinusoids of frequency ω is again a sinusoid of frequency ω. We also note that sine and cosine functions are identical, except for a $\pi/2$ radian difference in phase, that is,

$$\sin(\omega t + \phi) = \cos\left(\omega t + \phi - \frac{\pi}{2}\right) \tag{4-2}$$

as illustrated in Figure 4-2.

From the fact that, for all of the linear network elements defined in Chapter 1, the element voltage and current have the same frequency, and from the fact that the sum of any set of sinusoids is a sinusoid of the same frequency, it follows from KVL and KCL that *if the linear network is excited by a sinusoidal source of frequency ω, then all branch voltages and branch currents are sinusoids of the same frequency.* It is this most important property of linear networks that allows us to assume the functional forms of all the branch voltages and branch currents. This, in turn, allows us to evaluate the derivatives and integrals in the network differential equations, thereby reducing them to algebraic equa-

[1] The frequency in radians per second ω is related to the frequency in hertz f by $\omega = 2\pi f$.

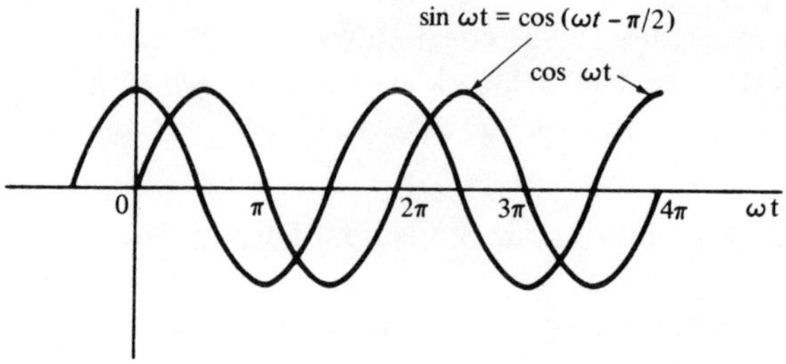

Figure 4-2 Sine and cosine waves.

tions in the unknown amplitudes and phases. This sinusoidal steady-state solution is the important solution for a large class of linear networks that are good models for useful circuits. In Chapter 10 we discuss other solutions, called transient solutions. The procedure is best illustrated by an example.

Example 4-1

Consider the series RL network presented in Figure 4-3. The problem is to compute the current $i(t)$ when the network is excited by the sinusoidal voltage source shown.

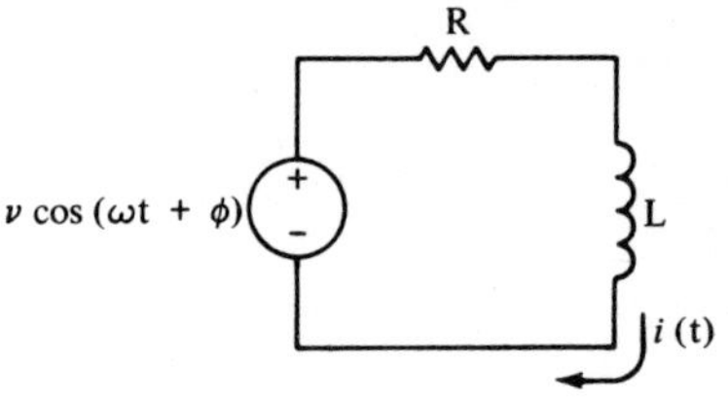

Figure 4-3 Example 4-1.

We know that $i(t)$ is given by the solution to the network (differential) equation

$$\nu \cos(\omega t + \phi) = Ri(t) + L\frac{d}{dt}[i(t)] \tag{4-3}$$

Since the forcing function is a sinusoid of frequency ω, we also know that the current $i(t)$ is a sinusoid of the same frequency. That is, the functional form of the particular solution to the differential equation is given by

$$i(t) = \iota \cos(\omega t + \theta) \tag{4-4}$$

where ω is known, but ι and θ are unknown constants. Therefore, we need only determine the amplitude and phase of the current to solve the problem. This can be accomplished by substituting Equation (4-4) into Equation (4-3) and solving for ι and θ.

Substituting Equation (4-4) into Equation (4-3) yields

$$\begin{aligned} \nu \cos(\omega t + \phi) &= R\iota \cos(\omega t + \theta) + L \frac{d}{dt}\{\iota \cos(\omega t + \theta)\} \\ &= R\iota \cos(\omega t + \theta) - \omega L\iota \sin(\omega t + \theta) \end{aligned} \tag{4-5}$$

Hence, knowing the functional form for $i(t)$ has allowed us to carry out the differentiation and thereby reduce the network differential equation to an algebraic equation. To solve Equation (4-5) for ι and θ, we use the trigometric identities

$$\begin{aligned} \cos(x + y) &= \cos x \cos y - \sin x \sin y \\ \sin(x + y) &= \sin x \cos y + \cos x \sin y \end{aligned} \tag{4-6}$$

According to Equations (4-6), we can write each of the sinusoids in Equation (4-5) in the form

$$\begin{aligned} \cos(\omega t + \phi) &= \cos\phi \cos\omega t - \sin\phi \sin\omega t \\ \cos(\omega t + \theta) &= \cos\theta \cos\omega t - \sin\theta \sin\omega t \\ \sin(\omega t + \theta) &= \sin\theta \cos\omega t + \cos\theta \sin\omega t \end{aligned} \tag{4-7}$$

Substituting Equations (4-7) into Equation (4-5), we obtain

$$\begin{aligned} \nu \cos\phi \cos\omega t - \nu \sin\phi \sin\omega t = {}& R\iota \cos\theta \cos\omega t - R\iota \sin\theta \sin\omega t \\ & - \omega L\iota \sin\theta \cos\omega t - \omega L\iota \cos\theta \sin\omega t \end{aligned} \tag{4-8}$$

Grouping all terms containing $\cos \omega t$ on the left-hand side and all terms containing $\sin \omega t$ on the right-hand side, we obtain

$$\begin{aligned} & (\nu \cos\phi - R\iota \cos\theta + \omega L\iota \sin\theta) \cos\omega t \\ & \qquad = (\nu \sin\phi - R\iota \sin\theta - \omega L\iota \cos\theta) \sin\omega t \end{aligned} \tag{4-9}$$

But $a \cos \omega t = b \sin \omega t$ if and only if $a = b = 0$. (This is obvious from Figure 4-2.) Hence Equation (4-9) requires

$$\begin{aligned} \nu \cos\phi - R\iota \cos\theta + \omega L\iota \sin\theta &= 0 \\ \nu \sin\phi - R\iota \sin\theta - \omega L\iota \cos\theta &= 0 \end{aligned} \tag{4-10}$$

Equations (4-10) are two equations in the two unknowns ι and θ, and their solution specifies the amplitude and phase of $i(t)$.

Note, however, that we have succeeded in reducing one linear [in the unknown $i(t)$] differential equation to two nonlinear [in the unknowns ι and θ] equations. Unfortunately, transcendental equations such as Equations (4-10) are quite difficult to solve. Furthermore, if we attempt to apply the same procedure to a set of n differential equations, we would obtain 2n transcendental equations in the 2n unknown amplitudes and phases. Clearly, this approach has little appeal because the resulting equations are hardly easier to solve than the original differential equations.

Fortunately, a slightly different approach will allow us to reduce any set of n linear differential equations to 2n *linear* algebraic equations in the 2n unknown amplitudes and phases. This technique requires that we use a complex exponential representation for sinusoids.

4-3 Complex Representation of Sinusoids

In this section we introduce a complex representation for sinusoidal waveforms. We further show that this notation allows us to reduce a set of n linear differential equations with sinusoidal forcing functions to a set of 2n linear algebraic equations — n linear equations in the n unknown amplitudes and n linear equations in the n unknown phases.

Sine and cosine functions can be written in terms of complex exponential functions via Euler's equations. Euler's equations are given by

$$\cos x = \frac{e^{jx} + e^{-jx}}{2} \qquad \sin x = \frac{e^{jx} - e^{-jx}}{j2} \tag{4-11}$$

where $j = \sqrt{-1}$. According to Equation (4-11), we can write

$$\begin{aligned} \nu \cos(\omega t + \phi) &= \frac{\nu e^{j(\omega t+\phi)} + \nu e^{-j(\omega t+\phi)}}{2} \\ &= \nu e^{j\phi}\frac{e^{j\omega t}}{2} + \nu e^{-j\phi}\frac{e^{-j\omega t}}{2} \\ &= V\frac{e^{j\omega t}}{2} + V^*\frac{e^{-j\omega t}}{2} \end{aligned} \tag{4-12}$$

where V and V^* are complex numbers defined by

$$\begin{aligned} V &= \nu e^{j\phi} = \nu\underline{/\phi} \\ V^* &= \nu e^{-j\phi} = \nu\underline{/-\phi} \end{aligned}$$

Note that the magnitude of V is the amplitude of the cosine function, and the angle of V is the phase of the cosine function. Note also that V^* is the conjugate of V.

We also note that, if the frequency ω is assumed known, then a one-to-one correspondence exists between the waveform $\nu \cos(\omega t + \phi)$ and the complex number V. That is, if we are given that $V = 2\underline{/\pi}$, then we can reconstruct the waveform $2e^{j\pi}(e^{j\omega t}/2) + 2e^{-j\pi}(e^{-j\omega t}/2) = 2\cos(\omega t + \pi)$. $V = \nu\underline{/\phi}$ is called the *complex number representation* for the waveform $v(t) = \nu \cos(\omega t + \phi)$. The complex number representation for a voltage waveform is called the *complex voltage amplitude*, or the *complex voltage*, and the complex number representation for a current waveform is called the *complex current amplitude*, or the *complex current*. In Example 4-1 we considered a voltage waveform $v(t) = \nu \cos(\omega t + \phi)$ and a current waveform $i(t) = \iota \cos(\omega t + \theta)$. These waveforms and their associated complex number representations are illustrated in Figure 4-4. Sine functions can be handled in an analogous fashion.

We conclude that any sinusoid of frequency ω can be expressed as a linear combination of two complex exponential functions, $e^{j\omega t}/2$ and its

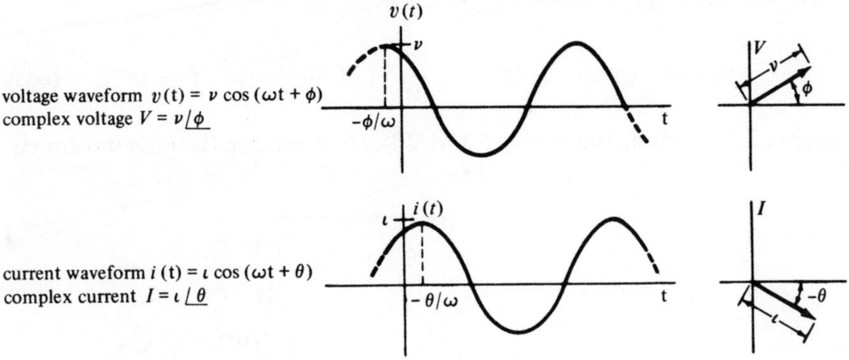

Figure 4-4 Complex number representation.

conjugate $e^{-j\omega t}/2$. The coefficient of $e^{j\omega t}/2$ is a complex number whose magnitude is that of the sinusoid and whose angle is the phase of the sinusoid; the coefficient of $e^{-j\omega t}/2$ is the conjugate of the first coefficient for cosine functions and the negative of the conjugate for sine functions.

Exponential functions possess a number of very convenient properties that the less well-behaved sinusoidal functions do not have. Therefore, it is sometimes more convenient to use the complex exponential representation rather than the sinusoidal function. The fact that the exponential functions are complex, whereas the sine and cosine functions are real, does not introduce any particular difficulty. And the fact that it takes two exponential functions to represent one sinusoidal function does not introduce any additional computation, since the coefficients of the two exponentials are complex conjugates (negative conjugates for the sine function).

Example 4-2

In this example we solve the same problem considered in Example 4-1, except that this time we represent all sinusoids with their complex exponential representations. Recall that the problem is to compute the current $i(t)$ in the network of Figure 4-5.

We again start by writing the differential equation

$$v(t) = Ri(t) + L\frac{d}{dt}[i(t)] \tag{4-13}$$

but this time we express $v(t)$ in the form

$$v(t) = \nu \cos(\omega t + \phi) = V\frac{e^{j\omega t}}{2} + V^*\frac{e^{-j\omega t}}{2}, \qquad V = \nu e^{j\phi} \tag{4-14}$$

Next, we write the assumed solution for $i(t)$ in the form

$$i(t) = \iota \cos(\omega t + \theta) = I\frac{e^{j\omega t}}{2} + I^*\frac{e^{-j\omega t}}{2}, \qquad I = \iota e^{j\theta} \tag{4-15}$$

where ι and θ (and, therefore, I and I^*) are constants to be determined.

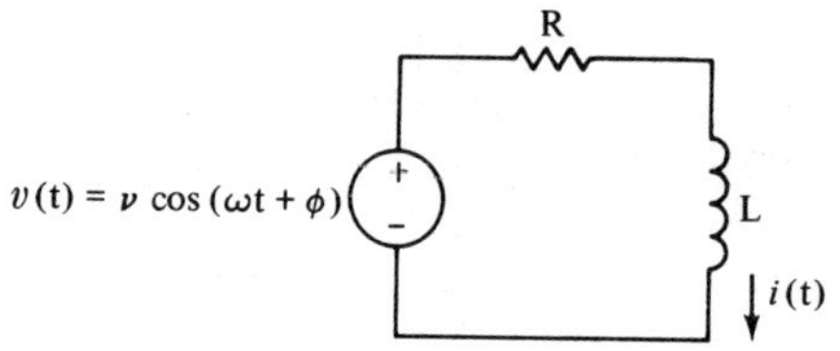

Figure 4-5 Example 4-2.

Substituting Equations (4-14) and (4-15) into Equation (4-13), we obtain

$$Ve^{j\omega t} + V^*e^{-j\omega t} = RIe^{j\omega t} + RI^*e^{-j\omega t} + L\frac{d}{dt}[Ie^{j\omega t} + I^*e^{-j\omega t j}] \tag{4-16}$$

$$= RIe^{j\omega t} + RI^*e^{-j\omega t} + j\omega LIe^{j\omega t} - j\omega LI^*e^{-j\omega t} \tag{4-17}$$

Or, grouping all the $e^{j\omega t}$ terms on the left-hand side and all the $e^{-j\omega t}$ terms on the right-hand side, we obtain

$$(V - RI - j\omega LI)e^{j\omega t} = (-V^* + RI^* - j\omega LI^*)e^{-j\omega t} \tag{4-18}$$

Equation (4-18) is of the form $Ae^{j\omega t} = A^*e^{-j\omega t}$, which is obviously true for all ωt if and only if $A = A^* = 0$. Hence, Equation (4-18) requires

$$\begin{aligned} V - RI - j\omega LI &= 0 \\ -V^* + RI^* - j\omega LI^* &= 0 \end{aligned} \tag{4-19}$$

Solving the first of Equations (4-19) for I we obtain

$$\begin{aligned} I &= \frac{V}{R + j\omega L} \\ &= \frac{\nu\underline{/\phi}}{\sqrt{R^2 + \omega^2L^2}\;\underline{/\tan^{-1}\left(\frac{\omega L}{R}\right)}} = \frac{\nu}{\sqrt{R^2 + \omega^2L^2}}\;\underline{/\phi - \tan^{-1}\left(\frac{\omega L}{R}\right)} \\ &= \frac{\nu}{\sqrt{R^2 + \omega^2L^2}}\,e^{j\left[\phi - \tan^{-1}\left(\frac{\omega L}{R}\right)\right]} \end{aligned} \tag{4-20}$$

But

$$I = \iota\underline{/\theta} = \frac{\nu}{\sqrt{R^2 + \omega^2 L^2}} \underline{\Big/ \phi - \tan^{-1}\left(\frac{\omega L}{R}\right)} \qquad \textbf{(4-21)}$$

so that

$$\begin{aligned} \iota &= \frac{\nu}{\sqrt{R^2 + \omega^2 L^2}} \\ \theta &= \phi - \tan^{-1}\frac{\omega L}{R} \end{aligned} \qquad \textbf{(4-22)}$$

Therefore, the current $i(t)$ is given by

$$i(t) = \frac{\nu}{\sqrt{R^2 - \omega^2 L^2}} \cos\left[\omega t + \phi - \tan^{-1}\left(\frac{\omega L}{R}\right)\right] \qquad \textbf{(4-23)}$$

The reader can verify by direct substitution that Equation (4-23) satisfies Equation (4-13). The reader is also invited to show that Equations (4-22) satisfy Equations (4-10).

In Example 4-2 we succeeded in reducing the linear differential equation in the current waveform $i(t)$ [Equation (4-13)] to a linear algebraic equation in the complex current I [Equation (4-19)], which in turn was reduced to two linear algebraic equations—one in the amplitude ι, and one in the phase θ [Equations (4-22)].

The same procedure can be followed for any set of n linear differential equations with constant coefficients and with one or more sinusoidal forcing functions of frequency ω. In every case the n linear differential equations in n unknown waveforms can be reduced to n linear algebraic equations in the corresponding n complex number representations. These n equations can then be solved for the n complex voltages and currents by any of the methods of linear algebra. The magnitude and angle of each of these gives the amplitude and phase of the corresponding waveforms.

After using this technique to solve a number of sets of differential equations with sinusoidal forcing functions, we begin to realize that it is not necessary to first write the n differential equations! The n linear algebraic equations in the n complex number representations can be written directly from the network diagram by the proper application of the algorithms of Section 3-6. However, before we can utilize this technique, we must master the concept of complex impedance.

4-4 Complex Impedance

Recall from Chapter 1 that, for any linear network element, a unique relationship exists between the element voltage and the element current. For example, the voltage-current relationships for a R-ohm resistance, an

L-henry inductance, and a C-farad capacitance are given by

$$\begin{aligned} v(t) &= Ri(t) \\ v(t) &= L\frac{d}{dt}[i(t)] \\ i(t) &= C\frac{d}{dt}[v(t)] \end{aligned} \tag{4-24}$$

In Section 4-1 we stated that if the element voltage is a sinusoid of frequency ω, then the element current is also a sinusoid of frequency ω. Hence, we can express the element voltage and the element current in the form

$$\begin{aligned} v(t) &= \nu \cos(\omega t + \phi) \\ i(t) &= \iota \cos(\omega t + \theta) \end{aligned} \tag{4-25}$$

In Section 4-3 we showed that a unique relationship exists between the voltage waveform $v(t) = \nu \cos(\omega t + \phi)$ and the complex voltage $V = \nu\underline{/\phi}$, and that a unique relationship exists between the current waveform $i(t) = \iota \cos(\omega t + \theta)$ and the complex current $I = \iota\underline{/\theta}$.

From the statements in the preceding three paragraphs it follows that, for any network element, a unique relationship must exist between the element complex voltage V and the element complex current I. This discussion is illustrated graphically in Figure 4-6.

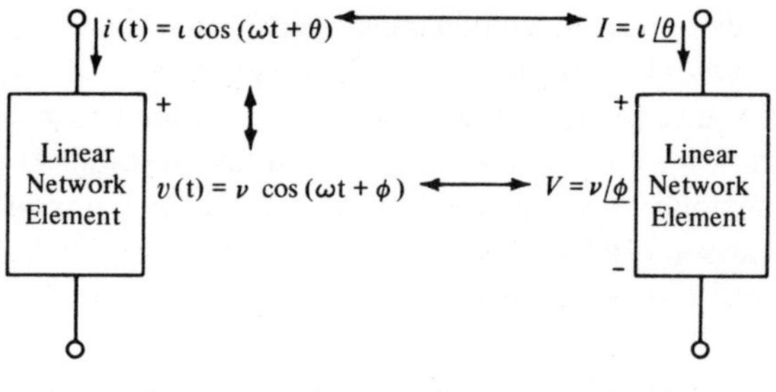

Figure 4-6 Transformation to complex number representation.

In this section we start with the known relationships between v and i, v and V, and i and I, and we derive the relationship between V and I for a resistance, an inductance, and a capacitance. In each case we will find that a linear relationship exists between V and I; that is, we shall find that the element complex voltage and complex current are related by an equation of the form $V = \mathfrak{z}I$, where $\mathfrak{z}$ is a complex number that depends only on the characteristics of the linear element. The complex number $\mathfrak{z}$ is called the *complex impedance* of the linear element. Hence, the *complex impedance* for any linear element is defined as the ratio of the element complex voltage to the element complex current; that is,

$$\mathfrak{z} = \frac{V}{I} \tag{4-26}$$

Resistance. For an R-ohm resistance, the element voltage and element current are related by

$$v(t) = Ri(t) \tag{4-27}$$

Substituting Equations (4-25) into Equation (4-27), we obtain

$$\nu \cos(\omega t + \phi) = R\iota \cos(\omega t + \theta) \tag{4-28}$$

The complex number representation for the left-hand side of (4-28) is $\nu\underline{/\phi}$, and the complex number representation for the right-hand side is $R\iota\underline{/\theta}$. If two sinusoidal waveforms are identical, their complex number representations must also be identical. Hence,

$$\nu\underline{/\phi} = R\iota\underline{/\theta} \tag{4-29}$$

or

$$V = RI \tag{4-30}$$

Therefore, the complex impedance for an R-ohm resistance is given by

$$\mathfrak{z} = \frac{V}{I} = R\underline{/0} = R \tag{4-31}$$

that is, a complex number of magnitude R and angle 0. Or, written as a sum of its real and imaginary parts, the complex impedance for an R-ohm resistance has a real part of R units and zero imaginary part.

Equation (4-31) states that the voltage and current complex number representations have the same angle, but differ in magnitude by a factor of R. Hence, the corresponding voltage and current waveforms have the same phase, but their magnitudes differ by a factor of R, as illustrated by Equation (4-28). These results are summarized in Figure 4-7, where we also interpret the complex numbers V, I, and $\mathfrak{z}$ as vectors in a plane. Such vectors are often referred to as *phasors*.

Inductance. The voltage-current relationship for an L-henry inductance is given by

$$v(t) = \frac{d}{dt}[Li(t)] \tag{4-32}$$

Using Equations (4-25) in Equation (4-32), we obtain

$$\begin{aligned}\nu \cos(\omega t + \phi) &= \frac{d}{dt}[L\iota \cos(\omega t + \theta)] \\ &= -\omega L\iota \sin(\omega t + \theta) \\ &= -\omega L\iota \cos\left(\omega t + \theta - \frac{\pi}{2}\right) = \omega L\iota \cos\left(\omega t + \theta + \frac{\pi}{2}\right)\end{aligned} \tag{4-33}$$

The complex number representation for the left-hand side of Equation (4-33) is $\nu\underline{/\phi}$, and for the right-hand side $\omega L\iota\underline{/\theta + \pi/2}$. Furthermore,

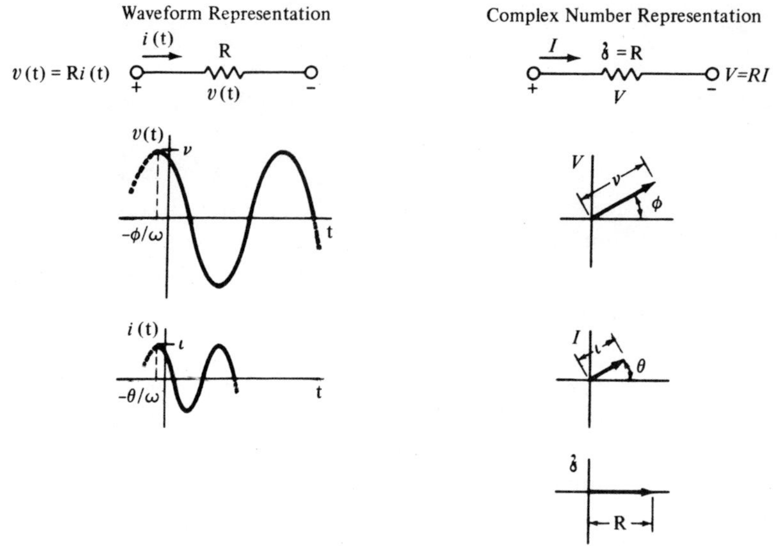

Figure 4-7 Impedance of a resistance.

Equation (4-33) implies

$$\begin{aligned} v\underline{/\phi} &= \omega L \iota \underline{/\theta + \frac{\pi}{2}} \\ &= \omega L \underline{/\frac{\pi}{2}} \cdot \iota\underline{/\theta} \end{aligned} \tag{4-34}$$

Therefore, the complex impedance for an L-henry inductance is given by

$$\mathfrak{z} = \frac{V}{I} = \omega L \underline{/\frac{\pi}{2}} = j\omega L \tag{4-35}$$

We conclude that the complex impedance associated with an L-henry inductance can be expressed as a complex number having magnitude ωL and angle $\pi/2$ radians, or as a complex number whose real part is zero and whose imaginary part is ωL. We can also think of $\mathfrak{z}$ as a vector (phasor), as illustrated in Figure 4-8.

From either Equation (4-33) or from Equation (4-34), we note that the magnitude of the inductance voltage is related to the magnitude of the inductance current by a factor of ωL; also, the phase of the inductance voltage is different from the phase of the inductance current by a factor of $\pi/2$ radians. Since $\phi = \theta + \pi/2$, we sometimes say that the inductance voltage "leads" the inductance current by $\pi/2$ radians. These results are summarized in Figure 4-8.

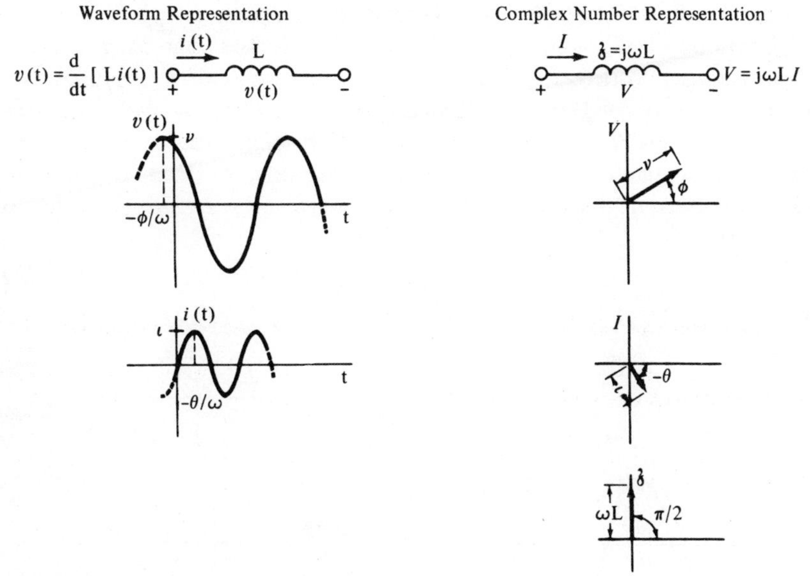

Figure 4-8 Impedance of an inductance.

Capacitance. For a C-farad capacitance, the capacitance voltage is related to the capacitance current by

$$i(t) = \frac{d}{dt}[Cv(t)] \tag{4-36}$$

For sinusoidal waveforms of frequency ω we have

$$\begin{aligned} \iota \cos(\omega t + \theta) &= \frac{d}{dt}[C\nu \cos(\omega t + \phi)] \\ &= -\omega C\nu \sin(\omega t + \phi) \\ &= \omega C\nu \cos\left(\omega t + \phi + \frac{\pi}{2}\right) \end{aligned} \tag{4-37}$$

Equation (4-37) implies that

$$\begin{aligned} \iota\angle\theta &= \omega C\nu \angle\phi + \frac{\pi}{2} \\ &= \omega C \angle\frac{\pi}{2} \cdot \nu\angle\phi \end{aligned} \tag{4-38}$$

Therefore,

$$\mathfrak{z} = \frac{V}{I} = \frac{1}{\omega C}\angle -\frac{\pi}{2} = -j\frac{1}{\omega C} = \frac{1}{j\omega C} \tag{4-39}$$

We conclude that the complex impedance associated with a C-farad capacitance has magnitude $1/\omega C$ volts per ampere and angle $-\pi/2$ radians. It can also be expressed as a complex number whose real part is 0 and whose imaginary part is $-1/\omega C$. It is also correct to say that the complex impedance associated with a C-farad capacitance is $1/j\omega C$.

Equations (4-37), (4-38), and (4-39) all state that the magnitude of the capacitance voltage is related to the magnitude of the capacitance current by a factor of $1/\omega C$ and that the phases are related by $\pi/2$ radians. Since $\theta = \phi + \pi/2$, we sometimes say that the capacitance current "leads" the capacitance voltage by $\pi/2$ radians, or that the voltage "lags" the current by $\pi/2$ radians. These results are summarized in Figure 4-9.

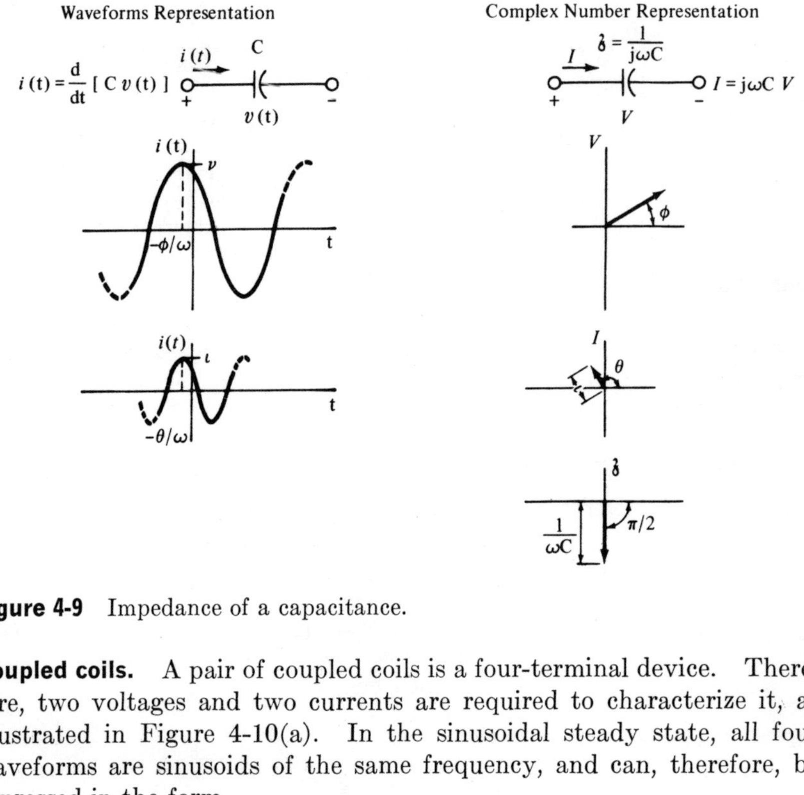

Figure 4-9 Impedance of a capacitance.

Coupled coils. A pair of coupled coils is a four-terminal device. Therefore, two voltages and two currents are required to characterize it, as illustrated in Figure 4-10(a). In the sinusoidal steady state, all four waveforms are sinusoids of the same frequency, and can, therefore, be expressed in the form

$$
\begin{aligned}
v_1(t) &= \nu_1 \cos(\omega t + \phi_1) = V_1 \frac{e^{j\omega t}}{2} + V_1^* \frac{e^{-j\omega t}}{2} \\
v_2(t) &= \nu_2 \cos(\omega t + \phi_2) = V_2 \frac{e^{j\omega t}}{2} + V_2^* \frac{e^{-j\omega t}}{2} \\
i_1(t) &= \iota_1 \cos(\omega t + \theta_1) = I_1 \frac{e^{j\omega t}}{2} + I_1^* \frac{e^{-j\omega t}}{2} \\
i_2(t) &= \iota_2 \cos(\omega t + \theta_2) = I_2 \frac{e^{j\omega t}}{2} + I_2^* \frac{e^{-j\omega t}}{2}
\end{aligned}
\tag{4-40}
$$

These voltages and currents are related by the element equations

$$v_1(t) = \frac{d}{dt}[L_1 i_1(t) + M i_2(t)]$$
$$v_2(t) = \frac{d}{dt}[M i_1(t) + L_2 i_2(t)] \qquad \textbf{(4-41)}$$

Substituting the exponential forms of Equations (4-40) into Equations (4-41) and following the procedure used in Example 4-2, we find that the linear differential equations given by Equations (4-41) reduce to the linear algebraic equations

$$V_1 = j\omega L_1 I_1 + j\omega M I_2$$
$$V_2 = j\omega M I_1 + j\omega L_2 I_2 \qquad \textbf{(4-42)}$$

The network model in terms of these complex voltages, currents, and impedances is shown in Figure 4-10(b).

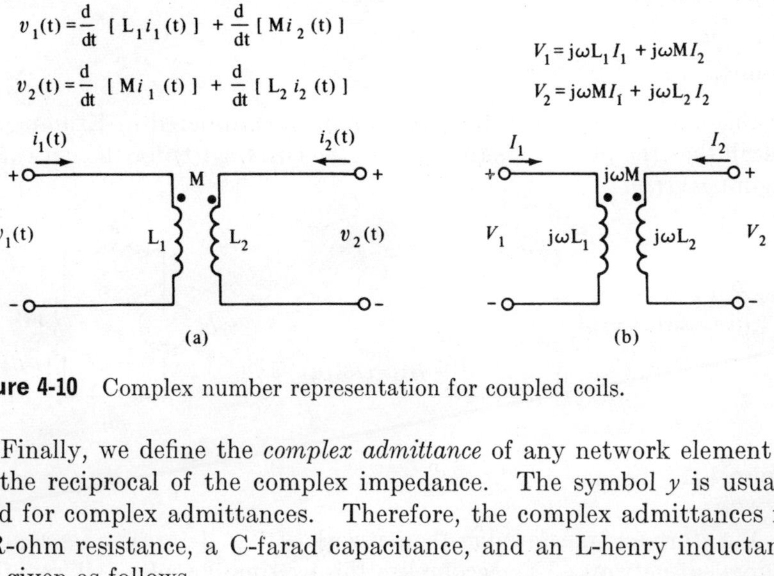

Figure 4-10 Complex number representation for coupled coils.

Finally, we define the *complex admittance* of any network element to be the reciprocal of the complex impedance. The symbol y is usually used for complex admittances. Therefore, the complex admittances for a R-ohm resistance, a C-farad capacitance, and an L-henry inductance are given as follows.

$$\text{resistance: } y = z^{-1} = \frac{1}{R}$$
$$\text{inductance: } y = z^{-1} = \frac{1}{j\omega L} \qquad \textbf{(4-43)}$$
$$\text{capacitance: } y = z^{-1} = j\omega C$$

We can also define an admittance representation for a pair of coupled coils by simply solving Equations (4-42) for I_1 and I_2. Impedance and admittance representations for multiterminal networks is discussed further in Chapters 5 and 6.

4-5 Sinusoidal Steady-State Network Analysis

In this section the results of Chapter 3 and Chapter 4 are combined into a method for analyzing linear networks with sinusoidal forcing functions.

The sinusoidal steady-state algorithm.

1. Transform the network into its complex equivalent network.
2. Write a set of linearly independent network equations using one of the two algorithms presented in Section 3-6.
3. Solve the set of equations.
4. Transform the resulting complex numbers into their equivalent sinusoidal waveforms.

This procedure is best illustrated by examples. The procedure can be justified by using the technique outlined in Example 4-2 to solve the same problem.

Example 4-3

In this example we solve the same problem considered in Example 4-2. Recall that the problem is to compute the current $i(t)$ for the network of Figure 4-11(a).

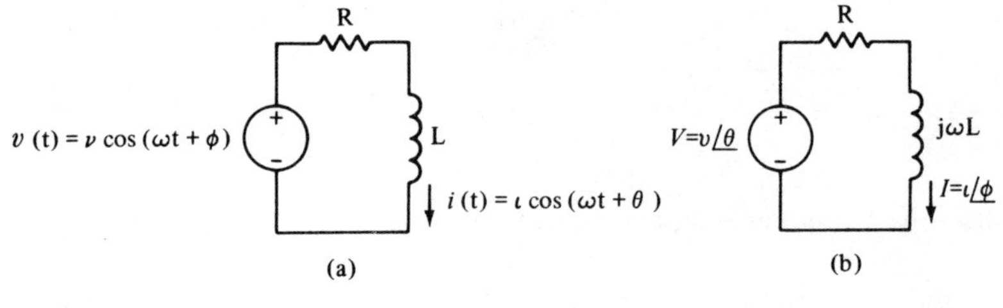

Figure 4-11 Example 4-3.

We start by transforming the network of Fig. 4-11(a) to its complex equivalent network. To accomplish this we simply replace all waveform sources with their corresponding complex number representations and all linear elements with their complex impedances. Hence, we replace the independent (waveform) voltage source $v(t)$ with the independent (complex voltage) source V. We also replace the R-ohm resistance and the L-henry inductance with their corresponding complex impedances. Finally, we replace the waveform current symbol $i(t)$ with its corresponding complex current symbol I. This completes step 1.

Step 2 requires that we write a set of network equations for the complex equivalent network of Figure 4-11(b). The straightforward way to accomplish this is to apply either of the algorithms of Section 3-6. However, the network of Figure 4-11(b) is simple enough so that the appropriate equation is obvious. Summing voltages by KVL around the loop, we write

$$V = RI + j\omega LI$$

The next step is to solve the network equation for the complex current I:

$$I = \frac{V}{R + j\omega L} \tag{4-44}$$

The magnitude and angle of I can be obtained by writing

$$I = \iota\underline{/\theta} = \frac{\nu\underline{/\phi}}{\sqrt{R^2 + \omega^2 L^2}\ \underline{/\tan^{-1}\frac{\omega L}{R}}} = \frac{\nu}{\sqrt{R^2 + \omega^2 L^2}}\ \underline{/\phi - \tan^{-1}\frac{\omega L}{R}} \tag{4-45}$$

or

$$\iota = \frac{\nu}{\sqrt{R^2 + \omega^2 L^2}}$$

$$\theta = \phi - \tan^{-1}\frac{\omega L}{R} \tag{4-46}$$

The final step is to transform the complex current I back to its waveform representation

$$i(t) = \frac{\nu}{\sqrt{R^2 + \omega^2 L^2}} \cos\left(\omega t + \phi - \tan^{-1}\frac{\omega L}{R}\right) \tag{4-47}$$

Equation (4-43), which was written directly from the complex equivalent network, is identical to Equation (4-19), which was obtained by first writing the network differential equation and reducing it by using the complex exponential representation for the sinusoids. The final result is, of course, the same as the result obtained in Example 4-2.

Example 4-4

In this example we compute the output voltage $v_2(t)$ for the amplifier circuit presented in Figure 4-12(a) for a sinusoidal input of peak amplitude 1 mA and radian frequency $\omega = \frac{1}{3} \times 10^6$ rad/sec (f = 53.1 kHz).

A model for the 2N3819 field-effect transistor (FET) is presented in Figure 4-12(b). Combining these two models, we find that the network to be analyzed is the one in Figure 4-12(c).

According to the sinusoidal steady-state algorithm, the first step is to construct the complex equivalent network. Hence, as illustrated in Figure 4-13(a), the sinusoidal current source is replaced by the complex current source given by $10^{-3}\underline{/0} = 10^{-3}$, the control voltage v is replaced

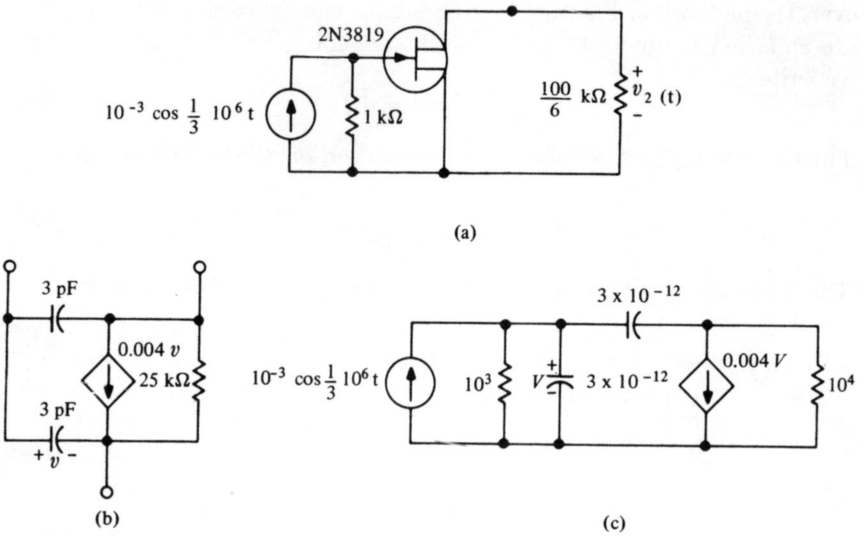

Figure 4-12 Example 4-4.

by the complex voltage V, each of the 3 pF capacitances is replaced by its complex admittance $j\omega C = j\frac{1}{3} \times 10^6 \times 3 \times 10^{-12} = j10^{-6}$, and for each resistance we write its complex admittance. We have chosen to replace the network elements with their complex admittances; we could as well have worked with complex impedances.

The second step is to write a set of network equations. Toward this end we construct a tree and cotree, as illustrated in Figure 4-13(b). Note that, as suggested in Chapter 3, we have put the two current sources in the cotree and the control voltage in the tree. Since our objective is to compute the output voltage, we also make the remaining branch across the output terminals a tree branch, and choose to write two equations in the two tree branch voltages. (Note that since two of the links correspond to current sources, two equations would also be required if we had chosen to write equations in the link currents.) From Figure

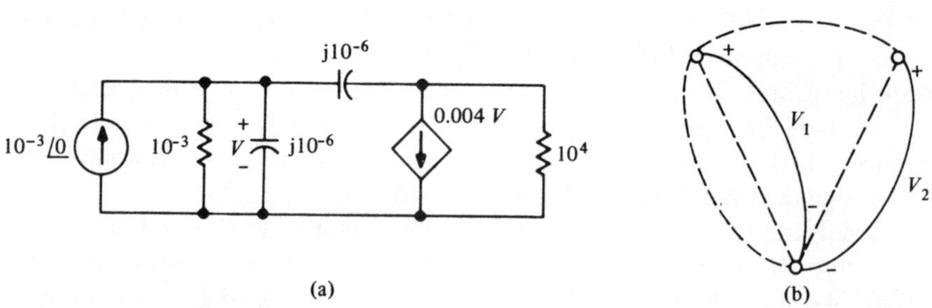

Figure 4-13 Complex equivalent network for Example 4-4.

4-13(b) we now write each tree branch current as a sum of link currents:

$$\begin{aligned} j10^{-6}V_1 &= 10^{-3} + 10^{-3}V_1 - j10^{-6}(V_1 - V_2) \\ 10^{-4}V_2 &= j10^{-6}(V_1 - V_2) - 0.004V_1 \end{aligned} \tag{4-48}$$

The result is two equations in the two complex voltages V_1 and V_2.

Step 3 dictates that we solve Equations (4-48). Since we are only interested in the output voltage, we need only solve for V_2. To accomplish this we first rearrange Equations (4-49) into the form

$$\begin{aligned} (10^{-3} + j2 \times 10^{-6})V_1 \qquad\qquad - (j10^{-6})V_2 &= 10^{-3} \\ (0.004 - j10^{-6})V_1 + (10^{-4} + j10^{-6})V_2 &= 0 \end{aligned} \tag{4-49}$$

We can now apply Cramer's rule (see Appendix II) to obtain V_2:

$$\begin{aligned} V_2 &= \frac{\begin{vmatrix} 10^{-3} + j2 \times 10^{-6} & 10^{-3} \\ 0.004 - j10^{-6} & 0 \end{vmatrix}}{\begin{vmatrix} 10^{-3} + j2 \times 10^{-6} & -j10^{-6} \\ 0.004 - j10^{-6} & 10^{-4} + j10^{-6} \end{vmatrix}} \\ &= \frac{-4 \times 10^{-6} + j10^{-9}}{(10^{-3} + j2 \times 10^{-6})(10^{-4} + j10^{-6}) - j10^{-6}(0.004 - j10^{-6})} \\ &\approx \frac{-10^{-6}(4 - j0.001)}{10^{-7}(1 + j0.052)} = -10\,\frac{4 - j0.001}{1 + j0.0025} \approx -10\,\frac{4\angle 0}{1\angle 0} \\ &= -40\angle 0 = 40\angle \pi \end{aligned} \tag{4-50}$$

We conclude that the complex voltage V_2 has magnitude 40 and angle π radians (180°).[2]

The final step is to construct the waveform $v_2(t)$ from the complex voltage V_2. The result is

$$v_2(t) = -40 \cos\left(\tfrac{1}{3} \times 10^6 t\right) = 40 \cos\left(\tfrac{1}{3} \times 10^6 t + \pi\right) \tag{4-51}$$

Hence, the output voltage is π radians (180°) out of phase with the input current, and the network gain is 40 V/A = 40,000 V/mA.

Actually, this result could have been obtained with less computation had we noted in Figure 4-13(a) that the complex admittances of the two capacitances were at least 1000 times smaller than the other two admittances. Hence, both capacitances are essentially open circuits and

[2] For complex number arithmetic we usually approximate terms when the approximation error is less than 1 percent in magnitude or 5° in phase. For example, $1 + j0.052 = 1.0027\angle 2.92°$, and so we use $1\angle 0$ as an approximation. Similarly, $4 + j0.001 \approx 4\angle 0$.

could have been omitted from the model. The same result is obtained if both capacitances are replaced by open circuits.

Example 4-5

In order to demonstrate the analysis of a network containing a pair of coupled coils, we next compute the output voltage $v_0(t)$ for the network presented in Figure 4-14(a).

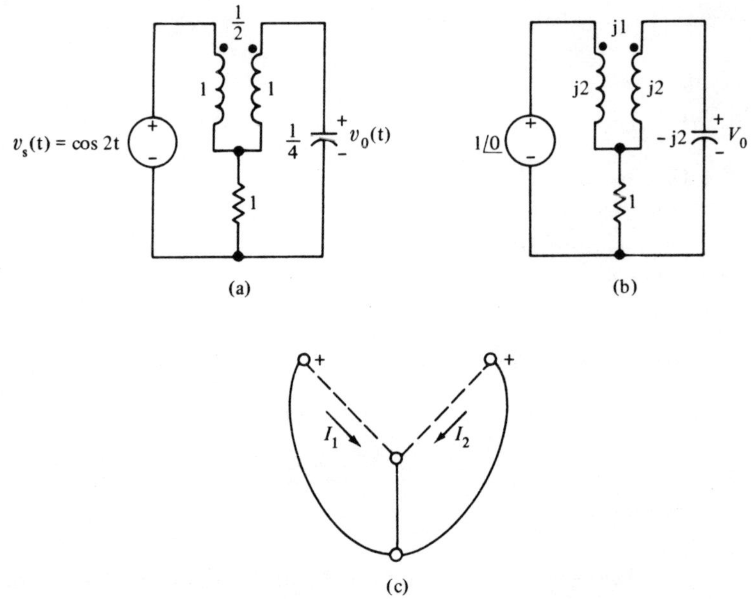

Figure 4-14 Example 4-5.

We start by transforming the time-domain network of Figure 14(a) into its frequency-domain equivalent, as shown in Figure 4-14(b). Next, we construct a tree according to the method suggested for coupled coils in Section 3-6, as shown in Figure 4-14(c). A set of network equations in terms of the two link (complex) currents I_1 and I_2 can now be written by applying the link current algorithm. Writing each link voltage as a sum of tree branch voltages in terms of I_1 and I_2, we obtain

$$\begin{aligned} j2I_1 + jI_2 &= 1 - (I_1 + I_2) \\ j2I_2 + jI_1 &= j2I_2 - (I_1 + I_2) \end{aligned} \tag{4-52}$$

Or, rearranging terms, we obtain

$$\begin{aligned} (j2 + 1)I_1 + (j + 1)I_2 &= 1 \\ (j + 1)I_1 \qquad + \quad I_2 &= 0 \end{aligned} \tag{4-53}$$

We next solve for I_2 by the method of determinants:

$$I_2 = \frac{\begin{vmatrix} j2+1 & 1 \\ j+1 & 0 \end{vmatrix}}{\begin{vmatrix} j2+1 & j+1 \\ j+1 & 1 \end{vmatrix}} = \frac{(j2+1)0 - (j+1)}{(j2+1) - (j+1)^2}$$

$$= \frac{-(j+1)}{1} = \sqrt{2}\,\underline{/-\pi/4} \tag{4-54}$$

The complex voltage V_0 is related to the complex current I_2 by

$$V_0 = j2I_2$$
$$= 2\,\underline{/\frac{\pi}{2}} \cdot \sqrt{2}\,\underline{/\frac{-\pi}{4}} = 2\sqrt{2}\,\underline{/\frac{\pi}{4}} \tag{4-55}$$

Finally, we transform back from the complex representation to the equivalent waveform representation by writing

$$v_0(t) = 2\sqrt{2}\cos\left(2t + \frac{\pi}{4}\right) \tag{4-56}$$

4-6 Power in the Sinusoidal Steady State

Instantaneous and average power. In Chapter 1 the instantaneous electric power delivered to a network was defined to be the product of the voltage and the current, provided that proper polarity conventions are assigned. If the current or the voltage varies with time, then the power varies with time. For the case of sinusoidal voltages and currents, the instantaneous power is given by

$$\begin{aligned} p(t) &= v(t)i(t) \\ &= \nu\cos(\omega t + \phi)\,\iota\cos(\omega t + \theta) \\ &= \tfrac{1}{2}\nu\iota\cos(\phi - \theta) + \tfrac{1}{2}\nu\iota\cos(2\omega t + \phi + \theta) \end{aligned} \tag{4-57}$$

Hence, the instantaneous power in the sinusoidal steady state consists of a constant plus a sinusoid at twice the frequency of the voltage and current.

In Chapter 1 the average power delivered to a network was defined as the time average of the instantaneous power. Hence, in the sinusoidal steady state, the average power is given by

$$\begin{aligned} \langle p(t)\rangle &= \langle v(t)i(t)\rangle \\ &= \langle \tfrac{1}{2}\nu\iota\cos(\phi - \theta) + \tfrac{1}{2}\nu\iota\cos(2\omega t + \phi + \theta)\rangle \\ &= \tfrac{1}{2}\nu\iota\cos(\phi - \theta) \end{aligned} \tag{4-58}$$

For an R-ohm resistance, we can use Equation (4-29) in Equation (4-57) to obtain the power delivered to the resistance

$$p_R(t) = \frac{1}{2}\frac{\nu^2}{R} + \frac{1}{2}\frac{\nu^2}{R}\cos(2\omega t + 2\phi) \tag{4-59}$$

$p_R(t)$ is sketched in Figure 4-15 for $R > 0$. Hence, the power dissipated in the resistance varies sinusoidally between a maximum value of ν^2/R watts and a minimum of 0 watts with an average power of $\nu^2/2R$.

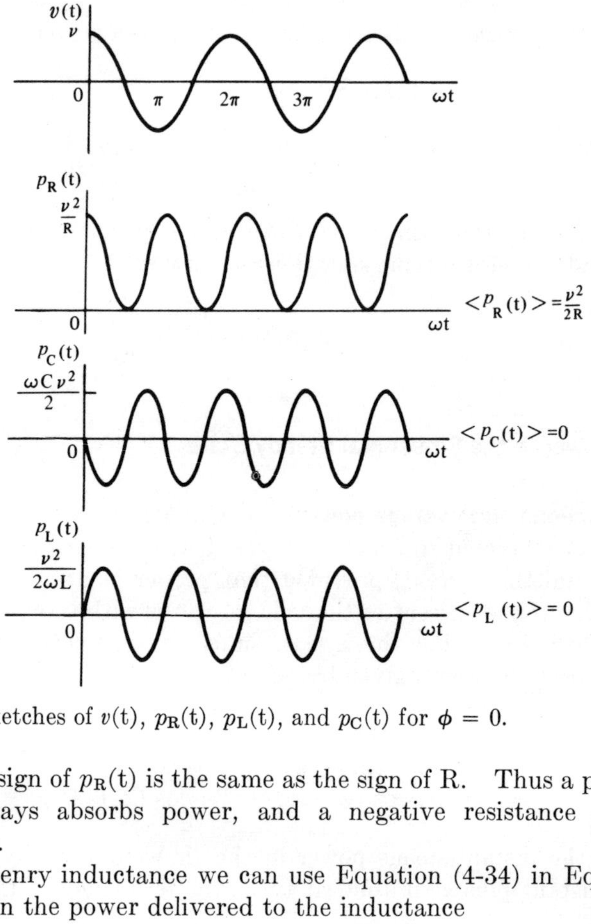

Figure 4-15 Sketches of $v(t)$, $p_R(t)$, $p_L(t)$, and $p_C(t)$ for $\phi = 0$.

Note that the sign of $p_R(t)$ is the same as the sign of R. Thus a positive resistance always absorbs power, and a negative resistance always delivers power.

For an L-henry inductance we can use Equation (4-34) in Equation (4-57) to obtain the power delivered to the inductance

$$p_L(t) = \frac{1}{2}\frac{\nu^2}{\omega L}\cos\left(\frac{\pi}{2}\right) + \frac{1}{2}\frac{\nu^2}{\omega L}\cos\left(2\omega t + 2\phi - \frac{\pi}{2}\right) \tag{4-60}$$

$p_L(t)$ is sketched in Figure 4-15. Note that the power varies sinusoidally about 0. The inductance absorbs power for $\frac{1}{4}$ cycle of the voltage, but returns it all during the next $\frac{1}{4}$ cycle. We conclude that, even though the stored energy varies, all that goes in comes out, and the average power absorbed is zero, that is,

$$\langle p_L(t)\rangle = 0 \tag{4-61}$$

For a C-farad capacitance, we use Equation (4-38) in Equation (4-57) to obtain the power delivered to the capacitance:

$$p_C(t) = \tfrac{1}{2}\omega C\nu^2 \cos\left(-\frac{\pi}{2}\right) + \tfrac{1}{2}\omega C\nu^2 \cos\left(2\omega t + \phi + \frac{\pi}{2}\right) \tag{4-62}$$

$p_C(t)$ is sketched in Figure 4-15. Again, the power varies sinusoidally at twice the frequency of the voltage, but the average power is zero:

$$\langle p_C(t)\rangle = 0 \tag{4-63}$$

These instantaneous-power and average-power formulas can also be written in terms of the complex voltage and complex current. Using Equation (4-12) in Equation (4-57), we obtain

$$\begin{aligned} p(t) &= \left[V\frac{e^{j\omega t}}{2} + V^*\frac{e^{-j\omega t}}{2}\right]\left[I\frac{e^{j\omega t}}{2} + I^*\frac{e^{-j\omega t}}{2}\right] \\ &= \tfrac{1}{4}[VI^* + V^*I + VIe^{j2\omega t} + V^*I^*e^{-j2\omega t}] \end{aligned} \tag{4-64}$$

Comparing Equation (4-64) to Equation (4-57), we note that $\frac{1}{4}[VI^* + V^*I] = \frac{1}{2}\nu\iota \cos(\phi - \theta)$ and $\frac{1}{4}[VIe^{j2\omega t} + V^*I^*e^{-j2\omega t}] = \frac{1}{2}\nu\iota \cos(2\omega t + \phi + \theta)$. Hence, in terms of the complex current and complex voltage, the average power is given by

$$\langle p(t)\rangle = \tfrac{1}{4}[VI^* + V^*I] \tag{4-65}$$

It is easy to show that (see Appendix III) $VI^* = (V^*I)^*$ so that $VI^* + V^*I = VI^* + (VI^*)^* = V^*I + (V^*I)^*$. It is also obvious from Appendix III-1 that $A + A^* = 2\,\mathrm{Re}\,\{A\}$, where Re $\{\cdot\}$ means the real part of the quantity inside the bracket. Therefore, we conclude that Equation (4-65) is equivalent to

$$\langle p(t)\rangle = \tfrac{1}{2}\,\mathrm{Re}\,\{VI^*\} = \tfrac{1}{2}\,\mathrm{Re}\,\{V^*I\} \tag{4-66}$$

Effective voltage, effective current, and power factor. In most applications of sinusoidal voltages and currents, from the 60-Hz power systems to microwave communications systems, power computations are often the most significant computations. To avoid the factor $\frac{1}{2}$ that appears in all the formulas above, quantities called effective voltages and effective currents are usually used. The *effective voltage* $\mathcal{V}$ is related to the voltage magnitude ν by

$$\mathcal{V} = \frac{1}{\sqrt{2}}\nu \approx 0.707\nu \tag{4-67}$$

Similarly, the *effective current* $\mathcal{I}$ is related to the current magnitude ι by

$$\mathcal{I} = \frac{1}{\sqrt{2}}\iota \approx 0.707\iota \tag{4-68}$$

These definitions for effective voltage and effective current are equivalent to the rms (root-mean-square) values. Furthermore, since the sinusoidal component of the power flow represents an exchange of power from source to load and back without loss, it is not of prime consideration. Thus, discussions of power flow usually center on the average power—the real power that must be delivered to the load, never to be returned. The average power delivered to any network is given by Equation (4-58)

$$\langle p(t)\rangle = \tfrac{1}{2}\nu\iota\cos(\phi - \theta) = \frac{\nu}{\sqrt{2}}\frac{\iota}{\sqrt{2}}\cos(\phi - \theta) = \mathcal{V}\mathcal{I}\cos(\phi - \theta) \tag{4-69}$$

The quantity $\cos(\phi - \theta)$ is called the *power factor*. Hence, *the average power dissipated in the network is the product of the effective voltage, the effective current, and the power factor.*

The power factor depends on the phase relationship of the voltage and current. If the voltage and current are in phase, $\phi = \theta$ and the power factor is one. On the other hand, if the voltage and current are $\pi/2$ radians (90°) out of phase (as for an inductance or capacitance), the power factor is zero.

Since power is a prime consideration, most volt-meters and ammeters are calibrated in effective values. The 110–120-volt ac line carries a 60-Hz (377 rad/sec) sine wave with effective value between 110 and 120. Thus the peak value of the ac line voltage is between 156 and 170 volts.

PROBLEMS

4-1 For the network shown in Figure P4-1

a. Write a differential equation in terms of $v(t)$.

b. Assume $i_s(t) = \cos t$ and use the sinusoidal steady-state analysis to solve for $v(t)$.

c. Check your answer by showing that your solution to part b satisfies the differential equation of part a.

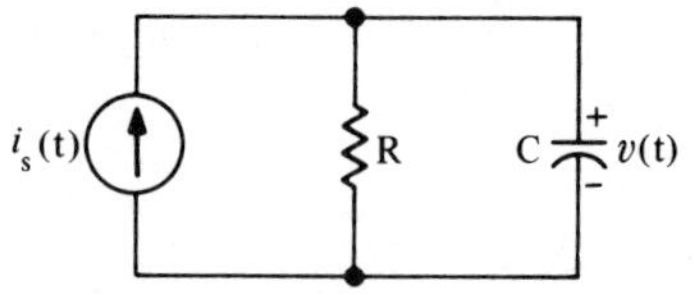

Figure P4-1

4-2 For each of the following voltages, find the complex voltage amplitude:

a. $v(t) = 10 \cos [377t + (\pi/3)]$

b. $v(t) = 25 \cos (6.28 \times 10^6 t + 3.14)$

c. $v(t) = 30 \cos 10^5 t + 40 \sin 10^5 t$

d. $v(t) = 75 \sin (10^5 t + 1.57)$

4-3 For each of the following complex voltage amplitudes, find the voltage as a function of time:

a. $V = 0.5 + j0.866$

b. $V = -27$

c. $V = j10$

d. $V = 10\underline{/45°}$

e. $V = 100\underline{/\pi/4}$

4-4 Compute $v(t)$ (see Figure P4-4). Use the sinusoidal steady-state analysis and use any reasonable approximations.

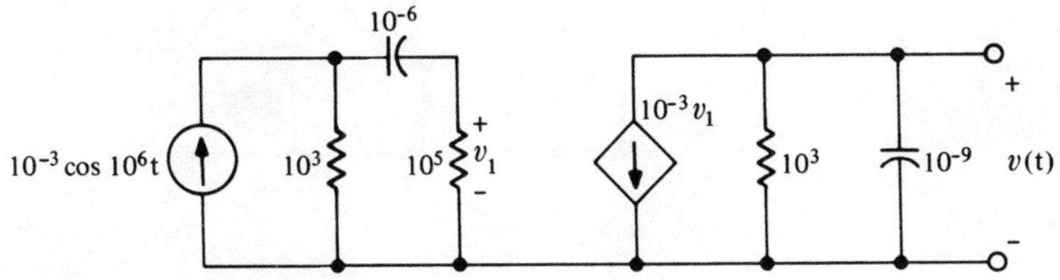

Figure P4-4

4-5 Compute the (complex) Norton and Thevenin equivalent networks for the network shown by Figure P4-5.

a. By computing the open-circuit voltage and short-circuit current.

b. By setting the independent source to zero and attaching a 1-ampere independent current source at the terminals.

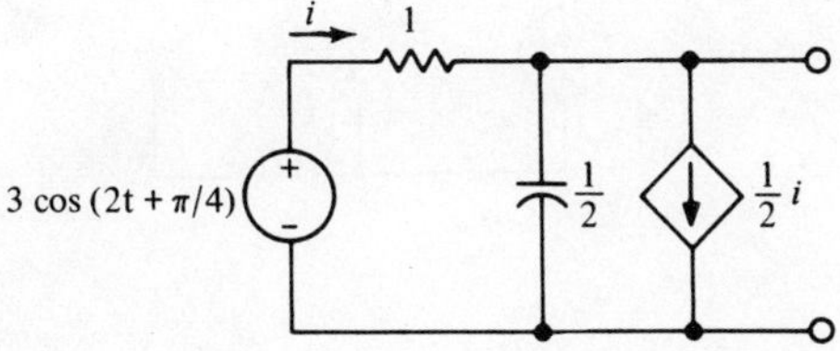

Figure P4-5

4-6 Compute the Thevenin and Norton equivalent networks with respect to the terminals indicated (see Figure P4-6).

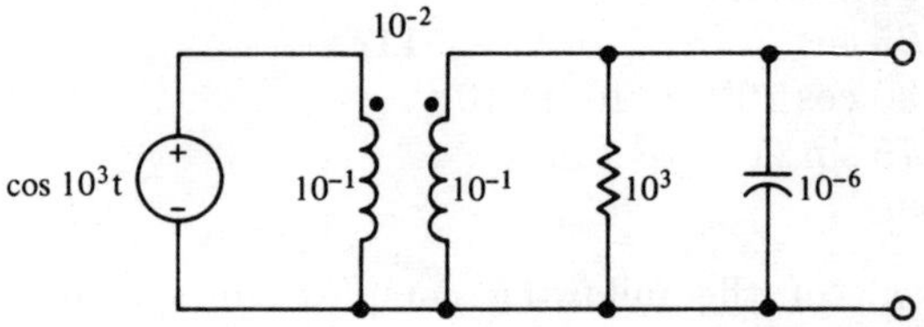

Figure P4-6

4-7 For the network shown in Figure P4-7

a. Compute $v(t)$ for $\omega = 10^4$.

b. Compute and plot the amplitude and phase of $v(t)$ versus ω for $10^3 \leq \omega \leq 10^5$. (Hint: Use the computer.)

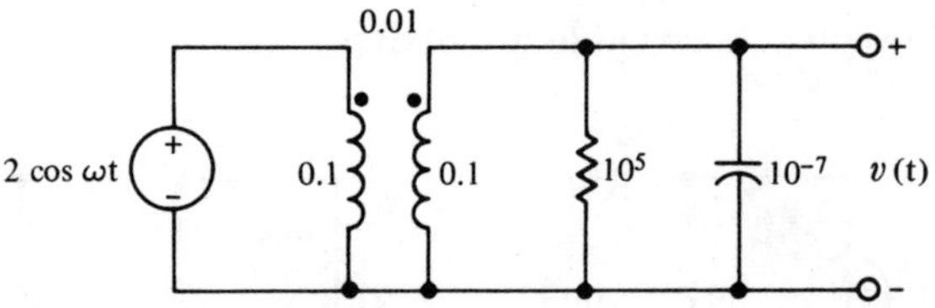

Figure P4-7

4-8 Consider the parallel R-C circuit shown in Figure P4-8. Compute the instantaneous and average power delivered to the resistance and to the capacitance for the following cases:

a. $\omega = 1, \phi = 0$

b. $\omega = 1, \phi = (\pi/2)$

c. $\omega = 0.1, \phi = 0$

d. $\omega = 10, \phi = 0$

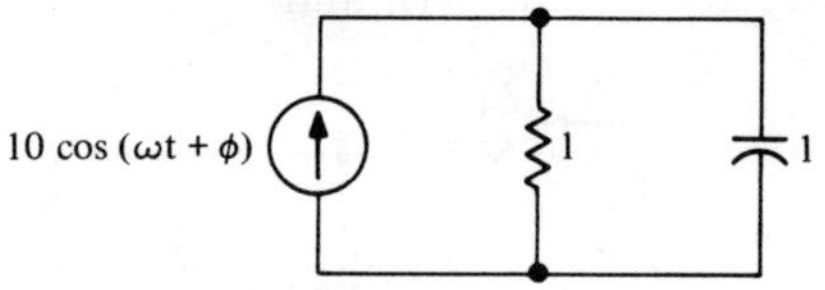

Figure P4-8

4-9 A three-phase, 60-cycle, 240-volt supply is connected to a three-phase motor connected to a compressor in an air-conditioning unit. What is the approximate BTU rating of the air-conditioner? (See Figure P4-9.)

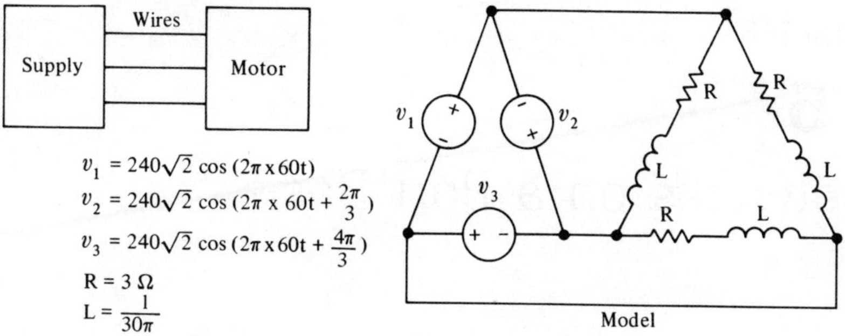

Figure P4-9

4-10 For the circuit shown in Figure P4-10, prove that the controlled source delivers power at low frequencies and absorbs power at high frequencies. Find the frequency where the power is zero in the controlled source.

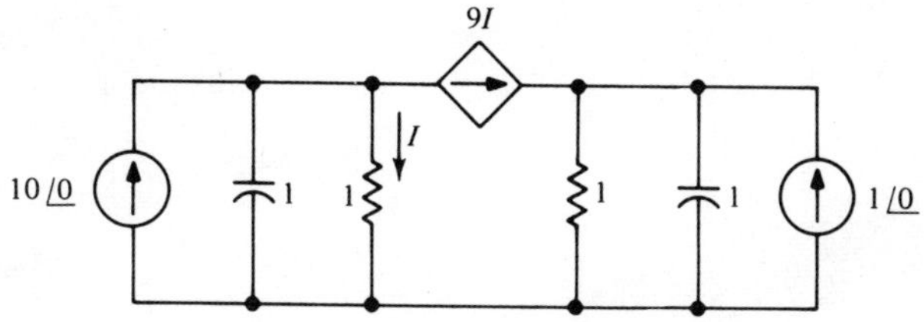

Figure P4-10

CHAPTER

5

Networks on a Port Basis

5-1 Introduction

The material in Chapter 3 pertained to the problem of formulating a set of linearly independent equations that describe an electronic network. By employing the algorithms presented there, we can write a set of equations whose solution yields all the branch voltages and currents in terms of the element parameters, that is, the resistances, inductances, capacitances, independent and controlled source parameters, and so on. However, we are not usually interested in solving for *all* the network voltages and currents; usually, we are interested in only a few of the 2b branch voltages and currents. In particular, we are usually interested in only those branch voltages and currents at those nodes chosen to be terminals (connecting points) to other networks.

Consider again the broadcast receiver (transistor radio) discussed briefly in Chapter 1. The circuit diagram and block diagram are shown in Figure 5-1. By replacing the transistors, the speaker, the antenna, and so on, in the circuit diagram with their models synthesized from the network elements described in Chapter 1, we find that each of the boxes in the block diagram corresponds to an electrical network. The boxes labeled mixer and detector contain nonlinear elements, and their analysis is beyond the scope of this text. Each of the remaining networks is a linear, lumped, time-invariant network of the sort we are concerned with. Note that some of the networks (boxes) have more terminals than others. For example, the antenna is a two-terminal device, that is, it has two connecting points for passing the signal developed across the antenna

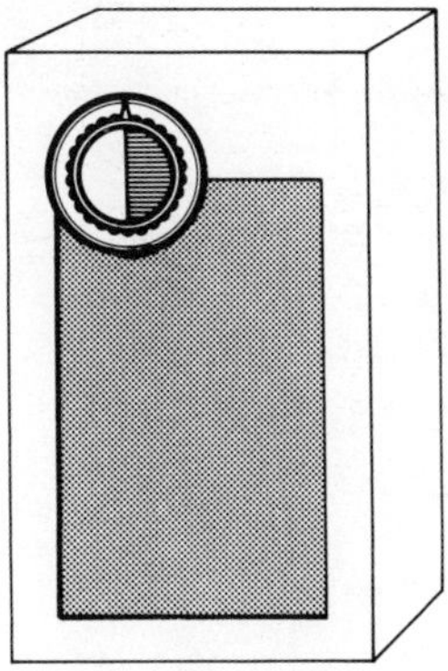

Trade Name:	Westinghouse Models H–903P8GP, H–904P8GP Chassis V–2403-7
Type Set:	2 Transistor AM Receiver
Power Supply:	3 volts dc
Rating:	15 mA@3V dc (No Signal Min. Volume)
	25 mA@3V dc (Signal, Normal Volume)
Tuning Range:	Broadcast 540 – 1625 kHz

(a)

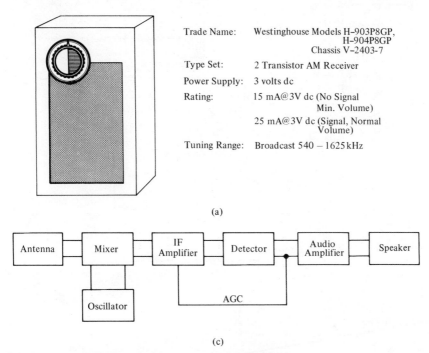

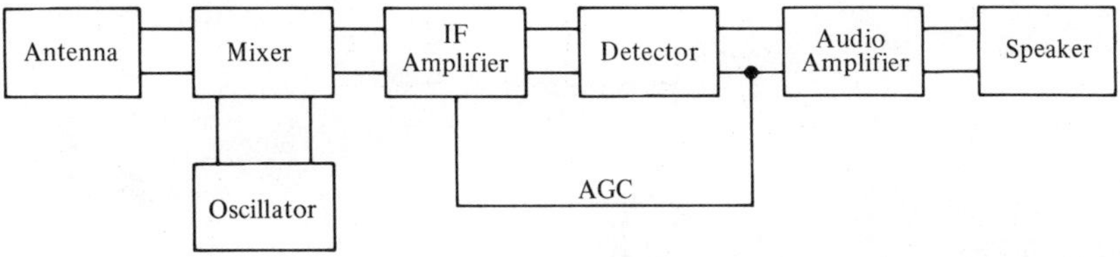

(c)

Figure 5-1 (a) Broadcast receiver. (c) Block diagram.

terminals to the mixer. On the other hand, the mixer has six terminals—two input terminals from the antenna, two input terminals from the oscillator, and two output terminals for passing the output signal to the IF (intermediate-frequency) amplifier.

The nodes of the networks that are internal to the blocks of the block diagram are of only secondary importance. The signal-processing function of each block can be characterized by the voltages and currents at its terminals. A terminal pair for which the current into one terminal is equal to the current out of the other terminal is called a *port*. Note that all terminal pairs are not ports. Whether or not a network terminal pair is a port does not always depend on the network; it may depend on terminal connections external to the network.

Recall that none of the electrical elements described in Chapter 2 is

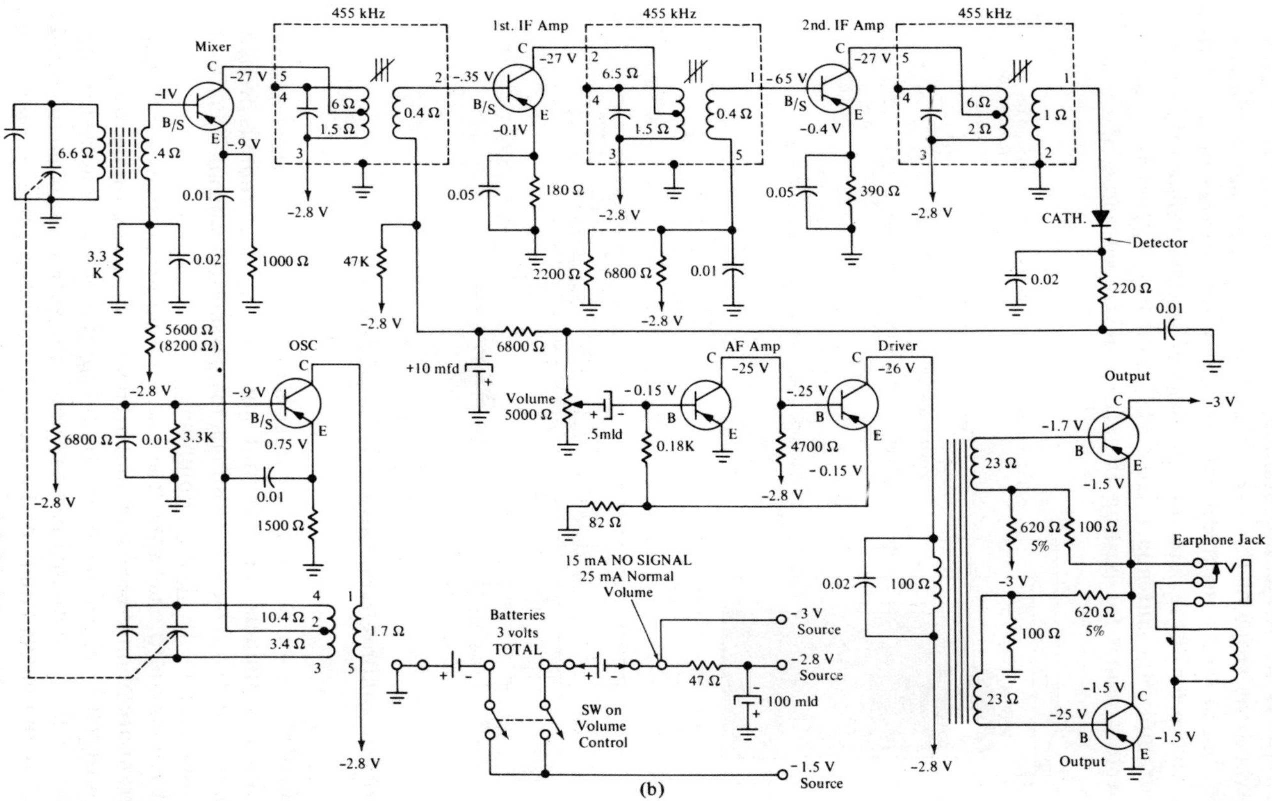

Figure 5-1 (b) Its circuit diagram.

capable of storing current. Therefore, networks constructed from these elements cannot store current. It follows that the net current into any network is zero. Now, consider a two-terminal network such as the antenna of Figure 5-1. Since the current into the network equals the current out of the network, we conclude that all two-terminal networks are 1-port networks, regardless of what is connected across their terminals. Next, consider the mixer in Figure 5-1. Since both input terminal pairs are ports (because the antenna and oscillator are both 1-port networks), the output terminal-pair must also be a port. It follows that the IF-amplifier input terminal pair is a port. Note, however, that if any current flows in the AGC (automatic gain control) line, the IF-amplifier output terminal pair is not a port. The same can be said for both detector terminal pairs. However, the audio-amplifier input terminal pair is a port because its output terminal pair is a port. In general, a network terminal pair is a port, provided no current-carrying paths exist between that terminal pair and any other terminal belonging to the same network.

As we have stated, we are usually interested in the network characteristics as seen from the network terminals. Therefore, it is convenient to characterize the network with equations involving only the terminal voltages and terminal currents. In this chapter we describe various schemes for characterizing n-port networks in terms of their port voltages and currents. We first consider 1-port networks, then 2-port networks, and, finally, n-port networks. We also consider the problem of connecting n-port networks into larger networks. The characterization of n-terminal networks (networks with terminals that may not be grouped into ports) is discussed in Chapter 6.

We shall consider only networks containing no independent sources. This restriction involves no loss in generality, since any independent sources can always be considered external to the network but connected to the network through an additional terminal pair. Furthermore, we characterize these networks in terms of their steady-state responses to independent sinusoidal sources. This assures that the sinusoidal steady-state analysis developed in Chapter 4 can be used to transform the network differential equations to algebraic equations. In these equations the port voltages V and the port currents I are complex numbers that represent the magnitudes and phases of the actual sinusoidal port voltages and port currents at the frequency of the independent source.

Finally, we state a port voltage polarity convention and a port current direction convention used throughout this chapter. Port voltage is always specified as the voltage at the upper terminal relative to the lower terminal, as illustrated in Figure 5-2. The port current is assumed to flow into the upper terminal (and out of the lower), as illustrated in Figure 5-2.

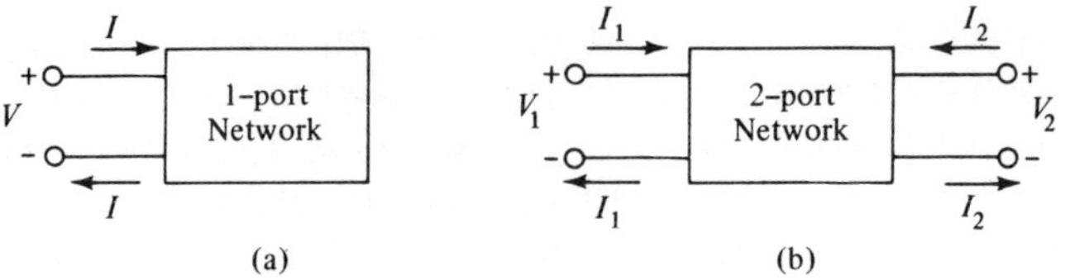

Figure 5-2 Port voltage polarity and port current direction conventions for (a) 1-port networks and (b) 2-port networks.

5-2 1-Port Networks

Consider the 1-port network of Figure 5-2(a). Since the network is assumed to be a linear network, a linear relationship exists between the port complex voltage V and the port complex current I. The most general linear algebraic equation relating these two variables is an equation of the form $V = \mathfrak{z}I + b$. This equation must be valid for all V and I that meet the constraints imposed by the network and any external connections. If a short circuit is connected across the terminals, $V = 0$. But, for any network containing no independent sources, $V = 0$ implies $I = 0$. Therefore, $b = 0$, and we can write

$$V = \mathfrak{z}I \tag{5-1}$$

Equation (5-1) requires a short discussion. First recall that the magnitude and angle of the complex voltage V are the magnitude and phase of the corresponding sinusoidal voltage waveform. [$V = \nu\angle\phi$ corresponds to $v = \nu \cos(\omega t + \phi)$.] Similarly, the magnitude and angle of the complex current $I = \iota\angle\theta$ are the magnitude and phase of the current waveform; that is, $i = \iota \cos(\omega t + \theta)$. Also note that, since V and I are complex numbers, $\mathfrak{z}$ must also be a complex number; for example, $\mathfrak{z} = |\mathfrak{z}|\angle\psi$. Equation (5-1) states that the port complex voltage V and the port complex current I are related in a very specific way. Indeed, we note that $|V| = |\mathfrak{z}|\,|I|$; that is, the complex voltage magnitude ν is related to the complex current magnitude ι by the factor $|\mathfrak{z}|$. Hence, the magnitude of the voltage waveform is $|\mathfrak{z}|$ multiplied by the magnitude of the current waveform; that is, $\nu = |\mathfrak{z}|\iota$. Furthermore, we note from Equation (5-1) that $\angle V = \angle\mathfrak{z} + \angle I$; that is, the angle of the complex voltage is different from the angle of the complex current I by a factor of $\angle\mathfrak{z}$. Hence, the phase of the voltage waveform ϕ is shifted from the phase of the current waveform θ by ψ radians. If $\psi > 0$, the voltage "leads" the current; if $\psi < 0$, the voltage "lags" the current; if $\psi = 0$, the voltage and current are "in phase." In conclusion, the complex number $\mathfrak{z}$ defines the relationship between the magnitudes and phases of the voltage and current waveforms.

Equation (5-1) further states that any 1-port network is completely characterized by the complex number $\mathfrak{z}$. Suppose that we are given a particular 1-port network and are asked to compute the complex number $\mathfrak{z}$ that characterizes that network. Solving Equation (5-1) for $\mathfrak{z}$, we have

$$\mathfrak{z} = \frac{V}{I} \tag{5-2}$$

Hence, one method for computing $\mathfrak{z}$ would be arbitrarily to choose I, say $I = 12\underline{/\pi}$, and constrain the port current to be this value by attaching a complex current source of this value to the port terminals. We could then compute the port voltage V and compute the ratio V/I. An alternative method would be to connect an arbitrary complex voltage source, say $V = 36\underline{/\pi/2}$, to the port terminals, compute the resulting port current I, and form the ratio V/I. In summary, $\mathfrak{z} = V/I = \nu\underline{/\phi}/\iota\underline{/\theta} = (\nu/\iota)\underline{/\phi - \theta} = |\mathfrak{z}|\underline{/\psi}$ depends only on the ratio of the voltage magnitude to the current magnitude and on the difference in their phases. Thus $\mathfrak{z}$ can be computed by fixing either the port voltage or the port current to any value. A particularly convenient choice is to set $I = 1\underline{/0} = 1 + j0 = 1$. For this case we have

$$\mathfrak{z} = \frac{V}{I} = V\bigg|_{I=1} \tag{5-3}$$

That is, for $I = 1$, the complex number that characterizes the 1-port network is identical to the complex port voltage.

In Equation (5-1) we recognize I as the independent variable and V as the dependent variable. By choosing V to be the independent variable and I the dependent variable, we obtain a linear equation of the form

$$I = yV \tag{5-4}$$

Hence, y is a complex number given by

$$y = \frac{I}{V} = I\bigg|_{V=1} \tag{5-5}$$

We conclude that any linear, lumped, 1-port network containing no independent sources can, as far as its behavior at its terminals is concerned, be uniquely represented by a single (complex) parameter. Depending on which variable is taken to be the independent variable, this parameter is either the parameter $\mathfrak{z}$ defined by Equation (5-2) or (5-3) or the parameter y defined by Equation (5-5). Obviously, a unique relationship exists between $\mathfrak{z}$ and y; that is, $\mathfrak{z} = y^{-1}$.

The complex number $\mathfrak{z}$ is called the *input impedance* for the 1-port network, and y is called the *input admittance*. Note that $\mathfrak{z}$ is a generalization of the Thevenin equivalent resistance defined in Chapter 2 for resistive networks. Indeed, $\mathfrak{z}$ is sometimes called the Thevenin equivalent impedance and y the Norton equivalent admittance. In general, any linear 1-port network can be characterized by its Thevenin equivalent network, which in general would consist of a Thevenin equivalent complex voltage source, say V_T, and a Thevenin equivalent (complex) impedance, say $\mathfrak{z}_T$, as in Figure 2-12(a). A Norton equivalent network is also possible. This network would be analogous to the network of Figure 2-12(b), except that the Norton equivalent current source would be a complex current I_N and the Norton admittance would be the input admittance y_N. Since we are restricting ourselves to networks containing no independent sources, the Thevenin equivalent voltage source (the open-circuit voltage) and the Norton equivalent current source (the short-circuit current) are zero for these networks. Hence, they are completely characterized by their input impedances (Thevenin equivalent impedance) or their input admittances (Norton equivalent admittance).

Example 5-1

In this example we compute the input impedance for the 1-port network of Figure 5-3(a). Using Equation (5-3) we start by connecting a 1-ampere current source across port 1, as indicated in Figure 5-3(b). To compute V, we first construct a tree and assign link currents, as illustrated in Figure 5-3(c). According to the link current algorithm of Section 3-5 (shortcut method) we write

$$R_3 I_b = -\mu R_2 I_b - I_b R_2 - R_1(-1 + I_b) \tag{5-6}$$

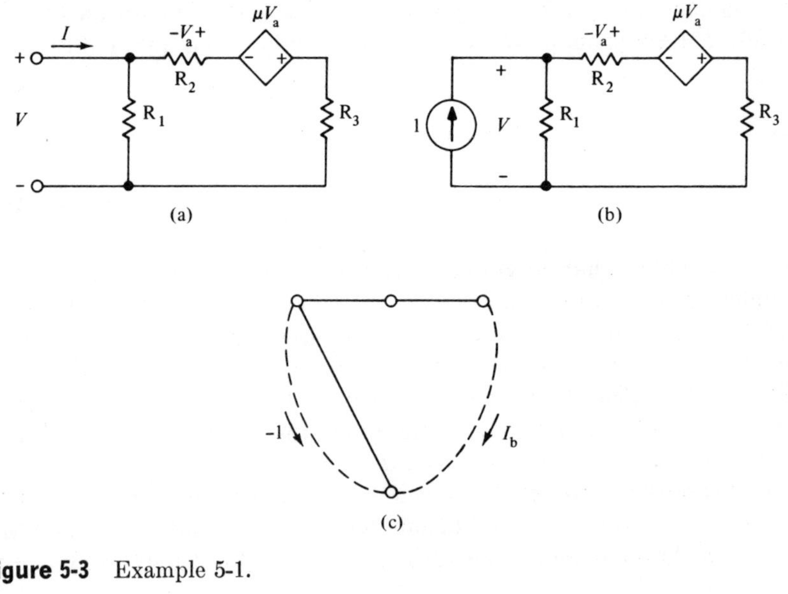

Figure 5-3 Example 5-1.

Solving for I_b, we have

$$I_b = \frac{R_1}{R_1 + R_3 + (1 + \mu)R_2} \tag{5-7}$$

Therefore

$$\mathfrak{z} = V\bigg|_{I=1} = -R_1(-1 + I_b) = \frac{R_1R_3 + (1 + \mu)R_1R_2}{R_1 + (1 + \mu)R_2 + R_3} \tag{5-8}$$

Since the network contained no inductances or capacitances, the input impedance is real. Hence, the port voltage and port current have the same phase, but are different in amplitude by the factor given by Equation (5-8).

5-3 2-Port Networks

Any 2-port network can be characterized in terms of its port voltages V_1 and V_2 and its port currents I_1 and I_2. The polarity of V_1 and V_2 and the direction of I_1 and I_2 are defined according to the convention indicated in Figure 5-4. Of the four variables V_1, V_2, I_1, I_2, any two

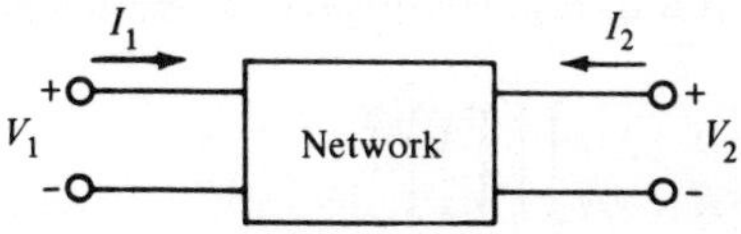

Figure 5-4 A two-port.

can be considered the independent variables, with the remaining two the dependent variables. For example, if V_1 and V_2 are fixed, then I_1 and I_2 are also determined; fixing V_1 and I_1 uniquely determines V_2 and I_2, and so on. Hence, from combinatorial analysis, we have a total of $4!/2!2! = 6$ (four items taken two at a time) possible choices for the pair of independent variables. For each choice we can write the dependent variables as linear combinations of the independent variables. Recall that a 1-port network can be characterized by either of two single (complex) parameters ($\mathfrak{z}$ or y). On the other hand, a 2-port network requires four (complex) parameters to characterize it. A total of six different sets of four parameters each are possible. These are presented in the next sections.

The short-circuit admittance parameters. Let the port voltages V_1 and V_2 be the independent variables and the port currents I_1 and I_2 the dependent variables. Then, writing each dependent variable as a linear combination of the independent variables, we characterize the 2-port network of Figure 5-4 with the two equations

$$\begin{aligned} I_1 &= y_{11}V_1 + y_{12}V_2 \\ I_2 &= y_{21}V_1 + y_{22}V_2 \end{aligned} \tag{5-9}$$

According to Equations (5-9) the four constant y_{ij} (i, j = 1, 2) are defined by

$$y_{11} = \frac{I_1}{V_1}\bigg|_{V_2=0} = I_1\bigg|_{V_1=1,V_2=0} \qquad y_{12} = \frac{I_1}{V_2}\bigg|_{V_1=0} = I_1\bigg|_{V_1=0,V_2=1}$$
$$y_{21} = \frac{I_2}{V_1}\bigg|_{V_2=0} = I_2\bigg|_{V_1=1,V_2=0} \qquad y_{22} = \frac{I_2}{V_2}\bigg|_{V_1=0} = I_2\bigg|_{V_1=0,V_2=1} \tag{5-10}$$

Hence, y_{21} can be computed (or measured) by short-circuiting port 2, connecting a 1-volt voltage source to port 1, and computing (or measuring) I_2. Similar statements can be made for the other three parameters. Note further that y_{11} is the input admittance at port 1, with port 2 short-circuited; y_{22} is the admittance seen looking into port 2 with port 1 short-circuited. The complex numbers y_{12} and y_{21} are transfer admittances (they have the physical dimensions of admittances, since they are ratios of currents to voltages, but the current and voltage correspond to different ports) with ports 1 and 2, respectively, short-circuited. Hence, the y_{ij}'s are called the *short-circuit admittance parameters*, or the *y parameters*.

Equations (5-9) can also be written in matrix notation (see Appendix I)

$$\begin{bmatrix} I_1 \\ I_2 \end{bmatrix} = \begin{bmatrix} y_{11} & y_{12} \\ y_{21} & y_{22} \end{bmatrix} \begin{bmatrix} V_1 \\ V_2 \end{bmatrix} \tag{5-11}$$

where the matrix of y_{ij}'s is called the *short-circuit admittance matrix*. We use **Y** to denote this matrix.

An equivalent 2-port network, analogous to the Norton equivalent for a 1-port network, can be synthesized from Equations (5-9). From the first equation, note that the current I_1 is expressed as the sum of two terms, each of which must be a current. Hence, we interpret $y_{11}V_1$ as the current through a y_{11}-mho admittance, with V_1 volts impressed across it, and $y_{12}V_2$ as the output of a voltage-controlled current source. The second equation can be interpreted in a similar fashion to yield the equivalent network of Figure 5-5. Other equivalent networks are also possible. See, for example, Problem 5-5.

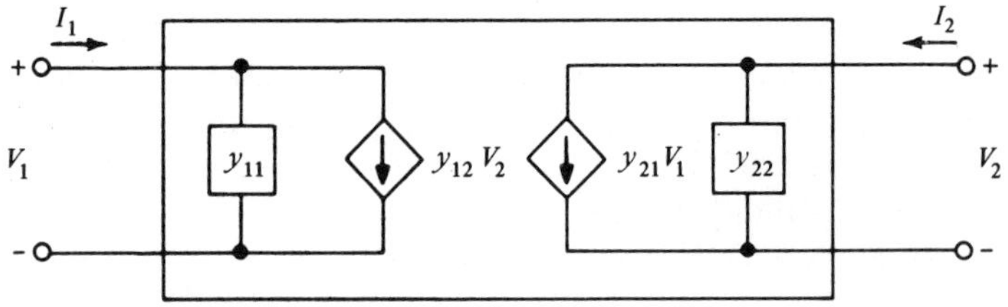

Figure 5-5 A y parameter 2-port equivalent network.

In summary, any lumped, linear, 2-port network containing no independent sources can be uniquely described by two equations in the 2-port voltages and the 2-port currents. It follows that any lumped, linear, 2-port network can, as far as its behavior at its terminals is concerned, be uniquely characterized by its short-circuit admittance parameter matrix.

Example 5-2

For a second example, we compute the short-circuit admittance parameter matrix for the network of Figure 5-6(a). Since the network contains capacitances, the y parameters are complex numbers that depend

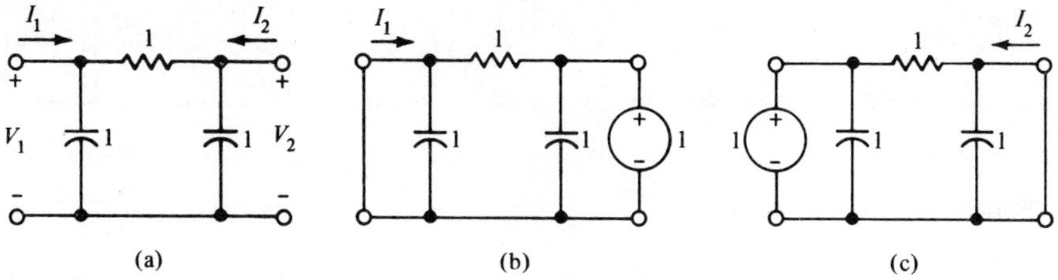

Figure 5-6 Example 5-2.

on the frequency ω as well as on the network parameters. y_{11} is the admittance seen looking into port 1, with port 2 short-circuited; that is,

$$y_{11} = j\omega + 1 \tag{5-12}$$

To compute y_{12}, we short-circuit port 1, connect a 1-volt voltage source to port 2, and compute the current I_1. From Figure 5-6(b) we observe that

$$y_{12} = I_1 \Big|_{V_1=0, V_2=1} = -1 \tag{5-13}$$

To compute y_{21}, we short-circuit port 2, connect a 1-volt voltage source to port 1, and compute the current I_2. From Figure 5-6(c) we observe that

$$y_{21} = I_2 \Big|_{V_1=1, V_2=0} = -1 \tag{5-14}$$

Finally, y_{22} is the admittance seen looking into port 2, with port 1 short-circuited; that is,

$$y_{22} = j\omega + 1 \tag{5-15}$$

Hence, the y parameter matrix for the network of Figure 5-6(a) is given by

$$\mathbf{Y} = \begin{bmatrix} j\omega + 1 & -1 \\ -1 & j\omega + 1 \end{bmatrix} \tag{5-16}$$

Example 5-3

For another example, we compute the short-circuit admittance parameter matrix for the common source field-effect transistor (FET) shown in Figure 5-7(a) along with its model in Figure 5-7(b).

The parameters y_{11} and y_{21} can be computed by short-circuiting port 2 and connecting a 1-volt voltage source to port 1, as illustrated in Figure 5-7(c), and computing I_1 and I_2, respectively. With port 2 short-circuited, the voltage across G and C_2 is zero, so that no current flows

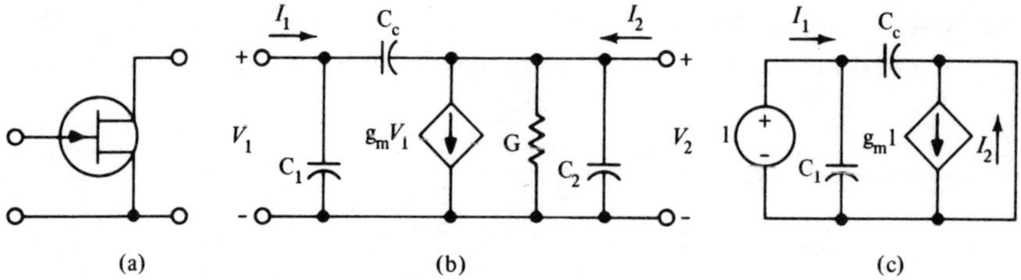

Figure 5-7 Example 5-3.

through these elements. Therefore, they have no effect on I_1 or I_2 and, consequently, have been omitted from the figure. Note, however, that the controlled current source must be included because the current in this branch is g_m amperes, independent of the voltage across it.

From Figure 5-7(c), we note that the current through C_1 is $1 \cdot j\omega C_1$, that the current through C_c is $1 \cdot j\omega C_c$, and that I_1 is the sum of these two currents so that

$$y_{11} = I_1 \Big|_{V_1=1, V_2=0} = j\omega(C_1 + C_c) \tag{5-17}$$

Furthermore, I_2 is the current through the controlled current source minus the current through C_c; that is,

$$y_{21} = I_2 \Big|_{V_1=1, V_2=0} = g_m - j\omega C_c \tag{5-18}$$

To compute y_{12} and y_{22} we first short-circuit port 1 and connect a 1-volt voltage source to port 2, as illustrated in Figure 5-8. Then we

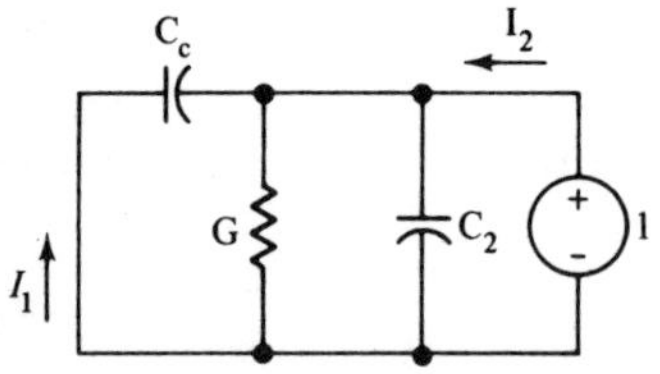

Figure 5-8 Network for computing y_{12} and y_{22}.

compute I_1 and I_2, respectively. Since $V_1 = 0$, the current through C_1 is zero, and the controlled current source current $g_m V_1 = 0$. Therefore, these two elements can be omitted. By inspection of Figure 5-8, we write

$$y_{12} = I_1 \Big|_{V_1=0, V_2=1} = -j\omega C_c \tag{5-19}$$

$$y_{22} = I_2 \Big|_{V_1=0, V_2=1} = G + j\omega(C_2 + C_c) \tag{5-20}$$

The short-circuit admittance parameter matrix for the common source FET of Figure 5-7 is given by

$$\mathbf{Y} = \begin{bmatrix} j\omega(C_1 + C_c) & -j\omega C_c \\ g_m - j\omega C_c & G + j\omega(C_2 + C_c) \end{bmatrix} \tag{5-21}$$

A typical good-quality FET has parameter values

$$C_1 \approx 2 \text{ pF},\ C_2 \approx 5 \text{ pF},\ C_c \approx 1 \text{ pF}$$
$$R = G^{-1} \approx 50 \text{ k}\Omega$$
$$g_m \approx 10^{-2}$$

Hence, a typical y parameter matrix for an FET is

$$\mathbf{Y} = \begin{bmatrix} j\omega 3 \cdot 10^{-12} & j\omega 1 \cdot 10^{-12} \\ 10^{-2} - j\omega 1 \cdot 10^{-12} & 2 \cdot 10^{-5} + j\omega 6 \cdot 10^{-12} \end{bmatrix} \tag{5-22}$$

At low frequencies, the terms involving ω are small compared to the other terms, and

$$\mathbf{Y} \approx \begin{bmatrix} 0 & 0 \\ 10^{-2} & 2 \cdot 10^{-5} \end{bmatrix} \tag{5-23}$$

Hence, the y parameters given by Equation (5-23) can be used in applications where the admittances y_{11} and y_{12} in Equation (5-22) are small compared to the admittances of the other networks connected across port 1 and where the frequency ω is low enough so that the y_{21} and y_{22} approximations are valid. For $\omega < 10^5$ the imaginary part of y_{22} is less than $6 \cdot 10^{-7}$, which can be neglected relative to the real part $2 \cdot 10^{-5}$. Note that neglecting the ω terms in Equations (5-22) is analogous to neglecting the capacitances in the FET model of Figure 5-7(b). At low frequencies, the admittances of the capacitances are sufficiently small compared to the admittances of the other branches so that we can assume that they are open circuits.

The open-circuit impedance parameters. By choosing the port currents to be the independent variables and the port voltages the dependent variables, we can write each port voltage as a linear combination of the port currents. The standard notation for this arrangement is

$$\begin{aligned} V_1 &= z_{11}I_1 + z_{12}I_2 \\ V_2 &= z_{21}I_1 + z_{22}I_2 \end{aligned} \tag{5-24}$$

From Equations (5-24) we observe that the z_{ij} (i, j = 1, 2) are given by

$$z_{11} = \frac{V_1}{I_1}\bigg|_{I_2=0} = V_1\bigg|_{I_1=1, I_2=0} \qquad z_{12} = \frac{V_1}{I_2}\bigg|_{I_1=0} = V_1\bigg|_{I_1=0, I_2=1}$$
$$z_{21} = \frac{V_2}{I_1}\bigg|_{I_2=0} = V_2\bigg|_{I_1=1, I_2=0} \qquad z_{22} = \frac{V_2}{I_2}\bigg|_{I_1=0} = V_2\bigg|_{I_1=0, I_2=1} \tag{5-25}$$

The four coefficients z_{ij} (i, j = 1, 2) are called the *open-circuit impedance parameters* or the *z parameters.* Equations (5-25) verify that all four z parameters are impedances (they are voltage-to-current ratios), and all are obtained with one of the two ports open-circuited. z_{11} and z_{22} can be interpreted as the input impedances seen looking into ports 1 and 2, respectively, with the other port open-circuited. z_{12} and z_{21} can be interpreted as open-circuited transfer impedances.

Equations (5-24) can also be written in the matrix form

$$\begin{bmatrix} V_1 \\ V_2 \end{bmatrix} = \begin{bmatrix} z_{11} & z_{12} \\ z_{21} & z_{22} \end{bmatrix} \begin{bmatrix} I_1 \\ I_2 \end{bmatrix} \tag{5-26}$$

where the matrix of z_{ij}'s denoted by **Z** is called the *open-circuit impedance matrix.*

An equivalent network in terms of the z parameters can be synthesized from Equations (5-24) if we interpret V_1 as the sum of the two voltages $z_{11}I_1$ and $z_{12}I_2$ and V_2 as the sum of $z_{21}I_1$ and $z_{22}I_2$. Here we interpret $z_{11}I_1$ as the voltage across a z_{11}-ohm impedance, with current I_1 through it, and $z_{22}I_2$ as the voltage across z_{22} with current I_2 through it. $z_{12}I_2$ and $z_{21}I_1$ can be interpreted as current-controlled voltage sources

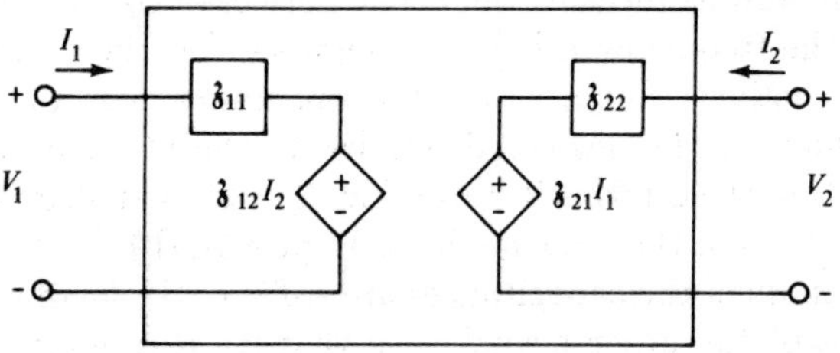

Figure 5-9 A z parameter 2-port equivalent network.

to obtain the equivalent network presented in Figure 5-9. Other equivalent networks in terms of the four z parameters are also possible. See, for example, Problem 5-12.

In summary, any linear, lumped, 2-port network containing no independent sources can be uniquely characterized by its open-circuit impedance parameter matrix.

Example 5-4

In this example we compute the open-circuit impedance parameter matrix for the network of Fig. 5-10(a). z_{11} is the impedance seen looking into port 1 with port 2 open-circuited; that is,

$$z_{11} = 1 + \frac{1}{j\omega} = \frac{j\omega + 1}{j\omega} \tag{5-27}$$

z_{22} is the impedance seen looking into port 2 with port 1 open-circuited; that is,

$$z_{22} = 1 + \frac{1}{j\omega} = \frac{j\omega + 1}{j\omega} \tag{5-28}$$

To compute z_{12} from Equations (5-25) we open-circuit port 1, connect a

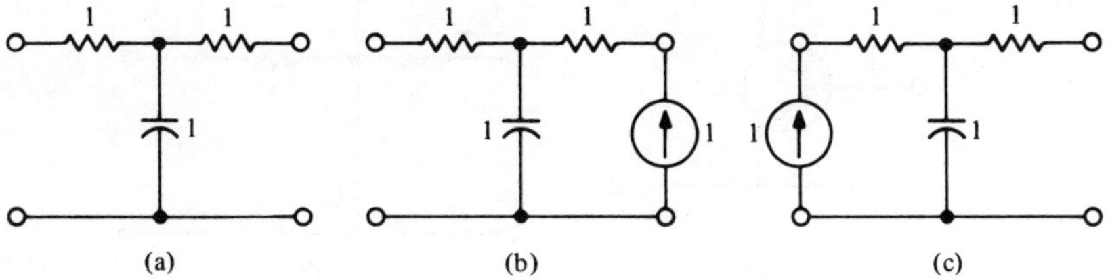

Figure 5-10 Example 5-4.

1-ampere current source to port 2, and compute V_1. By observing Figure 5-10(b), we write

$$z_{12} = V_1 \Big|_{I_1=0, I_2=1} = \frac{1}{j\omega} \tag{5-29}$$

Similarly, to compute z_{21} we observe Figure 5-10(c) and write

$$z_{21} = V_2 \Big|_{I_1=1, I_2=0} = \frac{1}{j\omega} \tag{5-30}$$

Consequently, the open-circuit impedance parameter matrix for the network of Figure 5-10(a) is given by

$$\mathbf{Z} = \begin{bmatrix} \dfrac{j\omega + 1}{j\omega} & \dfrac{1}{j\omega} \\ \dfrac{1}{j\omega} & \dfrac{j\omega + 1}{j\omega} \end{bmatrix} \tag{5-31}$$

Example 5-5

To give another example, we compute the open-circuit impedance matrix for the grounded emitter transistor of Figure 5-11(a), having the model shown in Figure 5-11(b).

To compute $\mathfrak{z}_{11}$ and $\mathfrak{z}_{21}$ from Equations (5-25), we open-circuit port 2 and connect a 1-ampere current source to port 1, as indicated in Figure 5-11(c). Observe that $I_2 = 0$ so that $I = -1$ and the controlled current source current is $-\alpha$. Hence, we write

$$\begin{aligned} \mathfrak{z}_{11} &= V_1 \Big|_{I_1=1, I_2=0} = (\mathrm{r_e} + \mathrm{r_b}) \cdot 1 = \mathrm{r_c} + \mathrm{r_b} \\ \mathfrak{z}_{21} &= V_2 \Big|_{I_1=1, I_2=0} = \mathrm{r_e} \cdot 1 - \alpha \mathrm{r_c} \cdot 1 = \mathrm{r_e} - \alpha \mathrm{r_c} \end{aligned} \qquad \textbf{(5-32)}$$

To compute $\mathfrak{z}_{12}$ and $\mathfrak{z}_{22}$, we open-circuit port 1 and connect a 1-ampere current source to port 2, as illustrated in Figure 5-11(d). Observe that

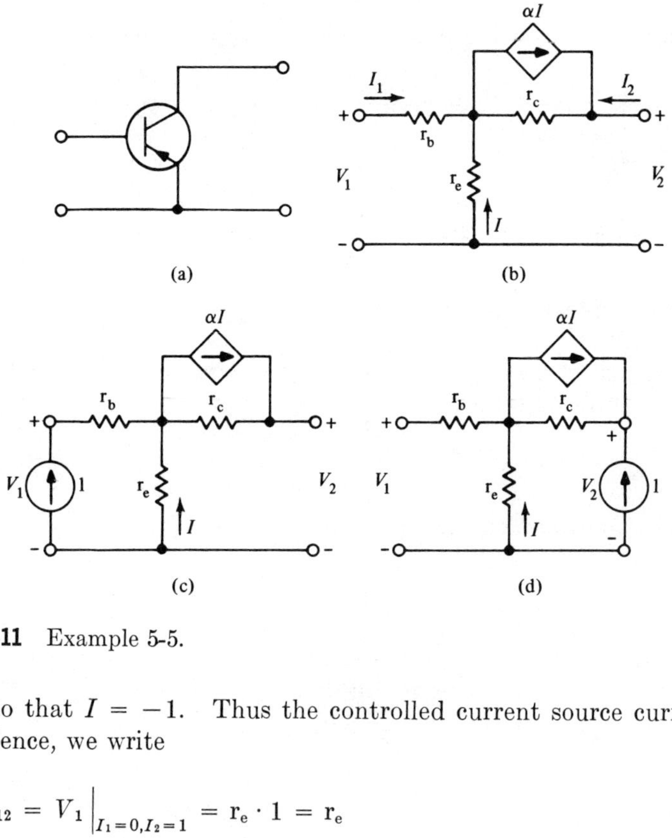

Figure 5-11 Example 5-5.

$I_1 = 0$ so that $I = -1$. Thus the controlled current source current is $-\alpha$. Hence, we write

$$\begin{aligned} \mathfrak{z}_{12} &= V_1 \Big|_{I_1=0, I_2=1} = \mathrm{r_e} \cdot 1 = \mathrm{r_e} \\ \mathfrak{z}_{22} &= V_2 \Big|_{I_1=0, I_2=1} = \mathrm{r_e} \cdot 1 + \mathrm{r_c} \cdot (1 - \alpha) = \mathrm{r_e} + (1 - \alpha)\mathrm{r_c} \end{aligned} \qquad \textbf{(5-33)}$$

We conclude that the open-circuit impedance parameter matrix for the grounded emitter transistor of Figure 5-11 is given by

$$\mathbf{Z} = \begin{bmatrix} \mathrm{r_b} + \mathrm{r_e} & \mathrm{r_e} \\ \mathrm{r_e} - \alpha \mathrm{r_c} & \mathrm{r_e} + (1 - \alpha)\mathrm{r_c} \end{bmatrix} \qquad \textbf{(5-34)}$$

The hybrid parameters. The hybrid parameters are obtained by choosing I_1 and V_2 to be the independent variables, with V_1 and I_2 the dependent variables. Writing the dependent variables as linear combinations of the independent variables, we obtain two equations of the form

$$\begin{aligned} V_1 &= \mathfrak{h}_{11}I_1 + \mathfrak{h}_{12}V_2 \\ I_2 &= \mathfrak{h}_{21}I_1 + \mathfrak{h}_{22}V_2 \end{aligned} \tag{5-35}$$

From Equations (5-35) we note that the four coefficients $\mathfrak{h}_{ij}$ (i, j = 1, 2) are given by

$$\begin{aligned} \mathfrak{h}_{11} &= \frac{V_1}{I_1}\bigg|_{V_2=0} = V_1\bigg|_{I_1=1,V_2=0} & \mathfrak{h}_{12} &= \frac{V_1}{V_2}\bigg|_{I_1=0} = V_1\bigg|_{I_1=0,V_2=1} \\ \mathfrak{h}_{21} &= \frac{I_2}{I_1}\bigg|_{V_2=0} = I_2\bigg|_{I_1=1,V_2=0} & \mathfrak{h}_{22} &= \frac{I_2}{V_2}\bigg|_{I_1=0} = I_2\bigg|_{I_1=0,V_2=1} \end{aligned} \tag{5-36}$$

$\mathfrak{h}_{11}$ can be interpreted as the input impedance seen looking into port 1 with port 2 short-circuited; $\mathfrak{h}_{12}$ is the reverse open-circuit voltage gain; $\mathfrak{h}_{21}$ is a short-circuit current gain; $\mathfrak{h}_{22}$ is the input admittance seen looking into port 2 with port 1 open-circuited. Hence the name *hybrid parameters*, or *h parameters*. The hybrid parameters came into common use with the bipolar transistor; for bipolar transistors at low frequencies, they are much easier to measure in the laboratory than are the y parameters or z parameters.

Equations (5-35) can also be written in matrix notation; that is,

$$\begin{bmatrix} V_1 \\ I_2 \end{bmatrix} = \begin{bmatrix} \mathfrak{h}_{11} & \mathfrak{h}_{12} \\ \mathfrak{h}_{21} & \mathfrak{h}_{22} \end{bmatrix} \begin{bmatrix} I_1 \\ V_2 \end{bmatrix} \tag{5-37}$$

where the matrix of $\mathfrak{h}_{ij}$'s is called the *hybrid parameter matrix.* We use **H** to denote this matrix.

An equivalent 2-port network in terms of the h parameters can be synthesized from Equations (5-35) by interpreting V_1 as a sum of two voltages and I_2 as a sum of two currents. $\mathfrak{h}_{11}I_1$ can be interpreted as a voltage developed across an $\mathfrak{h}_{11}$-ohm impedance due to current I_1; $\mathfrak{h}_{12}V_2$ can be interpreted as the output of a voltage-controlled voltage source. In the second equation, we interpret $\mathfrak{h}_{21}I_1$ as the output of a current-controlled current source and $\mathfrak{h}_{22}V_2$ as a current due to voltage V_2 across a $\mathfrak{h}_{22}$-mho admittance. The equivalent network is shown in Figure 5-12.

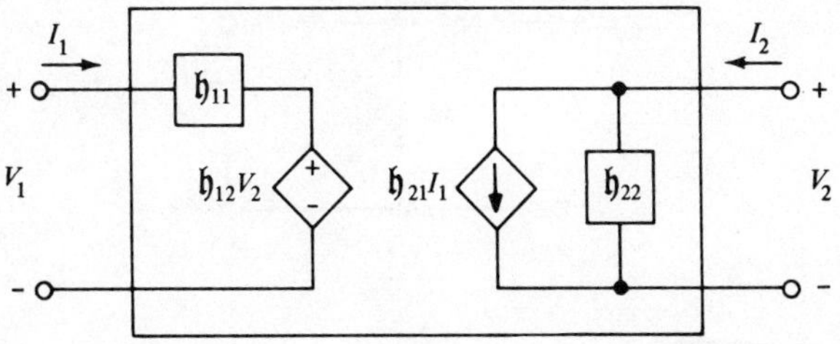

Figure 5-12 A 2-port equivalent network in terms of the hybrid parameter.

In summary, any linear, lumped, 2-port network containing no independent sources can be uniquely characterized by its hybrid parameter matrix. The hybrid parameters provide a convenient characterization for transistors and transistor networks.

Example 5-6

To give an example, we compute the hybrid parameters for the grounded base transistor presented in Figure 5-13(a) and (b). To compute $\mathfrak{h}_{11}$ and $\mathfrak{h}_{21}$ we short-circuit port 2, connect a 1-ampere current source to port 1, and compute V_1 and I_2, respectively, as dictated by Equations (5-36). Therefore, we first sketch the network of Figure 5-13(c) and construct a tree, as shown in Figure 5-13(d). The link current corresponding to the independent current source is 1 ampere, and the link current cor-

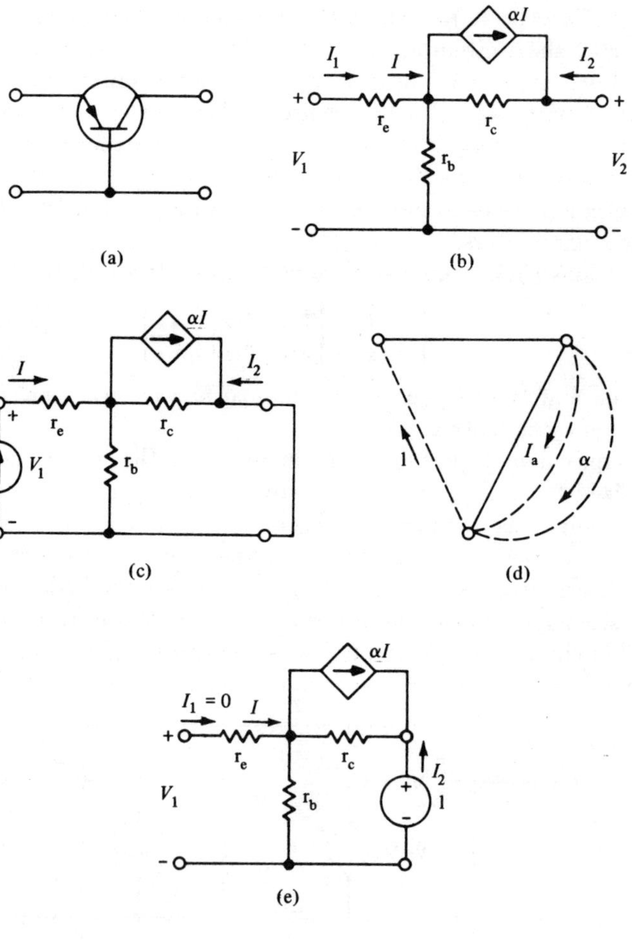

Figure 5-13 Example 5-6.

responding to the controlled current source is $\alpha \cdot I = \alpha \cdot 1 = \alpha$ amperes. Therefore, according to the link current algorithm of Section 3-5, we need only express the remaining link voltage as a sum of tree branch voltages (in terms of the three link currents) via KVL. We write

$$I_a r_c = (1 - I_a - \alpha) r_b \tag{5-38}$$

or

$$I_a = \frac{(1-\alpha) r_b}{r_b + r_c} \tag{5-39}$$

It follows that

$$\begin{aligned} \mathfrak{h}_{11} &= I_1 \Big|_{I_1=1, I_2=0} = r_e \cdot 1 + I_a r_c = r_e + \frac{(1-\alpha) r_b r_c}{r_b + r_c} \\ \mathfrak{h}_{21} &= I_2 \Big|_{I_1=1, I_2=0} = -I_a - \alpha = -\alpha - \frac{(1-\alpha) r_b}{r_b + r_c} \end{aligned} \tag{5-40}$$

To compute $\mathfrak{h}_{12}$ and $\mathfrak{h}_{22}$ we open-circuit port 1 and connect a 1-volt voltage source to port 2, as illustrated in Figure 5-13(e). Observe that $I_1 = I = 0$ so that the controlled current source is inactive. Hence, we can write

$$\begin{aligned} \mathfrak{h}_{12} &= V_1 \Big|_{I_1=0, V_2=1} = \frac{r_b}{r_b + r_c} \\ \mathfrak{h}_{22} &= I_2 \Big|_{I_1=0, V_2=1} = \frac{1}{r_b + r_c} \end{aligned} \tag{5-41}$$

Therefore, the hybrid parameter matrix for the grounded base transistor of Figure 5-13 is given by

$$\mathbf{H} = \begin{bmatrix} r_e + \dfrac{(1-\alpha) r_b r_c}{r_b + r_c} & \dfrac{r_b}{r_b + r_c} \\ -\alpha - \dfrac{(1-\alpha) r_b}{r_b + r_c} & \dfrac{1}{r_b + r_c} \end{bmatrix} \tag{5-42}$$

A good-quality transistor has $\alpha \approx 0.99$, $r_b \approx 10^3$, $r_e \approx 10$, and $r_c \approx 10^6$. The resulting h parameter matrix is given by

$$\mathbf{H} \approx \begin{bmatrix} 20 & 10^{-3} \\ -0.99 & 10^{-6} \end{bmatrix}$$

In many applications the input impedance of 20 ohms is negligible compared to the output impedance of the driving stage. Also, the output admittance of 10^{-6} mhos is usually negligible compared to input admittance of the succeeding stage. Furthermore, the reverse voltage gain of 10^{-3} has little effect. Hence, a good approximation in this case is

$$\mathbf{H} \approx \begin{bmatrix} 0 & 0 \\ -\alpha & 0 \end{bmatrix}$$

Thus, at low frequencies the grounded base transistor can be modeled by a current-controlled current source, as illustrated in Figure 5-14. At high frequencies, capacitances must be added to the model.

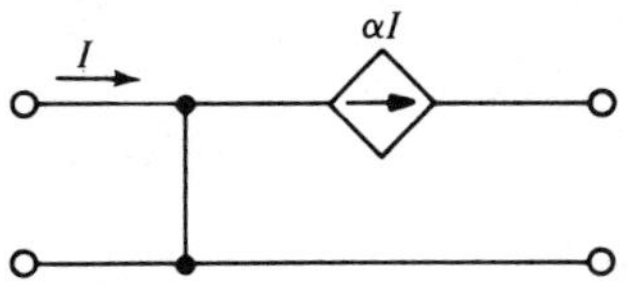

Figure 5-14 An equivalent network for a grounded-base transistor.

The transfer parameters. It is sometimes convenient to choose the port 2 voltage and current to be the independent variables and the port 1 voltage and current to be the dependent variables. This arrangement results in equations of the form

$$\begin{aligned} V_1 &= \mathfrak{a}V_2 - \mathfrak{b}I_2 \\ I_1 &= \mathfrak{c}V_2 - \mathfrak{d}I_2 \end{aligned} \tag{5-43}$$

where the four coefficients are given by

$$\begin{aligned} \mathfrak{a} &= \frac{V_1}{V_2}\bigg|_{I_2=0} = \frac{1}{V_2}\bigg|_{V_1=1,I_2=0} & \mathfrak{b} &= -\frac{V_1}{I_2}\bigg|_{V_2=0} = \frac{1}{I_2}\bigg|_{V_1=-1,V_2=0} \\ \mathfrak{c} &= \frac{I_1}{V_2}\bigg|_{I_2=0} = \frac{1}{V_2}\bigg|_{I_1=1,I_2=0} & \mathfrak{d} &= -\frac{I_1}{I_2}\bigg|_{V_2=0} = \frac{1}{I_2}\bigg|_{I_1=-1,V_2=0} \end{aligned} \tag{5-44}$$

From Equations (5-44) we observe that $\mathfrak{a}$ is the backward voltage gain with port 2 open-circuited; $\mathfrak{d}$ is the negative of the backward current gain with port 2 short-circuited; $\mathfrak{b}$ is a short-circuit transfer impedance; $\mathfrak{c}$ is an open-circuit transfer admittance. These four parameters are called the *transfer parameters* or the *t parameters.*

Equations (5-43) can also be written in matrix form

$$\begin{bmatrix} V_1 \\ I_1 \end{bmatrix} = \begin{bmatrix} \mathfrak{a} & \mathfrak{b} \\ \mathfrak{c} & \mathfrak{d} \end{bmatrix} \begin{bmatrix} V_2 \\ -I_2 \end{bmatrix} \tag{5-45}$$

where the matrix of coefficients is called the *transfer parameter matrix* and is denoted by $\mathbf{T}$.

An equivalent 2-port network in terms of the t parameters analogous to those defined in terms of the y, z, and h parameters is not possible.

In summary, any lumped, linear, n-port network containing no independent sources can be uniquely characterized by its transfer parameter matrix. The transfer parameters are very convenient for analyzing networks connected in cascade, as we shall see in Section 5-4. The reason for using two minus signs in the definitions will also become apparent in Section 5-4.

Example 5-7

In this example we compute the t parameter matrix for the network of Figure 5-15(a). To compute $\mathfrak{a}$ we open-circuit port 2 and connect a 1-volt voltage source to port 1, as illustrated in Figure 5-15(b), and write

$$\mathfrak{a} = \left.\frac{1}{V_2}\right|_{V_1=1, I_2=0} = \frac{R + 1/j\omega C}{1/j\omega C} = j\omega RC + 1 \tag{5-46}$$

To compute $\mathfrak{c}$ we replace the 1-volt voltage source with a 1-ampere current source and write

$$\mathfrak{c} = \left.\frac{1}{V_2}\right|_{I_1=1, I_2=0} = j\omega C \tag{5-47}$$

To compute $\mathfrak{b}$ we short-circuit port 2 and connect a -1-volt voltage

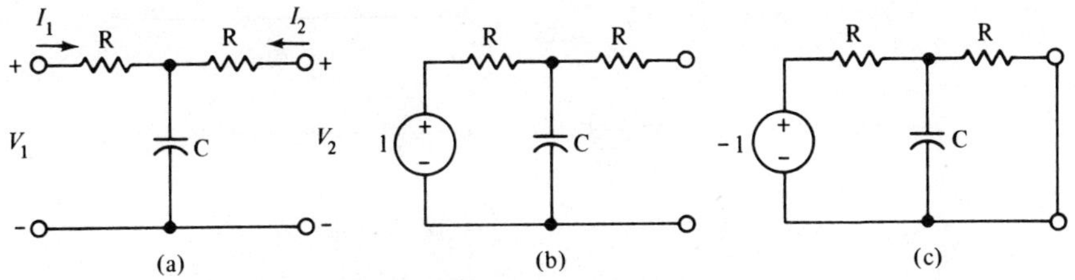

Figure 5-15 Example 5-7.

source to port 1, as illustrated in Figure 5-15(c), and write

$$\mathfrak{b} = \left.\frac{1}{I_2}\right|_{V_1=-1, V_2=0} = \frac{R + (1/j\omega C)}{2R + (1/j\omega C)} \cdot \frac{1}{R} = \frac{1 + j\omega RC}{R(1 + j\omega 2RC)} \tag{5-48}$$

To compute $\mathfrak{d}$ we replace the 1-volt voltage source with a -1-ampere current source and write

$$\mathfrak{d} = \left.\frac{1}{I_2}\right|_{I_1=-1, V_2=0} = \left(\frac{1/j\omega C}{R + 1/j\omega C}\right)^{-1} = j\omega RC + 1 \tag{5-49}$$

Therefore, the t parameter matrix is given by

$$\mathbf{T} = \begin{bmatrix} j\omega RC + 1 & \dfrac{1 + j\omega RC}{R(1 + j\omega 2RC)} \\ j\omega C & j\omega RC + 1 \end{bmatrix} \tag{5-50}$$

Example 5-8

In this example we compute the transfer parameter matrix for the grounded collector transistor of Figure 5-16(a). We assume the transistor model of Figure 5-16(b).

To compute $\mathfrak{a}$ we open-circuit port 2 and connect a 1-volt voltage source to port 1, as illustrated in Figure 5-16(c). We note that $I_2 = 0$ implies that $I = 0$ and $\alpha I = 0$. Therefore, by inspection of Figure 5-16(c), we write

$$\mathfrak{a} = \frac{1}{V_2}\bigg|_{V_1=1, I_2=0} = \frac{1}{r_c/(r_b + r_c)} = \frac{r_b + r_c}{r_c} \tag{5-51}$$

To compute $\mathfrak{c}$ we replace the 1-volt voltage source with a 1-ampere current source, as shown in Figure 5-16(d). We again have $I_2 = 0$ and

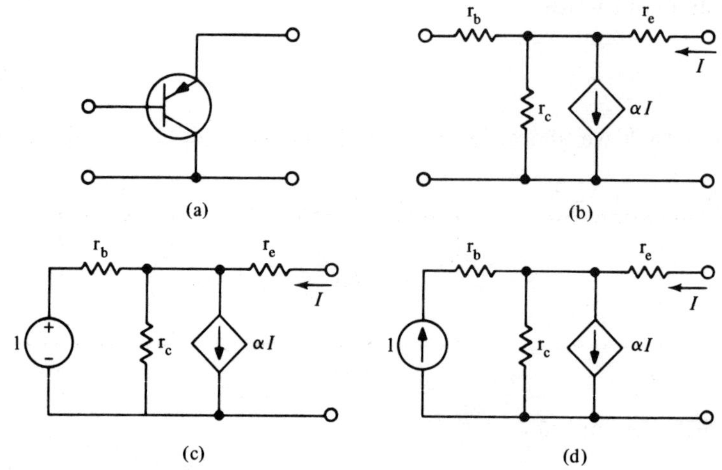

Figure 5-16 Example 5-8.

$I = \alpha I = 0$, and we write

$$\mathfrak{c} = \frac{1}{V_2}\bigg|_{I_1=1, I_2=0} = \frac{1}{r_c \cdot 1} = \frac{1}{r_c} \tag{5-52}$$

To compute $\mathfrak{b}$ we short-circuit port 2 and connect a -1-volt voltage source to port 1, as illustrated in Figure 5-17(a). For this case the solution is not obvious, so we construct a tree and assign tree branch voltage symbols, as in Figure 5-17(b), and write

$$\frac{V}{r_c} = \frac{-1 - V}{r_b} + \alpha\frac{V}{r_e} - \frac{V}{r_e}$$

or

$$V = \frac{-r_c r_e}{r_c r_e + r_b r_e + (1 - \alpha) r_b r_c} \tag{5-53}$$

Thus

$$I_2 = -\frac{V}{r_e} = \frac{r_c}{r_c r_e + r_b r_e + (1 - \alpha) r_b r_c} \tag{5-54}$$

and

$$\mathfrak{b} = \frac{1}{I_2}\bigg|_{V_1=-1,V_2=0} = \frac{\mathrm{r_c r_e + r_b r_e} + (1-\alpha)\mathrm{r_b r_c}}{\mathrm{r_c}} \tag{5-55}$$

To compute $\mathfrak{d}$ we replace the -1-volt voltage source with a

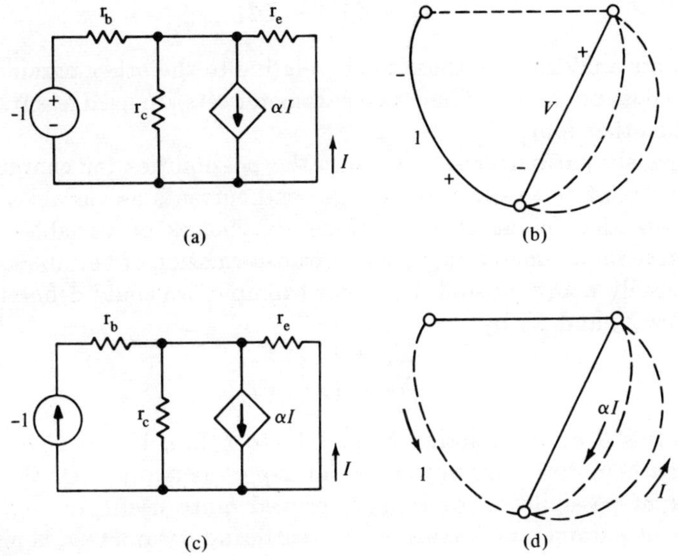

Figure 5-17 Networks for computing $\mathfrak{b}$ and $\mathfrak{d}$.

-1-ampere current source, as shown in Figure 5-17(c). We next construct a tree and assign link currents, as shown in Figure 5-17(d), and write

$$I\mathrm{r_e} = (1 + \alpha I - I)\mathrm{r_c}$$

or, solving for I, we obtain

$$I = \frac{\mathrm{r_c}}{\mathrm{r_e} + (1-\alpha)\mathrm{r_c}} \tag{5-56}$$

But $I_2 = I$ so that

$$\mathfrak{d} = \frac{1}{I_2}\bigg|_{I_1=-1,V_2=0} = \frac{\mathrm{r_e} + (1-\alpha)\mathrm{r_c}}{\mathrm{r_c}} \tag{5-57}$$

Therefore, the t parameter matrix for the grounded collector transistor of Figure 5-16 is given by

$$\mathbf{T} = \frac{1}{\mathrm{r_c}}\begin{bmatrix} \mathrm{r_b + r_c} & \mathrm{r_c r_e + r_b r_e} + (1-\alpha)\mathrm{r_b r_c} \\ 1 & \mathrm{r_e} + (1-\alpha)\mathrm{r_c} \end{bmatrix} \tag{5-58}$$

Other parameter sets for 2-ports. The remaining two of the six possible parameter sets are given by

g parameters
$$\begin{aligned} I_1 &= \mathfrak{g}_{11}V_1 + \mathfrak{g}_{12}I_2 \\ V_2 &= \mathfrak{g}_{21}V_1 + \mathfrak{g}_{22}I_2 \end{aligned} \tag{5-59}$$

$\hat{\mathfrak{t}}$ parameters
$$\begin{aligned} V_2 &= \mathfrak{A}V_1 - \mathfrak{B}I_1 \\ I_2 &= \mathfrak{C}V_1 - \mathfrak{D}I_1 \end{aligned} \tag{5-60}$$

Comments analogous to those made relative to the other parameter sets are also possible here. These two parameter sets are used less frequently than the other four.

These six parameter sets exhaust the possibilities for characterizing networks relative to complex voltages and currents as variables. From a mathematical point of view, these six choices of variables are not exhaustive for characterizing a linear transformation of variables. There are infinitely many possibilities. For example, we could define two new variables X_1 and X_2 by

$$\begin{aligned} X_1 &= I_1 + V_1 \\ X_2 &= I_1 - V_1 \end{aligned} \tag{5-61}$$

Then a 2×2 matrix relating X_1 and X_2 to I_2 and V_2 would be a legitimate mathematical description of our 2-port network. Of the infinite number of possibilities, some have proved quite useful in engineering. One set of parameters, known as the scattering parameters, is more useful in microwave circuits than any of the six sets of parameters defined above. The interested student should see Carlin and Giordono (Ref. [5]) for an elementary discussion of scattering parameters. A more advanced discussion is given in Kuh and Rohrer (Ref. [26]).

Relationships between the six sets of 2-port network parameters. Since each of the six 2-port parameter representations uniquely characterizes any particular network relative to its terminals, a unique relationship must exist between the six parameter sets. These thirty-six relationships can be derived by starting with one of the six pairs of equations and rearranging them into the form of another set. For example, to find the h parameters in terms of the y parameters, we start with the y parameter equations

$$\begin{aligned} I_1 &= y_{11}V_1 + y_{12}V_2 \\ I_2 &= y_{21}V_1 + y_{22}V_2 \end{aligned} \tag{5-62}$$

and rearrange them into the form of the h parameter equations:

$$\begin{aligned} V_1 &= \mathfrak{h}_{11}I_1 + \mathfrak{h}_{12}V_2 \\ I_2 &= \mathfrak{h}_{21}I_1 + \mathfrak{h}_{22}V_2 \end{aligned} \tag{5-63}$$

Apparently, the first y parameter equation can be put into the form of the first h parameter equation by simply solving the first y parameter equation for V_1, that is,

$$V_1 = \frac{1}{y_{11}} I_1 - \frac{y_{12}}{y_{11}} V_2 \tag{5-64}$$

To transform the second y parameter equation into an equation of the same form as the second h parameter equation, we must start with the second y parameter equation and express I_2 as a linear combination of I_1 and V_2. This can be accomplished by replacing V_1 in the second y parameter equation with Equation (5-64), that is, by solving the first y parameter equation for V_1 and substituting this into the second y parameter equation. Proceeding in this manner, we obtain

$$I_2 = \frac{y_{21}}{y_{22}} I_1 + \left(y_{22} - \frac{y_{12}y_{21}}{y_{11}}\right) V_2 \tag{5-65}$$

Now, comparing Equations (5-64) and (5-65) to Equations (5-63), we observe that

$$\begin{aligned} \mathfrak{h}_{11} &= \frac{1}{y_{21}} & \mathfrak{h}_{12} &= -\frac{y_{12}}{y_{11}} \\ \mathfrak{h}_{21} &= \frac{y_{21}}{y_{22}} & \mathfrak{h}_{22} &= y_{22} - \frac{y_{12}y_{21}}{y_{11}} \end{aligned} \tag{5-66}$$

To find the y parameters in terms of the h parameters, we could start with the h parameter equations and rearrange them into the form of the y parameter equations. Or, we could simply solve Equations (5-66) for the four y parameters in terms of the four h parameters.

The relationship between the z parameters and the y parameters is particularly significant because the roles of the independent variables and the dependent variables are interchanged. That is, in the y parameter equations, I_1 and I_2 are the dependent variables, while in the z parameter equations, V_1 and V_2 are the dependent variables. Equations (5-62) are easily solved for V_1 and V_2 in terms of I_1 and I_2 by Cramer's rule:

$$\begin{aligned} V_1 &= \frac{\begin{vmatrix} I_1 & y_{12} \\ I_2 & y_{22} \end{vmatrix}}{\begin{vmatrix} y_{11} & y_{12} \\ y_{21} & y_{22} \end{vmatrix}} = \frac{y_{22}}{y_{11}y_{22} - y_{12}y_{22}} I_1 + \frac{-y_{12}}{y_{11}y_{22} - y_{12}y_{22}} I_2 \\ V_2 &= \frac{\begin{vmatrix} y_{11} & I_1 \\ y_{21} & I_2 \end{vmatrix}}{\begin{vmatrix} y_{11} & y_{12} \\ y_{21} & y_{22} \end{vmatrix}} = \frac{-y_{21}}{y_{11}y_{22} - y_{12}y_{22}} I_1 + \frac{y_{11}}{y_{11}y_{22} - y_{12}y_{21}} I_2 \end{aligned} \tag{5-67}$$

Equations (5-67) are in z parameter form and so

$$\mathfrak{z}_{11} = \frac{y_{22}}{y_{11}y_{22} - y_{12}y_{21}} \qquad \mathfrak{z}_{12} = \frac{-y_{12}}{y_{11}y_{22} - y_{12}y_{21}}$$
$$\mathfrak{z}_{21} = \frac{-y_{21}}{y_{11}y_{22} - y_{12}y_{21}} \qquad \mathfrak{z}_{22} = \frac{y_{11}}{y_{11}y_{22} - y_{12}y_{21}} \tag{5-68}$$

We also note from Equation (5-68) that the $\mathfrak{z}_{ij}$'s are the elements of the inverse matrix $\mathbf{Y}^{-1}$. In general, we can show that

$$\mathbf{Z} = \mathbf{Y}^{-1} \tag{5-69}$$

Since the independent and dependent variables are also interchanged in g parameters and h parameters, and in the t parameters and $\hat{\mathrm{t}}$ parameters, similar statements can be made for these parameter sets.

Any of the six sets of 2-port parameters can be found in terms of any of the other five sets by following one of the procedures outlined here. Table 5-1 lists sixteen of these relationships. For some networks, one or more of these matrices may not exist. For example, neither the z or y parameters exist for the simplified transistor model of Figure 5-14.

Table 5-1 RELATIONSHIPS BETWEEN THE z, y, h, AND t PARAMETERS.

	Z	Y	H	T
Z	$\begin{bmatrix} \mathfrak{z}_{11} & \mathfrak{z}_{12} \\ \mathfrak{z}_{21} & \mathfrak{z}_{22} \end{bmatrix}$	$\begin{bmatrix} \frac{y_{22}}{\lvert\mathbf{Y}\rvert} & \frac{-y_{12}}{\lvert\mathbf{Y}\rvert} \\ \frac{-y_{21}}{\lvert\mathbf{Y}\rvert} & \frac{y_{11}}{\lvert\mathbf{Y}\rvert} \end{bmatrix}$	$\begin{bmatrix} \frac{\lvert\mathbf{H}\rvert}{\mathfrak{h}_{22}} & \frac{\mathfrak{h}_{12}}{\mathfrak{h}_{22}} \\ \frac{-\mathfrak{h}_{21}}{\mathfrak{h}_{22}} & \frac{1}{\mathfrak{h}_{22}} \end{bmatrix}$	$\begin{bmatrix} \frac{\mathfrak{a}}{\mathfrak{c}} & \frac{\lvert\mathbf{T}\rvert}{\mathfrak{c}} \\ \frac{1}{\mathfrak{c}} & \frac{\mathfrak{d}}{\mathfrak{c}} \end{bmatrix}$
Y	$\begin{bmatrix} \frac{\mathfrak{z}_{22}}{\lvert\mathbf{Z}\rvert} & \frac{-\mathfrak{z}_{12}}{\lvert\mathbf{Z}\rvert} \\ \frac{-\mathfrak{z}_{21}}{\lvert\mathbf{Z}\rvert} & \frac{\mathfrak{z}_{11}}{\lvert\mathbf{Z}\rvert} \end{bmatrix}$	$\begin{bmatrix} y_{11} & y_{12} \\ y_{21} & y_{22} \end{bmatrix}$	$\begin{bmatrix} \frac{1}{\mathfrak{h}_{11}} & \frac{-\mathfrak{h}_{12}}{\mathfrak{h}_{11}} \\ \frac{\mathfrak{h}_{21}}{\mathfrak{h}_{11}} & \frac{\lvert\mathbf{H}\rvert}{\mathfrak{h}_{11}} \end{bmatrix}$	$\begin{bmatrix} \frac{\mathfrak{d}}{\mathfrak{b}} & \frac{-\lvert\mathbf{T}\rvert}{\mathfrak{b}} \\ \frac{-1}{\mathfrak{b}} & \frac{\mathfrak{a}}{\mathfrak{b}} \end{bmatrix}$
H	$\begin{bmatrix} \frac{\lvert\mathbf{Z}\rvert}{\mathfrak{z}_{22}} & \frac{\mathfrak{z}_{12}}{\mathfrak{z}_{22}} \\ \frac{-\mathfrak{z}_{21}}{\mathfrak{z}_{22}} & \frac{1}{\mathfrak{z}_{22}} \end{bmatrix}$	$\begin{bmatrix} \frac{1}{y_{11}} & \frac{-y_{12}}{y_{11}} \\ \frac{y_{21}}{y_{11}} & \frac{\lvert\mathbf{Y}\rvert}{y_{11}} \end{bmatrix}$	$\begin{bmatrix} \mathfrak{h}_{11} & \mathfrak{h}_{12} \\ \mathfrak{h}_{21} & \mathfrak{h}_{22} \end{bmatrix}$	$\begin{bmatrix} \frac{\mathfrak{b}}{\mathfrak{d}} & \frac{\lvert\mathbf{T}\rvert}{\mathfrak{d}} \\ \frac{1}{\mathfrak{d}} & \frac{\mathfrak{c}}{\mathfrak{d}} \end{bmatrix}$
T	$\begin{bmatrix} \frac{\mathfrak{z}_{11}}{\mathfrak{z}_{21}} & \frac{\lvert\mathbf{Z}\rvert}{\mathfrak{z}_{21}} \\ \frac{1}{\mathfrak{z}_{21}} & \frac{\mathfrak{z}_{22}}{\mathfrak{z}_{21}} \end{bmatrix}$	$\begin{bmatrix} \frac{-y_{22}}{y_{21}} & \frac{-1}{y_{21}} \\ \frac{-\lvert\mathbf{Y}\rvert}{y_{21}} & \frac{-y_{11}}{y_{21}} \end{bmatrix}$	$\begin{bmatrix} \frac{-\lvert\mathbf{H}\rvert}{\mathfrak{h}_{21}} & \frac{-\mathfrak{h}_{11}}{\mathfrak{h}_{21}} \\ \frac{-\mathfrak{h}_{22}}{\mathfrak{h}_{21}} & \frac{-1}{\mathfrak{h}_{21}} \end{bmatrix}$	$\begin{bmatrix} \mathfrak{a} & \mathfrak{b} \\ \mathfrak{c} & \mathfrak{d} \end{bmatrix}$

The symbol | | denotes determinant; for example,

$$\lvert\mathbf{Z}\rvert = \begin{vmatrix} \mathfrak{z}_{11} & \mathfrak{z}_{12} \\ \mathfrak{z}_{21} & \mathfrak{z}_{22} \end{vmatrix} = \mathfrak{z}_{11}\mathfrak{z}_{22} - \mathfrak{z}_{12}\mathfrak{z}_{21}$$

See appendix I for a more detailed discussion of matrices and determinants.

Reciprocal networks. A network is said to be *reciprocal* if its open-circuit admittance parameter matrix and its short-circuit impedance parameter matrix are symmetrical. (A matrix having elements a_{ij} is said to be a *symmetrical matrix* if $a_{ij} = a_{ji}$ for all i and j. Thus, the symmetry is with respect to the main diagonal. See Appendix I.) Therefore, a 2-port network is reciprocal is $z_{12} = z_{21}$ and $y_{12} = y_{21}$. It can be shown that if $z_{12} = z_{21}$, then $y_{12} = y_{21}$, and vice versa.

We now state a very important theorem; a proof is given in Appendix II.

Reciprocity theorem: Any 2-port network containing only resistances, capacitances, inductances, and coupled coils is reciprocal.

The reader is invited to check the results of Examples 5-2 and 5-4, in which $z_{12} = z_{21}$ and $y_{12} = y_{21}$, as they should. More important, the reader is urged to check the results of Examples 5-3 and 5-5, which illustrate that, in general, networks containing controlled sources are not reciprocal.

One consequence of the reciprocity theorem is that reciprocal 2-port networks can be uniquely characterized by three parameters, which is one less than the four required for general 2-port networks. For example, since $z_{21} = z_{12}$, any reciprocal network can be characterized by the three parameters z_{11}, z_{12}, and z_{22}. Furthermore, equivalent networks in terms of the z parameters and the y parameters can be constructed for reciprocal networks without the use of controlled sources. A simplified equivalent for 2-port reciprocal networks in terms of the z parameters, called a T(tee) network is presented in Figure 5-18(a), and a simplified equivalent

Figure 5-18 Equivalent networks for reciprocal 2-ports. (a) T (tee) or Y (wye). (b) π (pi) or Δ (delta) network.

for 2-port reciprocal networks in terms of the y parameters, called a π(pi) network, is presented in Figure 5-18(b).

Both the T and π networks of Figure 5-18(a) and (b) are unique for reciprocal networks. That is, every reciprocal network has one and only one T equivalent, and one and only one π equivalent. It follows that any T network consisting of only nonsource elements has an equivalent π network representation consisting of only nonsource elements, and

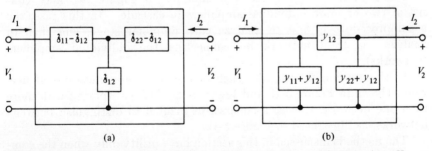

vice versa. The transformation giving the three T parameters in terms of the three π parameters is called the π-T transformation or, sometimes, the Δ-Y (delta-wye) transformation. This transformation is presented in Table 5-2.

It can also be shown that for a reciprocal network $\mathfrak{h}_{12} = -\mathfrak{h}_{21}$ and $\mathfrak{a}\mathfrak{d} = \mathfrak{b}\mathfrak{c} + 1$.

Table 5-2 THE T $-$ π (OR Δ $-$ Y) TRANSFORMATION

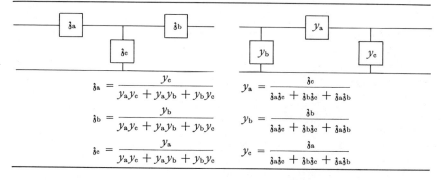

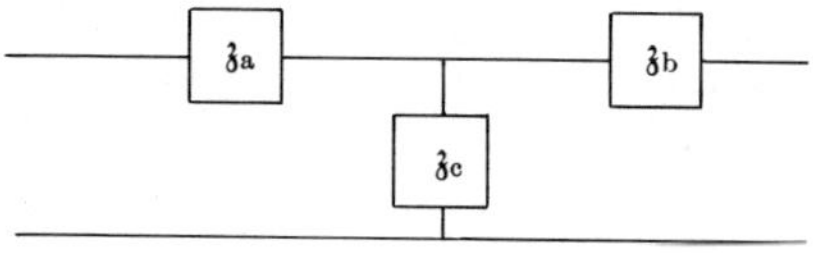

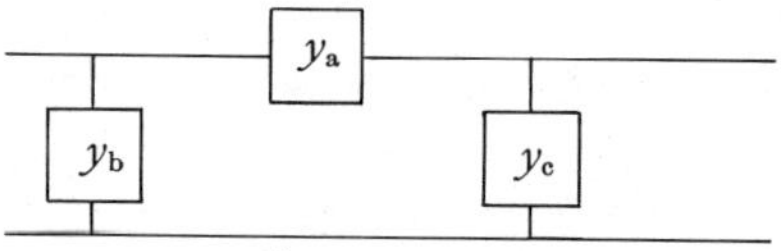

$\mathfrak{z}_a = \dfrac{y_c}{y_a y_c + y_a y_b + y_b y_c}$	$y_a = \dfrac{\mathfrak{z}_c}{\mathfrak{z}_a\mathfrak{z}_c + \mathfrak{z}_b\mathfrak{z}_c + \mathfrak{z}_a\mathfrak{z}_b}$
$\mathfrak{z}_b = \dfrac{y_b}{y_a y_c + y_a y_b + y_b y_c}$	$y_b = \dfrac{\mathfrak{z}_b}{\mathfrak{z}_a\mathfrak{z}_c + \mathfrak{z}_b\mathfrak{z}_c + \mathfrak{z}_a\mathfrak{z}_b}$
$\mathfrak{z}_c = \dfrac{y_a}{y_a y_c + y_a y_b + y_b y_c}$	$y_c = \dfrac{\mathfrak{z}_a}{\mathfrak{z}_a\mathfrak{z}_c + \mathfrak{z}_b\mathfrak{z}_c + \mathfrak{z}_a\mathfrak{z}_b}$

5-4 Interconnection of 2-Port Networks

Sometimes the most convenient way to analyze a network is first to divide it into a number of component networks, analyze each of the component networks separately, and then analyze the network of component networks. For example, the IF amplifier in Figure 5-1 may consist of two or more individual amplifiers in cascade. In this case, the usual procedure is first to analyze each amplifier stage, and then to analyze the total network by connecting the individual amplifiers (networks).

In this section we consider each of the six interconnections of networks shown in Figure 5-19 and Figure 5-23. When working with more than one network, we use a superscript symbol to distinguish between networks, as illustrated in Figure 5-19.

The methods discussed in this section have utility only when the component network terminal pairs remain ports after the interconnections have been made. (Recall that a port is defined as a terminal pair for which the current into one terminal is equal to the current out the other terminal.) If one or more terminal pairs do not retain their port properties after the interconnections are made, we must resort to a more general method of analysis; this more general method is the topic of Chapter 6.

2-port networks in parallel. Refer to Figure 5-19(a) and consider the parallel combination of networks N° and N^{Δ}, which we denote as network N. Our desideratum is to express the short-circuit admittance parameter matrix **Y** for network N in terms of the short-circuit admittance parameter matrices $\mathbf{Y}^\circ$ and $\mathbf{Y}^\Delta$ for networks N° and N$^\Delta$, respectively. For networks N°, N$^\Delta$, and N, we have

$$\text{network N}^\circ\text{:}\quad \begin{bmatrix} I_1^\circ \\ I_2^\circ \end{bmatrix} = \begin{bmatrix} y_{11}^\circ & y_{12}^\circ \\ y_{21}^\circ & y_{22}^\circ \end{bmatrix} \begin{bmatrix} V_1^\circ \\ V_2^\circ \end{bmatrix}$$
$$\text{network N}^\Delta\text{:}\quad \begin{bmatrix} I_1^\Delta \\ I_2^\Delta \end{bmatrix} = \begin{bmatrix} y_{11}^\Delta & y_{12}^\Delta \\ y_{21}^\Delta & y_{22}^\Delta \end{bmatrix} \begin{bmatrix} V_1^\Delta \\ V_2^\Delta \end{bmatrix} \tag{5-70}$$

and

$$\text{network N:}\quad \begin{bmatrix} I_1 \\ I_2 \end{bmatrix} = \begin{bmatrix} y_{11} & y_{12} \\ y_{21} & y_{22} \end{bmatrix} \begin{bmatrix} V_1 \\ V_2 \end{bmatrix} \tag{5-71}$$

If all terminal pairs remain ports after the connections are made, then

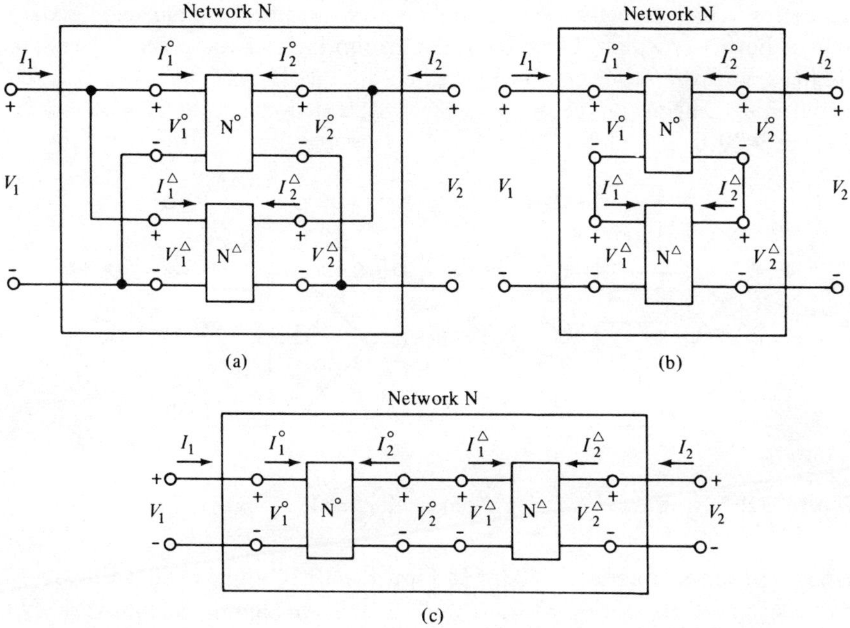

Figure 5-19 Networks connected in (a) parallel, (b) series, and (c) cascade.

Equations (5-70) remain valid after the connections are made. Furthermore, we observe from Figure 5-19(a) that

$$\begin{aligned} V_1 &= V_1^\circ = V_1^\Delta \\ V_2 &= V_2^\circ = V_2^\Delta \end{aligned} \quad \text{or} \quad \begin{bmatrix} V_1 \\ V_2 \end{bmatrix} = \begin{bmatrix} V_1^\circ \\ V_2^\circ \end{bmatrix} = \begin{bmatrix} V_1^\Delta \\ V_2^\Delta \end{bmatrix} \tag{5-72}$$

$$\begin{aligned} I_1 &= I_1^\circ + I_1^\Delta \\ I_2 &= I_2^\circ + I_2^\Delta \end{aligned} \quad \text{or} \quad \begin{bmatrix} I_1 \\ I_2 \end{bmatrix} = \begin{bmatrix} I_1^\circ \\ I_2^\circ \end{bmatrix} + \begin{bmatrix} I_1^\Delta \\ I_2^\Delta \end{bmatrix} \tag{5-73}$$

Now, we substitute Equations (5-72) into Equations (5-70) to eliminate V_1°, V_2°, V_1^Δ, and V_2^Δ. Next, we add the two Equations (5-70) and substitute Equation (5-73) into the left-hand side to obtain

$$\begin{aligned}\begin{bmatrix} I_1 \\ I_2 \end{bmatrix} &= \begin{bmatrix} \overset{\circ}{y}_{11} & \overset{\circ}{y}_{12} \\ \overset{\circ}{y}_{21} & \overset{\circ}{y}_{22} \end{bmatrix} \begin{bmatrix} V_1 \\ V_2 \end{bmatrix} + \begin{bmatrix} \overset{\Delta}{y}_{11} & \overset{\Delta}{y}_{12} \\ \overset{\Delta}{y}_{21} & \overset{\Delta}{y}_{22} \end{bmatrix} \begin{bmatrix} V_1 \\ V_2 \end{bmatrix} \\ &= \begin{bmatrix} \overset{\circ}{y}_{11} + \overset{\Delta}{y}_{11} & \overset{\circ}{y}_{12} + \overset{\Delta}{y}_{12} \\ \overset{\circ}{y}_{21} + \overset{\Delta}{y}_{21} & \overset{\circ}{y}_{22} + \overset{\Delta}{y}_{22} \end{bmatrix} \begin{bmatrix} V_1 \\ V_2 \end{bmatrix}\end{aligned} \tag{5-74}$$

Comparing Equation (5-74) to Equation (5-71), we conclude that

$$\mathbf{Y} = \mathbf{Y}^{\circ} + \mathbf{Y}^{\Delta} \tag{5-75}$$

That is, the y parameters add for networks in parallel, provided the terminal pairs remain ports.

Unfortunately, when networks are connected in parallel, their terminal pairs do not, in general, remain ports. However, there are two important cases for which the terminal pairs do remain ports. One case occurs when both networks have common grounds, as illustrated in Figure 5-20(a), and the other case occurs when one of the four terminal pairs is isolated by means of coupled coils or a transformer, as illustrated in Figure 5-20(b).

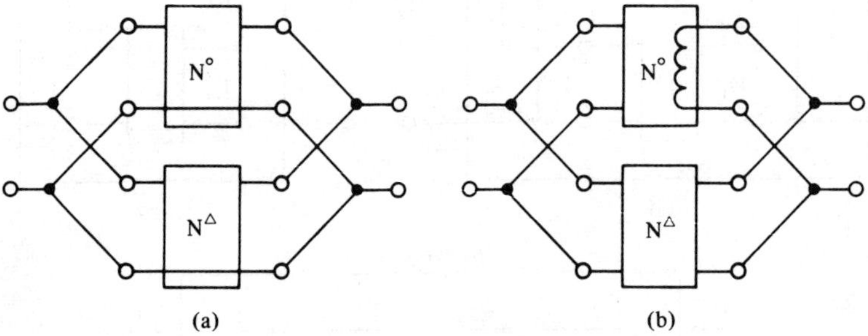

Figure 5-20 Networks for which y parameters add.

2-port networks in series. Refer to Figure 5-19(b) and consider the series connection of networks N° and N^{Δ}, which we denote as network N. Our desideratum is to express the open-circuit impedance parameter matrix **Z** for network N in terms of the open-circuit impedance matrices **Z**° and $\mathbf{Z}^{\Delta}$ for networks N° and N^{Δ}, respectively. We have

$$\begin{aligned}\text{for network N}^{\circ}: \begin{bmatrix} V_1^{\circ} \\ V_2^{\circ} \end{bmatrix} &= \begin{bmatrix} \overset{\circ}{z}_{11} & \overset{\circ}{z}_{12} \\ \overset{\circ}{z}_{21} & \overset{\circ}{z}_{22} \end{bmatrix} \begin{bmatrix} I_1^{\circ} \\ I_2^{\circ} \end{bmatrix} \\ \text{for network N}^{\Delta}: \begin{bmatrix} V_1^{\Delta} \\ V_2^{\Delta} \end{bmatrix} &= \begin{bmatrix} \overset{\Delta}{z}_{11} & \overset{\Delta}{z}_{12} \\ \overset{\Delta}{z}_{21} & \overset{\Delta}{z}_{22} \end{bmatrix} \begin{bmatrix} I_1^{\Delta} \\ I_2^{\Delta} \end{bmatrix}\end{aligned} \tag{5-76}$$

and

$$\text{for network N}: \begin{bmatrix} V_1 \\ V_2 \end{bmatrix} = \begin{bmatrix} z_{11} & z_{12} \\ z_{21} & z_{22} \end{bmatrix} \begin{bmatrix} I_1 \\ I_2 \end{bmatrix} \tag{5-77}$$

If all terminal pairs remain ports after the connections are made, then

Equations (5-76) remain valid and, furthermore, we note from Figure 5-19(b) that if the terminal pairs remain ports, then

$$\begin{aligned} I_1 &= I_1^\circ = I_1^\Delta \\ I_2 &= I_2^\circ = I_2^\Delta \end{aligned} \qquad \text{or} \qquad \begin{bmatrix} I_1 \\ I_2 \end{bmatrix} = \begin{bmatrix} I_1^\circ \\ I_2^\circ \end{bmatrix} = \begin{bmatrix} I_1^\Delta \\ I_2^\Delta \end{bmatrix} \tag{5-78}$$

Also note from Figure 5-19(b) that

$$\begin{aligned} V_1 &= V_1^\circ + V_1^\Delta \\ V_2 &= V_2^\circ + V_2^\Delta \end{aligned} \qquad \text{or} \qquad \begin{bmatrix} V_1 \\ V_2 \end{bmatrix} = \begin{bmatrix} V_1^\circ \\ V_2^\circ \end{bmatrix} + \begin{bmatrix} V_1^\Delta \\ V_2^\Delta \end{bmatrix} \tag{5-79}$$

Now, we substitute I_1 for I_1° and I_1^Δ, and I_2 for I_2° and I_2^Δ in Equations (5-76); then add the two Equations (5-76); and use Equations (5-79) to obtain

$$\begin{aligned} \begin{bmatrix} V_1 \\ V_2 \end{bmatrix} &= \begin{bmatrix} \overset{\circ}{\mathfrak{z}}_{11} & \overset{\circ}{\mathfrak{z}}_{12} \\ \overset{\circ}{\mathfrak{z}}_{21} & \overset{\circ}{\mathfrak{z}}_{22} \end{bmatrix} \begin{bmatrix} I_1 \\ I_2 \end{bmatrix} + \begin{bmatrix} \overset{\Delta}{\mathfrak{z}}_{11} & \overset{\Delta}{\mathfrak{z}}_{12} \\ \overset{\Delta}{\mathfrak{z}}_{21} & \overset{\Delta}{\mathfrak{z}}_{22} \end{bmatrix} \begin{bmatrix} I_1 \\ I_2 \end{bmatrix} \\ &= \begin{bmatrix} \overset{\circ}{\mathfrak{z}}_{11} + \overset{\Delta}{\mathfrak{z}}_{11} & \overset{\circ}{\mathfrak{z}}_{12} + \overset{\Delta}{\mathfrak{z}}_{12} \\ \overset{\circ}{\mathfrak{z}}_{21} + \overset{\Delta}{\mathfrak{z}}_{21} & \overset{\circ}{\mathfrak{z}}_{22} + \overset{\Delta}{\mathfrak{z}}_{22} \end{bmatrix} \begin{bmatrix} I_1 \\ I_2 \end{bmatrix} \end{aligned} \tag{5-80}$$

Comparing Equations (5-80) to Equations (5-77) we conclude that

$$\mathbf{Z} = \mathbf{Z}^\circ + \mathbf{Z}^\Delta \tag{5-81}$$

That is, the z parameters add for networks in series, provided the terminal pairs remain ports.

Unfortunately, when networks are connected in series, their terminal pairs do not, in general, remain ports. However, there are cases of some interest for which the terminal pairs remain ports and Equation (5-81) is valid. Both cases are illustrated in Figure 5-21.

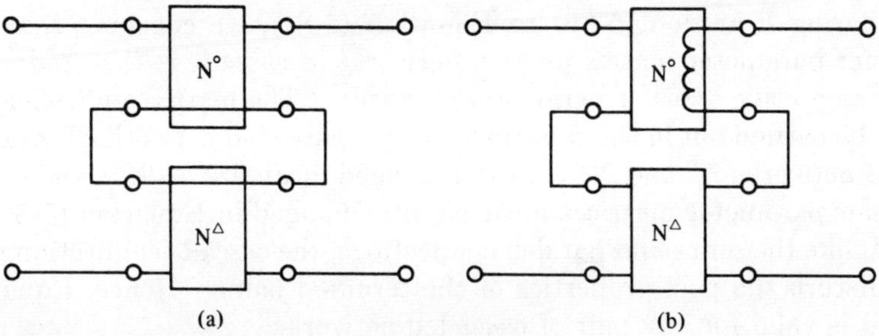

Figure 5-21 Networks for which z parameters add.

2-port networks in cascade. Networks connected end to end, as in Figure 5-19(c), are said to be connected in cascade. This interconnection has application when the output of one network (N°) is connected to the input of another network (NΔ). Note that all terminal pairs of both N° and NΔ remain ports after the connections are made, so that the results of this section are valid for all cascaded 2-port networks.

In this section we compute the transfer parameters **T** for network N in terms of the transfer parameters, $\mathbf{T}^\circ$ and $\mathbf{T}^\Delta$, for networks N° and N^Δ, respectively. For the three networks, we have

$$\text{network N}^\circ: \begin{bmatrix} V_1^\circ \\ I_1^\circ \end{bmatrix} = \begin{bmatrix} \mathfrak{a}^\circ & \mathfrak{b}^\circ \\ \mathfrak{c}^\circ & \mathfrak{d}^\circ \end{bmatrix} \begin{bmatrix} V_2^\circ \\ -I_2^\circ \end{bmatrix} \tag{5-82}$$

$$\text{network N}^\Delta: \begin{bmatrix} V_1^\Delta \\ I_1^\Delta \end{bmatrix} = \begin{bmatrix} \mathfrak{a}^\Delta & \mathfrak{b}^\Delta \\ \mathfrak{c}^\Delta & \mathfrak{d}^\Delta \end{bmatrix} \begin{bmatrix} V_2^\Delta \\ -I_2^\Delta \end{bmatrix} \tag{5-83}$$

$$\text{network N}: \begin{bmatrix} V_1 \\ I_1 \end{bmatrix} = \begin{bmatrix} \mathfrak{a} & \mathfrak{b} \\ \mathfrak{c} & \mathfrak{d} \end{bmatrix} \begin{bmatrix} V_2 \\ -I_2 \end{bmatrix} \tag{5-84}$$

From Figure 5-19(c) we observe that

$$\begin{aligned} \begin{bmatrix} V_1 \\ I_1 \end{bmatrix} &= \begin{bmatrix} V_1^\circ \\ I_1^\circ \end{bmatrix} \\ \begin{bmatrix} V_2^\circ \\ -I_2^\circ \end{bmatrix} &= \begin{bmatrix} V_1^\Delta \\ I_1^\Delta \end{bmatrix} \\ \begin{bmatrix} V_2^\Delta \\ I_2^\Delta \end{bmatrix} &= \begin{bmatrix} V_2 \\ I_2 \end{bmatrix} \end{aligned} \tag{5-85}$$

We now substitute the first equation of Equations (5-85) into the left-hand side of Equation (5-82); the second equation into the right-hand side of Equation (5-82); and the third equation into the right-hand side of Equation (5-83). We then substitute Equation (5-84) into Equation (5-82) to obtain

$$\begin{bmatrix} V_1 \\ I_1 \end{bmatrix} = \begin{bmatrix} \mathfrak{a}^\circ & \mathfrak{b}^\circ \\ \mathfrak{c}^\circ & \mathfrak{d}^\circ \end{bmatrix} \begin{bmatrix} \mathfrak{a}^\Delta & \mathfrak{b}^\Delta \\ \mathfrak{c}^\Delta & \mathfrak{d}^\Delta \end{bmatrix} \begin{bmatrix} V_2 \\ -I_2 \end{bmatrix} \tag{5-86}$$

Comparing Equation (5-86) to Equation (5-84), we conclude that the transfer parameter matrix for two networks in cascade is the product of their respective transfer parameter matrices. The matrix multiplication must be carried out in the same order as the cascaded networks; for example, if networks N° and N^Δ are interchanged in Figure 5-19(c), then their transfer parameter matrices must be interchanged in Equation (5-86).

Unlike the series and parallel connections, the cascade connection does not disturb the port properties of the terminal pairs. Hence, Equation (5-86) is valid for any pair of cascaded networks.

Example 5-9

In this example we compute the t parameter matrix for the network of Figure 5-22(a). The transistor model is given in Figure 5-22(b). We first divide the network of Figure 5-22(a) into two networks, as in Figure 5-22(b). Network N° consists of the transistor model and network N^Δ, the resistor network. The t parameter matrix for the grounded collector

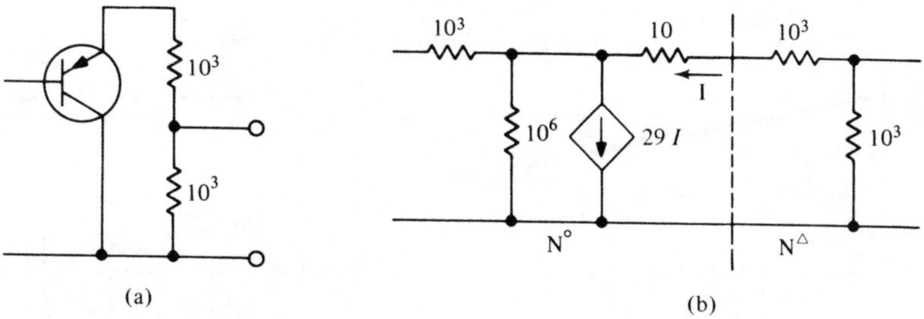

Figure 5-22 Example 5-9.

transistor was computed in Example 5-8 and is given by Equation (5-58). Substituting the typical transistor parameter values from Example 5-6 into Equation (5-58), we obtain the t parameter matrix for network N°:

$$\mathbf{T}^\circ = \begin{bmatrix} 1 & 20 \\ 10^{-6} & 10^{-2} \end{bmatrix} \tag{5-87}$$

The t parameter matrix for the resistor network N^Δ is given by

$$\mathbf{T}^\Delta = \begin{bmatrix} 2 & 10^3 \\ 10^{-3} & 1 \end{bmatrix} \tag{5-88}$$

Therefore, according to Equation (5-86), the t parameter matrix for the cascaded network is given by

$$\begin{aligned} \mathbf{T} &= \begin{bmatrix} 1 & 20 \\ 10^{-6} & 10^{-2} \end{bmatrix} \begin{bmatrix} 2 & 10^3 \\ 10^{-3} & 1 \end{bmatrix} \\ &= \begin{bmatrix} 2.02 & 1020 \\ 1.2 \times 10^{-5} & 0.011 \end{bmatrix} \end{aligned} \tag{5-89}$$

Other 2-port interconnections. In addition to the interconnections illustrated in Figure 5-19, the three interconnections illustrated by Figure 5-23 are also possible. In Figure 5-23(c), network N is constructed by connecting port 1 of networks N° and N^Δ in series, and port 2 in parallel. In Figure 5-23(b), we have connected ports 1 in parallel and ports 2 in series. Finally, in Figure 5-23(c), we have network N° cascaded behind network N^Δ. By following the same procedure used in the preceding paragraphs, it is easy to show that the h parameter matrix for network N in Figure 5-23(a) is the sum of the h parameter matrices of networks N° and N^Δ, provided that both terminal pairs remain ports; the g parameter matrix for network N of Figure 5-23(b) is the sum of the g parameter matrices for networks N° and N^Δ, provided the terminal pairs remain ports; the $\hat{t}$ parameter matrix for network N of Figure 5-23(c) is the product of the $\hat{t}$ parameter matrix of network N° premultiplied by the $\hat{t}$ parameter matrix for network N^Δ.

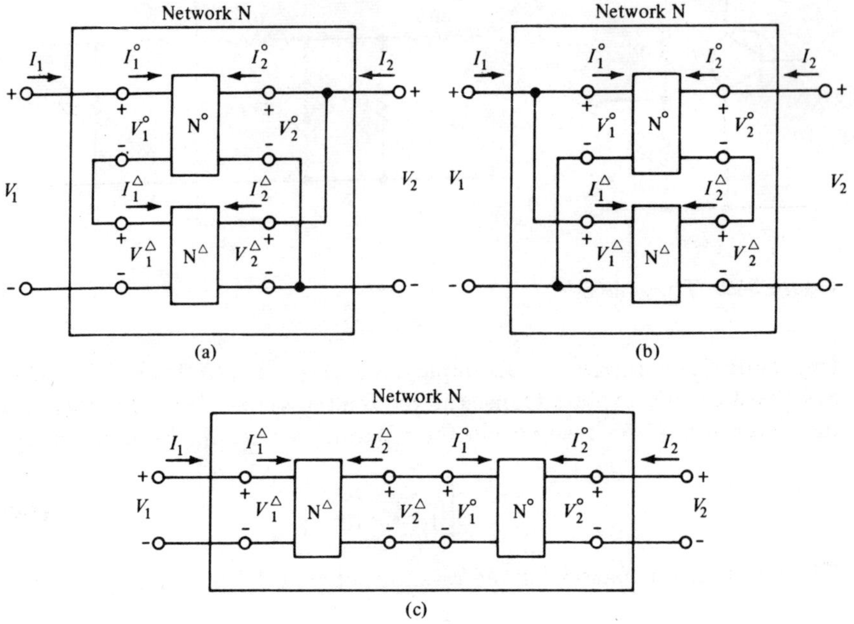

Figure 5-23 (a) Series-parallel, (b) parallel-series, (c) reverse cascade connections.

5-5 n-Port Networks

The results for 1-port and 2-port networks developed in Sections 5-2 and 5-3 can easily be extended to the case of n-port networks. To characterize an n-port network in terms of its n-port voltages and n-port currents, we first write a set of n linear equations in these 2n variables. Since n of the variables are independent variables, with the other n variables the dependent variables, $(2n)!/(n!)^2$ parameter matrices, which uniquely characterize the network, exist. For example, taking the n-port voltages to be the independent variables and the n-port currents to be the dependent variables, we write each port current as a linear combination of the n-port voltages,

$$\begin{aligned}
I_1 &= y_{11}V_1 + y_{12}V_2 + \cdots + y_{1n}V_n \\
I_2 &= y_{21}V_1 + y_{22}V_2 + \cdots + y_{2n}V_n \\
&\vdots \\
I_n &= y_{n1}V_1 + y_{n2}V_2 + \cdots + y_{nn}V_n
\end{aligned} \tag{5-90}$$

and observe that the n-port network can be uniquely characterized by the n × n short-circuit admittance parameter matrix

$$\mathbf{Y} = \begin{bmatrix} y_{11} & y_{12} & \cdots & y_{1n} \\ y_{21} & y_{22} & \cdots & y_{2n} \\ \cdot & \cdot & \cdot & \cdot \\ \cdot & \cdot & \cdot & \cdot \\ \cdot & \cdot & \cdot & \cdot \\ y_{n1} & y_{n2} & \cdots & y_{nn} \end{bmatrix} \tag{5-91}$$

where

$$y_{ij} = \frac{I_i}{V_j}\bigg|_{V_\ell=0,\ell\neq j} = I_i\bigg|_{V_j=1;V_\ell=0,\ell\neq j} \tag{5-92}$$

Note that for n = 1, Equations (5-90) reduce to Equations (5-3), and for n = 2, reduce to Equations (5-7).

If the n-port currents are taken to be the independent variables, we write

$$\begin{bmatrix} V_1 \\ V_2 \\ \cdot \\ \cdot \\ \cdot \\ V_n \end{bmatrix} = \begin{bmatrix} z_{11} & z_{12} & \cdots & z_{1n} \\ z_{21} & z_{22} & \cdots & z_{2n} \\ \cdot & \cdot & \cdot & \cdot \\ \cdot & \cdot & \cdot & \cdot \\ \cdot & \cdot & \cdot & \cdot \\ z_{n1} & z_{n2} & \cdots & z_{nn} \end{bmatrix} \begin{bmatrix} I_1 \\ I_2 \\ \cdot \\ \cdot \\ \cdot \\ I_n \end{bmatrix} \tag{5-93}$$

where the elements of the n × n open-circuit impedance parameter matrix are given by the open-circuit impedances

$$z_{ij} = \frac{V_i}{I_j}\bigg|_{I_j=0,\ell\neq j} = V_i\bigg|_{I_j=1;I_\ell=0,\ell\neq j} \tag{5-94}$$

A multiport network is said to be reciprocal if its open-circuit impedance matrix and/or its short-circuit admittance matrix is symmetrical. It can be shown that if **Z** is symmetrical, so is **Y**, and vice versa. Furthermore, the reciprocity theorem for n-port networks states that any n-port network consisting only of resistances, inductances, capacitances, coupled coils, and ideal transformers is reciprocal.

Finally, it can be shown that if two n-port networks are connected in parallel, their short-circuit admittance parameters add, as in Equation (5-75), provided that all n ports remain ports after the connections are made; if two n-port networks are connected in series, their open-circuit impedance parameters add, as in Equation (5-81), provided that all n ports remain ports after the connections are made; and if two n-port (n even) networks are connected in cascade, their transfer parameters always multiply, as in Equation (5-86).

5-6 Power on a Port Basis

In order to study power flow in networks described on a port basis, we recall the steady-state power formulas of Section 4-6 in terms of the complex voltage and current amplitudes. We found that the average power is given by [see Equation (4-66)]

$$\langle p(\mathrm{t})\rangle = \tfrac{1}{2}\,\mathrm{Re}\,\{VI^*\} = \tfrac{1}{2}\,\mathrm{Re}\,\{V^*I\} \tag{5-95}$$

where Re $\{\cdot\}$ means the real part of the quantity inside the brackets.

For a 1-port characterized by a complex impedance $\mathfrak{z}$, Equation (5-95) can be written

$$\begin{aligned}\tfrac{1}{2}\,\mathrm{Re}\,\{VI^*\} &= \tfrac{1}{2}\,\mathrm{Re}\,\{(\mathfrak{z}I)I^*\} = \tfrac{1}{2}\,\mathrm{Re}\,\{\mathfrak{z}(II^*)\} = \tfrac{1}{2}\,\mathrm{Re}\,\{\mathfrak{z}|I|^2\} \\ \langle p(\mathrm{t})\rangle &= \tfrac{1}{2}|I|^2\,\mathrm{Re}\,\{\mathfrak{z}\}\end{aligned} \tag{5-96}$$

When the complex impedance $\mathfrak{z}$ is written in rectangular or polar forms, and the effective current $\mathcal{I}$ is used instead of the complex current I, two very commonly used power formulas are readily derived from Equation (5-96). With the rectangular form

$$\mathfrak{z} = \mathrm{R} + \mathrm{jX} \tag{5-97}$$

we have

$$\langle p(\mathrm{t})\rangle = \tfrac{1}{2}|I|^2\,\mathrm{Re}\,\{\mathrm{R} + \mathrm{jX}\} = \tfrac{1}{2}|I|^2\mathrm{R} = \mathcal{I}^2\mathrm{R} \tag{5-98}$$

Although Equation (5-98) looks very much like the instantaneous power formula for a resistor [see Equation (1-21)], we must remember that $\mathcal{I}$ is an effective sinusoidal current and R is the real part of a complex impedance. With the polar form

$$\mathfrak{z} = |\mathfrak{z}|\underline{/\psi} \tag{5-99}$$

we have

$$\begin{aligned}\langle p(\mathrm{t})\rangle &= \tfrac{1}{2}|I^2|\,\mathrm{Re}\,\{|\mathfrak{z}|\underline{/\psi}\} = \tfrac{1}{2}|I^2|\,|\mathfrak{z}|\cos\psi \\ &= \mathcal{I}^2|\mathfrak{z}|\cos\psi\end{aligned} \tag{5-100}$$

The factor $\cos\psi$ in Equation (5-100) is exactly the power factor defined in Section 4-6. Consequently, one often characterizes a complex impedance by its magnitude and power factor rather than by its magnitude and angle. The sign of the angle is lost in such a description, but all the necessary information for power calculations is available.

For a 1-port characterized by a complex admittance y, Equation (5-95) can be written

$$\begin{aligned}\langle p(\mathrm{t})\rangle &= \tfrac{1}{2}\,\mathrm{Re}\,\{V^*I\} = \tfrac{1}{2}\,\mathrm{Re}\,\{V^*(Vy)\} = \tfrac{1}{2}\,\mathrm{Re}\,\{(V^*V)y\} \\ &= \tfrac{1}{2}\,\mathrm{Re}\,\{|V|^2y\} \\ &= \tfrac{1}{2}|V|^2\,\mathrm{Re}\,\{y\}\end{aligned} \tag{5-101}$$

To get formulas corresponding to Equations (5-98) and (5-100), we write

$$y = \mathrm{G} + \mathrm{jB} = |y|\underline{/\eta} \tag{5-102}$$

Then, in terms of the symbols of Equation (5-102) and the effective voltage $\mathcal{V}$, we have

$$\langle p(\mathrm{t}) \rangle = \mathcal{V}^2 \mathrm{G} = \mathcal{V}^2 |y| \cos \eta \tag{5-103}$$

The average power as given by Equation (5-103) is exactly the same as that given by Equations (5-98) and (5-100) when $\mathfrak{z}$ and y describe the same 1-port. The relations between the various quantities are

$$y = \frac{1}{\mathfrak{z}} \tag{5-104}$$

$$\mathrm{G} = \frac{\mathrm{R}}{\mathrm{R}^2 + \mathrm{X}^2} \tag{5-105}$$

$$\mathrm{R} = \frac{\mathrm{G}}{\mathrm{G}^2 + \mathrm{B}^2} \tag{5-106}$$

$$\eta = -\psi \tag{5-107}$$

$$\cos \eta = \cos \psi \tag{5-108}$$

As a matter of common terminology, R is called the *resistance*, X the *reactance*, G the *conductance*, and B the *susceptance* of the 1-port. From Equation (5-108), we see that the power factor is the same on both the admittance and the impedance basis.

Power transfer—passivity and activity. From Equations (5-96) and (5-101), we see that the average power delivered to a 1-port is positive if Re $|\mathfrak{z}|$ is positive, and the power is negative if Re $|\mathfrak{z}|$ is negative.[1] The complex amplitude of the independent source that drives the network does not affect the sign of the power, since the effect of the source is contained in $|I|^2$ or $|V|^2$ only. When the power delivered to a network is positive, the network is said to be *passive*. When the power is negative—the network is delivering average power back to the source—then the network is said to be *active*.

The concept of passivity and activity is readily extended to networks with more than one port, but the test to determine whether a network is active or passive becomes much more involved. We illustrate the problem for a 2-port. The extension to n-port networks is straightforward. For a 2-port, the total power delivered to the circuit is readily computed if the network equations are given on a g, h, z, or y parameter basis. Since the basic technique is the same for all four sets of equations, we can be specific and consider a circuit described by its h parameters.

The h parameter description of a 2-port is given by Equations (5-35) as

$$\begin{aligned} V_1 &= \mathfrak{h}_{11} I_1 + \mathfrak{h}_{12} V_2 \\ I_2 &= \mathfrak{h}_{21} I_1 + \mathfrak{h}_{22} V_2 \end{aligned} \tag{5-109}$$

[1] From Equation (5-105) it is clear that Re $\{\mathfrak{z}\}$ and Re $\{y\}$ have the same sign.

The average power delivered to the network at port 1 is $\frac{1}{2}$ Re $\{V_1 I_1^*\}$. The power delivered at port 2 is $\frac{1}{2}$ Re $\{V_2^* I_2\}$. The total average power is the sum of the two. Thus

$$
\begin{aligned}
2\langle p(t)\rangle &= \text{Re}\,\{V_1 I_1^*\} + \text{Re}\,\{V_2^* I_2\} \\
&= \text{Re}\,\{(\mathfrak{h}_{11} I_1 + \mathfrak{h}_{12} V_2) I_1^*\} + \text{Re}\,\{V_2^*)\mathfrak{h}_{21} I_1 + \mathfrak{h}_{22} V_2)\} \\
&= \text{Re}\,\{\mathfrak{h}_{11} |I_1|^2\} + \text{Re}\,\{\mathfrak{h}_{12} V_2 I_1^*\} + \text{Re}\,\{\mathfrak{h}_{21} V_2^* I_1\} + \text{Re}\,\{\mathfrak{h}_{22} |V_2|^2\} \\
&= |I_1|^2\,\text{Re}\,\{\mathfrak{h}_{11}\} + \text{Re}\,\{\mathfrak{h}_{12} I_1^* V_2 + \mathfrak{h}_{21} I_1 V_2^*\} + |V_2|^2\,\text{Re}\,\{\mathfrak{h}_{22}\} \\
2\langle p(\text{t})\rangle &= |I_1|^2\,\text{Re}\,\{\mathfrak{h}_{11}\} + \text{Re}\,\{(\mathfrak{h}_{12} + \mathfrak{h}_{21}^*) I_1^* V_2\} + |V_2|^2\,\text{Re}\,\{\mathfrak{h}_{22}\}
\end{aligned}
\qquad \textbf{(5-110)}
$$

The positive or negative nature of Equation (5-110) depends not only on the network parameters, but also on the excitations, that is, on I_1 and V_2. The examples below show that for some networks $\langle p(\text{t})\rangle$ can be positive or negative, depending on the magnitude and phase of I_1 and V_2. Others have $\langle p(t)\rangle$ positive for all I_1 and V_2. We say that a network is *passive* if it cannot deliver net average power for any excitations. If there is one condition for which $\langle p(t)\rangle$ in Equation (5-110) can be negative, then the network is *active*.

To show how the complex amplitudes of the excitations affect the power flow in a 2-port, let us reexamine Example 5-6. The h parameter matrix for the grounded base transistor was found to be

$$
\mathbf{H} = \begin{bmatrix} 20 & 10^{-3} \\ -0.99 & 10^{-6} \end{bmatrix} \qquad \textbf{(5-111)}
$$

Using these values in Equation (5-110), we find that the average power for this grounded base transistor is given by

$$
\langle p(\text{t})\rangle = 10|I_1|^2 - 0.495\ \text{Re}\,\{I_1^* V_2\} + 10^{-6}|V_2|^2 \qquad \textbf{(5-112)}
$$

If we make V_2 one volt and I_1 one milliampere, both with zero phase angle, then

$$
\langle p(\text{t})\rangle = 10^{-5} - 0.495 \times 10^{-3} + 10^{-6} = -4.84 \qquad \textbf{(5-113)}
$$

On the other hand, if V_2 is one volt and I_1 one milliampere, but with the angle of V_2 at zero and the angle of I_1 at $\pi/2$, then Re $\{I_1^* V_2\}$ is zero, and the net average power is positive.

We conclude that this network is active, since it is possible to get more power out than we put in.

A different set of parameters in Equation (5-111), that is, a set of parameters that doesn't make a very good transistor, can make the circuit passive. One possibility is a transistor with $r_b = r_e = r_c = 1000$, and $\alpha = 0.5$. Then $\mathfrak{h}_{11} = 1250$, $\mathfrak{h}_{12} = 500$, $\mathfrak{h}_{21} = -0.75$, and $\mathfrak{h}_{22} = 0.0005$, and

$$
\langle p(\text{t})\rangle = 625|I_1|^2 - \tfrac{1}{8}\ \text{Re}\,\{I_1^* V_2\} + \tfrac{1}{4000}|V_2|^2 \qquad \textbf{(5-114)}
$$

A thorough study of the right side of Equation (5-114) reveals that for these particular coefficients there are no values of I_1 and V_2 that make $\langle p(t)\rangle$ negative. Thus this network is passive.

In the discussion so far we have considered the frequency fixed and we have investigated power flow as a function of the complex amplitudes of the voltages and currents. As the frequency is varied, the network parameters vary. Consequently, a network may appear passive at one frequency and active at another. It is customary to use the word passive to describe a network only if the circuit is passive for all frequencies.

Example 5-10

As an example of a network that can deliver net average power at certain frequencies but not at others, consider the network presented in Figure

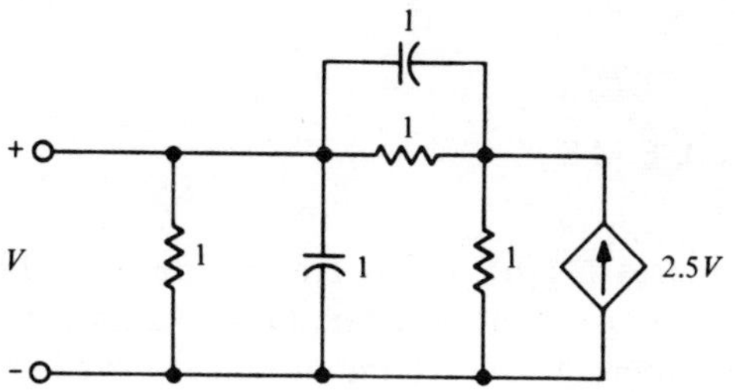

Figure 5-24 Example 5-10.

5-24. For this 1-port the input impedance (Thevenin equivalent impedance) $\mathfrak{z}$ is given by

$$\begin{aligned}\mathfrak{z} = \mathfrak{z}(j\omega) &= \frac{2 + j\omega}{(0.5 - \omega^2) + j1.5\omega} \\ &= \frac{1 - 0.5\omega^2}{(0.5 - \omega^2)^2 + 2.25\omega^2} + \frac{j[\omega(0.5 - \omega^2) - 3\omega]}{(0.5 - \omega^2)^2 + 2.25\omega^2}\end{aligned} \tag{5-115}$$

For $\omega < 2$, Re $\{\mathfrak{z}(j\omega)\}$ is positive, and for $\omega > 2$ it is negative. If the network is driven by a source of frequency $\omega > 2$, power will flow from network to source.

Consistency of power relationships—Tellegen's theorem. In Section 4-6, we saw that positive resistances absorb average power and that capacitances and inductances absorb no power on the average in the sinusoidal steady state. Thus, for the network of Example 5-10, the resistances absorb power, and the capacitances do not generate power. Thus the power that the network delivers back to the source for frequencies higher than 2 rad/sec must come from the controlled source. This reasoning, which is correct, is based on our concept of conservation of energy.

We assume the network equations are consistent with conservation of energy. That is, we assume that the power delivered to a network at its ports is equal to the sum of the power delivered to the individual elements. This assumption must be true if our network laws are to be consistent with physics. At the end of Section 2-3 above we stated Tellegen's Theorem for instantaneous power. The theorem[2] is also true for average power. Although the mathematical theorem was not proved until 1952, the fact of the theorem was so obvious from physical reasoning that it had been used for years without proof. The proof of Tellegen's theorem is given in Appendix II. With Tellegen's theorem, we can state that any network consisting of one or more positive resistances, inductances, and capacitances must be passive. To be active, a network must contain either a controlled source or a negative resistance.

PROBLEMS

5-1 Combine Figures 5-1(b) and 5-1(c) by drawing boxes around the elements of Figure 5-1(b) that correspond to the blocks of Figure 5-1(c). Note that one terminal of each terminal pair is grounded.

5-2 Find the input admittance for the network of Example 5-1 by using Equation (5-4). Check your answer with the result obtained in Example 5-1.

5-3 Find the admittance of the network given in Figure P5-3. The frequency ω is a free parameter.

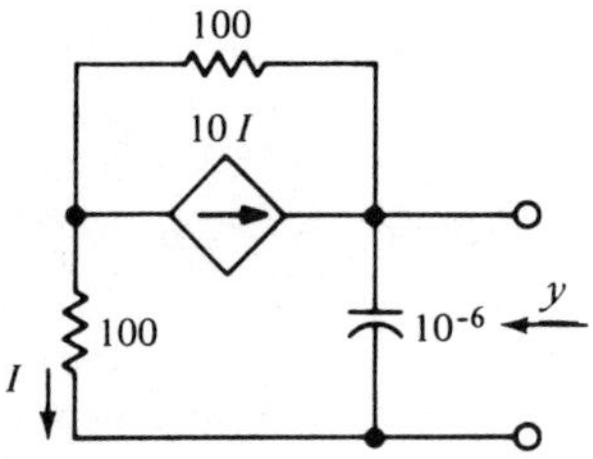

Figure P5-3

[2] B. D. H. Tellegen, "A General Network Theorem with Applications," *Philips Research Reports* 7, (1952) pp. 259–269. The theorem is more general than a statement of conservation of energy, but the generalizations are not needed for the material of this text.

5-4 Compute the y parameter matrix for the network shown in Figure P5-4. (Answers: 0, 0.023, 0, −0.02.)

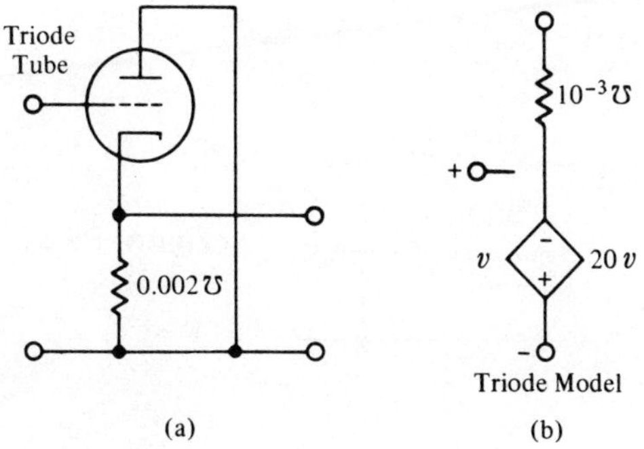

Figure P5-4

5-5 The y parameter matrix for the grounded base transistor is given by

$$\mathbf{Y} = \frac{1}{r_e(r_b + r_c) + (1 - \alpha)r_b r_c} \begin{bmatrix} r_b + r_c & -r_b \\ -(r_b + \alpha r_c) & r_e + r_b \end{bmatrix}$$

a. Verify any two of the four entries in this matrix.
b. A good-quality transistor has $\alpha = 0.99$, $r_b = 10^3$, $r_e = 10$, $r_c = 10^6$. Write the y parameter matrix for these values.

5-6 For the network shown in Figure P5-6, show that $|y_{21}|^2 = 1/(1 + \omega^4)$

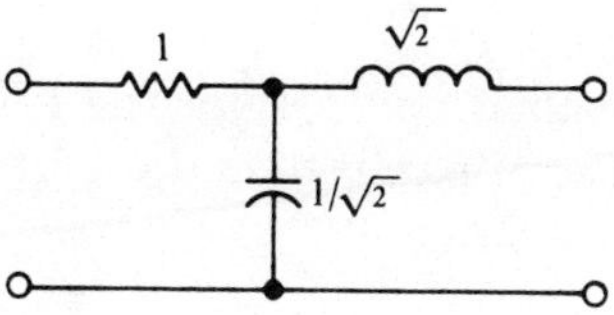

Figure P5-6

5-7 Compute the z parameter matrices for the networks shown in Figure P5-7.

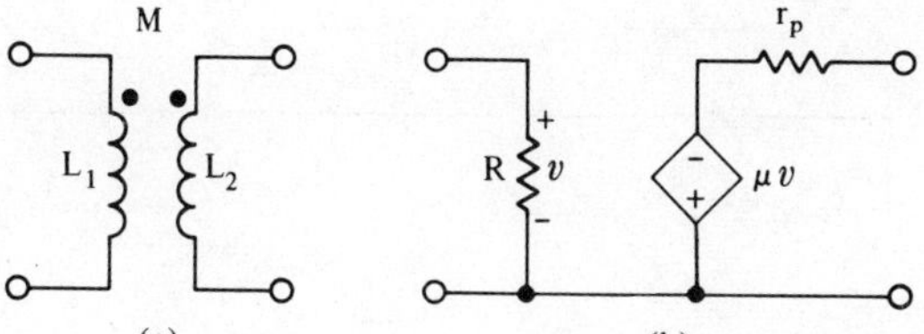

Figure P5-7

5-8 For the network shown in Figure P5-8

a. Compute $\mathfrak{z}_{21}$

b. Compute $\mathfrak{z}_{21}^*$ ($\mathfrak{z}_{21}$ conjugate)

c. Show that $|\mathfrak{z}_{21}|^2 = \mathfrak{z}_{21}\mathfrak{z}_{21}^* = 1/(1 + \omega^4)$

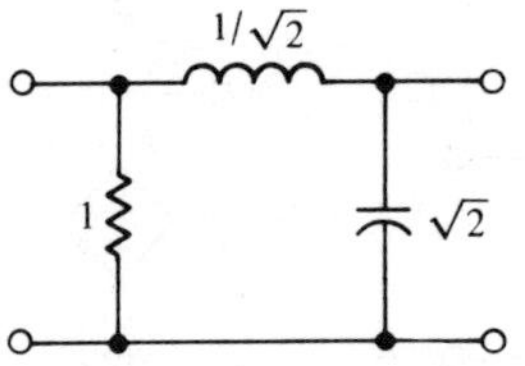

Figure P5-8

5-9 a. Show that the y parameter matrix for a grounded emitter transistor is given by

$$\begin{bmatrix} \dfrac{r_e + (1-\alpha)r_c}{r_er_c + r_br_e + (1-\alpha)r_br_c} & \dfrac{-r_e}{r_er_c + r_br_e + (1-\alpha)r_br_c} \\ \dfrac{-r_e + \alpha r_c}{r_er_c + r_br_e + (1-\alpha)r_br_c} & \dfrac{r_e + r_b}{r_er_c + r_br_e + (1-\alpha)r_br_c} \end{bmatrix}$$

b. A good-quality transistor has $\alpha \approx 0.99$, $r_b = 10^3$, $r_e \approx 10$, $r_c \approx 10^6$. Write the y parameter matrix for these values.

5-10 a. Show that the network shown in Figure P5-10 is equivalent to the network of Figure 5-5. Hint: Compute the y parameter matrix for both networks.

b. Show a low-frequency equivalent network for the FET discussed in Example 5-3; that is, assume the y parameters are given by Equation (5-23).

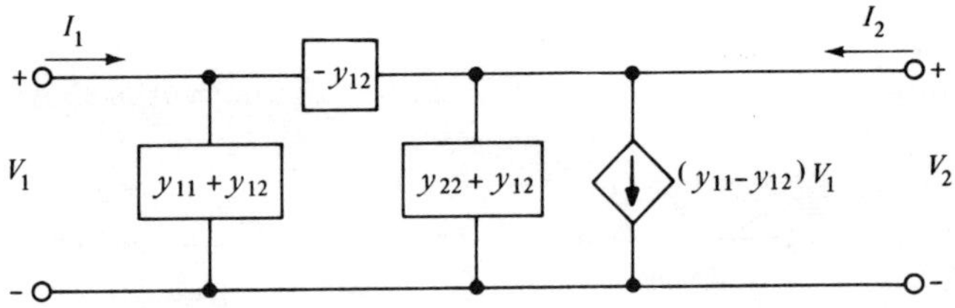

Figure P5-10

5-11 Compute the h parameter matrix for the grounded emitter transistor. Use the numbers given in Example 5-6 and compare your answer to the results obtained there for the grounded base transistor.

5-12 Compute the t parameters for the network shown in Figure P5-12.

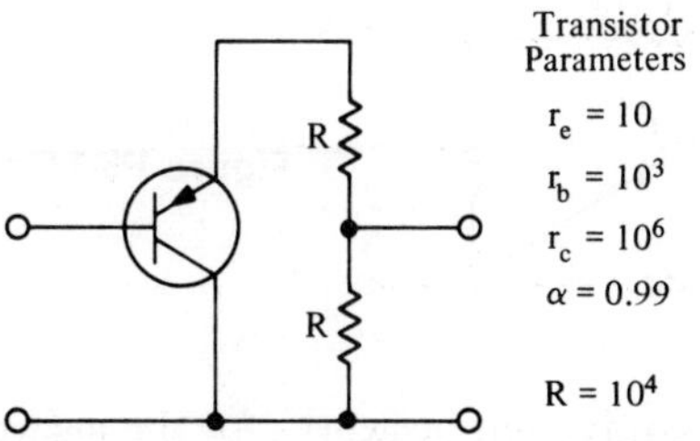

Figure P5-12

5-13 Is the network shown in Figure P5-13 equivalent to the network of Figure 5-9?

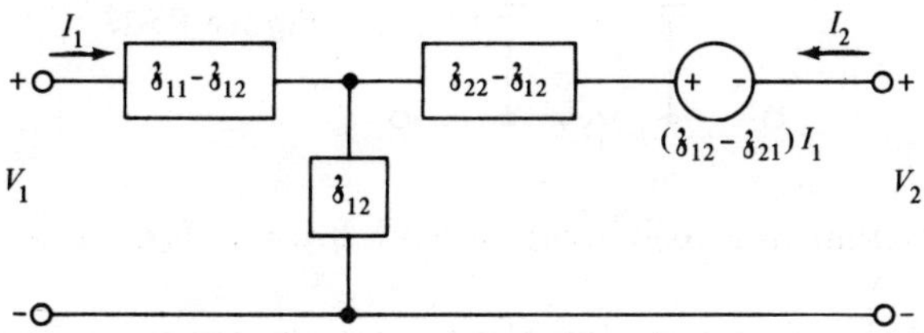

Figure P5-13

5-14 Compute the h parameters for the networks of Problem 5-7.

5-15 Compute the h parameter matrix for the grounded emitter transistor (see Figure P5-15).

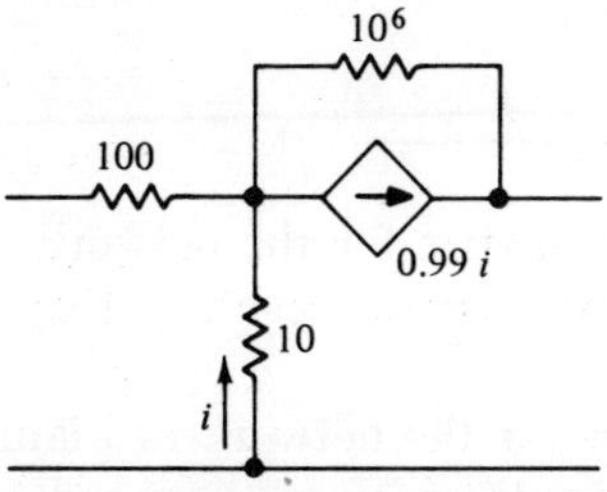

Figure P5-15

5-16 Compute the transfer parameter matrix for the network shown in Figure P5-16.

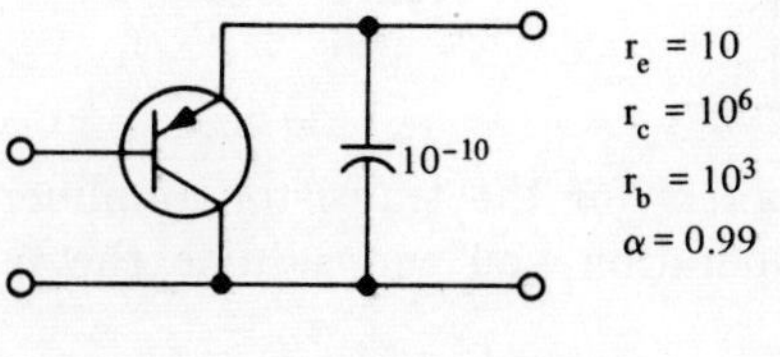

Figure P5-16

5-17 Find a T and π equivalent for the network shown in Figure P5-17.

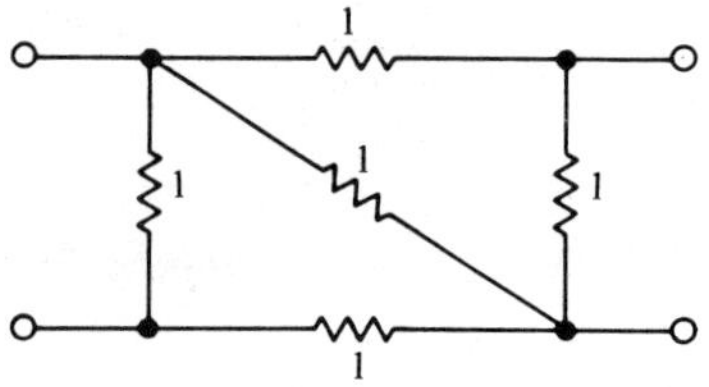

Figure P5-17

5-18 Compute the T and π equivalent networks for the network shown in Figure P5-18.

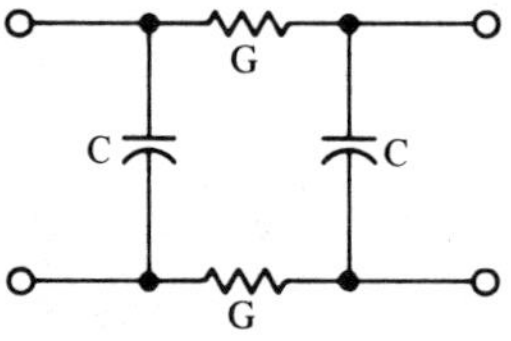

Figure P5-18

5-19 Show that in a reciprocal network $\mathfrak{h}_{12} = -\mathfrak{h}_{21}$.

5-20 Find the T and π equivalent networks for the network shown in Figure P5-20.

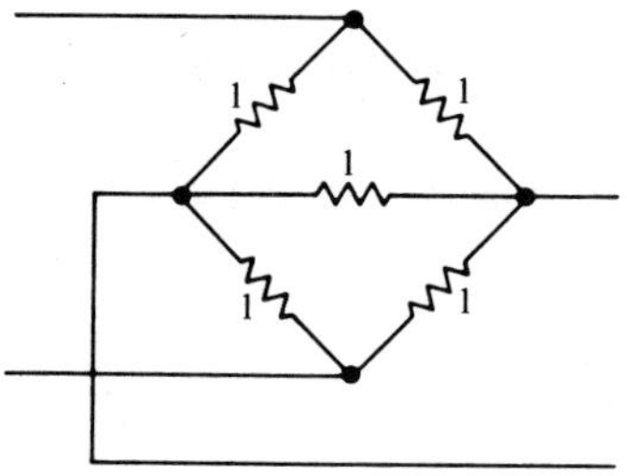

Figure P5-20

5-21 Compute the t parameter matrix for the network of Problem 5-12 by using the product rule. Hint: Use the result of Example 5-8.

5-22 **a.** Find the t parameters for the network in Figure P5-22.
b. Compute the t parameters for the network formed by connecting two such networks in cascade by the tandem rule.

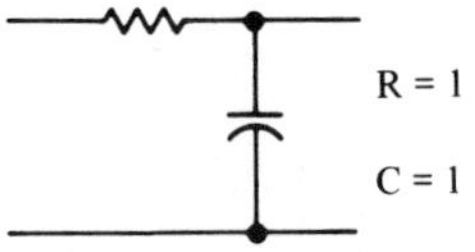

Figure P5-22

5-23 Compute the y parameters for the transistor amplifier shown in Figure P5-23. For signal operation you can assume the battery and capacitances are short circuits. Assume that the bias resistances R_{b_1} and R_{b_2} are chosen such that their parallel combination is 10 kΩ. The transistor has $r_e = 10$, $r_b = 100$, $r_c = 10^6$, $\alpha = 0.99$. Hint: Use the result of Example 5-5, Table 5-1, and Equation (5-75).

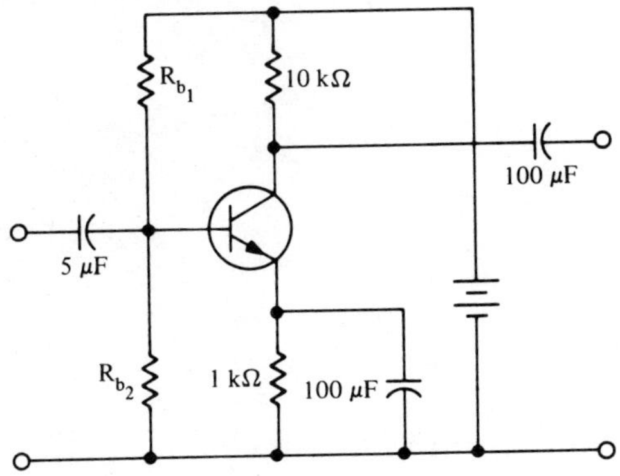

Figure P5-23

5-24 Two amplifiers identical to the one of Problem 5-12 are connected in cascade. Compute the t parameters for the 2-stage amplifier.

5-25 Compute y_{32} and y_{23} for the 3-port network shown in Figure P5-25.

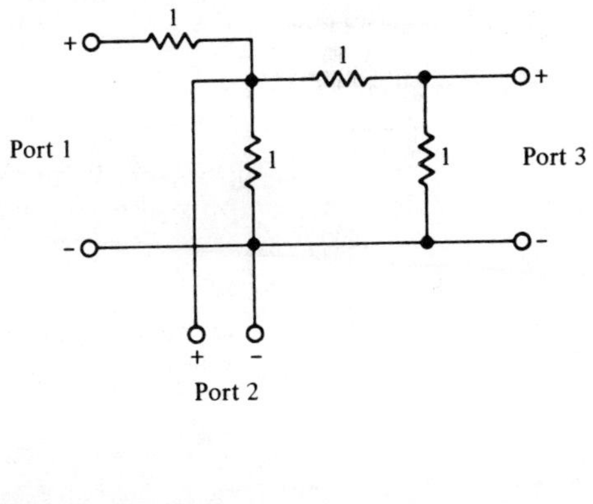

Figure P5-25

5-26 The symbol shown in Figure P5-26 is for a 1-ohm 3-port circulator. This device is used mainly in microwave circuits and has the following properties:

a. If a voltage source with a 1-ohm series resistance is connected to port 1, and a 1-ohm load is connected to port 2, then the voltage at port 3 is zero, independent of what load is connected to port 3.

b. If the voltage source with 1-ohm resistance is connected to port 2 and the 1-ohm load at port 3, then the voltage at port 1 is zero.

c. If the 1-ohm source is connected at port 3 and a 1-ohm load at port 1, then the port 2 voltage is zero.

Show that if such a device has the z parameter matrix given below, then properties a, b, and c are satisfied.

$$\mathbf{Z} = \begin{bmatrix} 0 & +1 & -1 \\ -1 & 0 & +1 \\ +1 & -1 & 0 \end{bmatrix}$$

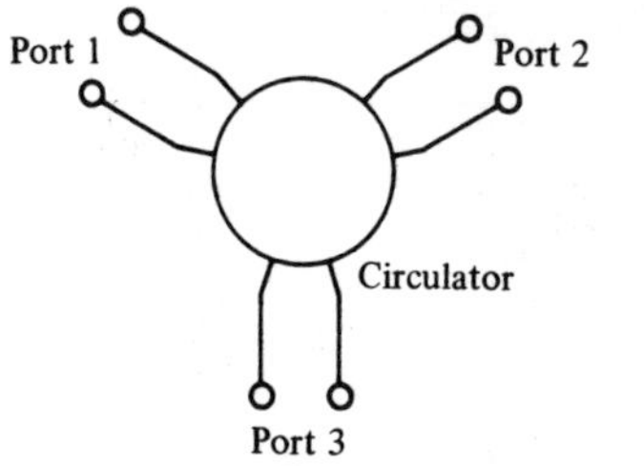

Figure P5-26

5-27 Show that $y_{23} = y_{32}$ (see Figure P5-27).

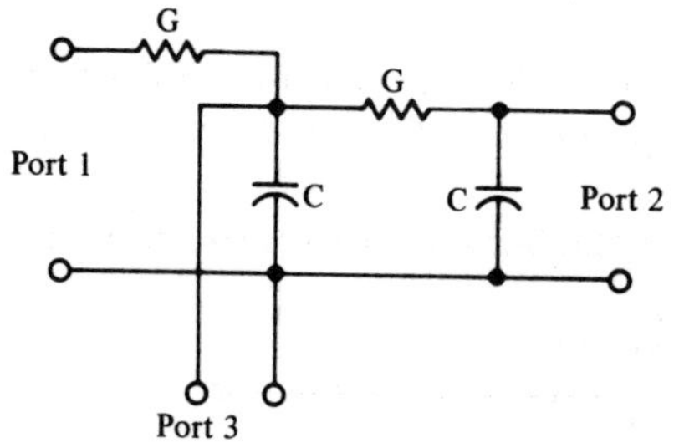

Figure P5-27

5-28 Consider the network of Problem 5-3. For one-watt input power at frequency one radian per second, compute the power delivered to each element.

5-29 Compute the power delivered to each element of the network shown in Figure P5-29 when one volt is applied at the input. Is the network active or passive?

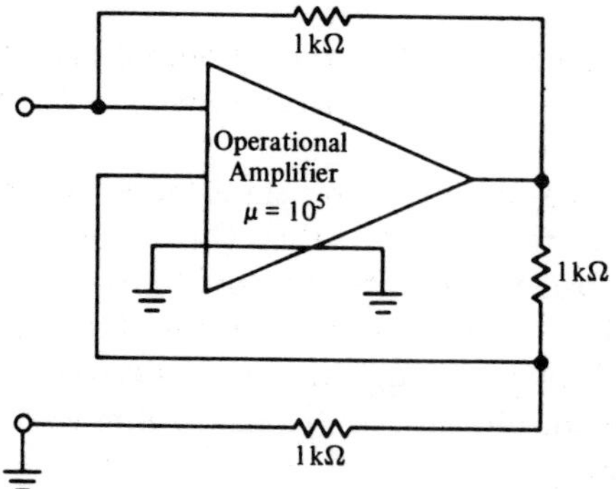

Figure P5-29

5-30 Consider the 2-port shown in Figure P5-30 (see also Problem 4-10).
a. For fixed frequency, show the relation between power flow and the relative phases of the sources (current sources) at the two ports.
b. Find a condition (frequency and phase of the sources at the ports) for which the network is passive.
c. Find a condition for which the network is active. Hint: Use voltage sources at the ports.

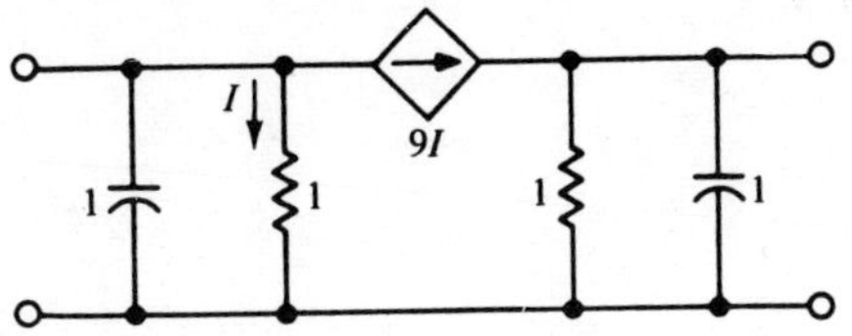

Figure P5-30

CHAPTER

6

Networks on a Terminal Basis

6-1 Introduction

As discussed in Section 5-1, it is not always possible to group the terminals of a network into ports. Furthermore, it is sometimes inconvenient to do so even when it is possible. For example, a transistor is basically a 3-terminal device. It is true that a transistor can be transformed into a 2-port device by making one of its terminals common to both ports, as illustrated in Figure 6-1, so that any of the 2-port characterizations discussed in Section 5-3 are applicable. However, a different parameter matrix is required for each configuration; for example, the grounded-base h parameter matrix is different from the grounded-emitter h parameter matrix. In many applications it is convenient to represent a transistor with a 3-terminal characterization that does not depend on the way the transistor is connected in the circuit.

In this chapter we introduce a network characterization that allows us to treat 3-terminal networks (such as transistor models) as 3-terminal networks without pairing their terminals into ports. This characterization, called the indefinite admittance matrix (IAM), has a num-

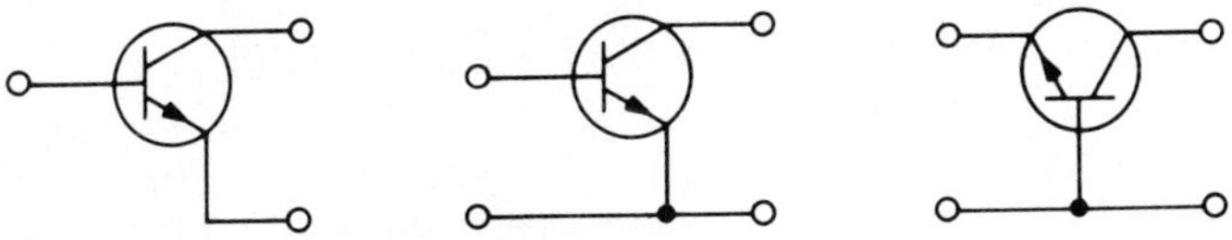

Figure 6-1 (a) A 3-terminal transistor. (b) In a 2-port grounded-emitter configuration. (c) In a 2-port grounded-base configuration.

ber of extremely convenient properties that make it a powerful tool in network analysis. It has the disadvantage of requiring more parameters to characterize a network than are required by the port characterization discussed in Chapter 5. For the same reasons that are listed in Section 5-1, we limit our discussion to networks containing no independent sources. Furthermore, we limit our discussion to the sinusoidal steady state and use complex voltages and complex currents to characterize these networks in terms of their response to sinusoidal sources of frequency ω rad/sec.

The network analysis technique developed in this chapter is based on the nodal analysis (node-to-datum equations) presented in Section 3-6. This technique is well suited for execution on a digital computer.

In Section 6-2 we define the IAM, and in Section 6-3 we list its important properties. In Sections 6-4 and 6-5 we show how to use the IAM as a tool for analyzing electronic networks. In Section 6-6 we show how a computer can be used to analyze networks.

6-2 The Indefinite Admittance Matrix (IAM)

In Section 5-5 we characterized an n-port linear network containing no independent sources by defining a set of n port voltages and n port currents, and by writing a set of equations in these 2n variables. For example, the y parameters were defined by writing each of the n port currents as linear combinations of the n port voltages. In contrast, we characterize an m-terminal linear network containing no independent sources by first defining a set of m terminal voltages and m terminal currents, and writing a set of m equations in these 2m variables. As pointed out in Chapter 1, currents flowing into (or out of) individual terminals are well defined. On the other hand, voltages are defined between terminal pairs. In the nodal analysis method of Section 3-6, we assigned a voltage to each node relative to one node in the network that was arbitrarily (conveniently) chosen as a reference node. In the indefinite admittance method, we establish an additional reference node external to the network and define voltages at various nodes (terminals) relative to this reference node. Using this external reference allows a freedom not readily available if a specific network node is chosen as a reference node.

An m-terminal network is presented in Figure 6-2(a). As illustrated in Figure 6-2(b), each terminal voltage is defined relative to a reference node external to the network; that is, V_i is the voltage at terminal i relative to the reference node. The terminal currents are the currents flowing from the reference node into the terminals; that is, I_i is the current

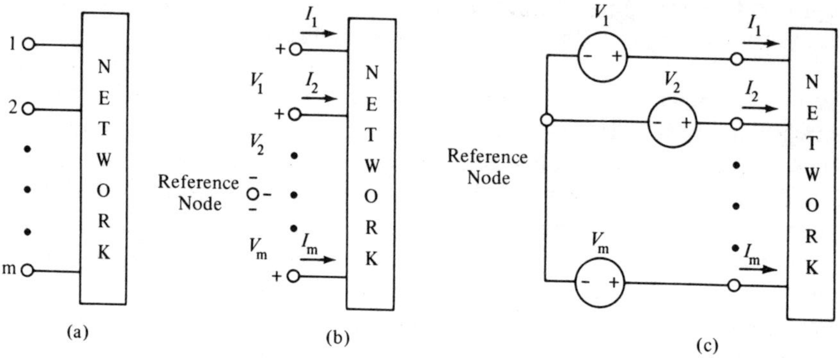

Figure 6-2 IAM definitions.

into terminal i from the reference node. Now, choosing the m terminal voltages to be the independent variables and the m terminal currents the dependent variables, we write each terminal current as a linear combination of the terminal voltages; that is,

$$\begin{aligned} I_1 &= y_{11}V_1 + y_{12}V_2 + \cdots + y_{1m}V_m \\ I_2 &= y_{21}V_1 + y_{22}V_2 + \cdots + y_{2m}V_m \\ &\;\;\vdots \\ I_m &= y_{m1}V_1 + y_{m2}V_2 + \cdots + y_{mm}V_m \end{aligned} \tag{6-1}$$

where the m^2 coefficients y_{ik} (i, k = 1, 2, $\cdots$, m) are defined by

$$y_{ik} = \left.\frac{I_i}{V_k}\right|_{V_\ell=0,\ell\neq k} = \left.I_i\right|_{V_k=1;V_\ell=0,\ell\neq k} \tag{6-2}$$

Equations (6-1) can also be written in matrix form:

$$\begin{bmatrix} I_1 \\ I_2 \\ \cdot \\ \cdot \\ \cdot \\ I_m \end{bmatrix} = \begin{bmatrix} y_{11} & y_{12} & \cdots & y_{1m} \\ y_{21} & y_{22} & \cdots & y_{2m} \\ \cdot & \cdot & \cdot & \cdot \\ \cdot & \cdot & \cdot & \cdot \\ \cdot & \cdot & \cdot & \cdot \\ y_{m1} & y_{m2} & \cdots & y_{mm} \end{bmatrix} \begin{bmatrix} V_1 \\ V_2 \\ \cdot \\ \cdot \\ \cdot \\ V_m \end{bmatrix} \tag{6-3}$$

The matrix of y_{ik}'s is called the indefinite admittance matrix (IAM).

This procedure can also be viewed in a slightly different manner. Clearly, the terminal voltages and terminal currents of an m-terminal network containing no independent sources are zero unless the network is excited by one or more external sources. Now, fixing any m of the 2m terminal voltages and terminal currents uniquely determines the remaining m quantities. Let us fix the m terminal voltages by connecting m independent voltage sources between the m network terminals and the reference node, as illustrated in Figure 6-2(c). Then, each terminal current is caused in part by each of the m independent sources. But, according to the superposition theorem (see Section 2-3), the currents due to the individual independent voltages sources are additive. Define $y_{ik}V_k$ to be the contribution to current I_i due to source V_k. Then the total ith terminal current due to all m independent voltage sources is

$$I_i = \sum_{k=1}^{m} y_{ik}V_k$$

An equation of this form can be written for each terminal current, that is, for i = 1, 2, · · · , m. The result is a set of equations identical to Equations (6-1).

Each y_{ik} is, in general, a complex number that depends only on the characteristics of the network and the frequency variable $j\omega$. That is, since $y_{ik}V_k = y_{ik}(j\omega)V_k$ is the contribution to current I_i due to the independent voltage source, V_k; y_{ik} simply represents the magnitude and phase relationship between V_k and that part of I_i produced by V_k, as discussed in Section 4-2. The set of m^2 complex numbers y_{ik} (i, k = 1, 2, · · · , m) uniquely characterizes the m-terminal network; that is, every m-terminal network has a unique IAM.

Finally, we note that Equations (6-1) are valid for any set of m terminal voltages and m terminal currents, and that since m of these 2m quantities are independent variables, m external constraints can be added. Indeed, these m equations, along with m constraint equations, would allow us to solve for the 2m terminal voltages and terminal currents.

Example 6-1

To give an example, we compute the IAM for the 3-terminal network presented in Figure 6-3(a). The IAM consists of the nine coefficients in the three equations.

$$\begin{aligned} I_1 &= y_{11}V_1 + y_{12}V_2 + y_{13}V_3 \\ I_2 &= y_{21}V_1 + y_{22}V_2 + y_{23}V_3 \\ I_3 &= y_{31}V_1 + y_{32}V_2 + y_{33}V_3 \end{aligned} \tag{6-4}$$

The independent variables V_1, V_2, and V_3 and the dependent variables I_1, I_2, and I_3 are defined in Figure 6-3(b).

A convenient procedure for determining the nine coefficients is to compute them by columns, since, if we set $V_1 = 1$, and $V_2 = V_3 = 0$, then $y_{11} = I_1$, $y_{21} = I_2$, and $y_{31} = I_3$. Hence, we refer to Figure 6-4(a)

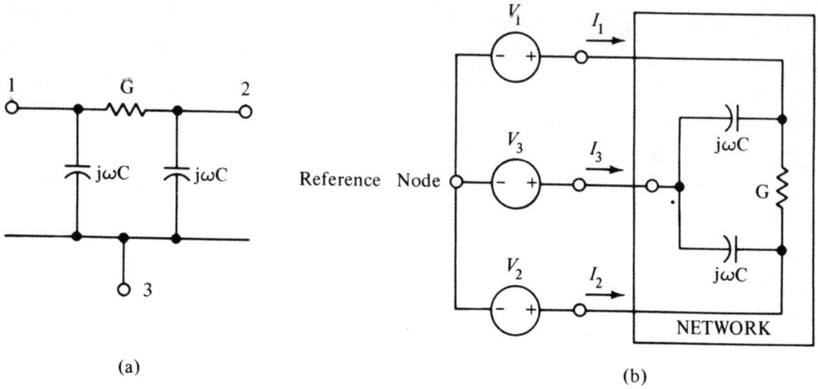

Figure 6-3 Example 6-1.

and compute the three terminal currents

$$
\begin{aligned}
y_{11} &= I_1 \Big|_{V_1=1, V_2=V_3=0} = G + j\omega C \\
y_{21} &= I_2 \Big|_{V_1=1, V_2=V_3=0} = -G \\
y_{31} &= I_3 \Big|_{V_1=1, V_2=V_3=0} = -j\omega C
\end{aligned}
\tag{6-5}
$$

To compute the second column of the IAM, we set $V_2 = 1$ and $V_1 = V_3 = 0$ and again compute the three terminal currents. Referring to Figure 6-4(b) and utilizing the symmetry, we can make use of the first column results and write

$$
\begin{aligned}
y_{12} &= I_1 \Big|_{V_2=1, V_1=V_3=0} = -G \\
y_{22} &= I_2 \Big|_{V_2=1, V_1=V_3=0} = G + j\omega C \\
y_{32} &= I_3 \Big|_{V_2=1, V_1=V_3=0} = -j\omega C
\end{aligned}
\tag{6-6}
$$

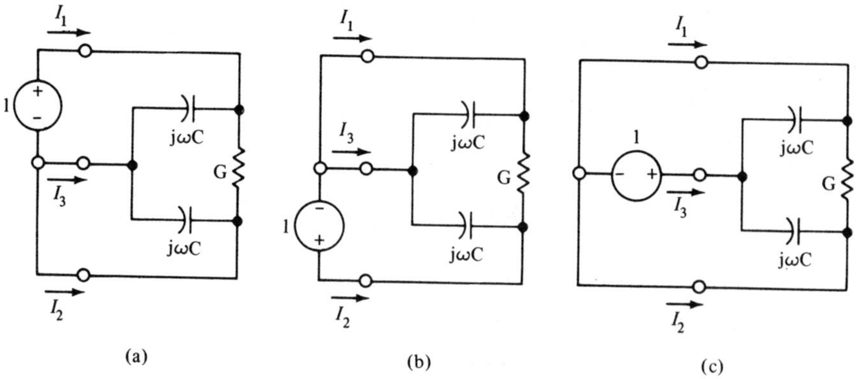

Figure 6-4 Networks for computing each column.

Finally, we compute the third column of IAM parameters by setting $V_3 = 1$ and $V_1 = V_2 = 0$, as illustrated in Figure 6-4(c), and again compute the three terminal currents

$$\begin{aligned} \mathbf{y}_{13} &= I_1 \Big|_{V_3=1, V_1=V_2=0} = -j\omega C \\ \mathbf{y}_{23} &= I_2 \Big|_{V_3=1, V_1=V_2=0} = -j\omega C \\ \mathbf{y}_{33} &= I_3 \Big|_{V_3=1, V_1=V_2=0} = j2\omega C \end{aligned} \tag{6-7}$$

Hence, the IAM for the network of Figure 6-3(a) is given by

$$\mathbf{I}_{1,2,3} = \begin{bmatrix} G + j\omega C & -G & -j\omega C \\ -G & G + j\omega C & -j\omega C \\ -j\omega C & -j\omega C & j2\omega C \end{bmatrix} \tag{6-8}$$

The subscripts on **I** indicate that the first row and column correspond to terminal 1, the second row and column correspond to terminal 2, and the third row and column correspond to terminal 3, respectively.

Example 6-2

For a second example, we compute the IAM for the transistor and the transistor model presented in Figure 6-5(a) and (b). $\mathfrak{z}$ represents a com-

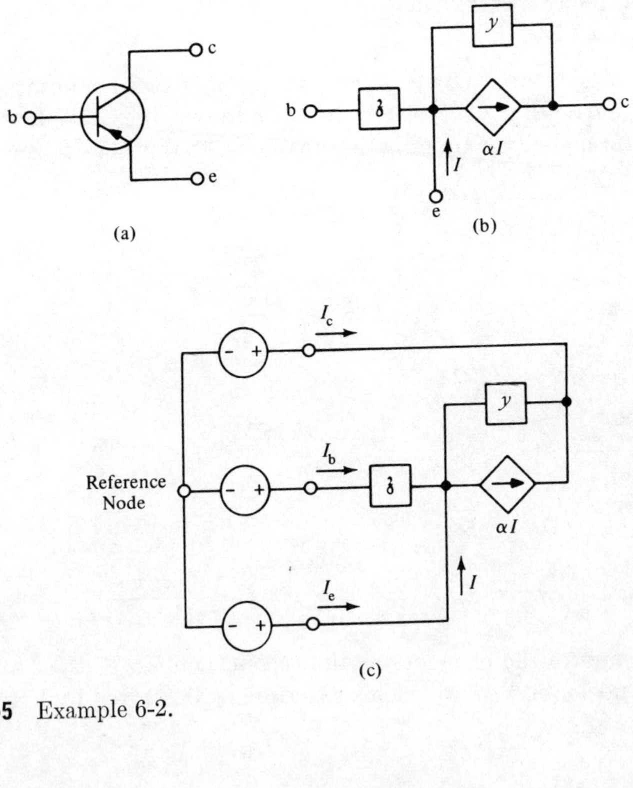

Figure 6-5 Example 6-2.

plex base impedance, and y a complex collector admittance. The emitter impedance is approximated by a short circuit. (See, however, Problem 6-4.) The IAM for this 3-terminal network consists of the $3^2 = 9$ coefficients in the equations

$$\begin{aligned} I_b &= y_{bb}V_b + y_{bc}V_c + y_{be}V_e \\ I_c &= y_{cb}V_b + y_{cc}V_c + y_{ce}V_e \\ I_e &= y_{eb}V_b + y_{ec}V_c + y_{ee}V_e \end{aligned} \tag{6-9}$$

These coefficients are defined by Equations (6-2) and are easily computed with the aid of Figure 6-5(c). To compute the first column of the IAM, we set $V_b = 1$ and $V_c = V_e = 0$ in Figure 6-5(c). This results in the situation shown in Figure 6-6, where we have labeled each of three terminal voltages relative to the reference node voltage. Since all of the

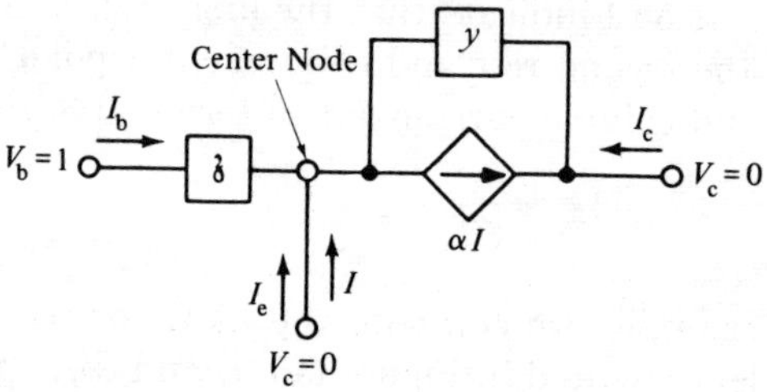

Figure 6-6 Network for column one.

branch voltages are known, it is a simple matter to compute the three terminal currents if we first determine I by writing a KCL equation at the "center node." Summing the currents into this node to zero, we write

$$\frac{1}{\mathfrak{z}} + I - \alpha I = 0 \tag{6-10}$$

or

$$I = \frac{-1}{\mathfrak{z}(1 - \alpha)} \tag{6-11}$$

We now have

$$\begin{aligned} y_{bb} &= I_b \Big|_{V_b=1, V_c=V_e=0} = \frac{1}{\mathfrak{z}} \\ y_{cb} &= I_c \Big|_{V_b=1, V_c=V_e=0} = -\alpha I = \frac{\alpha}{\mathfrak{z}(1-\alpha)} \\ y_{eb} &= I_e \Big|_{V_b=1, V_c=V_e=0} = I = \frac{-1}{\mathfrak{z}(1-\alpha)} \end{aligned} \tag{6-12}$$

To compute the elements in the second column of the IAM, we set $V_c = 1$ and $V_b = V_e = 0$. This situation is illustrated by Figure 6-7(a).

Again summing the three terminal currents to zero, we obtain

$$I + y - \alpha I = 0 \tag{6-13}$$

or

$$I = \frac{-y}{1 - \alpha} \tag{6-14}$$

Hence,

$$\begin{aligned} \mathsf{y}_{bc} &= I_b \Big|_{V_c=1, V_b=V_e=0} = 0 \\ \mathsf{y}_{cc} &= I_c \Big|_{V_c=1, V_b=V_e=0} = y - \alpha I = \frac{y}{1-\alpha} \\ \mathsf{y}_{ec} &= I_e \Big|_{V_c=1, V_b=V_e=0} = I = \frac{-y}{1-\alpha} \end{aligned} \tag{6-15}$$

To compute the third column of the IAM, we set $V_e = 1$ and $V_b = V_c = 0$, sketch the situation as in Figure 6-7(b), and again find I

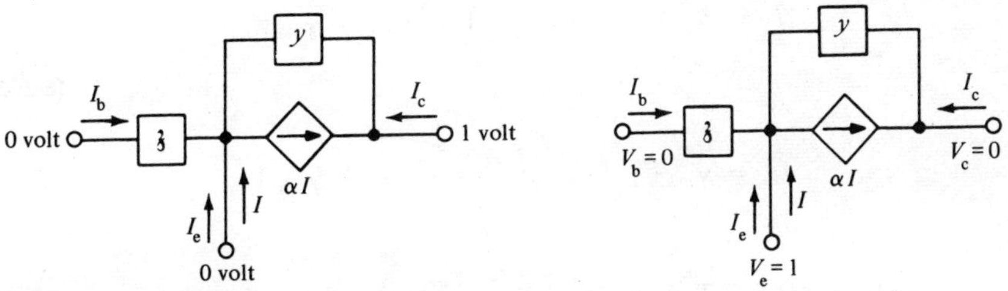

Figure 6-7 Networks for columns two and three.

by summing the terminal currents to zero; that is,

$$-\frac{1}{\mathfrak{z}} + I - \alpha I - y = 0 \tag{6-16}$$

or

$$I = \frac{1 + \mathfrak{z}y}{\mathfrak{z}(1 - \alpha)} \tag{6-17}$$

Therefore,

$$\begin{aligned} \mathsf{y}_{be} &= I_b \Big|_{V_e=1, V_b=V_c=0} = -\frac{1}{\mathfrak{z}} \\ \mathsf{y}_{ce} &= I_c \Big|_{V_e=1, V_b=V_c=0} = -y - \alpha I = \frac{-\alpha - y\mathfrak{z}}{\mathfrak{z}(1-\alpha)} \\ \mathsf{y}_{ee} &= I_e \Big|_{V_e=1, V_b=V_c=0} = I = \frac{1 + y\mathfrak{z}}{\mathfrak{z}(1-\alpha)} \end{aligned} \tag{6-18}$$

We conclude that the IAM for the 3-terminal transistor model of Figure 6-5(b) is given by

$$\mathbf{I}_{b,c,e} = \begin{bmatrix} \dfrac{1}{\mathfrak{z}} & 0 & -\dfrac{1}{\mathfrak{z}} \\ \dfrac{\alpha}{\mathfrak{z}(1-\alpha)} & \dfrac{y}{1-\alpha} & \dfrac{-\alpha - y\mathfrak{z}}{\mathfrak{z}(1-\alpha)} \\ \dfrac{-1}{\mathfrak{z}(1-\alpha)} & \dfrac{-y}{1-\alpha} & \dfrac{1+y\mathfrak{z}}{\mathfrak{z}(1-\alpha)} \end{bmatrix} \tag{6-19}$$

6-3 Properties of the Indefinite Admittance Matrix

The IAM has a number of interesting and easily provable properties that account for its great utility. Since we shall refer to Equations (6-1) a number of times, we rewrite them here for the reader's convenience.

$$\begin{aligned} I_1 &= \mathsf{y}_{11}V_1 + \mathsf{y}_{12}V_2 + \cdots + \mathsf{y}_{1m}V_m \\ I_2 &= \mathsf{y}_{21}V_1 + \mathsf{y}_{22}V_2 + \cdots + \mathsf{y}_{2m}V_m \\ \cdot &= \cdot \qquad \cdot \qquad \cdot \qquad \cdot \\ &\vdots \\ I_m &= \mathsf{y}_{m1}V_1 + \mathsf{y}_{m2}V_2 + \cdots + \mathsf{y}_{mm}V_m \end{aligned} \tag{6-20}$$

PROPERTY 1

The sum of the elements in each column of the IAM is zero.

Proof: To show that the kth column sums to zero, we set $V_k = 1$ and $V_\ell = 0$, $\ell \neq k$. Then, summing Equations (6-20), we have

$$\sum_{i=1}^{m} I_i = \sum_{i=1}^{m} \mathsf{y}_{ik}$$

But

$$\sum_{i=1}^{m} I_i$$

is the sum of the currents out of the reference node (see Figure 6-2(c)), which, by KCL, must equal zero.

Comment: A quick check of Equations (6-8) and (6-19) shows property 1 for the IAM's computed in Examples 1 and 2. Note that the use of property 1 in Examples 1 and 2 would have saved $\frac{1}{3}$ of the computations performed there. One element of each column could have been obtained from the other two by applying property 1. A more conservative approach is to compute all the elements of the IAM via Equation (6-2) and to use property 1 as a check for errors.

PROPERTY 2

The sum of the elements of each row of the IAM is zero.

Proof: To show that the ith-row elements of the IAM sum to zero, we set all m terminal voltages equal to one volt; that is, $V_k = 1$ (k = 1, 2, · · · , m). Then the ith equation of Equations (6-20) becomes

$$I_i = \sum_{k=1}^{m} \mathbf{y}_{ik}$$

But all m terminal currents (including I_i) are zero, because the potential difference between all terminal pairs is zero.

Comment: Checking Equations (6-8) and (6-19), we verify property 2 for the IAM's computed in Examples 1 and 2. Note that by making use of both properties 1 and 2, only $m^2 - (2m - 1)$ of the m^2 IAM parameters need be computed from Equation (6-2). In Example 1 as well as Example 2, five of the nine parameters could have been obtained via properties 1 and 2.

Example 6-3

In this example we compute the IAM for the 2-terminal admittance presented in Figure 6-8(a). The IAM for this network is given by

$$\mathbf{I}_{13} = \begin{bmatrix} \mathbf{y}_{11} & \mathbf{y}_{13} \\ \mathbf{y}_{31} & \mathbf{y}_{33} \end{bmatrix} \tag{6-21}$$

where the elements $\mathbf{y}_{ij}$ (i, j = 1, 3) are defined by Equation (6-2). Setting $V_1 = 1$ and $V_3 = 0$, we refer to Figure 6-8(b) and write

$$\mathbf{y}_{11} = I_1 \Big|_{V_1=1, V_3=0} = y \tag{6-22}$$

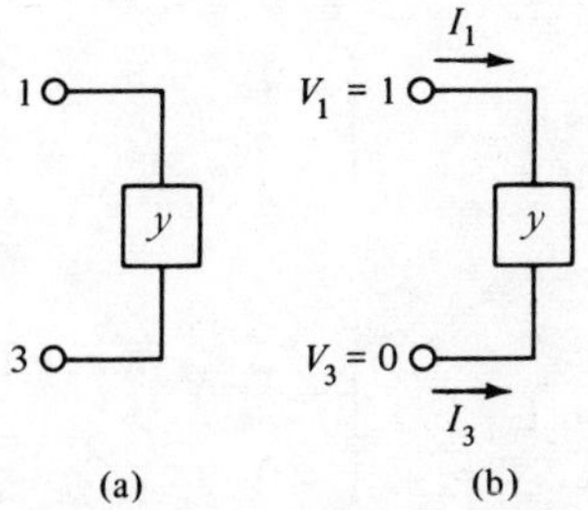

Figure 6-8 Example 6-3.

Entering this element into the IAM, we have

$$\mathbf{I}_{1,3} = \begin{bmatrix} y & (\ ) \\ (\ ) & (\ ) \end{bmatrix} \tag{6-23}$$

Now using property 2, we obtain the upper right-hand element and enter it into the IAM to obtain

$$\mathbf{I}_{1,3} = \begin{bmatrix} y & -y \\ (\ ) & (\ ) \end{bmatrix} \tag{6-24}$$

Finally, property 1 allows us to fill in the second row; that is,

$$\mathbf{I}_{1,3} = \begin{bmatrix} y & -y \\ -y & y \end{bmatrix} \tag{6-25}$$

PROPERTY 3

Reordering network terminals is equivalent to reordering rows and columns in the IAM; for example, if

$$\mathbf{I}_{a,b,c} = \begin{bmatrix} y_{aa} & y_{ab} & y_{ac} \\ y_{ba} & y_{bb} & y_{bc} \\ y_{ca} & y_{cb} & y_{cc} \end{bmatrix} \tag{6-26}$$

then

$$\mathbf{I}_{b,a,c} = \begin{bmatrix} y_{bb} & y_{ba} & y_{bc} \\ y_{ab} & y_{aa} & y_{ac} \\ y_{cb} & y_{ca} & y_{cc} \end{bmatrix} \tag{6-27}$$

Proof: Reordering network terminals simply corresponds to reordering Equations (6-20).

Example 6-4

If

$$\begin{aligned} I_a &= 1V_a + 2V_b - 3V_c \\ I_b &= 4V_a + 5V_b - 9V_c \\ I_c &= -5V_a - 7V_b + 12V_c \end{aligned} \qquad \mathbf{I}_{a,b,c} = \begin{bmatrix} 1 & 2 & -3 \\ 4 & 5 & -9 \\ -5 & -7 & 12 \end{bmatrix} \tag{6-28}$$

then

$$\begin{aligned} I_a &= 1V_a - 3V_c + 2V_b \\ I_c &= -5V_a + 12V_c - 7V_b \\ I_b &= 4V_a - 9V_c + 5V_b \end{aligned} \qquad \mathbf{I}_{a,c,b} = \begin{bmatrix} 1 & -3 & 2 \\ -5 & 12 & -7 \\ 4 & -9 & 5 \end{bmatrix} \tag{6-29}$$

and

$$\begin{aligned} I_c &= 12V_c - 5V_a - 7V_b \\ I_a &= -3V_c + 1V_a + 2V_b \\ I_b &= -9V_c + 4V_a + 5V_b \end{aligned} \qquad \mathbf{I}_{c,a,b} = \begin{bmatrix} 12 & -5 & -7 \\ -3 & 1 & 2 \\ -9 & 4 & 5 \end{bmatrix} \tag{6-30}$$

Note that $\mathbf{I}_{a,c,b}$ can be obtained from $\mathbf{I}_{a,b,c}$ by interchanging rows c and b and interchanging columns c and b.

PROPERTY 4

If r isolated terminals are added to an m-terminal network to form an (m + r)-terminal network, the IAM for the (m + r)-terminal network can be formed from the IAM for the m-terminal network by adding r rows and columns of zeros to the m-terminal network IAM.

Proof: Consider the two-terminal resistance presented in Figure 6-9 and its IAM given by (see Example 6-3)

$$\mathbf{I}_{1,3} = \begin{bmatrix} G & -G \\ -G & G \end{bmatrix} \tag{6-31}$$

We now compute the IAM for the 4-terminal network of Figure 6-9(b) by using Equations (6-2). To compute the second column of the

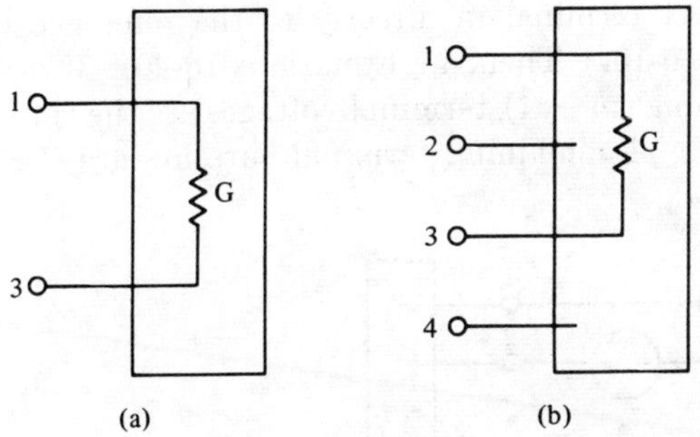

Figure 6-9 (a) A two-terminal network consisting of a resistance. (b) A four-terminal network consisting of a resistance and two isolated terminals.

IAM, we connect a 1-volt independent voltage source to terminal 2 and compute the four terminal currents. Obviously, they are all zero. A similar argument holds for the fourth column, which corresponds to terminal 4. Now consider the second row of the IAM. These elements, from left to right, are the four terminal 2 currents as the 1-volt independent voltage source is connected to terminals 1, 2, 3, and 4, respectively. Obviously, the second-row elements of the IAM are zero. A similar argument holds for the fourth row, which corresponds to terminal 4. The y_{ij}'s for i, j = 1, 3 are identical to those in Equation (6-31), so that we have

$$\mathbf{I}_{1,2,3,4} = \begin{bmatrix} G & 0 & -G & 0 \\ 0 & 0 & 0 & 0 \\ -G & 0 & G & 0 \\ 0 & 0 & 0 & 0 \end{bmatrix} \tag{6-32}$$

Although we have treated only a specific example, the extension to the general case is obvious.

Comment: If r isolated terminals are added to an m-terminal network, the (m + r) × (m + r) IAM for the (m + r)-terminal network is constructed by (1) inserting r columns and r rows of zeros corresponding to the isolated terminals, (2) entering each of the remaining elements (which are identical to the elements of the m × m IAM) in the row and column corresponding to its terminal number.

PROPERTY 5

If an m-terminal network is made into an (m − 1)-port network by making terminal m common to all ports, then the short-circuit admittance parameter matrix for the (m − 1)-port network can be obtained from the m-terminal IAM by deleting the mth row and mth column.

Proof: Connect terminal m directly to the reference node, as illustrated in Figure 6-10. Then, in Equations (6-20), $V_m = 0$. Furthermore, the remaining (m − 1) terminal voltages are the (m − 1) port voltages, and the m − 1 remaining terminal currents are the (m − 1) port

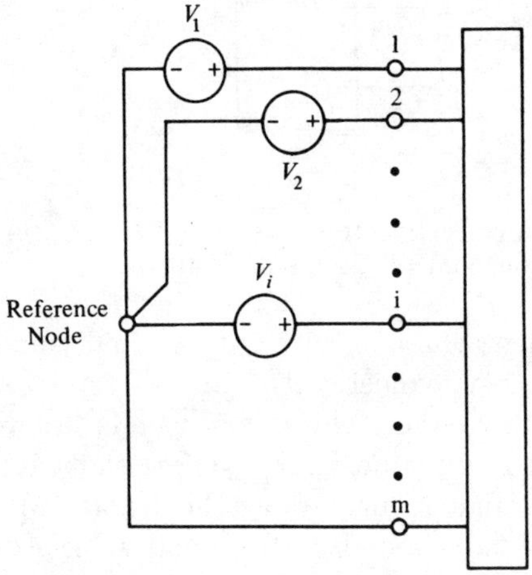

Figure 6-10 Conversion of m-terminal to (m − 1) port network.

currents. Hence, by deleting the mth equation of Equations (6-20) and by setting $V_m = 0$ in the remaining m − 1 equations, we obtain a set of equations identical to Equations (5-90), with n = m − 1.

Comment: Let us illustrate this property by considering the three terminal transistor presented in Figure 6-11(a). The IAM, $\mathbf{I}_{e,c,b}$, for this

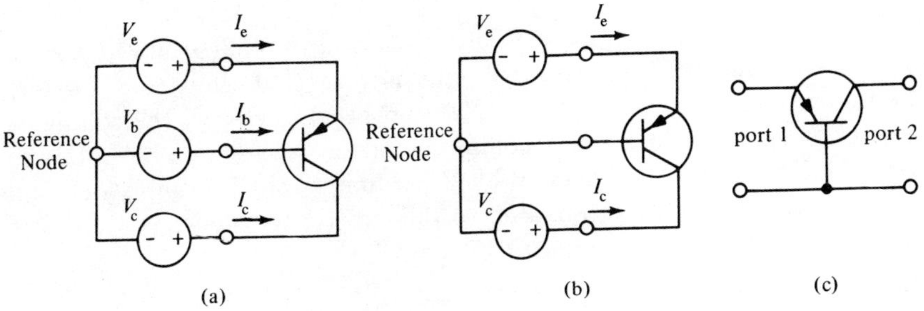

Figure 6-11 The 3-terminal transistor connected in the common-base configuration.

network consists of the coefficients in the three equations

$$\begin{aligned} I_e &= y_{ee}V_e + y_{ec}V_c + y_{eb}V_b \\ I_c &= y_{ce}V_e + y_{cc}V_c + y_{cb}V_b \\ I_b &= y_{be}V_e + y_{bc}V_c + y_{bb}V_b \end{aligned} \qquad \textbf{(6-33)}$$

Now, consider the network of Figure 6-11(b), where we have connected the base terminal directly to the reference node. The terminal equations for this network are given by Equations (6-33), with $V_b = 0$, that is,

$$\begin{aligned} I_e &= y_{ee}V_e + y_{ec}V_c \\ I_c &= y_{ce}V_e + y_{cc}V_c \\ I_b &= y_{be}V_e + y_{bc}V_c \end{aligned} \qquad \textbf{(6-34)}$$

Next, refer to Figure 6-11(b) and note that if we consider this network to be a 2-port network, as illustrated in Figure 6-11(c), V_e and I_e are the port 1 voltage and current, and V_c and I_c are the port 2 voltage and current. Hence, the first two equations of Equations (6-34) are also the port equations for the common-base 2-port, and the four coefficients of these equations are the short-circuit admittance parameters for the common-base 2-port. Therefore, by deleting the row and column corresponding to the base terminal from the 3-terminal IAM defined by Equations (6-33), we obtain the common-base 2-port y parameter matrix defined by the first two equations of Equations (6-34).

To find the common-emitter 2-port y parameter matrix from the 3-terminal IAM defined by Equations (6-33), we first connect the emitter terminal directly to the reference node, as illustrated in Figure 6-12(a). The equations for this arrangement are given by Equations (6-33), with $V_e = 0$, that is,

$$\begin{aligned} I_e &= y_{ec}V_c + y_{eb}V_b \\ I_c &= y_{cc}V_c + y_{cb}V_b \\ I_b &= y_{bc}V_c + y_{bb}V_b \end{aligned} \qquad \textbf{(6-35)}$$

But note that the networks of Figure 6-12(a) and 6-12(b) are identical, and note further that the last two equations of Equations (6-35) define the y parameters for the 2-port of Figure 6-11(b). Hence, the common-emitter y parameters for the common-emitter 2-port can be obtained from the 3-terminal IAM defined by Equations (6-33) by simply deleting the row and column corresponding to the emitter terminal.

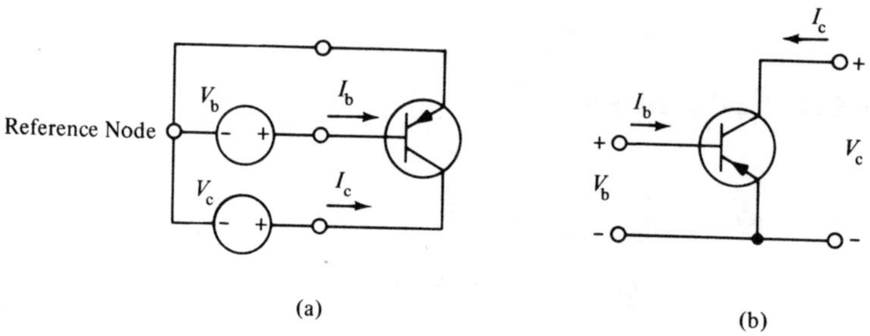

Figure 6-12 The 3-terminal transistor connected in the common-emitter configuration.

By following a similar procedure, one can show that the common-collector 2-port y parameter matrix consists of the four elements in the 3-terminal IAM that remain after the row and column corresponding to the collector terminal are deleted.

In summary, if the 3-terminal IAM is given by

$$\mathbf{I}_{e,c,b} = \begin{bmatrix} y_{ee} & y_{ec} & y_{eb} \\ y_{ce} & y_{cc} & y_{cb} \\ y_{be} & y_{bc} & y_{bb} \end{bmatrix} \tag{6-36}$$

then the three 2 × 2 y parameter matrices are given by

$$\begin{aligned} \mathbf{Y}_{\text{common emitter}} &= \begin{bmatrix} \cancel{y_{ee}} & \cancel{y_{ec}} & \cancel{y_{eb}} \\ \cancel{y_{ce}} & y_{cc} & y_{cb} \\ \cancel{y_{be}} & y_{bc} & y_{bb} \end{bmatrix} = \begin{bmatrix} y_{cc} & y_{cb} \\ y_{bc} & y_{bb} \end{bmatrix} \\ \mathbf{Y}_{\text{common base}} &= \begin{bmatrix} y_{ee} & y_{ec} & \cancel{y_{eb}} \\ y_{ce} & y_{cc} & \cancel{y_{cb}} \\ \cancel{y_{be}} & \cancel{y_{bc}} & \cancel{y_{bb}} \end{bmatrix} = \begin{bmatrix} y_{ee} & y_{ec} \\ y_{ce} & y_{cc} \end{bmatrix} \\ \mathbf{Y}_{\text{common collector}} &= \begin{bmatrix} y_{ee} & \cancel{y_{ec}} & y_{eb} \\ \cancel{y_{ce}} & \cancel{y_{cc}} & \cancel{y_{cb}} \\ y_{be} & \cancel{y_{bc}} & y_{bb} \end{bmatrix} = \begin{bmatrix} y_{ee} & y_{eb} \\ y_{be} & y_{bb} \end{bmatrix} \end{aligned} \tag{6-37}$$

Finally, note that the 3 × 3 IAM can be reconstructed from any of the y parameter equations defined by Equations (6-37) by using properties 1 and 2 to reconstruct the missing row and column. (See Example 6-5.)

Example 6-5

In this example we construct the IAM for the field-effect transistor (FET) presented in Figure 6-13(a) from the grounded-source y parameters. A typical grounded-source FET (as in Figure 6-13(b)) has y parameters given by

$$\mathbf{Y}^{\text{grounded source}} = \begin{bmatrix} j\omega \cdot 10^{-10} & -j\omega \cdot 10^{-11} \\ 10^{-2} & 10^{-4} + j\omega \cdot 10^{-9} \end{bmatrix} \tag{6-38}$$

Let

$$\mathbf{I}_{g,d,s} = \begin{bmatrix} y_{gg} & y_{gd} & y_{gs} \\ y_{dg} & y_{dd} & y_{ds} \\ y_{sg} & y_{sd} & y_{ss} \end{bmatrix} \tag{6-39}$$

represent the IAM for the 3-terminal FET. Then, according to property 5, the grounded-source y parameter matrix for the network of Figure 6-13(b) is given by deleting the row and column corresponding

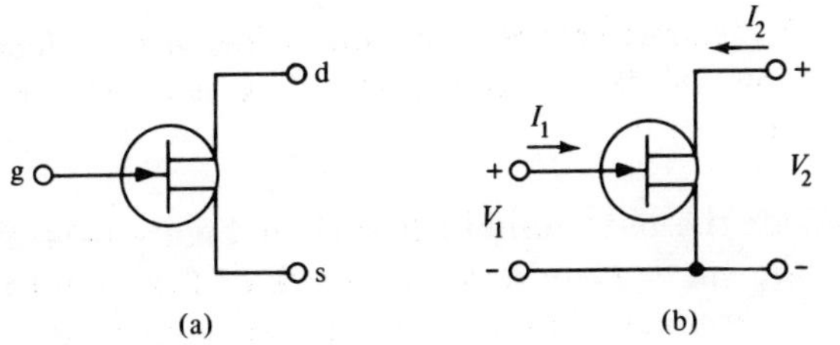

Figure 6-13 (a) A field-effect transistor with gate, drain, and source terminals. (b) FET in grounded-source configuration.

to the source terminal; that is,

$$\mathbf{Y}^{\text{grounded source}} = \begin{bmatrix} y_{gg} & y_{gd} & \cancel{y_{gs}} \\ y_{dg} & y_{dd} & \cancel{y_{ds}} \\ \cancel{y_{sg}} & \cancel{y_{sd}} & \cancel{y_{ss}} \end{bmatrix} = \begin{bmatrix} j\omega \cdot 10^{-10} & -j\omega \cdot 10^{-11} \\ 10^{-2} & 10^{-4} + j\omega \cdot 10^{-9} \end{bmatrix} \tag{6-40}$$

Entering these elements into the IAM, we have

$$\mathbf{I}_{g,d,s} = \begin{bmatrix} j\omega \cdot 10^{-10} & -j\omega \cdot 10^{-11} & (\quad) \\ 10^{-2} & 10^{-4} + j\omega \cdot 10^{-9} & (\quad) \\ (\quad) & (\quad) & (\quad) \end{bmatrix} \tag{6-41}$$

The third row and third column can now be filled in by applying properties 1 and 2 to obtain

$$\mathbf{I}_{g,d,s} = \begin{bmatrix} j\omega \cdot 10^{-10} & -j\omega \cdot 10^{-11} & -j\omega 0.9 \cdot 10^{-10} \\ 10^{-2} & 10^{-4} + j\omega \cdot 10^{-9} & -9.9 \cdot 10^{-3} - j\omega \cdot 10^{-9} \\ -10^{-2} - j\omega \cdot 10^{-10} & -10^{-4} + j\omega \cdot 9.9 \cdot 10^{-10} & 9.9 \cdot 10^{-3} + j\omega 1.09 \cdot 10^{-9} \end{bmatrix} \tag{6-42}$$

PROPERTY 6

If an (m − 1)-terminal network is generated from an m-terminal network by suppressing the mth terminal of the m-terminal network, the resulting (m − 1) × (m − 1) IAM is given by

$$\mathbf{I} = \mathbf{I}_* - \frac{1}{y_{mm}} \mathbf{C}_* \mathbf{R}_* \tag{6-43}$$

where

$\mathbf{I}_*$ = (m − 1) × (m − 1) matrix formed from the m × m IAM for the m-terminal network by deleting the mth row and mth column.

y_{mm} = mth row, mth-column element of the m-terminal IAM.

$\mathbf{C}_*$ = m − 1 element column matrix consisting of the elements in the mth column of the m-terminal IAM with the mth element deleted.

$\mathbf{R}_*$ = m − 1 element row matrix consisting of the elements in the mth row of the m-terminal IAM with the mth element deleted.

Proof: Consider the m-terminal network of Figure 6-14(a), which we transform into the (m − 1)-terminal network of Figure 6-14(b) by suppressing the mth terminal; that is, the network node that was formerly

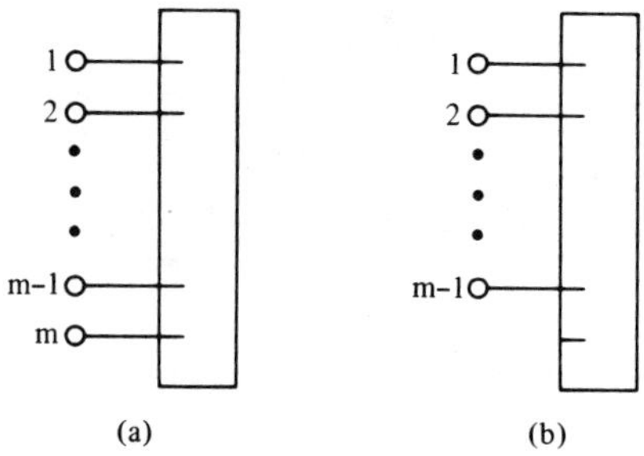

Figure 6-14 (a) m-terminal network. (b) (m − 1)-terminal network obtained by suppressing the mth terminal.

labeled the mth terminal is no longer considered to be a terminal. The terminal equations for the m-terminal network, given by Equations (6-20), can be transformed into the terminal equations for the (m − 1)-terminal network by setting $I_m = 0$ (since the current into the terminal m from the reference node is now zero), solving the mth equation for V_m, and substituting this value for V_m into the remaining m − 1 equations. That is, the mth equation of Equations (6-20) is now given by

$$I_m = 0 = y_{m1}V_1 + y_{m2}V_2 + \cdots + y_{m,m-1}V_{m-1} + y_{mm}V_m \tag{6-44}$$

Solving for V_m we have

$$V_m = -\left[\frac{y_{m1}}{y_{mm}} V_1 + \frac{y_{m2}}{y_{mm}} V_2 + \cdots + \frac{y_{m,m-1}}{y_{mm}} V_{m-1}\right] \tag{6-45}$$

Substituting Equation (6-45) into the first m-1 equations of Equations (6-20) results in

$$I_1 = \left(y_{11} - \frac{y_{1m}y_{m1}}{y_{mm}}\right) V_1 + \left(y_{12} - \frac{y_{1m}y_{m2}}{y_{mm}}\right) V_2 + \cdots + \left(y_{1,m-1} - \frac{y_{1m}y_{m,m-1}}{y_{mm}}\right) V_{m-1}$$

$$I_2 = \left(y_{21} - \frac{y_{2m}y_{m1}}{y_{mm}}\right) V_1 + \left(y_{22} - \frac{y_{2m}y_{m2}}{y_{mm}}\right) V_2 + \cdots + \left(y_{2,m-1} - \frac{y_{2m}y_{m,m-1}}{y_{mm}}\right) V_{m-1} \tag{6-46}$$

$$\vdots$$

$$I_{m-1} = \left(y_{m-1,1} - \frac{y_{m-1,m}y_{m1}}{y_{mm}}\right) V_1 + \left(y_{m-1,2} - \frac{y_{m-1,m}y_{m2}}{y_{mm}}\right) V_2 + \cdots + \left(y_{m-1,m-1} - \frac{y_{m-1,m}y_{m,m-1}}{y_{mm}}\right) V_{m-1}$$

The (m − 1) × (m − 1) array of coefficients in Equations (6-46) constitutes the elements in the (m − 1) × (m − 1) IAM for the (m − 1)-terminal network of Figure 6-14(b). These coefficients are precisely those defined by Equation (6-43).

Comment: Equation (6-43) can be illustrated graphically by first partitioning the m-terminal IAM in the form

$$\text{m-terminal IAM} = \begin{bmatrix} [\; \mathbf{I}_* \;] & [\; \mathbf{C}_* \;] \\ [\; \mathbf{R}_* \;] & y_{mm} \end{bmatrix} \tag{6-47}$$

and then writing

$$[\; \mathbf{I} \;] = [\; \mathbf{I}_* \;] - \frac{1}{y_{mm}} [\; \mathbf{C}_* \;][\; \mathbf{R}_* \;] \tag{6-48}$$

Example 6-6

In this example we first compute the 4 × 4 IAM for the 4-terminal network of Figure 6-15(a), and then we suppress terminal 4 to obtain the 3 × 3 IAM for the 3-terminal network of Figure 6-15(b).

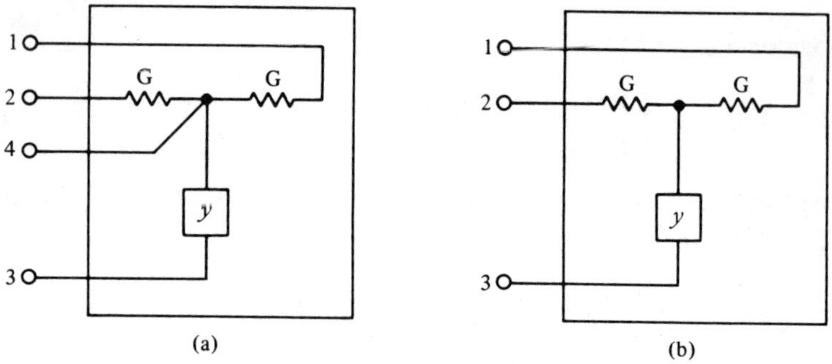

Figure 6-15 (a) A 4-terminal network. (b) The resulting 3-terminal network after suppressing terminal 4.

The four first-column elements of the 4 × 4 IAM for the 4-terminal network of Figure 6-15(a) are given by the four terminal currents I_1, I_2, I_4 and I_3, with $V_1 = 1$ and $V_2 = V_4 = V_3 = 0$. By inspection of Figure 6-15(a), we enter these four elements; that is,

$$\mathbf{I}_{1,2,4,3} = \begin{bmatrix} G & (\quad) & (\quad) & (\quad) \\ 0 & (\quad) & (\quad) & (\quad) \\ -G & (\quad) & (\quad) & (\quad) \\ 0 & (\quad) & (\quad) & (\quad) \end{bmatrix} \tag{6-49}$$

The four elements of the second column are the four terminal currents I_1, I_2, I_4, and I_3, with $V_2 = 1$ and $V_1 = V_4 = V_3 = 0$. Again, by inspection, we enter these elements; that is,

$$\mathbf{I}_{1,2,4,3} = \begin{bmatrix} G & 0 & (\quad) & (\quad) \\ 0 & G & (\quad) & (\quad) \\ -G & -G & (\quad) & (\quad) \\ 0 & 0 & (\quad) & (\quad) \end{bmatrix} \tag{6-50}$$

The four third-column elements are the same four terminal currents, with $V_4 = 1$ and $V_1 = V_2 = V_3 = 0$. We enter the third column

$$\mathbf{I}_{1,2,4,3} = \begin{bmatrix} G & 0 & -G & (\quad) \\ 0 & G & -G & (\quad) \\ -G & -G & 2G + y & (\quad) \\ 0 & 0 & -y & (\quad) \end{bmatrix} \tag{6-51}$$

Finally, we enter the four terminal currents, with $V_3 = 1$ and $V_1 = V_2 = V_4 = 0$ in the fourth column

$$\mathbf{I}_{1,2,4,3} = \begin{bmatrix} G & 0 & -G & 0 \\ 0 & G & -G & 0 \\ -G & -G & 2G + y & -y \\ 0 & 0 & -y & y \end{bmatrix} \tag{6-52}$$

As a check on our work we verify that all four rows and all four columns sum to zero.

We next suppress terminal 4 by applying property 6. According to Equation (6-43), we write for the 3 × 3 IAM for the three-terminal network of Figure 6-15(b)

$$\mathbf{I}_{1,2,3} = \begin{bmatrix} G & 0 & 0 \\ 0 & G & 0 \\ 0 & 0 & y \end{bmatrix} - \frac{1}{2G+y}\begin{bmatrix} -G \\ -G \\ -y \end{bmatrix}\begin{bmatrix} -G & -G & -y \end{bmatrix}$$

$$= \begin{bmatrix} G & 0 & 0 \\ 0 & G & 0 \\ 0 & 0 & y \end{bmatrix} - \frac{1}{2G+y}\begin{bmatrix} G^2 & G^2 & Gy \\ G^2 & G^2 & Gy \\ Gy & Gy & y^2 \end{bmatrix}$$

$$= \begin{bmatrix} G - \dfrac{G^2}{2G+y} & -\dfrac{G^2}{2G+y} & -\dfrac{Gy}{2G+y} \\ -\dfrac{G^2}{2G+y} & G - \dfrac{G^2}{2G+y} & -\dfrac{Gy}{2G+y} \\ -\dfrac{Gy}{2G+y} & -\dfrac{Gy}{2G+y} & y - \dfrac{y^2}{2G+y} \end{bmatrix} \tag{6-53}$$

$$= \begin{bmatrix} \dfrac{G^2 - Gy}{2G+y} & -\dfrac{G^2}{2G+y} & -\dfrac{Gy}{2G+y} \\ \dfrac{-G^2}{2G+y} & -\dfrac{G^2+Gy}{2G+y} & -\dfrac{Gy}{2G+y} \\ \dfrac{-Gy}{2G+y} & -\dfrac{Gy}{2G+y} & \dfrac{2Gy}{2G+y} \end{bmatrix}$$

The 3 × 3 IAM for the network of Figure 6-15(b) can also be obtained directly from Figure 6-15(b); roughly the same amount of computation is required.

PROPERTY 7

If an (m − 1)-terminal network is generated from an m-terminal network by connecting (shorting) terminal m − 1 to terminal m, the resulting (m − 1) × (m − 1) IAM is given by the following: All elements other than those corresponding to the (m − 1)st and mth row and (m − 1)st and mth column are the same as the corresponding elements in the m-terminal network IAM; each element of the (m − 1)st row is obtained by summing the corresponding (m − 1)st- and mth-row elements of the m-terminal IAM; each element of the (m − 1)st column is the sum of the corresponding elements of the (m − 1)st and mth column of the m-terminal IAM.

Proof: The proof is simple and straightforward. See Problem 6-20.

Comment: Property 7 is illustrated graphically in Equation (6-54), where the dots represent the elements of the m-terminal IAM.

(m − 1)-terminal IAM

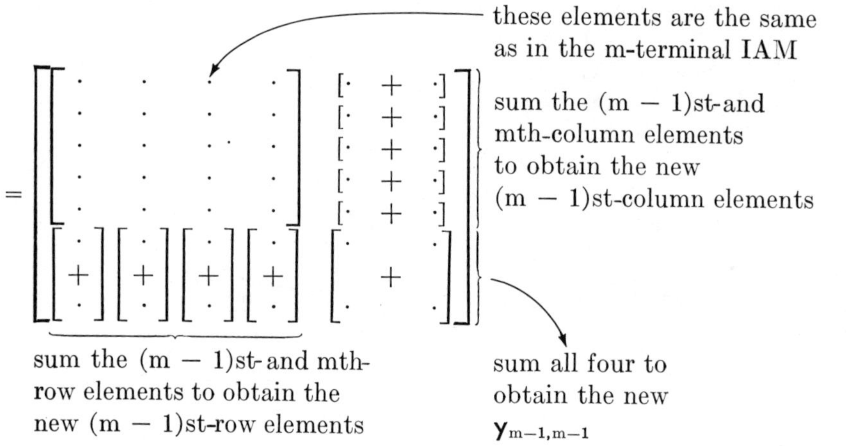

(6-54)

Example 6-7

In this example we first compute the 4 × 4 IAM for the network of Figure 6-16(a) and then use property 7 to obtain the 3 × 3 IAM for the network of Figure 6-16(b).

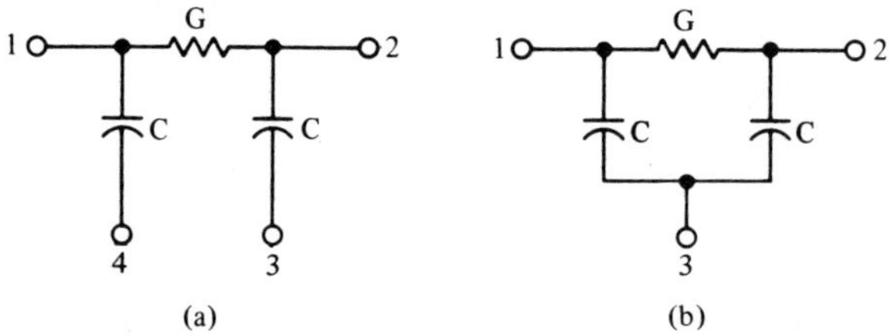

Figure 6-16 Example 6-7.

The 4 × 4 IAM for the 4-terminal network of Figure 6-16(a) can be written by inspection of the figure:

$$\mathbf{I}_{1,2,3,4} = \begin{bmatrix} G + j\omega C & -G & 0 & -j\omega C \\ -G & G + j\omega C & -j\omega C & 0 \\ 0 & -j\omega C & j\omega C & 0 \\ -j\omega C & 0 & 0 & j\omega C \end{bmatrix} \qquad \textbf{(6-55)}$$

We now short terminals 3 and 4 of the 4-terminal network of Figure 6-16(a) to obtain the 3-terminal network of Figure 6-16(b). According to Equation (6-54), the 3 × 3 IAM for the 3-terminal network can be constructed by writing

$$\mathbf{I}_{1,2,3} = \begin{bmatrix} G + j\omega C & -G & (0 + -j\omega C) \\ -G & G + j\omega C & (-j\omega C + 0) \\ \begin{pmatrix} 0 \\ + \\ -j\omega C \end{pmatrix} & \begin{pmatrix} -j\omega C \\ + \\ 0 \end{pmatrix} & \begin{pmatrix} j\omega C + 0 \\ +0 + j\omega C \end{pmatrix} \end{bmatrix}$$

$$= \begin{bmatrix} G + j\omega C & -G & -j\omega C \\ -G & G + j\omega C & -j\omega C \\ -j\omega C & -j\omega C & j2\omega C \end{bmatrix} \tag{6-56}$$

This is the same result as obtained in Example 6-1, where the IAM for the 3-terminal network of Figure 6-16(b) was computed directly. [See Equation (6-8).]

The results of this section are summarized in Table 6-1.

Table 6-1

PROPERTY	OPERATION
1. Rows sum to zero	$\sum_k y_{ik} = 0$
2. Columns sum to zero	$\sum_i y_{ik} = 0$
3. Reorder terminals	Interchange rows and interchange columns
4. Isolated terminals	To add an isolated terminal add a row and column of zeros
5. Port parameters	Cross out row and column corresponding to common terminal
6. Suppress terminal m	$\mathbf{I} = \mathbf{I}_* - \frac{1}{y_{mm}} \mathbf{C}_* \mathbf{R}_*$
7. Short terminals m − 1 and m	Add (m − 1)st- and mth-row elements and add (m − 1)st- and mth-column elements

6-4 Interconnection of m-Terminal Networks

The real advantage of the IAM description of networks is that it allows us to analyze large networks by first dividing the large network into a number of smaller component networks. Each of these component networks is characterized by its IAM, and these IAM's are combined to characterize the large network. If we use the seven IAM properties previously defined, the interconnection rules become so simple that even a digital computer can execute them.

In this section we show that if two m-terminal networks are connected in parallel by joining like-numbered terminals, the IAM of the resulting network is the sum of the IAM's of the component networks, provided that like-numbered terminals correspond to the same rows and columns in both IAM's.

Consider the parallel connection of the two m-terminal networks N°

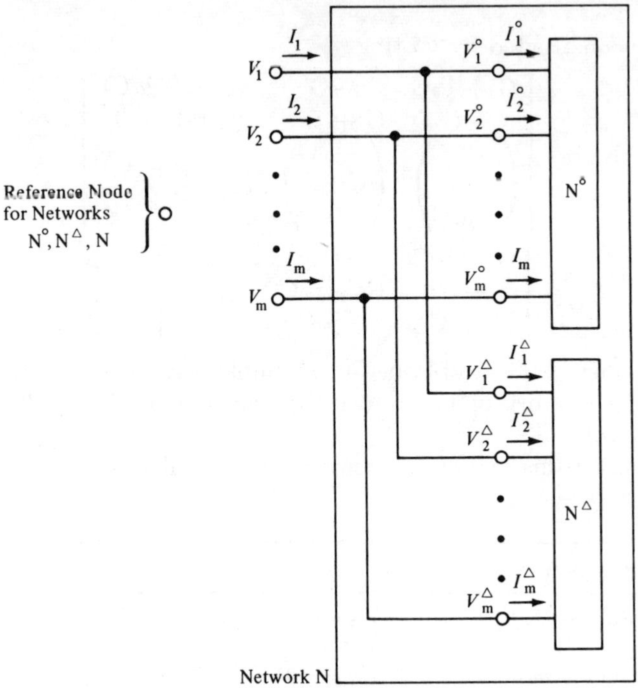

Figure 6-17 Two m-terminal networks connected in parallel.

and N^Δ, as illustrated in Figure 6-17. The equation governing the behavior of networks N° and N^Δ relative to their terminals are given by

$$\text{network } N^\circ: \begin{bmatrix} I_1^\circ \\ I_2^\circ \\ \cdot \\ \cdot \\ \cdot \\ I_m^\circ \end{bmatrix} = \begin{bmatrix} y_{11}^\circ & y_{12}^\circ & \cdots & y_{1m}^\circ \\ y_{21}^\circ & y_{22}^\circ & \cdots & y_{2m}^\circ \\ \cdot & \cdot & \cdot & \cdot \\ \cdot & \cdot & \cdot & \cdot \\ \cdot & \cdot & \cdot & \cdot \\ y_{m1}^\circ & y_{m2}^\circ & \cdots & y_{mm}^\circ \end{bmatrix} \begin{bmatrix} V_1^\circ \\ V_2^\circ \\ \cdot \\ \cdot \\ \cdot \\ V_m^\circ \end{bmatrix}$$

$$\text{network } N^\Delta: \begin{bmatrix} I_1^\Delta \\ I_2^\Delta \\ \cdot \\ \cdot \\ \cdot \\ I_m^\Delta \end{bmatrix} = \begin{bmatrix} y_{11}^\Delta & y_{12}^\Delta & \cdots & y_{1m}^\Delta \\ y_{21}^\Delta & y_{22}^\Delta & \cdots & y_{2m}^\Delta \\ \cdot & \cdot & \cdots & \cdot \\ \cdot & \cdot & \cdot & \cdot \\ \cdot & \cdot & \cdot & \cdot \\ y_{m1}^\Delta & y_{m2}^\Delta & & y_{mm}^\Delta \end{bmatrix} \begin{bmatrix} V_1^\Delta \\ V_2^\Delta \\ \cdot \\ \cdot \\ \cdot \\ V_m^\Delta \end{bmatrix} \tag{6-57}$$

and the behavior of network N relative to its terminal is governed by the equations

$$\text{network } N: \begin{bmatrix} I_1 \\ I_2 \\ \cdot \\ \cdot \\ \cdot \\ I_m \end{bmatrix} = \begin{bmatrix} y_{11} & y_{12} & \cdots & y_{1m} \\ y_{21} & y_{22} & \cdots & y_{2m} \\ \cdot & \cdot & \cdot & \cdot \\ \cdot & \cdot & \cdot & \cdot \\ \cdot & \cdot & \cdot & \cdot \\ y_{m1} & y_{m2} & \cdots & y_{mm} \end{bmatrix} \begin{bmatrix} V_1 \\ V_2 \\ \cdot \\ \cdot \\ \cdot \\ V_m \end{bmatrix} \tag{6-58}$$

Furthermore, it is apparent from Figure 6-17 that

$$\begin{bmatrix} V_1 \\ V_2 \\ \cdot \\ \cdot \\ \cdot \\ V_m \end{bmatrix} = \begin{bmatrix} V_1^{\circ} \\ V_2^{\circ} \\ \cdot \\ \cdot \\ \cdot \\ V_m^{\circ} \end{bmatrix} = \begin{bmatrix} V_1^{\Delta} \\ V_2^{\Delta} \\ \cdot \\ \cdot \\ \cdot \\ V_m^{\Delta} \end{bmatrix} \tag{6-59}$$

$$\begin{bmatrix} I_1 \\ I_2 \\ \cdot \\ \cdot \\ \cdot \\ I_m \end{bmatrix} = \begin{bmatrix} I_1^{\circ} \\ I_2^{\circ} \\ \cdot \\ \cdot \\ \cdot \\ I_m^{\circ} \end{bmatrix} + \begin{bmatrix} I_1^{\Delta} \\ I_2^{\Delta} \\ \cdot \\ \cdot \\ \cdot \\ I_m^{\Delta} \end{bmatrix} \tag{6-60}$$

We now substitute $\begin{bmatrix} V_1 \\ V_2 \\ \cdot \\ \cdot \\ \cdot \\ V_m \end{bmatrix}$ for $\begin{bmatrix} V_1^{\circ} \\ V_2^{\circ} \\ \cdot \\ \cdot \\ \cdot \\ V_m^{\circ} \end{bmatrix}$ and $\begin{bmatrix} V_1^{\Delta} \\ V_1^{\Delta} \\ \cdot \\ \cdot \\ \cdot \\ V_m^{\Delta} \end{bmatrix}$ in the right-hand side of Equations (6-57), and then substitute Equations (6-57) into Equation (6-60) to obtain

$$\begin{bmatrix} I_1 \\ I_2 \\ \cdot \\ \cdot \\ \cdot \\ I_m \end{bmatrix} = \begin{bmatrix} y_{11}^{\circ} & y_{12}^{\circ} & \cdots & y_{1m}^{\circ} \\ y_{21}^{\circ} & y_{22}^{\circ} & \cdots & y_{2m}^{\circ} \\ \cdot & \cdot & \cdot & \cdot \\ \cdot & \cdot & \cdot & \cdot \\ \cdot & \cdot & \cdot & \cdot \\ y_{m1}^{\circ} & y_{m2}^{\circ} & \cdots & y_{mm}^{\circ} \end{bmatrix} \begin{bmatrix} V_1 \\ V_2 \\ \cdot \\ \cdot \\ \cdot \\ V_m \end{bmatrix} + \begin{bmatrix} y_{11}^{\Delta} & y_{12}^{\Delta} & \cdots & y_{1m}^{\Delta} \\ y_{21}^{\Delta} & y_{22}^{\Delta} & \cdots & y_{2m}^{\Delta} \\ \cdot & \cdot & \cdot & \cdot \\ \cdot & \cdot & \cdot & \cdot \\ \cdot & \cdot & \cdot & \cdot \\ y_{m1}^{\Delta} & y_{m2}^{\Delta} & \cdots & y_{mm}^{\Delta} \end{bmatrix} \begin{bmatrix} V_1 \\ V_2 \\ \cdot \\ \cdot \\ \cdot \\ V_m \end{bmatrix}$$

$$= \left\{ \begin{bmatrix} y_{11}^{\circ} & y_{12}^{\circ} & \cdots & y_{1m}^{\circ} \\ y_{21}^{\circ} & y_{22}^{\circ} & \cdots & y_{2m}^{\circ} \\ \cdot & \cdot & \cdot & \cdot \\ \cdot & \cdot & \cdot & \cdot \\ \cdot & \cdot & \cdot & \cdot \\ y_{m1}^{\circ} & y_{m2}^{\circ} & \cdots & y_{mm}^{\circ} \end{bmatrix} + \begin{bmatrix} y_{11}^{\Delta} & y_{12}^{\Delta} & \cdots & y_{1m}^{\Delta} \\ y_{21}^{\Delta} & y_{22}^{\Delta} & \cdots & y_{2m}^{\Delta} \\ \cdot & \cdot & \cdot & \cdot \\ \cdot & \cdot & \cdot & \cdot \\ \cdot & \cdot & \cdot & \cdot \\ y_{m1}^{\Delta} & y_{m2}^{\Delta} & \cdots & y_{mm}^{\Delta} \end{bmatrix} \right\} \begin{bmatrix} V_1 \\ V_2 \\ \cdot \\ \cdot \\ \cdot \\ V_m \end{bmatrix}$$

$$= \begin{bmatrix} y_{11}^{\circ} + y_{11}^{\Delta} & y_{12}^{\circ} + y_{12}^{\Delta} & \cdots & y_{1m}^{\circ} + y_{1m}^{\Delta} \\ y_{21}^{\circ} + y_{21}^{\Delta} & y_{22}^{\circ} + y_{22}^{\Delta} & \cdots & y_{2m}^{\circ} + y_{2m}^{\Delta} \\ \cdot & \cdot & \cdot & \cdot \\ \cdot & \cdot & \cdot & \cdot \\ \cdot & \cdot & \cdot & \cdot \\ y_{m1}^{\circ} + y_{m1}^{\Delta} & y_{m2}^{\circ} + y_{m2}^{\Delta} & \cdots & y_{mm}^{\circ} + y_{mm}^{\Delta} \end{bmatrix} \begin{bmatrix} V_1 \\ V_2 \\ \cdot \\ \cdot \\ \cdot \\ V_m \end{bmatrix} \tag{6-61}$$

Comparing Equations (6-58) and (6-61), we conclude that if two m-terminal networks having IAM's $\mathbf{I}^{\circ}_{1,2,\ldots,m}$ and $\mathbf{I}^{\Delta}_{1,2,\ldots,m}$ are connected in parallel by joining like-numbered terminals, the IAM of the resulting network is given by

$$\mathbf{I}_{1,2,\ldots,m} = \mathbf{I}^{\circ}_{1,2,\ldots,m} + \mathbf{I}^{\Delta}_{1,2,\ldots,m} \tag{6-62}$$

Example 6-8

In this example we use the IAM concept to compute the short-circuit admittance matrix for the network presented in Figure 6-18(a). We assume that the field-effect transistor (FET) has grounded-source y

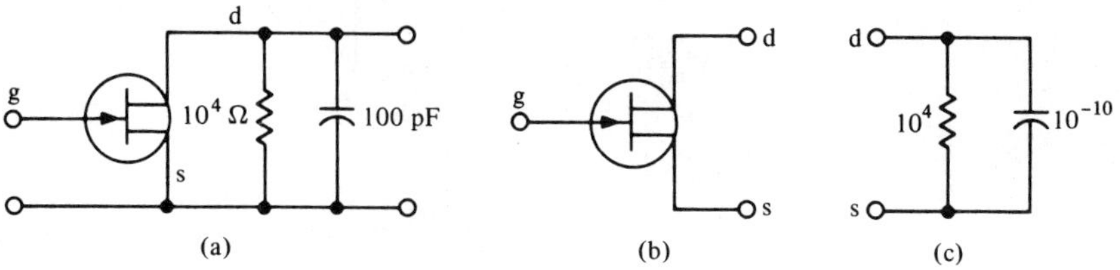

Figure 6-18 Example 6-8.

parameters given by

$$\mathbf{Y}^{(\text{grounded-source FET})} = \begin{bmatrix} j\omega \times 10^{-10} & -j\omega \times 10^{-11} \\ 10^{-2} & 10^{-4} + j\omega \times 10^{-9} \end{bmatrix} \tag{6-63}$$

Our strategy is first to compute the IAM for the FET, then to compute the IAM for the parallel RC network, then to combine these to construct the IAM for the complete network, and, finally, to recognize the network y parameters from the network IAM.

Starting with the grounded-source y parameters for the FET, we construct the 3 × 3 IAM for the 3-terminal FET of Figure 6-18(b). This is easily accomplished by applying property 5 and property 1. As discussed in Example 6-5, we obtain (see Equation 6-42)

$$\mathbf{I}^{(\text{FET})}_{g,d,s} = \begin{bmatrix} j\omega \times 10^{-10} & -j\omega \times 10^{-11} & -j\omega 0.9 \times 10^{-10} \\ 10^{-2} & 10^{-4} + j\omega \times 10^{-9} & -j\omega \times 10^{-9} \\ -10^{-2} - j\omega \times 10^{-10} & -10^{-4} - j\omega \times 9.9 \times 10^{-10} & 9.9 \times 10^{-3} + j\omega 1.09 \times 10^{-9} \end{bmatrix} \tag{6-64}$$

Next, we compute the 3 × 3 IAM for the parallel RC network. We start by computing the 2 × 2 IAM for the 2-terminal RC network of Figure 6-18(c). Here we make use of a result obtained in Example 6-3 to write

$$\mathbf{I}^{(\text{RC})}_{d,s} = \begin{bmatrix} 10^{-4} + j\omega \times 10^{-10} & -10^{-4} - j\omega \times 10^{-10} \\ -10^{-4} - j\omega \times 10^{-10} & 10^{-4} + j\omega \times 10^{-10} \end{bmatrix} \tag{6-65}$$

where we have simply let $y = R^{-1} + j\omega C = 10^{-4} + j\omega \times 10^{-10}$ in Equation (6-25). Next, we add an isolated terminal to the RC network in order to obtain the 3-terminal network shown in Figure 6-19(a).

According to property 4, the 3 × 3 IAM for the 3-terminal network of Figure 6-19(a) can be constructed by adding a row and a column of zeros to the 2 × 2 IAM given by Equation (6-65). Hence, we write

$$\mathbf{I}^{(RC)}_{g,d,s} = \begin{bmatrix} 0 & 0 & 0 \\ 0 & 10^{-4} + j\omega \times 10^{-10} & -10^{-4} - j\omega \times 10^{-10} \\ 0 & -10^{-4} - j\omega \times 10^{-10} & 10^{-4} + j\omega \times 10^{-10} \end{bmatrix} \tag{6-66}$$

The next step is to connect the 3-terminal FET and the 3-terminal RC network in parallel by connecting like-numbered terminals, as shown in Figure 6-19(b). The result of Section 6-4 stated that the IAM

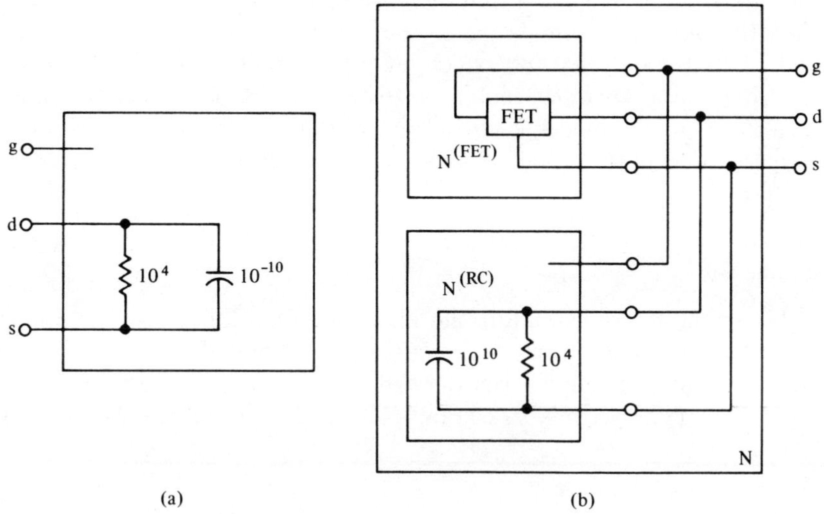

Figure 6-19 Interconnection for Example 6-8.

for the parallel combination of two networks is the sum of the IAM's for the two networks. Therefore, the IAM for the 3-terminal network designated as network N in Figure 6-19(b) is given by

$$\begin{aligned} \mathbf{I}^{(N)}_{g,d,s} &= \mathbf{I}^{(FET)}_{g,d,s} + \mathbf{I}^{(RC)}_{g,d,s} \\ &= \begin{bmatrix} j\omega \times 10^{-10} & -j\omega \times 10^{-11} & -j\omega.9 \times 10^{-10} \\ 10^{-2} & 2 \times 10^{-4} + j\omega 1.1 \times 10^{-9} & -10^{-2} - j\omega 1.1 \times 10^{-9} \\ -10^{-2} - j\omega \times 10^{-10} & -2 \times 10^{-4} - j\omega 1.1 \times 10^{-9} & 10^{-2} + j\omega 1.2 \times 10^{-9} \end{bmatrix} \end{aligned} \tag{6-67}$$

Finally, we note that network N of Figure 6-19(b) can be converted into the 2-port network of Figure 6-18(a) by making terminal s common to both ports. Therefore, by property 5, the y parameters for the network presented in Figure 6-18(a) can be obtained by deleting the row and column of the matrix $\mathbf{I}^{(N)}_{g,d,s}$ corresponding to the terminal common to both ports. Deleting the third row and third column of $\mathbf{I}^{(N)}_{g,d,s}$ we obtain

$$\mathbf{Y}^{(N)} = \begin{bmatrix} j\omega \times 10^{-10} & -j\omega \times 10^{-11} \\ 10^{-2} & 2 \times 10^{-4} + j\omega 1.1 \times 10^{-9} \end{bmatrix} \tag{6-68}$$

Comparing Equations (6-63) and (6-68), we note that the addition of the parallel RC network affects only the short-circuit output admittance y_{dd}. Actually, this result is obvious by inspection of Figure 6-18(a). y_{gg} is the input admittance with the output terminals short-circuited; hence, because the parallel RC network is shorted, it has no effect on y_{gg}. y_{dg} is the transfer admittance, again with the output terminal short-circuited. Hence, y_{dg} is not affected by the RC network. y_{gd} is the short-circuit gate current caused by a 1-volt independent voltage source connected across the output terminals; since the RC network does not alter the terminal voltage, it has no effect on y_{gd}. Finally, we note that y_{dd} represents the admittance looking into the output port with the input port a short circuit; hence, y_{dd} is the sum of the admittance of the parallel RC network and the admittance seen looking into the output port of the FET.

Example 6-9

In this example we compute the short-circuit admittance parameter matrix for the 2-port network presented in Figure 6-20(a). We assume that the transistor can be represented by the model shown in Figure 6-20(b). Our strategy is to compute the IAM for the transistor, then

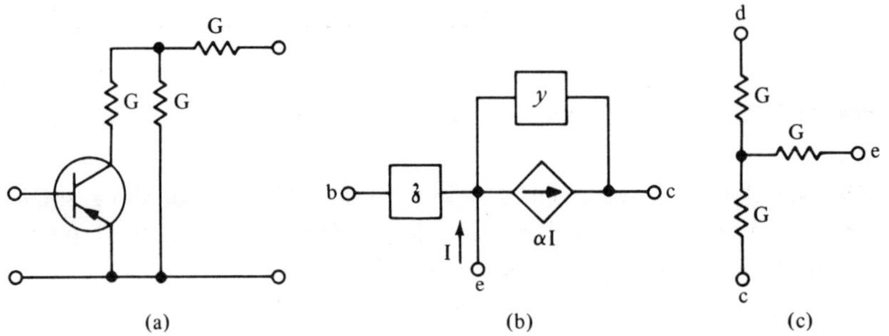

Figure 6-20 Example 6-9.

to compute the IAM for the Tee network, and, finally, to construct the IAM for the network of Figure 6-20(a). This strategy is quite tedious for a human being, but preferable for a computer. Our ultimate objective is to have a computer handle the tedium, but we must first do a few examples by hand so that we understand the procedure well enough to use the computer effectively.

The IAM for the 3-terminal transistor of Figure 6-20(b) was computed in Example 6-2. The result, given by Equation (6-19), is

$$\mathbf{I}_{b,c,e}^{(\text{transistor})} = \begin{bmatrix} \frac{1}{\mathfrak{z}} & 0 & -\frac{1}{\mathfrak{z}} \\ \frac{\alpha}{\mathfrak{z}(1-\alpha)} & \frac{y}{1-\alpha} & \frac{-\alpha - y\mathfrak{z}}{\mathfrak{z}(1-\alpha)} \\ \frac{-1}{\mathfrak{z}(1-\alpha)} & \frac{-y}{1-\alpha} & \frac{1+y\mathfrak{z}}{\mathfrak{z}(1-\alpha)} \end{bmatrix} \tag{6-69}$$

Furthermore, the 3-terminal resistance network of Figure 6-20(c) is a special case of the Tee network considered in Example 6-6. Therefore, the IAM for the 3-terminal resistance network of Figure 6-20(c) is given by Equation (6-53), with $y = G$; that is,

$$\mathbf{I}_{d,c,e}^{(R)} = \begin{bmatrix} \frac{2G}{3} & \frac{-G}{3} & \frac{-G}{3} \\ \frac{-G}{3} & \frac{2G}{3} & \frac{-G}{3} \\ \frac{-G}{3} & \frac{-G}{3} & \frac{2G}{3} \end{bmatrix} \tag{6-70}$$

The result of Section 6-4 states that if two networks are connected in parallel, the IAM of the resulting network is the sum of the IAM's of the two component networks, provided that (1) like-numbered terminals are connected and (2) like-numbered terminals correspond to the same rows and columns in the IAM's. Therefore, in the problem at hand, we must add a terminal d to the transistor network, as illustrated in Figure 6-21(a), and a terminal b to the resistor network, as illustrated in Figure 6-21(b). We are now ready to construct the IAM's for the two 4-terminal networks of Figure 6-21(a) and (b). First, we use property 4 to construct the IAM for the 4-terminal transistor network by adding a row and column of zeros corresponding to the isolated terminal. Since terminal d already corresponds to the first row and column in the resistance network IAM, we make the first row and first column of the IAM for the transistor network correspond to terminal d. Hence, we write

$$\mathbf{I}_{d,b,c,e}^{(\text{transistor})} = \begin{bmatrix} 0 & 0 & 0 & 0 \\ 0 & \frac{1}{\mathfrak{z}} & 0 & -\frac{1}{\mathfrak{z}} \\ 0 & \frac{\alpha}{\mathfrak{z}(1-\alpha)} & \frac{y}{1-\alpha} & \frac{-\alpha - y\mathfrak{z}}{\mathfrak{z}(1-\alpha)} \\ 0 & \frac{-1}{\mathfrak{z}(1-\alpha)} & \frac{-y}{1-\alpha} & \frac{1+y\mathfrak{z}}{\mathfrak{z}(1-\alpha)} \end{bmatrix} \tag{6-71}$$

We next construct the IAM for the 4-terminal resistance network of

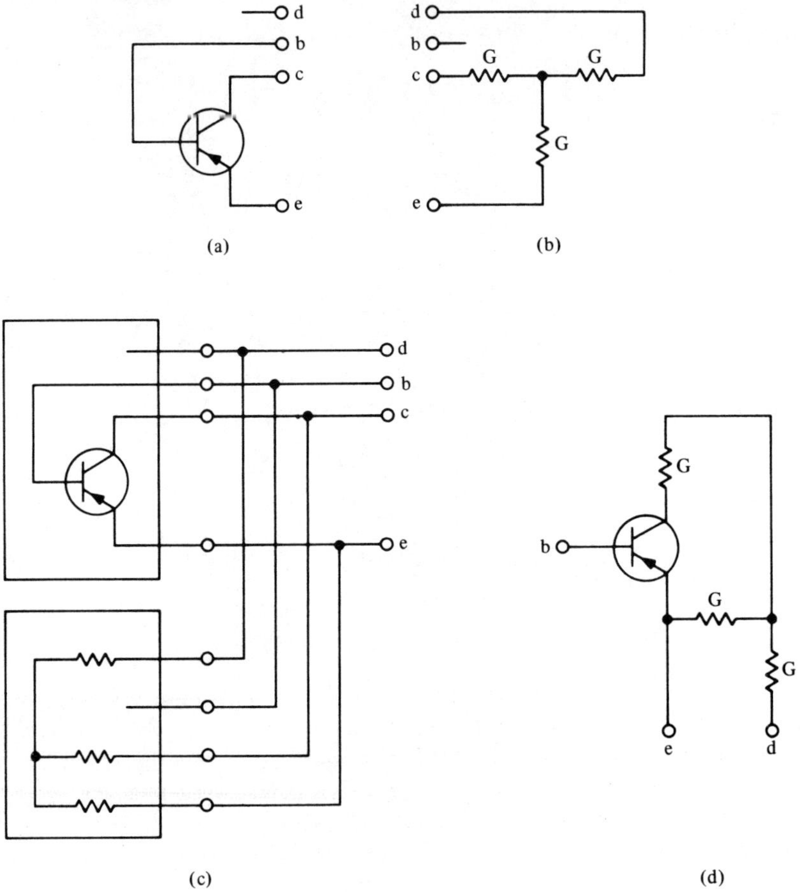

Figure 6-21 Interconnection of Example 6-9.

Figure 6-21(b). In the transistor IAM, terminals d, b, c, and e correspond to rows and columns 1, 2, 3, and 4, respectively. Since the same must be true of the resistance network IAM, the isolated terminal b must correspond to the second row and second column of the resistance network IAM. Again using property 4, we write the IAM for the 4-terminal network of Figure 6-21(b).

$$\mathbf{I}^{(R)}_{d,b,c,e} = \begin{bmatrix} \frac{2G}{3} & 0 & \frac{-G}{3} & \frac{-G}{3} \\ 0 & 0 & 0 & 0 \\ \frac{-G}{3} & 0 & \frac{2G}{3} & \frac{-G}{3} \\ \frac{-G}{3} & 0 & \frac{-G}{3} & \frac{2G}{3} \end{bmatrix} \tag{6-72}$$

We are now ready to connect the two 4-terminal networks of Figure

6-21(a) and Figure 6-21(b) in parallel by connecting like-lettered terminals. The IAM, say $\mathbf{I}_{d,b,c,e}$, for the parallel combination illustrated in Figure 6-21(c) can be expressed as the sum of the IAM's for the networks of Figure 6-21(a) and (b). Therefore, we have

$$\mathbf{I}_{d,b,c,e} = \mathbf{I}^{(\text{transistor})}_{d,b,c,e} + \mathbf{I}^{(R)}_{d,b,c,e}$$

$$= \begin{bmatrix} 0 & 0 & 0 & 0 \\ 0 & \frac{1}{z} & 0 & -\frac{1}{z} \\ 0 & \frac{\alpha}{z(1-\alpha)} & \frac{y}{1-\alpha} & \frac{-\alpha - yz}{z(1-\alpha)} \\ 0 & \frac{-1}{z(1-\alpha)} & \frac{-y}{1-\alpha} & \frac{1+yz}{z(1-\alpha)} \end{bmatrix} + \begin{bmatrix} \frac{2G}{3} & 0 & \frac{-G}{3} & \frac{-G}{3} \\ 0 & 0 & 0 & 0 \\ \frac{-G}{3} & 0 & \frac{2G}{3} & \frac{-G}{3} \\ \frac{-G}{3} & 0 & \frac{-G}{3} & \frac{2G}{3} \end{bmatrix}$$

$$= \begin{bmatrix} \frac{2G}{3} & 0 & \frac{-G}{3} & \frac{-G}{3} \\ 0 & \frac{1}{z} & 0 & -\frac{1}{z} \\ \frac{-G}{3} & \frac{\alpha}{z(1-\alpha)} & \frac{y}{1-\alpha} + \frac{2G}{3} & \frac{-\alpha - yz}{z(1-\alpha)} - \frac{G}{3} \\ \frac{-G}{3} & \frac{-1}{z(1-\alpha)} & \frac{-y}{1-\alpha} - \frac{G}{3} & \frac{1+yz}{z(1-\alpha)} + \frac{2G}{3} \end{bmatrix} \tag{6-73}$$

If terminal c in the network of Figure 6-21(c) is suppressed, the resulting 3-terminal network can be transformed into the 2-port network of Figure 6-20(a) by simply making terminal e common to both. Hence, the next step is to suppress terminal c and construct the IAM for the resulting 3-terminal network. This is easily accomplished by utilization of property 6. Applying property 6, we write the IAM for the 3-terminal network of Figure 6-21(d).

$$\mathbf{I}_{d,b,e} = \begin{bmatrix} \frac{2G}{3} & 0 & \frac{-G}{3} \\ 0 & \frac{1}{z} & -\frac{1}{z} \\ \frac{-G}{3} & \frac{-1}{z(1-\alpha)} & \frac{3(1+yz) + 2Gz(1-\alpha)}{3z(1-\alpha)} \end{bmatrix}$$

$$- \frac{\begin{bmatrix} \frac{-G}{3} \\ 0 \\ \frac{3y - G(1-\alpha)}{3(1-\alpha)} \end{bmatrix} \begin{bmatrix} -\frac{G}{3} & \frac{\alpha}{z(1-\alpha)} & \frac{-3\alpha + yz - Gz(1-\alpha)}{3z(1-\alpha)} \end{bmatrix}}{3y + 2G(1-\alpha)} \tag{6-74}$$

$$\mathbf{I}_{d,b,e} = \begin{bmatrix} \frac{2G}{3} & 0 & (\ ) \\ 0 & \frac{1}{\mathfrak{z}} & (\ ) \\ (\ ) & (\ ) & (\ ) \end{bmatrix} - \frac{3(1-\alpha)}{3\mathfrak{y} + 2G(1-\alpha)} \begin{bmatrix} \frac{G^2}{9} & \frac{-G\alpha}{3\mathfrak{z}(1-\alpha)} & (\ ) \\ 0 & 0 & (\ ) \\ (\ ) & (\ ) & (\ ) \end{bmatrix}$$

$$= \begin{bmatrix} \frac{2G}{3} - \frac{G^2}{9\mathfrak{y} + 6G(1-\alpha)} & \frac{3G\alpha}{3\mathfrak{z}[3\mathfrak{y} + 2G(1-\alpha)]} & (\ ) \\ 0 & \frac{1}{\mathfrak{z}} & (\ ) \\ (\ ) & (\ ) & (\ ) \end{bmatrix} \qquad \text{(6-75)}$$

We have not bothered to compute the third row and the third column of the IAM because we are interested only in the y parameters for the common-emitter terminal 2-port. According to property 5 these parameters are given by the four elements we did compute; that is,

$$\mathbf{Y}_{(\text{network N})} = \begin{bmatrix} \frac{2G}{3} - \frac{G^2}{9\mathfrak{y} + 6G(1-\alpha)} & \frac{3G\alpha}{3\mathfrak{z}[3\mathfrak{y} + 2G(1-\alpha)]} \\ 0 & \frac{1}{\mathfrak{z}} \end{bmatrix} \qquad \text{(6-76)}$$

6-5 Applications of the IAM

In the examples at the end of Section 6-4, we illustrated the application of the IAM to some relatively simple network analysis problems. For more complicated networks, the principles are the same. The network analyst constructs IAM's for each of the network elements (or small groups of network elements) in the network. Then properties 3 and 4 of Section 6-3 and the interconnection rule of Section 6-4 are used to combine them into a single IAM for the complete network. Next, terminals of no interest are removed from the analysis, with consequent reduction in the degree of the IAM matrix via properties 6 and 7. (Property 6 can be viewed as a generalization of connecting impedances in series, and property 7 a generalization of combining two elements in parallel.) After the IAM is reduced until only the terminals of interest are left (say, m terminals, so that the resulting IAM is m $\times$ m) we have, according to Equation (6-1), m equations in these m terminal voltages and m terminal currents. At this point an additional m equations can be written (one constraint equation for each terminal that depends on what is connected to the terminal). For example, if terminal 1 is not connected to anything, then $I_1 = 0$; if a V-volt voltage source is connected across terminals 2 and 3, then we can write $V_2 - V_3 = V$ and $I_2 = -I_3$.

Since the analysis of any network via the IAM consists of successive applications of properties 1 to 7 and the interconnection rule, it is ideally suited for execution on a digital computer. Computer subroutines that perform the sequence of steps required for application of each of the properties and for the application of the interconnection rule are available. (See Appendix IV.) Hence, the network analyst need only write a simple program instructing the computer of the network to be analyzed, the order for calling the subroutines, and a program for the few algebraic steps that specify the connection of inputs and outputs to the terminals. Standard programs are available for solving the resulting set of algebraic equations.

For a final example of this procedure, we analyze a differential amplifier. Consider the differential amplifier of Figure 6-22. The

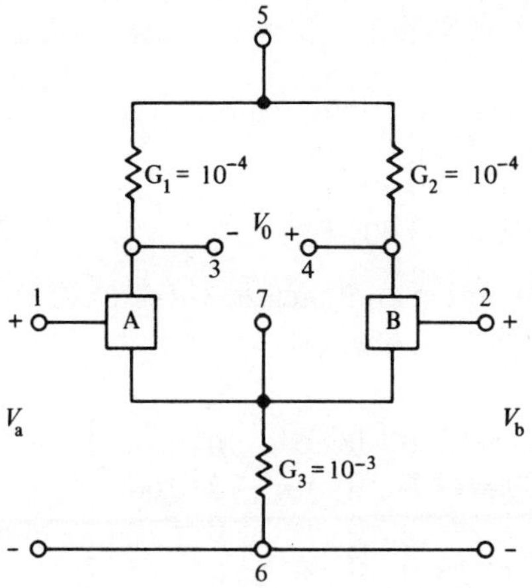

Figure 6-22 A differential amplifier.

devices labeled A and B are 3-terminal amplifying devices. In normal operation, terminal 5 is connected to a dc supply for bias, input voltage sources are connected to the terminal pairs (1, 6) and (2, 6), and a load is connected to (3, 4). We shall show that the output voltage V_0 is proportional to the difference between the input voltages V_a and V_b [$V_0 = K(V_a - V_b)$] when the network is balanced. The balanced condition occurs when devices A and B are identical and when $G_1 = G_2$. Balance is the important consideration. The type of amplifying devices (bipolar transistors, operational amplifiers, FET's, and so on) is not significant.

Suppose that both amplifying devices are FET's characterized by the

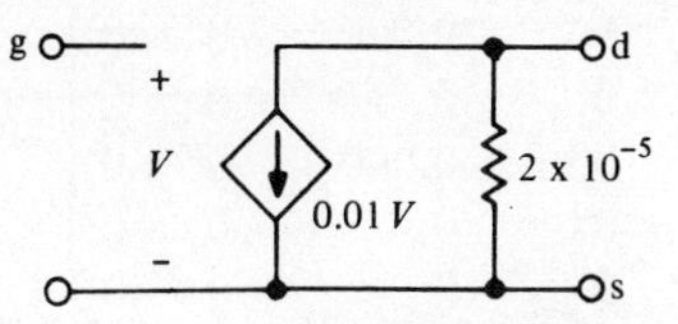

Figure 6-23 Low-frequency FET model.

models presented in Figure 6-23. The gates are connected to terminals 1 and 2, and the sources at terminal 7.

The first step is to construct the IAM's for each of the network components. From Figure 6-23 we note that the IAM for the FET is given by

$$\mathbf{I}_{gds} = \begin{bmatrix} 0 & 0 & 0 \\ 0.01 & 0.00002 & -0.01002 \\ -0.01 & -0.00002 & 0.01002 \end{bmatrix} \tag{6-77}$$

Hence, the IAM's for devices A and B are given by

$$\mathbf{I}^{(A)}_{1,3,7} = \mathbf{I}^{(B)}_{2,4,7} = \begin{bmatrix} 0 & 0 & 0 \\ 0.01 & 0.00002 & -0.01002 \\ -0.01 & -0.00002 & 0.01002 \end{bmatrix} \tag{6-78}$$

The IAM's for G_1 and G_2 are given by

$$\mathbf{I}^{(G_1)}_{5,3} = \mathbf{I}^{(G_2)}_{5,4} = \begin{bmatrix} 0.0001 & -0.0001 \\ -0.0001 & 0.0001 \end{bmatrix} \tag{6-79}$$

and for G_3 we write

$$\mathbf{I}^{(G_3)}_{7,6} = \begin{bmatrix} 0.001 & -0.001 \\ -0.001 & 0.001 \end{bmatrix} \tag{6-80}$$

The next step is to use property 4 to expand each of these IAM's to 7×7's by adding in rows and columns of zeros. Hence, we obtain

$$\mathbf{I}^{(A)}_{1,2,3,4,5,6,7} = \begin{bmatrix} 0 & 0 & 0 & 0 & 0 & 0 & 0 \\ 0 & 0 & 0 & 0 & 0 & 0 & 0 \\ 0.01 & 0 & 0.00002 & 0 & 0 & 0 & -0.01002 \\ 0 & 0 & 0 & 0 & 0 & 0 & 0 \\ 0 & 0 & 0 & 0 & 0 & 0 & 0 \\ 0 & 0 & 0 & 0 & 0 & 0 & 0 \\ -0.01 & 0 & -0.00002 & 0 & 0 & 0 & 0.01002 \end{bmatrix} \tag{6-81}$$

$$\mathbf{I}^{(B)}_{1,2,3,4,5,6,7} = \begin{bmatrix} 0 & 0 & 0 & 0 & 0 & 0 & 0 \\ 0 & 0 & 0 & 0 & 0 & 0 & 0 \\ 0 & 0 & 0 & 0 & 0 & 0 & 0 \\ 0 & 0.01 & 0 & 0.00002 & 0 & 0 & -0.01002 \\ 0 & 0 & 0 & 0 & 0 & 0 & 0 \\ 0 & 0 & 0 & 0 & 0 & 0 & 0 \\ 0 & -0.01 & 0 & -0.00002 & 0 & 0 & 0.01002 \end{bmatrix} \tag{6-82}$$

$$\mathbf{I}^{(G_1)}_{1,2,3,4,5,6,7} = \begin{bmatrix} 0 & 0 & 0 & 0 & 0 & 0 & 0 \\ 0 & 0 & 0 & 0 & 0 & 0 & 0 \\ 0 & 0 & 0.0001 & 0 & -0.0001 & 0 & 0 \\ 0 & 0 & 0 & 0 & 0 & 0 & 0 \\ 0 & 0 & -0.0001 & 0 & 0.0001 & 0 & 0 \\ 0 & 0 & 0 & 0 & 0 & 0 & 0 \\ 0 & 0 & 0 & 0 & 0 & 0 & 0 \end{bmatrix} \tag{6-83}$$

$$\mathbf{I}^{(G_2)}_{1,2,3,4,5,6,7} = \begin{bmatrix} 0 & 0 & 0 & 0 & 0 & 0 & 0 \\ 0 & 0 & 0 & 0 & 0 & 0 & 0 \\ 0 & 0 & 0 & 0 & 0 & 0 & 0 \\ 0 & 0 & 0 & 0.0001 & -0.0001 & 0 & 0 \\ 0 & 0 & 0 & -0.0001 & 0.0001 & 0 & 0 \\ 0 & 0 & 0 & 0 & 0 & 0 & 0 \\ 0 & 0 & 0 & 0 & 0 & 0 & 0 \end{bmatrix} \quad \text{(6-84)}$$

$$\mathbf{I}^{(G_3)}_{1,2,3,4,5,6,7} = \begin{bmatrix} 0 & 0 & 0 & 0 & 0 & 0 & 0 \\ 0 & 0 & 0 & 0 & 0 & 0 & 0 \\ 0 & 0 & 0 & 0 & 0 & 0 & 0 \\ 0 & 0 & 0 & 0 & 0 & 0 & 0 \\ 0 & 0 & 0 & 0 & 0 & 0 & 0 \\ 0 & 0 & 0 & 0 & 0 & 0.001 & -0.001 \\ 0 & 0 & 0 & 0 & 0 & -0.001 & 0.001 \end{bmatrix} \quad \text{(6-85)}$$

According to the interconnection rule of Section 6-4, the IAM for the complete network is the sum of these IAM's. Adding these, we obtain

$$\mathbf{I}_{1,2,3,4,5,6,7} = \begin{bmatrix} 0 & 0 & 0 & 0 & 0 & 0 & 0 \\ 0 & 0 & 0 & 0 & 0 & 0 & 0 \\ 0.01 & 0 & 0.00012 & 0 & -0.0001 & 0 & -0.01002 \\ 0 & 0.01 & 0 & 0.00012 & -0.0001 & 0 & -0.01002 \\ 0 & 0 & -0.0001 & -0.0001 & 0.0002 & 0 & 0 \\ 0 & 0 & 0 & 0 & 0 & 0.001 & -0.001 \\ -0.01 & -0.01 & -0.00002 & -0.00002 & 0 & -0.001 & 0.02104 \end{bmatrix} \quad \text{(6-86)}$$

We now have the 7×7 IAM for the 7-terminal network of Figure 6-22. The next step is to eliminate the terminals of no interest, that is, terminals 5 and 7. Terminal 5 is usually connected to a direct-current bias power supply. This could be modeled in Figure 6-22 by connecting a direct-current independent voltage source between terminals 5 and 6. As a consequence, the output voltage V_0 (by superposition) consists of a direct-current term due to this bias, as well as a signal component due to the input signals. We are interested only in the signal component, which, by superposition (Section 2-3), can be computed by setting the direct-current bias independent voltage source to zero. This is equivalent to connecting terminal 5 to terminal 6. (See also Section 2-4.) Hence, we eliminate terminal 5 via property 7 to obtain the 6×6 IAM.

$$\mathbf{I}_{1,2,3,4,6,7} = \begin{bmatrix} 0 & 0 & 0 & 0 & 0 & 0 \\ 0 & 0 & 0 & 0 & 0 & 0 \\ 0.01 & 0 & 0.00012 & 0 & -0.0001 & -0.01002 \\ 0 & 0.01 & 0 & 0.00012 & -0.0001 & -0.01002 \\ 0 & 0 & -0.0001 & -0.0001 & 0.0012 & -0.001 \\ -0.01 & -0.01 & -0.00002 & -0.00002 & -0.001 & 0.02104 \end{bmatrix} \quad \text{(6-87)}$$

Terminal 7 can be suppressed by the application of property 6. This results in the 5 × 5 IAM:

$$\mathbf{I}_{1,2,3,4,6} = \begin{bmatrix} 0 & 0 & 0 & 0 & 0 \\ 0 & 0 & 0 & 0 & 0 \\ 0.00524 & -0.00476 & 0.00011 & -0.00001 & -0.00058 \\ -0.00476 & 0.00524 & -0.00001 & 0.00011 & -0.00058 \\ -0.00048 & -0.00048 & -0.00010 & -0.00010 & 0.00115 \end{bmatrix} \quad \textbf{(6-88)}$$

One further reduction can be accomplished by using property 5 to convert from a terminal to a port basis. Hence, the y parameter matrix

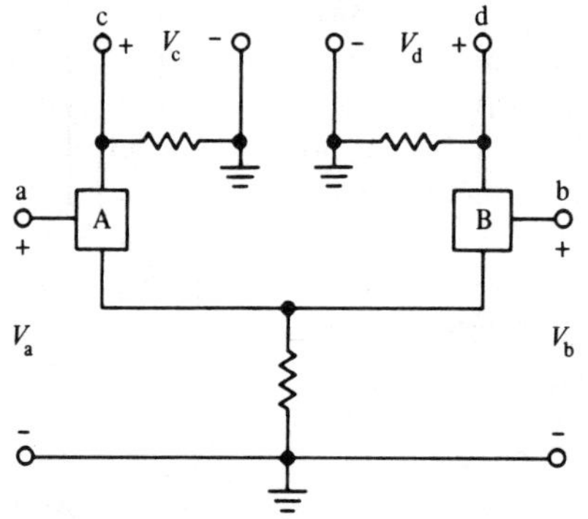

Figure 6-24 4-port network.

for the 4-port network of Figure 6-24 is obtained from Equation (6-88) by setting $V_6 = 0$ and discarding the last row.

$$\mathbf{Y}_{a,b,c,d} = \begin{bmatrix} 0 & 0 & 0 & 0 \\ 0 & 0 & 0 & 0 \\ 0.00524 & -0.00476 & 0.00011 & -0.00001 \\ -0.00476 & 0.00524 & -0.00001 & 0.00011 \end{bmatrix} \quad \textbf{(6-89)}$$

Hence, the equations relating the port voltages and the port currents for the network of Figure 6-24 are

$$\begin{aligned} I_a &= 0 \\ I_b &= 0 \\ I_c &= 0.00524V_a - 0.00476V_b + 0.00011V_c - 0.00001V_d \\ I_d &= -0.00476V_a + 0.00524V_b - 0.00001V_c + 0.00011V_d \end{aligned} \quad \textbf{(6-90)}$$

The 4-port network of Figure 6-24 can also be interpreted as a 3-port network of Figure 6-25. To get the 3-port equations for the 3-port network from the 4-port equations for the 4-port networks, we note that $V_0 = V_d - V_c$ and that $I_0 = I_c = -I_d$ so that $I_0 = \frac{1}{2}(I_c - I_d)$. Hence, in Equations (6-90), we subtract the fourth equation from the third equation:

$$\begin{aligned} I_c - I_d &= 0.01000V_a - 0.01000V_b + 0.00012V_c - 0.00012V_d \\ &= 0.01(V_a - V_b) + 0.00012(V_c - V_d) \end{aligned} \quad \textbf{(6-91)}$$

$$2I_0 = 0.01(V_a - V_b) + 0.00012V_0 \quad \textbf{(6-92)}$$

Hence, the three equations for the 3-port of Figures 6-25 are given by

$$\begin{aligned} I_a &= 0 \\ I_b &= 0 \\ I_0 &= 0.005V_a - 0.005V_b - 0.00006V_0 \end{aligned} \tag{6-93}$$

The easiest way to see that the network is a differential amplifier is to make a Norton (or Thevenin) equivalent network with respect to the output port. To accomplish this we assume that an independent voltage

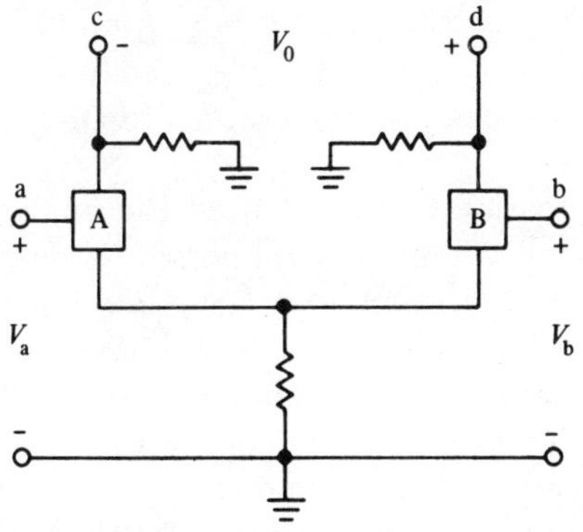

Figure 6-25 3-port differential amplifier.

source V_{s_1} is connected to port a and a second independent voltage source V_{s_2} is connected to port b, and we compute the open-circuit voltage and short-circuit current at the output port. To compute the open-circuit voltage, we use Equations (6-93) along with the constraint equations

$$\begin{aligned} V_a &= V_{s_1} \\ V_b &= V_{s_2} \\ I_0 &= 0 \end{aligned} \tag{6-94}$$

The solution is

$$V_{\text{open-circuit}} = V_0 \Big|_{I_0=0} = \frac{0.005}{0.00006}(V_{s_1} - V_{s_2}) = 83.33(V_{s_1} - V_{s_2}) \tag{6-95}$$

To compute the short-circuit current, we use the constraint equations

$$\begin{aligned} V_a &= V_{s_1} \\ V_b &= V_{s_2} \\ V_0 &= 0 \end{aligned} \tag{6-96}$$

and solve for I_0:

$$I_{\text{short-circuit}} = I_0 \Big|_{V_0=0} = 0.005(V_{s_1} - V_{s_2}) \tag{6-97}$$

Hence, the Thevenin equivalent impedance is given by

$$\mathfrak{z}^{(\text{Thevenin})} = \frac{V_{\text{open-circuit}}}{I_{\text{short-circuit}}} = \frac{83.3(V_{s_1} - V_{s_2})}{0.005(V_{s_1} - V_{s_2})} = 16{,}666 \tag{6-98}$$

The Norton and Thevenin equivalent networks with respect to the output port with independent sources V_{s_1} and V_{s_2} connected at ports a and b

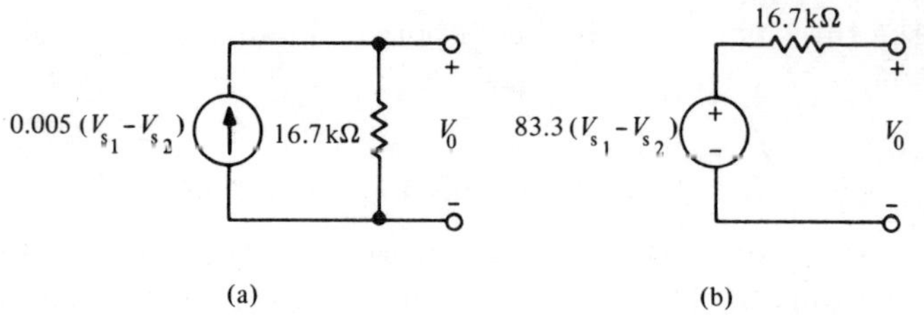

Figure 6-26 (a) Norton's and (b) Thevenin's equivalent networks for the network of Figure 6-22.

are presented in Figure 6-26. The load sees a source that is proportional to $(V_{s_1} - V_{s_2})$ and a 16.7 kΩ resistance.

This example illustrates the application of the IAM to the analysis of a moderately complicated network. For this numerical example there are short cuts that would allow us to arrive at the answer more quickly. Usually, however, the network analyst is not interested in numerical examples. A thorough analysis of the differential amplifier would normally include an investigation of the dependence of the output on variations in the pertinent parameter values. The effect of various amounts of imbalance should be investigated; for example, what happens if the g_m's of the FET's are not quite identical—just how well matched must they be? Could the effect of mismatched FET's be lessened by changing the resistors? For such a thorough analysis of the more general case, the IAM is an effective approach.

We also point out that for more complicated networks, the procedure is no more difficult; it is simply more tedious—but the tedium can be delegated to a digital computer. In the differential-amplifier example, we could have stopped after writing Equations (6-78, 79, and 80) and had the computer do the rest, including plotting curves of the important input-output quantities versus the various element values and the frequency.

PROBLEMS

6-1 Construct the IAM for the network shown in Figure P6-1. (The y's are complex admittances.)

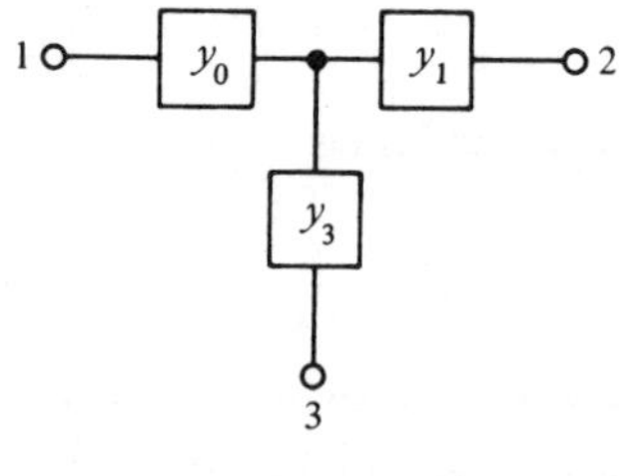

Figure P6-1

6-2 Write the IAM for the network shown in Figure P6-2. This is a special case of Problem 6-1.

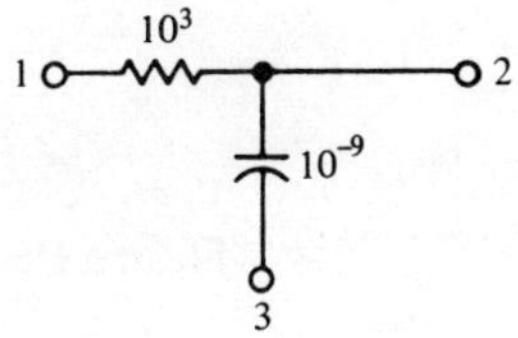

Figure P6-2

6-3 Compute the IAM for the field-effect transistor (see Figure P6-3). Answer:

$$\mathbf{I}_{gds} = \begin{bmatrix} j\omega(C_1 + C_c) & -j\omega C_c & -j\omega C_1 \\ g_m - j\omega C_c & G + j\omega C_c & -g_m - G \\ -g_m - j\omega C_1 & -G & g_m + G + j\omega C_1 \end{bmatrix}$$

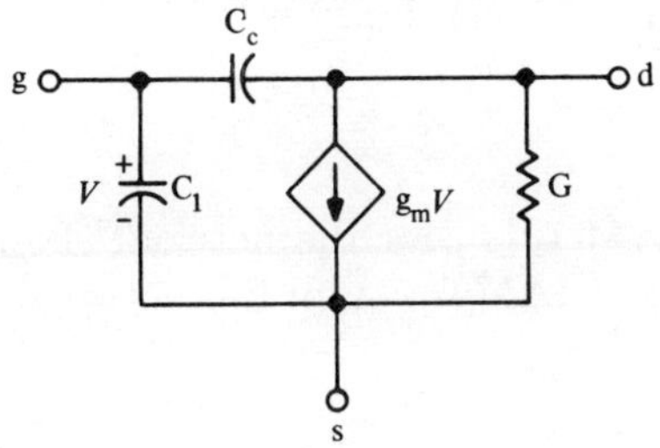

Figure P6-3

6-4 Construct the IAM for the transistor model shown in Figure P6-4.

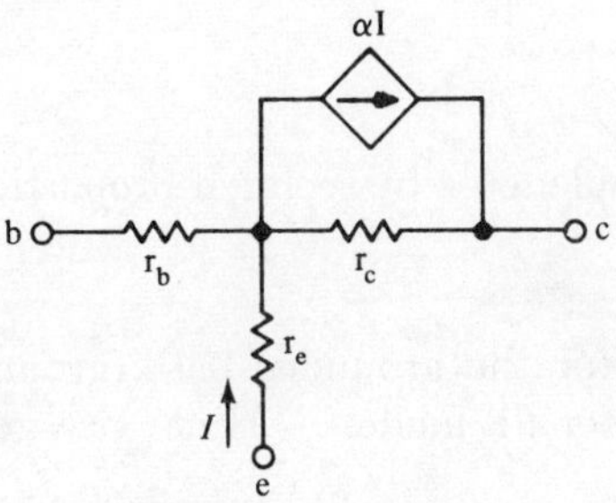

Figure P6-4

6-5 **a.** Construct the IAM for the network shown in Figure P6-5.
b. Show that if $\beta = \alpha/1 - \alpha$, then the IAM is the same as that of Problem 6-4. $r_d = r_c(1 - \alpha)$

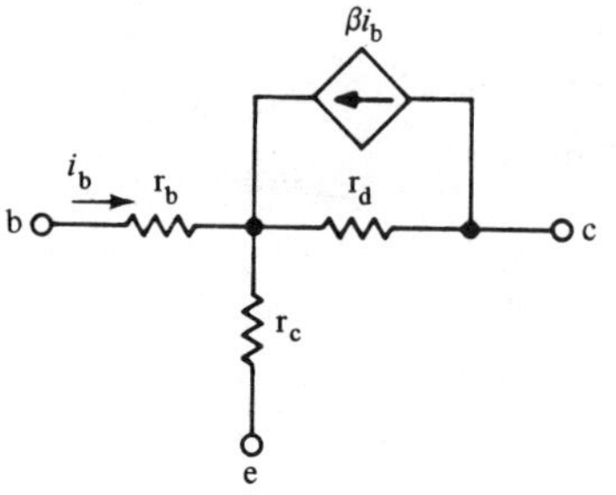

Figure P6-5

6-6 Write the IAM's $\mathbf{I}_{1,2,3}$ and $\mathbf{I}_{2,1,3}$ for the network shown in Figure P6-6.

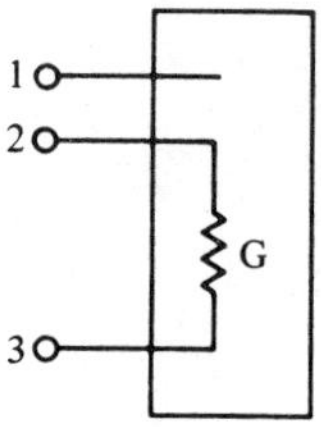

Figure P6-6

6-7 Construct the 4 × 4 IAM $\mathbf{I}_{1,2,3,4}$ for the network shown in Figure P6-7. Also construct the 4 × 4 IAM $\mathbf{I}_{1,3,2,4}$ for this network and compare to $\mathbf{I}_{1.2\ 3.4}$.

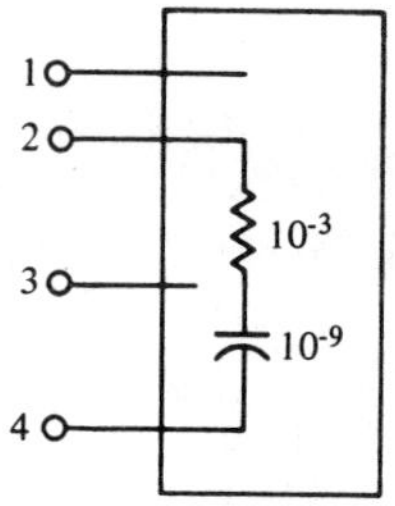

Figure P6-7

6-8 Check your answer to Problem 6-4 by seeing if properties 1 and 2 are satisfied.

6-9 Write the y parameters for the grounded-base, grounded-emitter, and grounded-collector transistor (T model). Hint: Use your result for Problem 6-4 and property 5.

6-10 Start with the 3 × 3 IAM constructed in Problem 6-2, and suppress terminal 2 to obtain the 2 × 2 IAM for the network shown in Figure P6-10. Check your answer by writing the 2 × 2 IAM directly from the network shown.

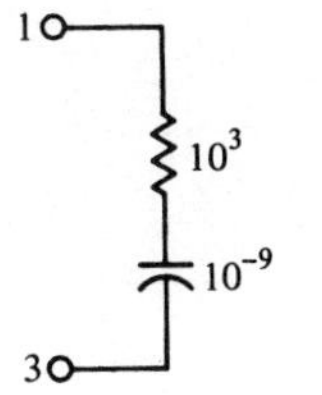

Figure P6-10

6-11 Construct the 4 × 4 IAM for the network shown in Figure P6-11. Then use property 6 to suppress terminal a to find the 3 × 3 IAM for the transistor. Which method is easier—this one or the direct method of Problem 6-5?

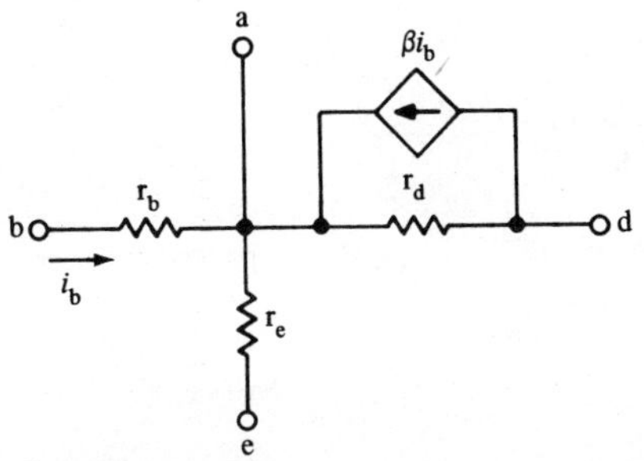

Figure P6-11

6-12 Compute the 3 × 3 IAM for the FET by first writing the 4 × 4 IAM for the four-terminal network shown in Figure P6-12 and then using property 7. Which way is easier?

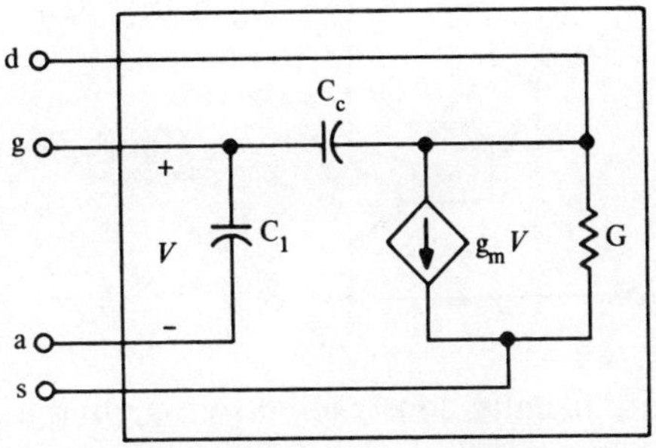

Figure P6-12

6-13 Compute the y parameters for each of the networks shown in Figure P6-13. Use the IAM techniques, but think before you start writing. You may assume all batteries and capacitors are short circuits.

The transistor is a 2N3705 with $\alpha = -0.99$, $r_c = 10$ kΩ, $r_e = 30$ Ω, $r_b = 1$ kΩ. Also assume $R_c = 1$ kΩ, $R_e = 100$ Ω, $R_B = 20$ kΩ (R_B is a parallel combination of R_{B1} and R_{B2}).

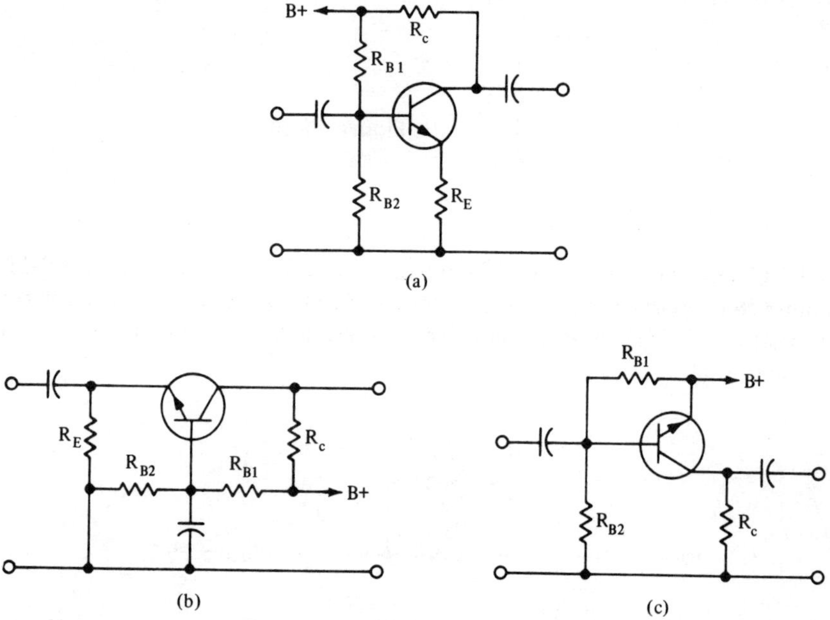

Figure P6-13

6-14 Compute the y parameters for the Darlington compound transistor shown in Figure P6-14. Both transistors are 2N3705's.

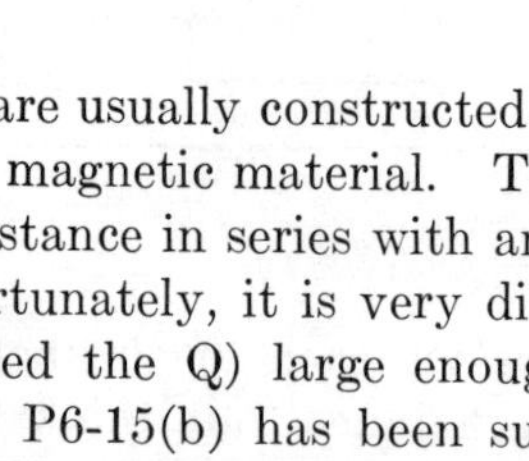

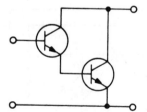

Figure P6-14

6-15 Inductors are usually constructed by winding a number of turns of wire on a core of magnetic material. The equivalent network of such an inductor is a resistance in series with an inductance, as shown in Figure P6-15(a). Unfortunately, it is very difficult (or impossible) to get the ratio $\omega L/R$ (called the Q) large enough for many applications. The circuit of Figure P6-15(b) has been suggested for providing a high Q network. Use $R = 100$, $C = 100$ pF. For the transistor $r_e = 10$,

$r_b = 10^3$, $r_c = 10^6$, $\alpha = 0.99$. Verify that $\mathfrak{z}_{in}$ is of the form $\mathfrak{z}_{in} = R + jX$. If $Q = X/R$, what is the Q?

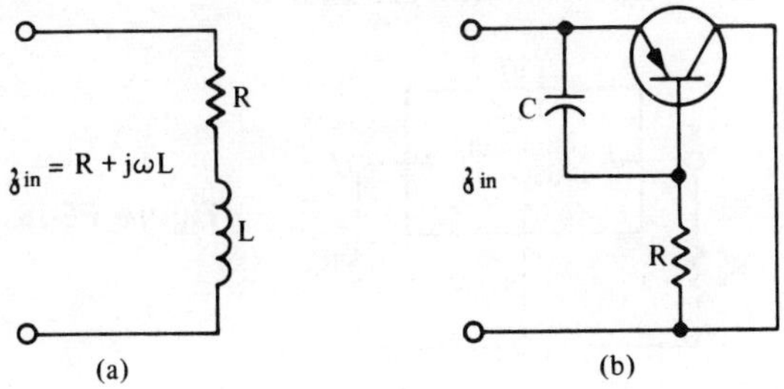

Figure P6-15

6-16 For the bridge circuit shown in Figure P6-16

a. Construct the 4 × 4 IAM.

b. Suppose the circuit is to be used as a 2-port with terminals 1 and 2 as port 1, and 3 and 4 as port 2. Relate the terminal voltages, V_{T1}, V_{T2}, V_{T3}, V_{T4}, and the terminal currents I_1, I_2, I_3, and I_4, in the IAM to the port voltages V_1 and V_2 and port currents J_1 and J_2.

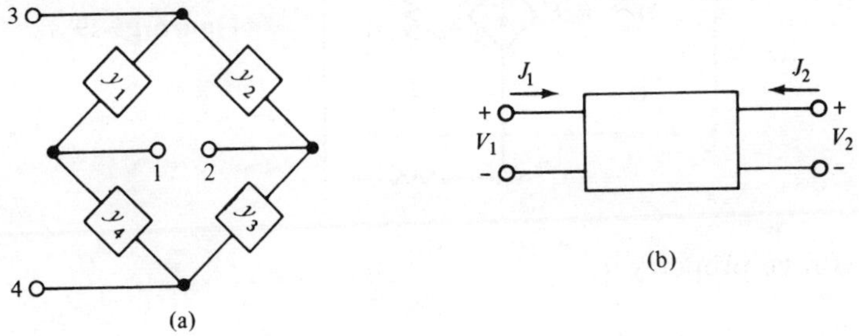

Figure P6-16

6-17 For the "Twin-T" network shown in Figure P6-17

a. Compute the IAM.

b. Obtain the y parameter for the above "Twin-T" network with terminal b grounded.

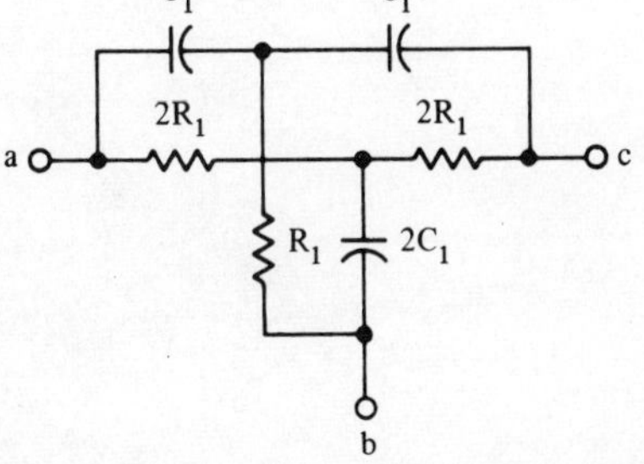

Figure P6-17

6-18 A properly biased grounded-source FET amplifier would have the ac equivalent circuit of Figure P6-18. Find the total IAM, I_{gds}, and the y parameter with terminal s grounded.

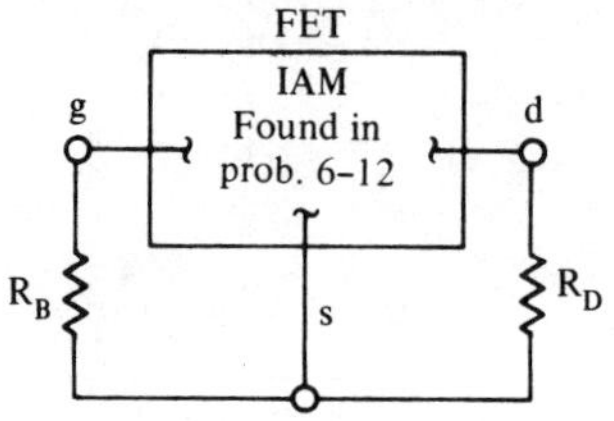

Figure P6-18

6-19 A properly biased grounded-drain (sometimes called a source-follower) FET amplifier would have the low-frequency, ac equivalent circuit of Figure P6-19. Find the total IAM, I_{gds}, and the y parameter with terminal d grounded.

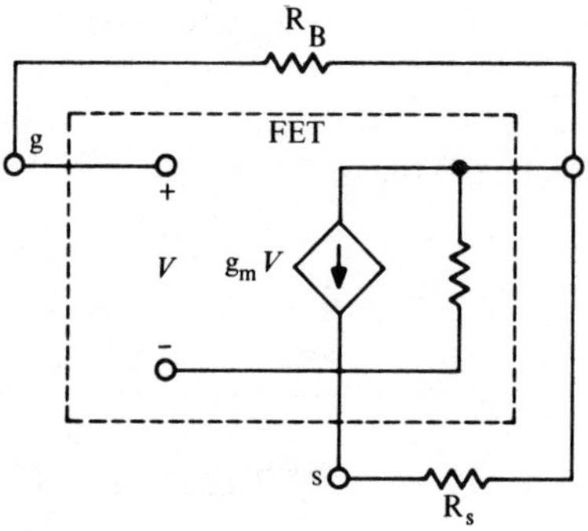

Figure P6-19

6-20 Prove property 7.

CHAPTER

7

Linear Amplifiers

7-1 Introduction

The most common application of linear electronic circuits is for signal processing. The most common signal-processing operations are amplification and filtering. This chapter is devoted to a discussion of linear amplifiers. Linear filters and frequency selective amplifiers are discussed in Chapter 9.

The term "amplifier" is used in a rather unrestrictive sense in engineering. Amplifiers have input terminals and output terminals. When a signal is applied at the input terminal, a signal is produced at the output terminal. Several amplifier characteristics that are frequently used to specify amplifier performance are voltage gain, current gain, power gain, insertion power gain, input impedance, output impedance, isolation, and bandwidth. In Section 7-2 we define these quantities. In Section 7-3 we use these concepts to analyze a bipolar transistor amplifier, and in Section 7-4 we analyze a field-effect transistor amplifier. The method used in these sections is a very general and very powerful one that can be used for any 2-port network. In Section 7-3 we also show how to analyze a simple transistor circuit by a straightforward application of the analysis techniques presented in Chapters 2 and 3. We again utilize the sinusoidal steady-state analysis developed in Chapter 4 to discuss amplifier performance for sinusoidal input signals.

7-2 Amplifier Characteristics

An amplifier with one input port and one output port can be modeled by the 2-port network presented in Figure 7-1(a). We denote the amplifier input voltage and current by V_1 and I_1, and the amplifier output voltage and current by V_2 and I_2. An amplifier is usually "driven" by a "driving source" and "loaded" by a "load," as illustrated in Figure 7-1(b). The

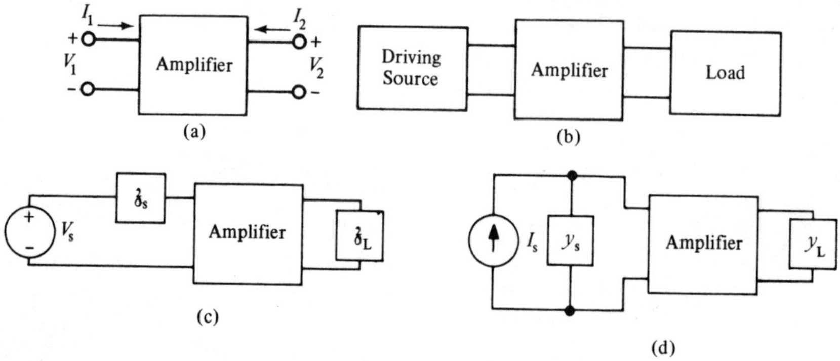

Figure 7-1 Amplifier connections.

driving source may represent the signal source (such as an antenna) or the output of the preceding stage if there are other signal processing operations between the signal source and the amplifier. Similarly, the load impedance usually represents the input to the succeeding stage. (See Problem 7-1.) The driving source can always be represented by its Thevenin equivalent, as illustrated in Figure 7-1(c). In many applications, no independent sources are contained in the succeeding stages so that the load has a Thevenin equivalent network consisting only of an impedance, as shown in Figure 7-1(c). An equally valid representation is presented in Figure 7-1(d), where I_s is an independent current source, and y_s and y_L represent the source and load admittances. It is apparent from Figure 7-1(c) that both the input terminal pair and output terminal pair are ports. Hence, any of the 2-port characterizations discussed in Section 5-3 can be used to characterize the amplifier. One convenient characterization for present-day 2-port amplifiers is the short-circuit admittance parameter matrix.[1] Hence, the amplifier input port and output port voltages and currents are related by the two equations

$$\left.\begin{aligned} I_1 &= y_{11}V_1 + y_{12}V_2 \\ I_2 &= y_{21}V_1 + y_{22}V_2 \end{aligned}\right\} \text{amplifier equations} \qquad \begin{aligned}&(7\text{-}1)\\&(7\text{-}2)\end{aligned}$$

where, for any particular amplifier, the y parameters are assumed known. By referring to Figure 7-1(d), we can also write a driving-source (or input) constraint equation of the form

$$I_1 = I_s - y_s V_1\} \text{ input constraint equation} \qquad (7\text{-}3)$$

and a load (or output) constraint equation of the form

$$I_2 = -y_L V_2\} \text{ output constraint equation} \qquad (7\text{-}4)$$

If we assume the short-circuit driving-source current I_s to be a known quantity, Equations (7-1) through (7-4) are four equations in the four unknown quantities V_1, I_1, V_2, and I_2. One way to solve for the terminal voltages and currents is to substitute Equation (7-3) into Equation (7-1) and Equation (7-4) into Equation (7-2), which gives

$$\begin{aligned} I_s - y_s V_1 &= y_{11}V_1 + y_{12}V_2 \\ -y_L V_2 &= y_{21}V_1 + y_{22}V_2 \end{aligned} \qquad (7\text{-}5)$$

or

$$\begin{aligned} I_s &= (y_{11} + y_s)V_1 + (y_{12})V_2 \\ 0 &= (y_{21})V_1 + (y_{22} + y_L)V_2 \end{aligned} \qquad (7\text{-}6)$$

Now, using Cramer's rule, we write

$$V_1 = \frac{\begin{vmatrix} I_s & y_{12} \\ 0 & y_{22} + y_L \end{vmatrix}}{\begin{vmatrix} y_{11} + y_s & y_{12} \\ y_{21} & y_{22} + y_L \end{vmatrix}} = \frac{y_{22} + y_L}{(y_{11} + y_s)(y_{22} + y_L) - y_{12}y_{21}} I_s \qquad (7\text{-}7)$$

$$V_2 = \frac{\begin{vmatrix} y_{11} + y_L & I_s \\ y_{21} & 0 \end{vmatrix}}{\begin{vmatrix} y_{11} + y_s & y_{12} \\ y_{21} & y_{22} + y_L \end{vmatrix}} = \frac{-y_{21}}{(y_{11} + y_s)(y_{22} + y_L) - y_{12}y_{21}} I_s \qquad (7\text{-}8)$$

[1] The most convenient characterization for an amplifier depends on the physical properties of the devices used in the amplifier. In the days of vacuum tubes, the $\mathfrak{z}$ parameters were easiest to measure in the laboratory; in the early days of transistors, the $\mathfrak{h}$ parameters were considered easiest to measure; in the present day of the field-effect transistor, the y parameters are easiest to measure. However, we choose the y parameters not for ease of measurement, but for ease of computation. The y parameters can always be obtained from any other parameter set from Table 5-1. The student should be able to compute the desired quantities starting from any given 2-port description.

Finally, using Equations (7-7) and (7-8) in Equations (7-3) and (7-4), we obtain

$$I_1 = \frac{y_{11}(y_{22} + y_L) - y_{12}y_{21}}{(y_{11} + y_s)(y_{22} + y_L) - y_{12}y_{21}} I_s \qquad \textbf{(7-9)}$$

$$I_2 = \frac{y_{21}y_L}{(y_{11} + y_s)(y_{22} + y_L) - y_{12}y_{21}} I_s \qquad \textbf{(7-10)}$$

These four quantities (V_1, V_2, I_1, and I_2) in themselves are not of particular interest in amplifier circuits. However, one or more of the various combinations of these quantities is usually used to specify amplifier performance.

Voltage gain (V_2/V_1). The *voltage gain* of an amplifier is the ratio of output voltage to input voltage. This quantity is easily computed from Equations (7-7) and (7-8):

$$\frac{V_2}{V_1} = \frac{-y_{21}}{y_{22} + y_L} \qquad \textbf{(7-11)}$$

Hence, the voltage gain of the amplifier depends on the load admittance as well as the amplifier characteristics, but does not depend on either of the driving source parameters. The voltage gain is a complex number whose magnitude gives the factor by which the output voltage amplitude is larger than the input voltage amplitude and whose angle gives the phase difference between input and output voltages.

Current gain (I_2/I_1). The *current gain* of an amplifier is the ratio of the output current to the input current. This quantity can be computed from Equations (7-9) and (7-10):

$$\frac{I_2}{I_1} = \frac{y_{21}y_L}{y_{11}(y_{22} + y_L) - y_{12}y_{21}} \qquad \textbf{(7-12)}$$

Therefore, the current gain depends on the load admittance as well as the amplifier characteristics, but it does not depend on either of the driving source parameters. Its magnitude gives the magnitude of the current gain, and its angle the phase shift of the amplifier.

Power gain (P_L/P_1). The *power gain* of an amplifier is defined as the ratio of the average power delivered to the load admittance to the average power delivered to the amplifier input. The average power delivered to the load admittance is given by

$$P_L = \tfrac{1}{2}|V_2|^2 \operatorname{Re}\{y_L\} \qquad \textbf{(7-13)}$$

while the average power delivered to the amplifier input is given by

$$\mathrm{P}_1 = \tfrac{1}{2}|V_1|^2 \operatorname{Re}\left\{\frac{I_1}{V_1}\right\} \tag{7-14}$$

where I_1/V_1, given by the ratio of Equation (7-9) to (7-8), is later interpreted as the amplifier input admittance. The power gain is the ratio of Equation (7-13) to (7-14):

$$\begin{aligned}\frac{\mathrm{P_L}}{\mathrm{P}_1} &= \frac{\frac{1}{2}|V_2|^2 \operatorname{Re}\{y_\mathrm{L}\}}{\frac{1}{2}|V_1|^2 \operatorname{Re}\{I_1/V_1\}} = \left|\frac{V_2}{V_1}\right|^2 \frac{\operatorname{Re}\{y_\mathrm{L}\}}{\operatorname{Re}\{I_1/V_1\}} \\ &= \left|\frac{y_{21}}{y_{22}+y_\mathrm{L}}\right|^2 \frac{\operatorname{Re}\{y_\mathrm{L}\}}{\operatorname{Re}\left\{y_{11} - \dfrac{y_{12}y_{21}}{y_{22}+y_\mathrm{L}}\right\}}\end{aligned} \tag{7-15}$$

Often one sees the statement that the power gain is the product of the voltage gain and current gain. This is true only if the input impedance of the amplifier and the load impedance are real. So long as the input impedance is real we may write[2]

$$\frac{\mathrm{P_L}}{\mathrm{P}_1} = \operatorname{Re}\left\{\left(\frac{V_2}{V_1}\right)\left(\frac{I_2}{I_1}\right)^*\right\} \qquad \frac{V_1}{I_1} \text{ real} \tag{7-16}$$

The power gain is always a real number.

Insertion power gain ($\mathrm{P_L}/\mathrm{P'_L}$). The *insertion power gain* of an amplifier is defined as the ratio of the average power delivered to the load admittance with the amplifier connected between the driving source and the load admittance, as in Figure 7-2(a), to the average power delivered to

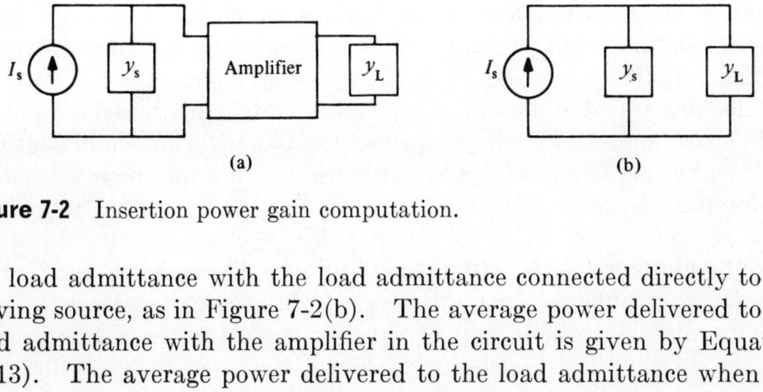

Figure 7-2 Insertion power gain computation.

the load admittance with the load admittance connected directly to the driving source, as in Figure 7-2(b). The average power delivered to the load admittance with the amplifier in the circuit is given by Equation (7-13). The average power delivered to the load admittance when the

[2] This formula is easily related to the average power formulae of Chapter 4. The average power carried by a voltage and current at a port is $\frac{1}{2}$ Re $[VI^*]$.

load admittance is connected directly to the driving source is given by

$$P_L' = \frac{1}{2}\left|\frac{y_L}{y_L + y_s} I_s\right|^2 \text{Re}\{y_L^{-1}\} \tag{7-17}$$

The insertion power gain can now be computed from Equations (7-13) and (7-17):

$$\begin{aligned}\frac{P_L}{P_L'} &= \frac{\frac{1}{2}|V_2|^2 \text{Re}\{y_L\}}{\frac{1}{2}\left|\frac{y_L}{y_L + y_s}\right|^2 |I_s|^2 \text{Re}\{y_L^{-1}\}} \\ &= \left|\frac{-y_{21}}{(y_{11} + y_s)(y_{22} + y_L) - y_{12}y_{21}} \times \frac{y_L + y_s}{y_L}\right|^2 \frac{\text{Re}\{y_L\}}{\text{Re}\{y_L^{-1}\}}\end{aligned} \tag{7-18}$$

Input admittance (I_1/V_1 with source disconnected). The *input admittance* of an amplifier is defined as the ratio of the input current to the input voltage, with the driving source disconnected. Hence, the input admittance is not constrained by the driving source or the input constraint equation that characterizes it. The input admittance is computed by substituting Equation (7-4) into Equation (7-2), solving the resulting equation for V_2, and substituting this result into Equation (7-1). We find

$$\frac{I_1}{V_1} = y_{11} - \frac{y_{12}y_{21}}{y_{22} + y_L} \tag{7-19}$$

The input admittance depends on the characteristics of the load admittance, as well as the characteristics of the amplifier, and gives the amplitude and phase relationships between the input current and the input voltage.

Note that the same result is obtained by dividing Equation (7-9) by Equation (7-7). Even though the driving source constrains the input current and the input voltage, their ratio is independent of this constraint.

The amplifier input admittance is the Norton equivalent admittance seen looking into the amplifier input port. Since neither the amplifier or its load contains any independent sources, the Norton equivalent network for the amplifier and load relative to its input terminals consists of the input admittance only.

Output admittance (I_2/V_2 with load admittance disconnected and $I_s = 0$). The *output admittance* of an amplifier is defined as the ratio of the output current to the output voltage, with the load admittance disconnected and the driving source set to zero. This is the same computation used to compute the input admittance, except that the roles of ports 1 and 2 are interchanged and y_L replaced by y_s. (Set $I_s = 0$ in Figure 7-1(d) and note the symmetry.) Therefore, in Equation (7-19), we interchange subscripts 1 and 2 and replace y_L by y_s to obtain

$$\left.\frac{I_2}{V_2}\right|_{y_L = I_s = 0} = y_{22} - \frac{y_{21}y_{12}}{y_{11} + y_s} \tag{7-20}$$

The output admittance depends on the driving source admittance as well as the amplifier characteristics.

The amplifier output admittance can also be computed by disconnecting the load admittance, computing the output short-circuit current and the output open-circuit voltage, and taking the ratio of the two.

The amplifier output admittance is the Norton equivalent admittance seen looking into its output terminals. The Norton equivalent network of the amplifier and its driving source relative to its output terminals consist of an independent current source in parallel with an admittance. The value of the independent current source is the output short-circuit current, and the value of the admittance is given by Equation (7-20).

Voltage isolation. The *voltage isolation* of an amplifier is defined as the ratio of the amplifier voltage gain to the voltage gain obtained with the amplifier input port and output port interchanged. The amplifier voltage gain is given by Equation (7-11). The amplifier backward voltage gain is computed by disconnecting the current source I_s in Figure 7-1(d) from the input terminals, reconnecting to the output terminals, and computing the ratio V_1/V_2. Because of the symmetry, this is given by Equation (7-11), with y_L replaced by y_s and the port subscripts interchanged. Hence, we obtain

$$\left(\frac{V_1}{V_2}\right)' = \left.\frac{V_1}{V_2}\right|_{I_s \text{ connected at output}} = \frac{-y_{12}}{y_{11} + y_s} \tag{7-21}$$

and the voltage isolation is given by

$$\frac{V_2/V_1}{(V_1/V_2)'} = \frac{y_{21}(y_{11} + y_s)}{y_{12}(y_{22} + y_L)} \tag{7-22}$$

The voltage isolation is a measure of the effectiveness of the amplifier in isolating the load from the driving source.

For example, it is important that radio receivers do not act as transmitters. Hence, the ratio of the forward to backward gain of the RF (radio frequency) amplifier must be high in order to keep the various signals generated in the mixer (see Figures 1-1 and 5-1) from propagating back to the antenna where they would be radiated. Finally, we point out that we could also define a "current isolation" and a "power isolation" in a similar manner. However, isolation is almost always defined as voltage isolation.

For the reader's convenience, the results of this section are summarized in Table 7-1.

Table 7-1

Voltage gain	$= \dfrac{y_{21}}{y_{22} + y_L}$
Current gain	$= \dfrac{y_{21} y_L}{y_{11}(y_{22} + y_L) - y_{12} y_{21}}$
Power gain	$= \left\lvert \dfrac{y_{21}}{y_{22} + y_L} \right\rvert^2 \dfrac{\text{Re}\{y_L\}}{\text{Re}\{y_{11} - [y_{12} y_{21}/(y_{22} + y_L)]\}}$
Insertion power gain	$= \left\lvert \dfrac{-y_{21}}{(y_{11} + y_s)(y_{22} + y_L) - y_{12} y_{21}} \left(\dfrac{y_L + y_s}{y_L} \right) \right\rvert^2 \dfrac{\text{Re}\{y_L\}}{\text{Re}\{y_L^{-1}\}}$
Input admittance	$= y_{11} - \dfrac{y_{12} y_{21}}{y_{22} + y_L}$
Output admittance	$= y_{22} - \dfrac{y_{12} y_{21}}{y_{11} + y_s}$
Voltage isolation	$= \dfrac{y_{21}(y_{11} + y_s)}{y_{12}(y_{22} + y_L)}$

7-3 Bipolar Transistor Amplifiers

One common low-frequency amplifying device is the bipolar transistor. Bipolar transistors are 3-terminal devices and, therefore, can be used in three distinct connections as 2-port amplifiers, grounded emitter, grounded collector, or grounded base. To illustrate the amplifier characteristics discussed in Section 7-2, we consider all three connections for the Delco Radio 2N174 bipolar power transistor. This low-frequency transistor is characterized by a very simple model. For high-frequency transistors with more complicated linear circuit models, the same procedure with more lengthy computations still applies. Since we intend to investigate all three connections, we first construct the IAM for the 3-terminal transistor and then obtain the three sets of y parameters from the IAM.

Like most bipolar junction transistor manufacturers, Delco Radio publishes h parameter data for the grounded-emitter connection. The h parameter $\mathfrak{h}_{21}$ is stated to be between 25 and 50. Let us use 35. The input and output voltage-current characteristics are specified by the curves of Figure 7-3. From the slope of the input characteristics curves we obtain, nominally, $\mathfrak{h}_{11} = 0.1$ volt/0.1 ampere $= 1$ ohm. From the output characteristics curves, we obtain $\mathfrak{h}_{22} = 0.1$ ampere/10 volts $=$ 0.01 mho. Sufficient data for obtaining $\mathfrak{h}_{12}$ are not given. However, for most applications of this transistor, $\mathfrak{h}_{12}$ is negligible; therefore, we use $\mathfrak{h}_{12} = 0$. The only other usable number for linear circuit calculations is the current amplification cutoff frequency, which is specified as 10 kHz. This number will be introduced into our calculations later in this section.

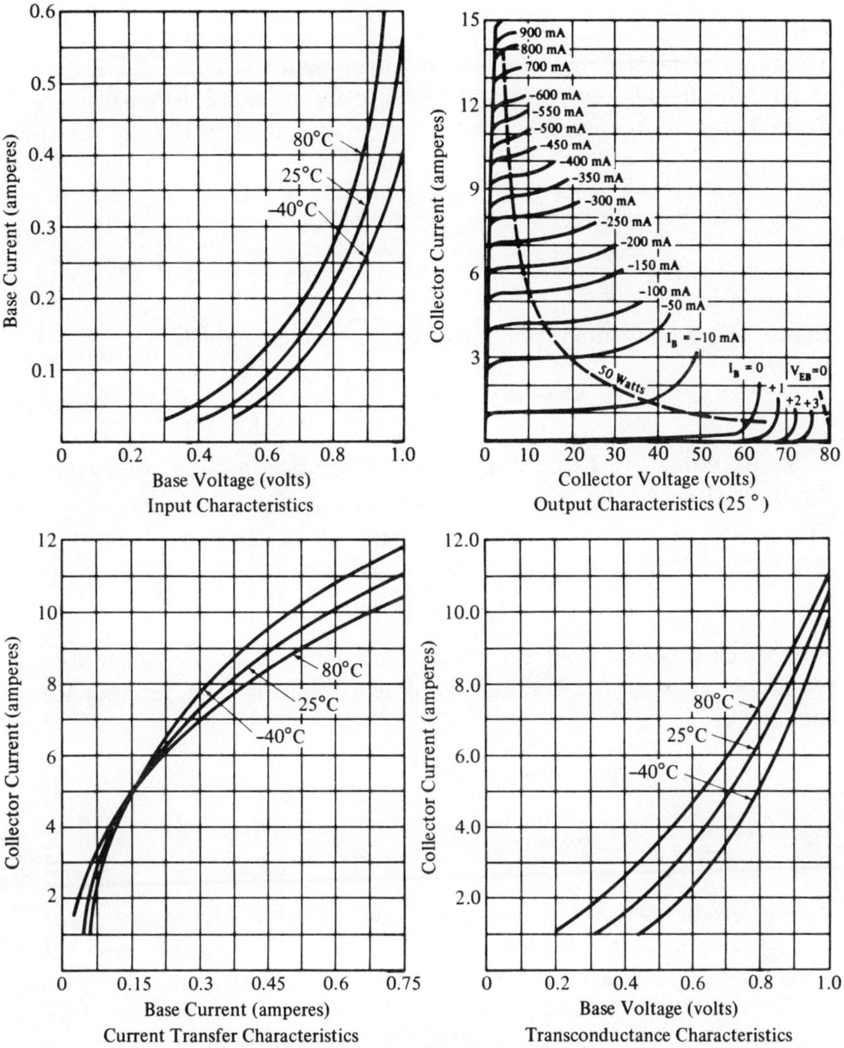

Figure 7-3 Characteristics of the 2N174.

With these data we can construct a linear mathematical model that is a good approximation to the operation of the transistor for the processing of comparatively small signals varying about an appropriately chosen quiescent operating point.

$$\mathbf{H}^{(\text{grounded emitter})} = \begin{bmatrix} 1 & 0 \\ 35 & 0.01 \end{bmatrix} \tag{7-23}$$

From this matrix we can compute all the important amplifier quantities for all three connections. Since we have derived formulas based on the y parameters, it is just as easy to compute the y parameters from the h parameter matrix and use our previous results. Table 5-1 gives the results of our previous computations to convert from h to y parameters. From Table 5-1 we can now write, for the grounded-emitter connection,

$$y_{11} = \frac{1}{\mathfrak{h}_{11}} = 1 \qquad y_{12} = \frac{-\mathfrak{h}_{12}}{\mathfrak{h}_{11}} = 0$$
$$y_{21} = \frac{\mathfrak{h}_{21}}{\mathfrak{h}_{11}} = 35 \qquad y_{22} = \frac{\mathfrak{h}_{11}\mathfrak{h}_{22} - \mathfrak{h}_{12}\mathfrak{h}_{21}}{\mathfrak{h}_{11}} = 0.01 \tag{7-24}$$

Hence, the grounded-emitter y parameter matrix is given by

$$\mathbf{Y}^{(\text{grounded emitter})} = \begin{bmatrix} 1 & 0 \\ 35 & 0.01 \end{bmatrix} \tag{7-25}$$

Using properties 1 and 2 of the IAM (see Section 6-3), we construct the transistor IAM:

$$\mathbf{I}_{b,c,e} = \begin{bmatrix} 1 & 0 & -1 \\ 35 & 0.01 & -35 \\ -36 & -0.01 & 36 \end{bmatrix} \tag{7.26}$$

Now using property 5 of the IAM (see Section 6-3), we can write the y parameter matrices for the grounded-collector and grounded-base connections

$$\mathbf{Y}^{(\text{grounded collector})} = \begin{bmatrix} 1 & -1 \\ -36 & 36 \end{bmatrix} \tag{7-27}$$

$$\mathbf{Y}^{(\text{grounded base})} = \begin{bmatrix} 36 & -0.01 \\ -35 & 0.01 \end{bmatrix} \tag{7-28}$$

For the moment let us assume a 1-mho source admittance ($y_s = 1$) and a 0.01-mho load admittance ($y_L = 0.01$). Then, using these values along with the transistor y parameters specified by Equations (7-25), (7-27), and (7-28) in the equations of Table 7-1, we compute the bipolar transistor amplifier characteristics listed in Table 7-2.

Grounded-emitter connection. From Table 7-2 we get the impression that the grounded-emitter transistor is "best," provided that voltage gain, current gain, power gain, or insertion power gain is taken as the criterion of goodness. The reason is that the source admittance is matched to the amplifier input admittance, and the load admittance is matched[3] to the amplifier output admittance for this particular connec-

[3] In Chapter 2 we saw that when a source with an internal resistance R was connected to a load resistor of value R, then maximum power was transferred from source to load. In the sinusoidal steady state, the corresponding condition of maximum-power transfer occurs between source and load when one is the complex conjugate of the other. This condition is known as having source and load matched.

tion. By comparing $y_s = 1$ to the input admittance and $y_L = 0.01$ to the output admittance in Table 7-2, we see that for the other connections, both the input and output are badly mismatched.

Note also that the voltage isolation is very high for the grounded-emitter connection (it is infinite because $y_{12} = 0$) compared to the grounded-collector connection.

Grounded-collector connection. The insertion power gain is a maximum when the source admittance is matched to the amplifier input admittance and the load admittance is matched to the amplifier output admittance. Then the source delivers maximum power to the transistor, and the transistor delivers maximum power to the load. The source admittance will be matched to the amplifier input admittance if

$$y_s = \left(y_{11} - \frac{y_{12}y_{21}}{y_{22} + y_L}\right)^* \tag{7-29}$$

and the load admittance will be matched to the amplifier output admittance if

$$y_L = \left(y_{22} - \frac{y_{12}y_{21}}{y_{11} + y_s}\right)^* \tag{7-30}$$

Now substituting the grounded-collector y parameters into Equations (7-29) and (7-30) and solving for y_s and y_L, we find

$$y_s = 0.017, \qquad y_L = 0.60 \tag{7-31}$$

Using these values for y_s and y_L and the y parameters given by Equations (7-25), (7-27), (7-28) in the equations listed in Table 7-1, we obtain the results presented in Table 7-3.

From Table 7-3 we note that the grounded-collector voltage gain is again less than one. Indeed, using the grounded-collector y parameters in the voltage gain formula, we find

$$\text{voltage gain} = \frac{36}{36 + y_L} \tag{7-32}$$

so that the voltage gain is one or less for all source and load conditions. Therefore, the grounded-collector amplifier has no value as a voltage amplifier.

Now using the grounded-collector y parameters in the formula for current gain, we obtain

$$\text{current gain} = \frac{-36y_L}{0.01 + y_L} \tag{7-33}$$

so that for large y_L ($y_L \gg 0.01$) the current gain is -36. Thus the grounded-collector amplifier is very useful for delivering a large current

Table 7-2 ($y_s = 1$, $y_L = 0.01$)

	VOLTAGE GAIN	CURRENT GAIN	POWER GAIN	INSERTION POWER GAIN	INPUT ADMITTANCE	OUTPUT ADMITTANCE	ISOLATION
Grounded emitter	−1750	17.5	30,600	783,000	1	0.01	∞
Grounded collector	1	−18	18	1.02	0.00056	18.0	−2
Grounded base	1750	−0.947	1650	8100	18.5	0.00055	63,000

Table 7-3 ($y_s = 0.017$, $y_L = 0.60$)

	VOLTAGE GAIN	CURRENT GAIN	POWER GAIN	INSERTION POWER GAIN	INPUT ADMITTANCE	OUTPUT ADMITTANCE	ISOLATION
Grounded emitter	−57.4	34.4	1980	1150	1	0.01	∞
Grounded collector	0.98	−36	36	320	0.016	0.6	1
Grounded base	57.4	−0.975	57.6	0.98	35.4	0.0003	207,000

to a large admittance load. From both Equations (7-32) and (7-33) we note that if y_L is large compared to 0.01, but small compared to 36, the power gain is essentially the same as the current gain. [See Equation (7-16).]

We conclude that the primary utility of the grounded-collector amplifier is an impedance transformer with some power gain.

Grounded-base connection. Grounded-base transistor amplifiers are not used as often as the other two connections. For this connection the input admittance is very high, and the output admittance is very low, so that it does have application when a high internal admittance source must be converted to a low internal admittance source. That is, it can be used to convert an almost ideal voltage source into an almost ideal current source.

High-frequency performance. The frequency limitation of a bipolar transistor such as the 2N174 is primarily due to capacitance between the base and collector.[4] Since the base to emitter admittance is quite large, the same capacitance appears between the collector and emitter. Therefore, in the grounded-emitter connection of this transistor (for which the specifications are given), the effect of this capacitance can be incorporated into the transistor model by including an admittance $j\omega C$ in y_{22}. Hence, the grounded-emitter y parameter matrix becomes

$$\mathbf{Y}_{(\text{grounded emitter})} = \begin{bmatrix} 1 & 0 \\ 35 & 0.01 + j\omega C \end{bmatrix} \tag{7-34}$$

where C must be chosen to give the current amplification cutoff frequency of 10 kHz listed in the 2N174 specifications. If we use the y parameters given in Equation (7-34), the grounded-emitter current gain is given by

$$\text{current gain} = \frac{-35y_L}{0.01 + j\omega C + y_L} \tag{7-35}$$

which depends on y_L. The transistor specifications on current amplification cutoff are given for y_L small compared to y_{22}. Hence, the current gain is down by a factor of 0.707 (3 dB) when $\omega C = 0.01$ or

$$C = \frac{0.01}{\omega} = \frac{0.01}{2\pi \times 10^4} = 0.16\ \mu\text{F} \tag{7-36}$$

[4] In high-frequency transistors the base-collector capacitance is still the first limitation of performance, but other parasitic effects such as inductance of the lead wires and capacitances between other electrodes must also be considered. Such transistors require a more complex though quite straightforward circuit model to predict their small signal performance.

Using this value for C in Equation (7-34), we can now construct the high-frequency IAM for the 2N174 transistor:

$$\mathbf{I}_{b,c,e} = \begin{bmatrix} 1 & 0 & -1 \\ 35 & 0.01 + j\omega 1.6 \times 10^{-7} & -35 - j\omega 1.6 \times 10^{-7} \\ -36 & -0.01 - j\omega 1.6 \times 10^{-7} & 36 + j\omega 1.6 \times 10^{-7} \end{bmatrix} \tag{7-37}$$

The effect of this capacitance on the voltage gain, current gain, and so on, for the three connections can be determined by computing these quantities by the same method used to analyze the low-frequency case.

Approximate analysis of the low-frequency bipolar transistor. The amplifier analysis technique presented in Section 7-2 is a very powerful technique that can be used for a detailed analysis of any 2-port network with a driving source connected to one of the ports and a load connected to the other port. In Section 7-3 we used this method to analyze three bipolar amplifiers. In some cases the method may be too powerful. For example, sometimes we would rather take a quick look at the approximate behavior of an amplifier. In this section we present a technique for obtaining an approximate but quick look at bipolar transistor amplifiers.

Consider the common-emitter transistor amplifier presented in Figure 7-4(a). The transistor model is given in Figure 7-4(b), and the combined network in Figure 7-4(c). For signal analysis we assume that the blocking capacitors are short circuits, and we set the bias supply voltage zero. (See Section 2-4.) Furthermore, for many transistors, a reasonable approximation for a first-order approximate solution is to assume $r_d = \infty$. This results in the model presented in Figure 7-4(d). We have used R_e in place of r_e, since in some transistor amplifiers, R_4 is not bypassed. Hence, if R_4 is bypassed, $R_e = r_e$; if not, $R_e = r_e + R_4$.

We start by computing the input admittance. From the illustration in Figure 7-4(d), we note that y_{in} is $(1/R_b) + (1/z''_{in})$ (the admittance $1/R_b$ in parallel with the admittance $1/z'_{in}$) and $z'_{in} = r_b + z''_{in}$ (z'_{in} is r_b in series with z''_{in}). Hence,

$$\begin{aligned} y_{in} &= \frac{1}{R_b} + \frac{1}{z'_{in}} \\ &= \frac{1}{R_b} + \frac{1}{r_b + z''_{in}} \end{aligned} \tag{7-38}$$

But, from Figure 7-4(d), z''_{in} is simply the voltage across R_e divided by I_b. Applying KCL at the upper R_e terminal, we note that the current down through R_e is $I_b + \beta I_b = (\beta + 1)I_b$. It follows that the voltage across R_e is $(\beta + 1)I_bR_e$ and that $z''_{in} = (\beta + 1)R_e$. Hence, Equation (7-38) becomes

$$y_{in} = \frac{1}{R_b} + \frac{1}{r_b + (\beta + 1)R_e} \tag{7-39}$$

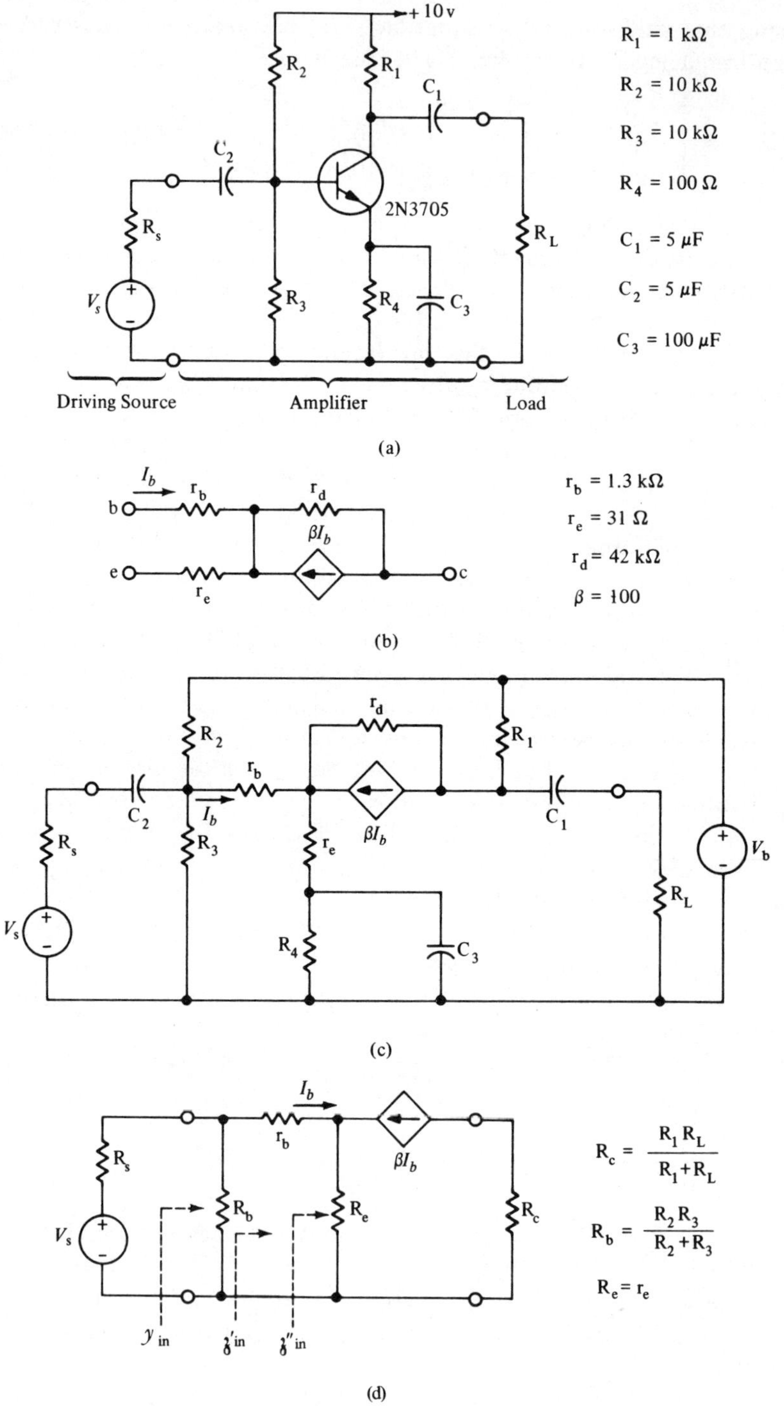

Figure 7-4 Common emitter amplifier.

To compute the voltage gain, we note that if the voltage across R_b is V_1 volts, then I_b is given by

$$I_b = \frac{V_1}{z'_{in}} = \frac{V_1}{r_b + (\beta + 1)R_e} \tag{7-40}$$

Hence, the voltage across R_c is given by

$$V_2 = -\beta I_b R_c = \frac{-\beta R_c}{r_b + (\beta + 1)R_e} \tag{7-41}$$

The voltage gain is

$$\frac{V_2}{V_1} = \frac{-\beta R_c}{r_b + (\beta + 1)R_e} \tag{7-42}$$

The other amplifier characteristics can be computed by a similar procedure. Grounded-collector and grounded-base transistor amplifiers can also be handled in this manner.

7-4 Field-Effect Transistor Amplifiers

General models. As a device with a slightly more complex high-frequency model, we consider a typical n-channel planar silicon field-effect transistor (FET)—the 2N3819. The small-signal equivalent network usually assumed for an FET is presented in Figure 7-5. At moderate frequencies, two capacitances must be considered. Thus this device has a

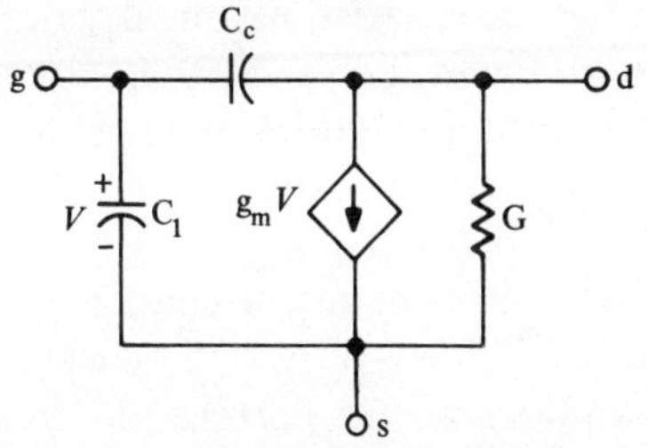

Figure 7-5 FET equivalent network.

slightly more complex frequency dependence in its normal operating range than the power transistor of the previous section. The common-source specifications for the 2N3819 as given by the manufacturer are listed in Table 7-4. These data are sufficient to determine the y parameter matrix for the grounded-source FET.

The y parameter matrix for the grounded-source FET is given by (see Example 5-3)

$$\mathbf{Y}_{\text{(grounded source)}} = \begin{bmatrix} j\omega(C_1 + C_c) & -j\omega C_c \\ g_m - j\omega C_c & G + j\omega C_c \end{bmatrix} \tag{7-43}$$

Table 7-4

Forward transfer admittance at 1 kHz	min 2×10^{-3}, max 6.5×10^{-3} mhos
Output admittance at 1 kHz	max 5×10^{-5} mhos
Input capacitance (output shorted)	max 8×10^{-12} farads
Reverse transfer capacitance (input shorted)	max 4×10^{-12} farads
Forward transfer admittance at 100 MHz	min 1.6×10^{-3} mhos

The third specification gives 8 pF as a maximum for $C_1 + C_c$; a realistic value is 6 pF. The fourth specification gives 4 pF as a maximum for C_c; let us use 3 pF. The first specification states that at low frequencies, $0.002 < y_{21} < 0.0065$; let us use 0.004. Finally, the second specification states that the output admittance G is less than 5×10^{-5}; let us set $G = 4 \times 10^{-5}$. Then

$$\mathbf{Y}_{(\text{grounded source})} = \begin{bmatrix} j\omega 6 \times 10^{-12} & -j\omega 3 \times 10^{-12} \\ 0.004 - j\omega 3 \times 10^{-12} & 4 \times 10^{-5} + j\omega 3 \times 10^{-12} \end{bmatrix} \tag{7-44}$$

and the IAM for the FET is given by

$$\mathbf{I}_{g,d,s} = \begin{bmatrix} j\omega 6 \times 10^{-12} & -j\omega 3 \times 10^{-12} & -j\omega 3 \times 10^{-12} \\ 0.004 - j\omega 3 \times 10^{-12} & 4 \times 10^{-5} + j\omega 3 \times 10^{-12} & -0.004 \\ -0.004 - j\omega 3 \times 10^{-12} & -4 \times 10^{-5} & 0.004 + j\omega 3 \times 10^{-12} \end{bmatrix} \tag{7-45}$$

The y parameters for the grounded-drain and grounded-gate FET's are easily obtained from $\mathbf{I}_{g,d,s}$. These quantities, along with assumed values for y_s and y_L, can now be used in the equations of Table 7-1 to evaluate the grounded-source, grounded-drain, and grounded-gate amplifier characteristics. For $y_s = y_L = 10^{-4}$, we obtain the results listed in Table 7-5.

Low-frequency models. The low-frequency performance of the FET in its various connections can be obtained from Table 7-5 by setting $\omega = 0$ in all the formulas. Such a procedure would give the correct answers, but it would not give much of a physical feeling for what is going on. A more instructive approach is to set up an equivalent circuit for the FET, plus source and load, and with all capacitances replaced by open circuits. With the FET of Figure 7-5, a 10-kΩ source and a 10-kΩ load, the three equivalent networks are those of Figure 7-6. For each of these equivalent networks we can write V_1, V_2, I_1, and I_2 by inspection. There is no need for the matrix formalism.

For the grounded-source configuration, the input is an open circuit. Thus $I_1 = 0$, and current and power gains are meaningless. The output admittance is given by the 25-kΩ internal resistance. The voltage gain is $g_m = 0.004$ times the parallel combination of the load and internal

Table 7-5

GROUND	V_2/V_1	I_2/I_1	I_1/V_1	I_2/V_2
s	$\dfrac{-40 + 3 \times 10^{-8}j\omega}{1.4 + 3 \times 10^{-9}j\omega}$	$\dfrac{4 - 3 \times 10^{-5}j\omega}{3 \times 10^{-8}j\omega[4.28 + 3 \times 10^{-9}j\omega]}$	$6 \times 10^{-12}j\omega \left[1 + \dfrac{4 - 3 \times 10^{-9}j\omega}{0.28 + 6 \times 10^{-9}j\omega}\right]$	$4 \times 10^{-5} + 3 \times 10^{-12}j\omega \left[1 + \dfrac{40 - 3 \times 10^{-8}j\omega}{1 + 6 \times 10^{-8}j\omega}\right]$
d	$\dfrac{4 + 3 \times 10^{-9}j\omega}{4.14 + 3 \times 10^{-9}j\omega}$	$\dfrac{-(4 + 3 \times 10^{-9}j\omega)}{3 \times 10^{-8}j\omega[4.14 + 3 \times 10^{-9}j\omega]}$	$6 \times 10^{-12}j\omega \left[1 - \dfrac{2 + 1.5 \times 10^{-9}j\omega}{4.14 + 3 \times 10^{-9}j\omega}\right]$	$4.04 \times 10^{-3} + 3 \times 10^{-12}j\omega \left[1 - \dfrac{40 + 3 \times 10^{-8}j\omega}{1 + 6 \times 10^{-8}j\omega}\right]$
g	$\dfrac{40.4}{1.4 + 3 \times 10^{-8}j\omega}$	$\dfrac{-4.04}{4.04 + 1.254 \times 10^{-7}j\omega + 9 \times 10^{-17}(j\omega)^2}$	$4.04 \times 10^{-3}\left(1 - \dfrac{4}{14 + 3 \times 10^{-8}j\omega}\right) + 3 \times 10^{-12}j\omega$	$4 \times 10^{-5}\left(1 - \dfrac{4.04}{4.14 + 3 \times 10^{-9}j\omega}\right) + 3 \times 10^{-12}j\omega$

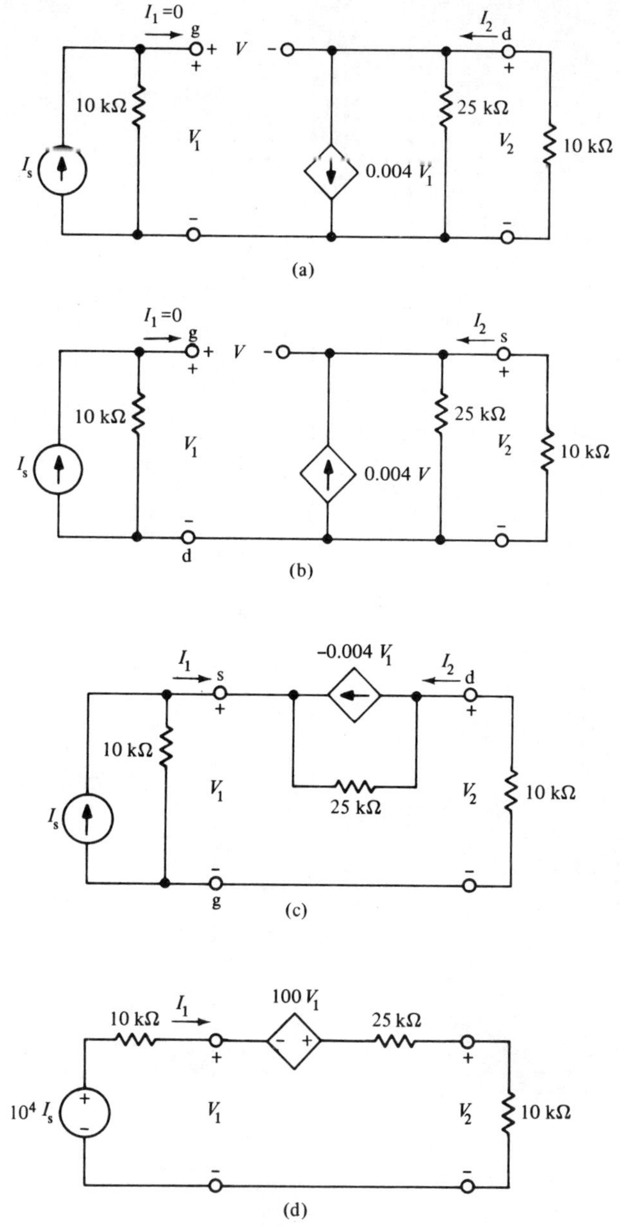

Figure 7-6 Low-frequency FET amplifiers. (a) Grounded source. (b) Grounded drain. (c) Grounded gate. (d) Series grounded gate.

resistances. In this case

$$\frac{V_2}{V_1} = -\frac{40}{1.4} = -28.6 \tag{7-46}$$

Thus, at very low frequencies, the grounded-source FET has very high input impedance, moderately high output impedance, and a moderate voltage gain. The voltage isolation is infinite with this simple model.

For the grounded-drain configuration, Figure 7-6(b), the input is again an open circuit for this simple model. Again the only significant quantities are voltage gain and output impedance. The output voltage, V_2, is $g_m V$ times the parallel combination of the load and the internal resistance of the FET. That is,

$$V_2 = \frac{4 \times 10^{-3}V}{4 \times 10^{-5} + 10^{-4}} = 28.6V \tag{7-47}$$

To relate V_2 to V_1, we must relate V to V_1. By KVL

$$V = V_1 - V_2 \tag{7-48}$$

Thus

$$V_2 = \frac{28.6}{29.6} V_1 \tag{7-49}$$

The voltage gain is slightly less than one. Computation of the output admittance requires some minor manipulation, since V is not zero when V_1 is zero. With $V_1 = 0$

$$I_2 = 4 \times 10^{-5}V_2 + 0.004V_2 = 4.04 \times 10^{-3}V_2 \tag{7-50}$$

Thus the output admittance is essentially equal to g_m. For very low frequencies, the advantage of the grounded-drain FET is its low output impedance.

The grounded-gate FET of Figure 7-6(c) is most easily analyzed by making source transformations to give the circuit of Figure 7-6(d). For this configuration $I_1 = -I_2$, so the current gain is always one in magnitude. The current I_1 is related to V_1 by

$$I_1 = \frac{V_1(101)}{(10 + 25) \times 10^3} \tag{7-51}$$

Thus

$$\frac{V_2}{V_1} = \frac{10^4 I_1}{V_1} = \frac{101}{3.5} = 28.9 \tag{7-52}$$

The voltage gain of the grounded-gate FET is essentially the same (1 percent higher) as that of the grounded-source stage. The input admittance is found immediately from Equation (7-51). It is

$$\frac{I_1}{V_1} = \frac{101}{35 \times 10^3} \tag{7-53}$$

Thus the input admittance is large. It is 101 times as large as the admittance of the series combination of the load and the internal output resistor. The output admittance is found by setting $I_2 = 0$ and fixing V_2 by a source in Figure 7-6(d). Then $V_1 = 10^4 I_2$, and KVL gives

$$101 \times 10^4 I_2 + 2.5 \times 10^4 I_2 = V_2 \tag{7-54}$$

or

$$\frac{V_2}{I_2} = 1.035 \times 10^6 \tag{7-55}$$

Thus the output impedance is very high — essentially the source impedance times 101.

General considerations. With the dc characteristics of the various FET connections as a starting point and the general formulas of Table 7-1 for the complete description, we can summarize the applications of FET's as follows.

The grounded source is the basic amplifying connection for the FET. The other two connections are used only for their special input and output impedance properties. At low frequencies, the input to both the grounded-source and the grounded-drain connection is essentially capacitive. The grounded-source transistor looks like 90 pF, whereas the grounded drain looks like 3 pF. Since the isolation of both is very high at low frequencies, the grounded-drain FET makes an excellent input element on a voltmeter or oscilloscope probe, where low capacitance and good isolation are of prime importance. The gain can be taken care of by later stages.

The low-frequency output admittance of the grounded-source and grounded-drain FET differ by a factor of 100 for this particular transistor. Since the output admittance of the grounded-drain transistor is large, it makes a good output stage for an amplifier where a low output impedance is desired. Again, the gain can be provided by earlier stages.

The grounded-gate transistor does not isolate the source from the load. The input admittance depends on the load admittance, and the output admittance depends on the source admittance. By proper selection of the load, the input admittance can be made much higher than that of the other two connections. Correspondingly, proper source selection can give a much lower output admittance than the other two connections.

A general discussion of the high-frequency performance of this FET requires material from the next chapter. For the present we can ascertain the cutoff frequency, that frequency where the voltage gain is down to 0.707 of its zero frequency value, for the grounded-source connection. Since the real part of the numerator of the voltage gain is so much larger than that of the denominator, and since the imaginary parts are the same, the denominator determines the cutoff frequency. This occurs when the real and imaginary parts are equal, or when $\omega = 4.67 \times 10^9$ rad/sec. The corresponding frequency is 743 MHz. Thus for this FET, the range of application of the very low frequency formulas extends well up into the megahertz region.

PROBLEMS

7-1 The circuit shown in Figure P7-1 is claimed to be a low-impedance microphone preamplifier, provided that one of the three transistors listed are used. Reference: GE Transistor Manual, 7th ed., p. 375. Choose one of the three transistors, look up its characteristics, and compute the input impedance. (It should be something like 150 Ω.) What should the input impedance of the next stage be if it is to be matched to the output of the preamplifier?

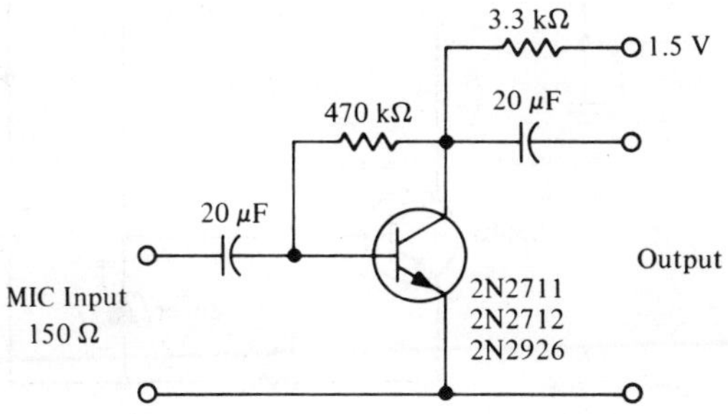

Figure P7-1

7-2 Compute each of the quantities of Table 7-1 for each of the four amplifiers of Problems 6-13 and 6-14. Assume $R_S = R_L = 1$ kΩ. Enter the results in a table similar to Table 7-2. Comment briefly on the utility of such an amplifier, the impedance match between input and output, and so on. (Hint: Use the computer.)

7-3 a. Calculate the voltage gain V_{cb}/V_{ab} for the "Twin-T" circuit given in Problem 6-17. Assume a load y_L is attached between terminals b and c.

b. With $R_1 = 5$ kΩ, $C_1 = 1$ μF, and $y_L = 10^{-6}$ mho, plot the *magnitude* of the gain V_{cb}/V_{ab} as a function of frequency ω, using the computer. Let ω vary from 50 rad/sec to 150 rad/sec in steps of 1 rad/sec or less.

7-4 Most FET's have wide parameter variations from transistor to transistor. These variations make biasing for a specific bias current somewhat difficult. One method of overcoming this problem is to bias FET's with a constant current source, that is, force the bias current to be a specific value. Furthermore, if this technique is employed for a grounded-drain (source follower) amplifier, the voltage gain becomes very close to one. Typically a bipolar transistor circuit is used to approximate the desired current source. Such a circuit for a common-drain (source-follower) amplifier is shown in Figure P7-4(a).

a. Using the model shown in Figure P7-4(b) for the bipolar transistor, show that the bipolar transistor circuit does approximate an ideal dc current source between s and the negative terminal of E_2. What is the dc value of the current source?

b. Assuming[5] the bipolar transistor is an ideal dc current source for the

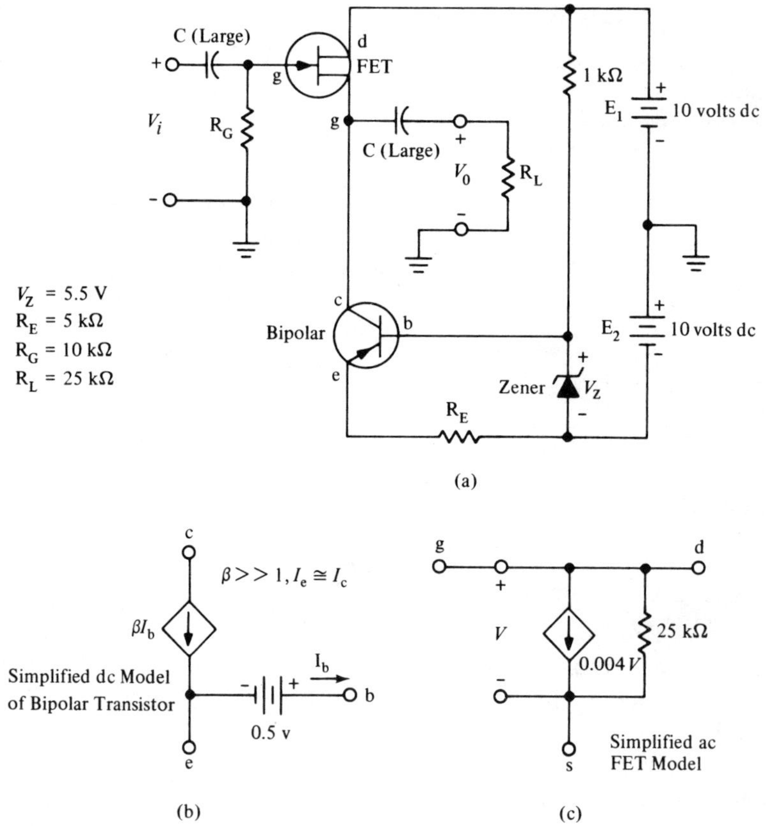

Figure P7-4

[5] If you cannot determine the ac equivalent circuit [Problem 7-4**b**.(1)], ask your instructor to give it to you and do the rest of **b**.

FET (1) determine the resulting ac equivalent circuit, (2) determine the gain, V_0/V_i, (3) determine the input impedance, (4) determine the output impedance, and (5) using the voltage gain as a starting point, determine the current gain. (See Figure P7-4(c).

7-5 What is the gain V_0/V_i of the circuit in Figure P7-5 (use the FET model of Problem 7-4)?

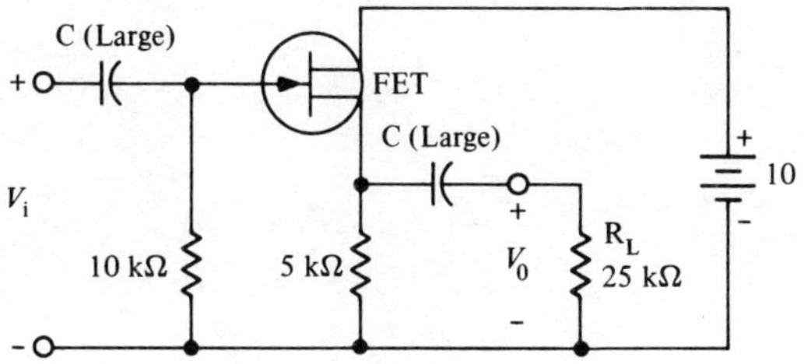

Figure P7-5

7-6 A voltage amplifier of fixed gain, when operated from a low impedance source, can be constructed from a very high gain operational amplifier, as shown in Figure P7-6. Compute the gain, input admittance (output open) and output admittance (input shorted) as a function of the resistances.

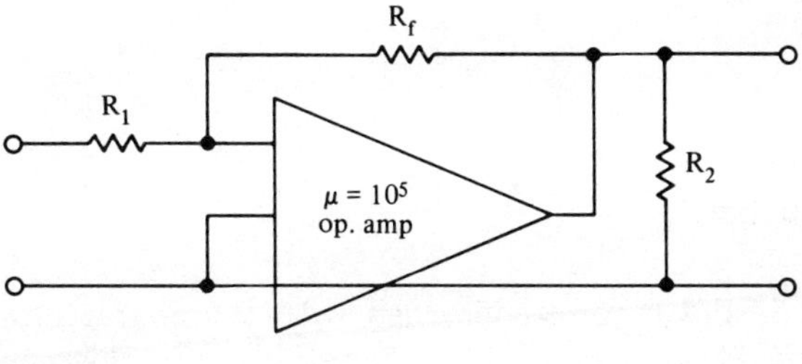

Figure P7-6

CHAPTER

8

Network Functions and Transfer Functions

8-1 Introduction

Chapters 4, 5, and 6 were concerned with the characterization of networks for steady-state sinusoidal excitations. In this chapter we continue this study of the response of linear networks to sinusoidal inputs, but we shift our viewpoint to the frequency domain. That is, we shall develop several techniques for describing the response of linear networks as a function of the frequency of the input sinusoid. These techniques are a necessary prerequisite to filter design, which is the topic of Chapter 9.

In Section 5-2 we considered 1-port networks and concluded that any linear 1-port network containing no independent sources can be completely characterized by a single parameter called the input admittance y (or, its reciprocal, the input impedance). In Section 5-3 we showed that any 2-port network can be characterized by a set of four parameters, for example, the h parameters $\mathfrak{h}_{11}$, $\mathfrak{h}_{12}$, $\mathfrak{h}_{21}$, and $\mathfrak{h}_{22}$. In Chapter 6 we introduced the indefinite admittance matrix, which, for an m-terminal network, contains the m^2 elements $\mathbf{y}_{ij}$ $(i, j = 1, 2, \cdots, m)$. Finally, in

Chapter 7 we considered 2-port amplifiers and defined various amplifier characteristics, such as voltage gain, current gain, insertion power gain, input impedance, and so on. All of these quantities—input admittances, h parameters, indefinite admittance parameters, insertion power gain, and others—have certain properties in common. In order that we can study these basic properties without specifying whether the parameter is an input impedance, current gain, and so on, we introduce the concept of a network function and a transfer function. Any ratio of two complex voltages and/or complex currents, such as voltage gain, output impedance, and so on, is called a *network function.*

8-2 Properties of Network Functions, Complex Frequency

Network functions, whether they are voltage gains, output admittances, or others, possess a number of interesting properties. In this section we outline some of their most significant properties. We use the symbol $\mathcal{H}(j\omega)$ to denote all network functions.

Properties of network functions.

PROPERTY 1

All network functions can be written as a ratio of two polynomials in $j\omega$ with real coefficients. That is, all network functions can be expressed in the form

$$\mathcal{H}(j\omega) = \frac{a_n(j\omega)^n + a_{n-1}(j\omega)^{n-1} + \cdots + a_1(j\omega) + a_0}{b_m(j\omega)^m + b_{m-1}(j\omega)^{m-1} + \cdots + b_1(j\omega) + b_0} \tag{8-1}$$

where the a_i $(i = 0, 1, \cdots, n)$ and the b_i $(i = 0, 1, \cdots, m)$ are real numbers.

Example 8-1

In this example we compute the voltage gain for the network presented in Figure 8-1. Using the voltage-divider technique, we write the voltage

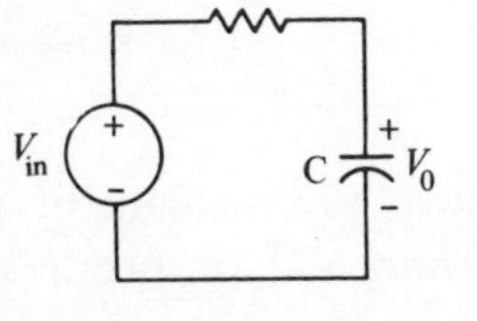

Figure 8-1 Example 8-1.

gain by inspection:

$$\mathcal{H}(j\omega) = \frac{V_0}{V_{in}} = \frac{(1/j\omega C)}{R + (1/j\omega C)} \tag{8-2}$$

Multiplying numerator and denominator by jωC, we obtain

$$\mathcal{H}(j\omega) = \frac{1}{RC(j\omega) + 1} \tag{8-3}$$

The numerator of Equation (8-3) is a zeroth-degree polynomial in jω, and the denominator is a first-degree polynomial in jω. All of the coefficients, $a_0 = 1$, $b_1 = RC$, $b_0 = 1$ are real numbers.

Example 8-2

In this example we compute the input admittance for the network of Figure 8-2. The input admittance is the sum of the admittances of the

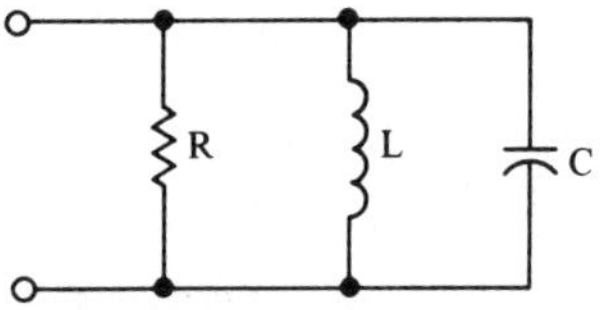

Figure 8-2 Example 8-2.

three parallel branches. We find

$$\mathcal{H}(j\omega) = y_{in} = \frac{1}{R} + \frac{1}{j\omega L} + j\omega C \tag{8-4}$$

Putting the three terms of Equation (8-4) over a least common denominator, we obtain

$$\begin{aligned}\mathcal{H}(j\omega) &= \frac{j\omega L + R + Rj\omega Cj\omega L}{Rj\omega L} \\ &= \frac{RCL(j\omega)^2 + L(j\omega) + R}{RL(j\omega)}\end{aligned} \tag{8-5}$$

Equation (8-5) is the ratio of a second-degree polynomial in jω to a first-degree polynomial in jω. All of the coefficients in both polynomials depend only on the network quantities R, L, and C, and are, therefore, real numbers.

PROPERTY 2

For all network functions, the roots of both the numerator polynomial and the denominator polynomial are either real or occur in complex conjugate pairs. This property follows directly from the fact that the coefficients in both polynomials are real.

Example 8-3

In Example 8-1 we found that the voltage gain for the RC network of Figure 8-1 is given by

$$\mathcal{H}(j\omega) = \frac{V_0}{V_{in}} = \frac{1}{RC(j\omega) + 1} \tag{8-6}$$

An nth-degree polynomial has n roots. Since the numerator of Equation (8-6) is a zeroth-degree polynomial in $j\omega$, it has no roots. The denominator of Equation (8-6) is a first-degree polynomial in $j\omega$ and has one root. The root is given by the value of $j\omega$, for which the polynomial is zero. Hence we set

$$RC(j\omega) + 1 = 0 \tag{8-7}$$

and solve for $j\omega$

$$j\omega = -\frac{1}{RC} \tag{8-8}$$

Obviously, the root is real.

Example 8-4

In Example 8-2 we found that the input admittance for the parallel RLC network of Figure 8-2 is given by

$$\mathcal{H}(j\omega) = y_{in} = \frac{RCL(j\omega)^2 + L(j\omega) + R}{RL(j\omega)} \tag{8-9}$$

In this example we compute the roots of both the numerator polynomial in $j\omega$ and the denominator polynomial in $j\omega$.

The denominator polynomial is simplest, so we tackle that first. Obviously, the value of $j\omega$ for which this function of $j\omega$ is zero is given by

$$j\omega = 0 \tag{8-10}$$

The result is a real number, as required by property 2.

The numerator polynomial requires more thought. The values of $j\omega$ for which the numerator polynomial is zero are given by the solutions to the equation

$$RCL(j\omega)^2 + L(j\omega) + R = 0 \tag{8-11}$$

The solutions to Equation (8-11) are known to be

$$j\omega = \frac{-L \pm \sqrt{L^2 - 4R^2CL}}{2RCL} \tag{8-12}$$

Hence, if $L^2 \geq 4R^2CL$, both roots are real; if $L^2 < 4R^2CL$, the roots are given by the complex conjugate pair

$$j\omega = \frac{-L \pm j\sqrt{4R^2CL - L^2}}{2RCL} \tag{8-13}$$

PROPERTY 3

For any network function $\mathcal{H}(j\omega)$, it can be shown that the magnitude of $\mathcal{H}(j\omega)$, that is, $|\mathcal{H}(j\omega)| = \sqrt{\mathcal{H}(j\omega)\mathcal{H}^*(j\omega)}$ is an even function of ω, and that the angle of $\mathcal{H}(j\omega)$, that is, $\underline{/\mathcal{H}(j\omega)}$, is an odd function of ω.[1]

Example 8-5

In Example 8-1 we showed that the voltage gain for the RC network of Figure 8-1 is given by

$$\mathcal{H}(j\omega) = \frac{V_0}{V_{\text{in}}} = \frac{1}{RC(j\omega) + 1} \tag{8-14}$$

In this example we show that the magnitude of $\mathcal{H}(j\omega)$ is an even function of ω, while the angle of $\mathcal{H}(j\omega)$ is an odd function of ω.

The magnitude of $\mathcal{H}(j\omega)$ is given by

$$\begin{aligned} |\mathcal{H}(j\omega)| &= \sqrt{\mathcal{H}(j\omega)\mathcal{H}^*(j\omega)} \\ &= \sqrt{\frac{1}{RC(j\omega) + 1} \cdot \frac{1}{RC(-j\omega) + 1}} \\ &= \sqrt{\frac{1}{R^2C^2\omega^2 + 1}} \end{aligned} \tag{8-15}$$

Obviously $|\mathcal{H}(j\omega)| = |\mathcal{H}(-j\omega)|$ so that $|\mathcal{H}(j\omega)|$ is an even function of ω. The angle of $\mathcal{H}(j\omega)$ is given by

$$\begin{aligned} \underline{/\mathcal{H}(j\omega)} &= \frac{1}{\underline{/RC(j\omega) + 1}} = -\underline{/RC(j\omega) + 1} \\ &= -\tan^{-1}\omega RC \end{aligned} \tag{8-16}$$

which is an odd function of ω.

Complex frequency. Recall that ω is the frequency variable; it represents the frequency of the sinusoidal excitation and is, therefore, a real number. If ω is a real number, then $j\omega$ is a pure imaginary number. We have also

---

[1] $f(\omega)$ is an even function of ω if its value at $+\omega$ is the same as its value at $-\omega$, that is, $f(\omega) = f(-\omega)$. $f(\omega)$ is an odd function of ω if its value at $+\omega$ is the negative of its value at $-\omega$, that is, $f(\omega) = -f(-\omega)$ and $f(0) = 0$.

shown that network functions $\mathcal{H}(j\omega)$ are ratios of polynomials in the variable $j\omega$, and we have determined that the roots of these polynomials are either real or occur in complex conjugate pairs. For example, in Example 8-3 we found that the roots of the denominator polynomial is real and given by $j\omega = -1/RC$. For this equation to hold, we have $\omega = j/RC$, which, for real R and C, requires that ω be imaginary.

Although we appear to be in a dilemma, actually, there is no problem. Equation (8-8) merely states that the denominator polynomial of Equation (8-6) is zero if $\omega = -j/RC$—it does not say that other than real values of ω must occur in nature. We simply conclude that there is no real positive value of ω for which the denominator polynomial is zero. (This is obvious when we recall that $\mathcal{H}(j\omega)$ for this case represents the voltage gain for the RC network of Figure 8-1. If some real value of ω were to make the denominator zero, the voltage gain would be infinite at this frequency.)

Nevertheless, we sometimes find it convenient to extend the use of the term "frequency" to complex values such as those in Equations (8-8) and (8-13). Hence, we define a *complex frequency* variable p, and we write the network function in the form

$$\mathcal{H}(p) = \frac{a_n p^n + a_{n-1} p^{n-1} + \cdots + a_1 p + a_0}{b_m p^m + b_{m-1} p^{m-1} + \cdots + b_1 p + b_0} \tag{8-17}$$

When the network function $\mathcal{H}(j\omega)$ is written in terms of the complex-frequency variable p, as in Equation (8-17), it is called a *transfer function* or *system function.* We also point out that the symbol "s" is sometimes used for the complex frequency rather than "p."

In the next section we show that, by allowing p to take on complex values, it is easy to visualize the frequency characteristic of transfer functions for real frequencies. To evaluate the transfer function $\mathcal{H}(p)$ for any real frequency ω, we simply set $p = j\omega$; that is, we constrain p to purely imaginary numbers. Hence, in the work that follows, we treat p as a complex variable. ω is always the real frequency of the sinusoidal excitation.

8-3 Graphical Representations of Transfer Functions

Most transfer functions are rather complicated complex valued functions of the complex frequency variable p. Hence, in general, we gain little insight into the frequency characteristics of the network by simply observing the functional form of the transfer function. Fortunately, several graphical techniques for displaying transfer functions as functions of frequency have been developed.

Perhaps the most convenient method for observing the frequency behavior of transfer functions is to examine the roots of the numerator and denominator polynomials. Before attempting this, however, we pause to review a basic property of polynomials: Any nth-degree polynomial can be written in factored form:

$$c_n p^n + c_{n-1} p^{n-1} + \cdots + c_1 p + c_0 = c_n(p - p_1)(p - p_2) \cdots (p - p_n) \tag{8-18}$$

where $p_1, p_2, \cdots, p_n$ are the n roots of the polynomial. For example, the polynomial $2p^2 + 6p + 4$ can be factored into the form

$$\begin{aligned} 2p^2 + 6p + 4 &= 2(p^2 + 3p + 2) = 2(p + 1)(p + 2) \\ &= 2(p - p_1)(p - p_2); \qquad p_1 = -1, p_2 = -2 \end{aligned} \tag{8-19}$$

If $p = p_1 = -1$, or if $p = p_2 = -2$, both sides of the equation are zero, which verifies that p_1 and p_2 are the roots of the polynomial. In general, the roots may be complex. For example, the second-degree polynomial $p^2 + 2p + 4$ has roots given by

$$\begin{Bmatrix} p_1 \\ p_2 \end{Bmatrix} = -1 \pm \sqrt{1 - 4} = -1 \pm j\sqrt{3} \tag{8-20}$$

Therefore, we can write

$$\begin{aligned} p^2 + 2p + 4 &= [p - p_1][p - p_2] \\ &= [p - (-1 + j\sqrt{3})][p - (-1 - j\sqrt{3})] \end{aligned} \tag{8-21}$$

The reader should multiply out the right-hand side of Equation (8-21) to verify the equality. Also note that if the roots did not occur in complex conjugate pairs, the $+j$ and $-j$ terms would not cancel, and the resulting polynomial would have complex coefficients.

Finally, we point out that for first- and second-degree polynomials, it is a relatively simple task to extract the roots of the polynomial. Formulas for extracting the roots of a third-degree polynomial are also available, but they are too complicated to be of use. Therefore, for third-degree polynomials, trial-and-error methods with hand computation are generally used. For fourth- and higher-degree polynomials, one usually resorts to a digital computer. Standard computer routines for extracting the roots of a polynomial are available in most program libraries.

Since any transfer function can be expressed as a ratio of two polynomials, and since any polynomial can be written in the factored form given by Equation (8-18), it follows that any transfer function can be written in the form

$$\begin{aligned} \mathcal{H}(p) &= \frac{a_n(p)^n + \cdots + a_1(p) + a_0}{b_m(p)^m + \cdots + b_1(p) + b_0} \\ &= \frac{a_n(p - z_1)(p - z_2) \cdots (p - z_n)}{b_m(p - p_1)(p - p_2) \cdots (p - p_m)} \end{aligned} \tag{8-22}$$

where $z_1, z_2, \cdots, z_n$ are the n roots of the nth-degree numerator polynomial, and $p_1, p_2, \cdots, p_m$ are the m roots of the mth-degree denominator polynomial. The z_i (i = 1, 2, · · · , n) are called the *zeros* of the transfer function because they are the values of p for which the transfer function is zero; that is,

$$\mathcal{H}(p) = 0; \quad p = z_1, z_2, \cdots, z_n \tag{8-23}$$

Note that if $p = p_1, p_2, \cdots, p_m$, the denominator polynomial is zero, and the transfer function is infinite; that is,

$$\mathcal{H}(p) = \infty; \qquad p = p_1, p_2, \cdots, p_m \tag{8-24}$$

The p_i (i = 1, 2, · · · , m) are called the *poles* of the transfer function.

In the next subsection we show that knowledge of the poles and zeros of a transfer function yields insight into its behavior as a function of the real frequency ω.

Pole-zero diagrams. The most convenient tool for analyzing transfer functions is the pole-zero diagram. A pole-zero diagram can be constructed by simply marking the poles and zeros of the transfer function on a complex plane. Circles are used to indicate zeros, and x's are used to indicate poles.

Example 8-6

In this example we make a pole-zero diagram for the voltage gain of the RC network presented in Figure 8-3(a). We have already shown that

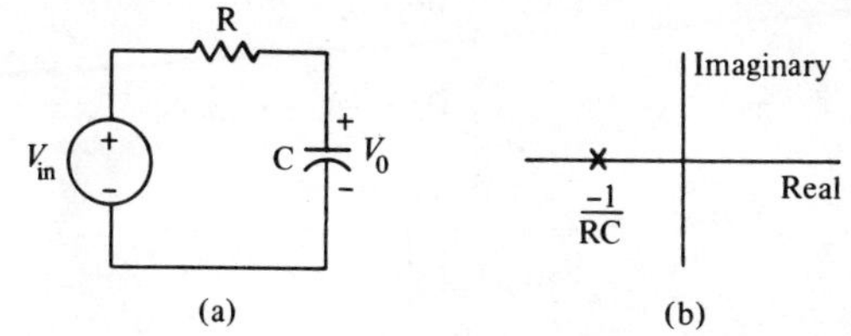

Figure 8-3 Example 8-6.

this transfer function is given by (see Example 8-1)

$$\mathcal{H}(p) = \frac{V_0}{V_{in}} = \frac{1}{RCp + 1} = \frac{1}{RC(p - p_1)} \qquad p_1 = -\frac{1}{RC} \tag{8-25}$$

The single pole has real part $-1/RC$ and imaginary part 0. This number is plotted in Figure 8-3(b).

Example 8-7

For a second example, we construct the pole-zero diagram for the input admittance for the parallel RLC network of Figure 8-4(a).

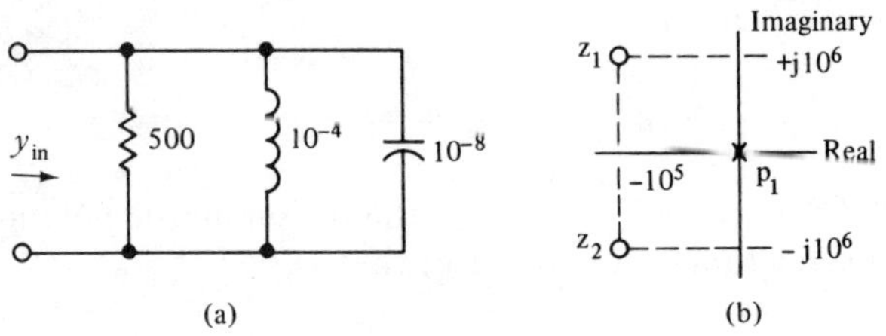

Figure 8-4 Example 8-7.

The input admittance for this network was derived in Example 8-2. Substituting $R = 500$, $L = 10^{-4}$, and $C = 10^{-8}$ into Equation (8-5), we obtain

$$\mathcal{H}(p) = y_{in} = \frac{5 \times 10^{-10}p^2 + 10^{-4}p + 500}{5 \times 10^{-2}p} = 10^{-8}\frac{p^2 + 2 \times 10^5 p + 10^{12}}{p} \tag{8-26}$$

The zeros are given by the roots of $p^2 + 2 \times 10^5 p + 10^{12}$:

$$\begin{Bmatrix} z_1 \\ z_2 \end{Bmatrix} = -10^5 \pm \sqrt{10^{10} - 10^{12}} \approx -10^5 \pm j10^6 \tag{8-27}$$

and the pole is given by

$$p_1 = 0 \tag{8-28}$$

Hence, the pole-zero diagram for this transfer function is as illustrated in Figure 8-4(b).

We also note that the transfer function can be expressed in the form

$$\mathcal{H}(p) = 10^{-8}\frac{[p - (-10^5 + j10^6)][p - (-10^5 - j10^6)]}{[p - 0]} \tag{8-29}$$

In the next subsection we show how to plot the magnitude and phase of transfer functions as functions of frequency from the pole-zero diagram.

Magnitude and phase versus frequency plots. One way to visualize the frequency characteristics of a network is to plot both the magnitude and the phase of the transfer function as a function of the frequency ω. In this section we show how to construct magnitude and phase plots from the pole-zero diagram. This procedure also develops insight into the relationship between the frequency characteristics of a network and the locations of the poles and zeros of its transfer function.

Before proceeding, however, let us consider a simple example with which we have already had some experience. In Example 8-5 we computed the magnitude and angle of the voltage gain for the network of Figure 8-5(a). The results are presented graphically in Figures 8-5(b) and (c).

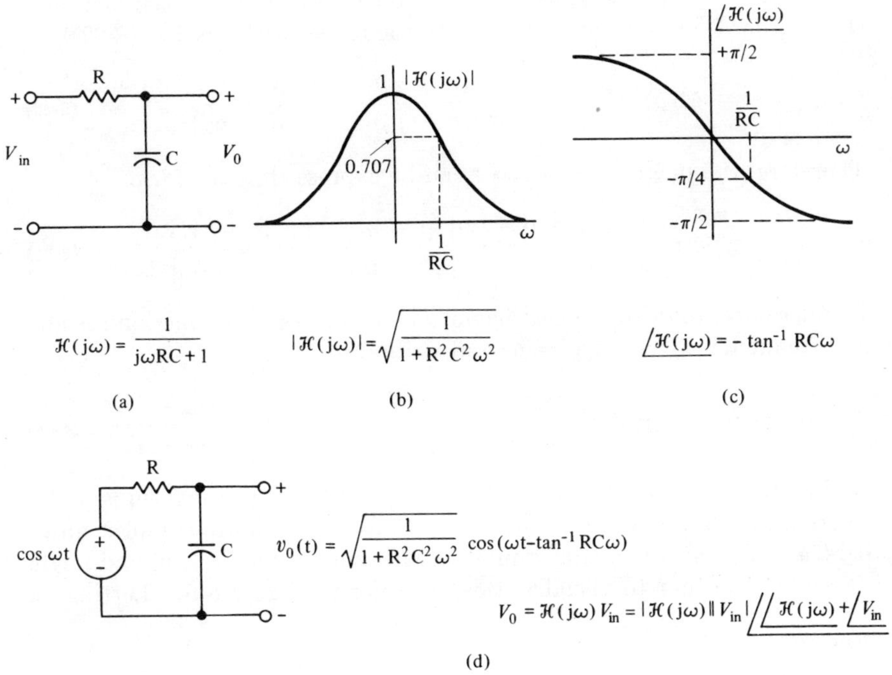

Figure 8-5 A frequency-selective network.

From the equation $|\mathcal{H}(j\omega)| = \sqrt{1/(1 + R^2C^2\omega^2)}$ it is obvious that $|\mathcal{H}(j0)| = 1$, $|\mathcal{H}[j(1/RC)]| = \sqrt{\frac{1}{2}} \approx 0.707$, $|\mathcal{H}(j\infty)| = 0$, and that $|\mathcal{H}(j\omega)|$ is a monotonically decreasing function of ω for $\omega > 0$. From the equation $\angle \mathcal{H}(j\omega) = -\tan^{-1} RC\omega$, it is apparent that $\angle \mathcal{H}(j0) = 0$, $\angle \mathcal{H}[j(1/RC)] = -\pi/4$, $\angle \mathcal{H}(j\infty) = -\pi/2$ and that $\angle \mathcal{H}(j\omega)$ is a monotonically decreasing function of ω. Next recall that V_{in} and V_0 are complex voltages related by the equation $\mathcal{H}(j\omega) = V_0/V_{in}$, or $V_0 = \mathcal{H}(j\omega)V_{in}$. Therefore, as illustrated in Figure 8-5(d), the magnitude of V_0 is different from the magnitude of V_{in} by a factor of $|\mathcal{H}(j\omega)|$, and the angle of V_0 is different from the angle of V_{in} by the amount $\angle \mathcal{H}(j\omega)$. Hence, back in the time domain, the input and output voltage waveforms are cosine waves of the same frequency but different amplitudes and phases. The amplitude of the output is scaled by a factor of $|\mathcal{H}(j\omega)|$ relative to the amplitude of the input, and the phase of the output is shifted by $\angle \mathcal{H}(j\omega)$ radians from the phase of the input. We conclude that by making a plot of $|\mathcal{H}(j\omega)|$ versus ω, as in Figure 8-5(b), and a plot of $\underline{/\mathcal{H}(j\omega)}$ versus ω, as in Figure 8-5(c), we can visualize the effect of the RC network on the input waveform as the frequency of the input is varied.

To obtain the magnitude-versus-frequency and angle-versus-frequency plots directly from the pole-zero diagram, we make use of the fact that the transfer function can be written in the form [see Equation (8-22)]

$$\mathcal{H}(p) = \frac{a_n(p - z_1)(p - z_2) \cdots (p - z_n)}{b_m(p - p_1)(p - p_2) \cdots (p - p_m)} \tag{8-30}$$

Therefore, the magnitude of $\mathcal{H}(p)$ can be expressed in the form

$$|\mathcal{H}(p)| = \left|\frac{a_n}{b_m}\right| \frac{|p - z_1|\,|p - z_2| \cdots |p - z_n|}{|p - p_1|\,|p - p_2| \cdots |p - p_m|} \tag{8-31}$$

Furthermore, to evaluate the magnitude of the transfer function for any real frequency ω, we set $p = j\omega$ and write

$$|\mathcal{H}(j\omega)| = |\mathcal{H}(p = j\omega)| = \left|\frac{a_n}{b_m}\right| \frac{|j\omega - z_1|\,|j\omega - z_2| \cdots |j\omega - z_n|}{|j\omega - p_1|\,|j\omega - p_2| \cdots |j\omega - p_m|} \tag{8-32}$$

Equation (8-32) is made up of a number of terms of the form $|a - b|$, where a and b are complex numbers. Since a and b are complex numbers, $a - b$ is also a complex number with magnitude $|a - b|$ and angle $\underline{/a - b}$. In order to visualize this, we refer to Figure 8-6. Letting a_R

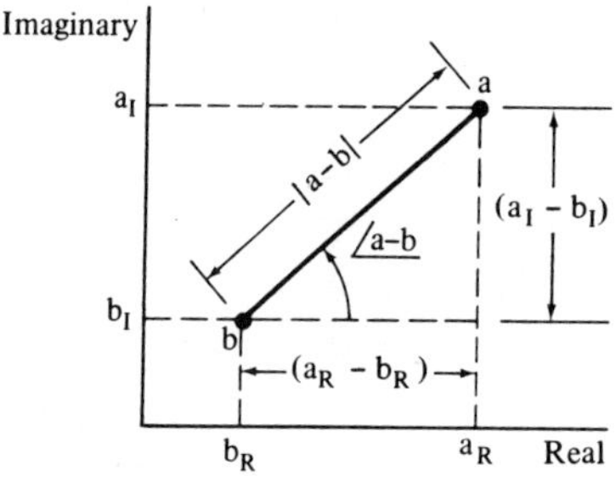

Figure 8-6 Difference of two complex numbers. Note: $a = a_R + ja_I$, $b = b_R + jb_I$, $a - b = (a_R - b_R) + j(a_I - b_I)$,
$|a - b| = \sqrt{(a_R - b_R)^2 + (a_I - b_I)^2}$,
$\underline{/a - b} = \tan^{-1}[(a_I - b_I)/(a_R - b_R)]$.

and a_I be the real and imaginary parts of a, respectively ($a = a_R + ja_I$), and b_R and b_I the real and imaginary parts of b, respectively ($b = b_R + jb_I$), we find $a - b = (a_R - b_R) + j(a_I - b_I)$. Hence, $a - b$ is a complex number with real part $(a_R - b_R)$ and imaginary part $(a_I - b_I)$. Consequently, $|a - b| = \sqrt{(a_R - b_R)^2 + (a_I - b_I)^2}$ and $\underline{/a - b}$ is the angle whose tangent is $(a_I - b_I)/(a_R - b_R)$. Hence, in Figure 8-6, we plot the two complex numbers a and b (according to their real and imaginary parts) and draw a line between them. The number $|a - b|$ is the length of this line. Furthermore, if we think of the complex number a − b as a vector with head at a and tail at b, $\underline{/a - b}$ is the angle of this vector.

Returning to Equation (8-32), we now interpret the number $|j\omega - z_1|$ to be the magnitude of a vector whose head is at the point $j\omega$ and whose tail is at the point z_1; $|j\omega - z_2|$ can be interpreted as the magnitude of a vector with head at $j\omega$ and tail at z_2, and so forth. Similarly, the number $|j\omega - p_1|$ can be interpreted as the magnitude of a vector whose head is at $j\omega$ and whose tail is at p_1; $|j\omega - p_2|$ can be interpreted as the magnitude of a vector with head at $j\omega$ and tail at p_2, and so forth.

Assuming that we have available a pole-zero diagram, the points $z_1, z_2, \cdots, z_n$ and the points $p_1, p_2, \cdots, p_m$ are already plotted. The only remaining point that we require is the point $j\omega$. But the point $j\omega$ is easily plotted on the complex plane, since its real part is zero, and its imaginary part is ω. Hence, we plot this point, draw in the vectors, and use only simple geometry to compute $|\mathcal{H}(j\omega)|$.

Example 8-8

The pole-zero plot for the voltage gain of the RC network of Figure 8-7(a) is presented in Figure 8-7(b). In this example we shall use the pole-zero diagram to construct a plot of $|\mathcal{H}(j\omega)|$ vs. ω. We start by writing the

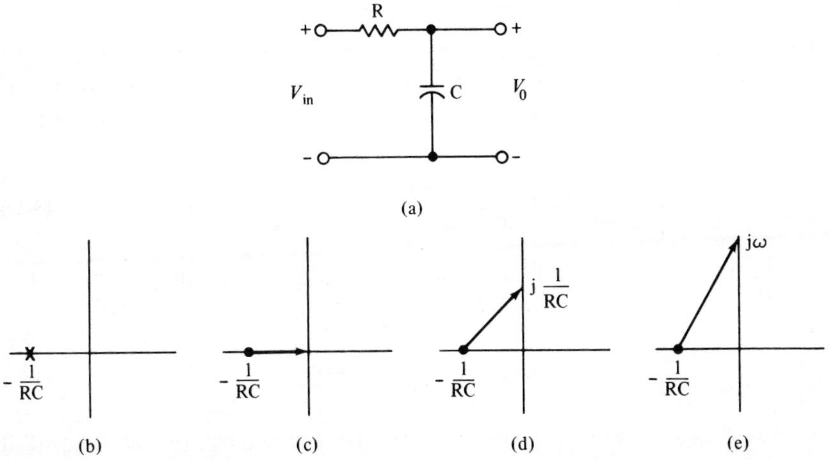

Figure 8-7 Example 8-8.

transfer function in the form

$$\mathcal{H}(p) = \frac{1}{RC(p - p_1)} \qquad p_1 = -\frac{1}{RC} \tag{8-33}$$

so that

$$|\mathcal{H}(p)| = \frac{1}{RC}\frac{1}{|p - p_1|} \tag{8-34}$$

Next, set $p = j\omega$ to obtain

$$|\mathcal{H}(j\omega)| = \frac{1}{RC} \frac{1}{|j\omega - p_1|} \tag{8-35}$$

To evaluate the magnitude of the transfer function at $\omega = 0$, we write

$$|\mathcal{H}(j0)| - \frac{1}{RC} \frac{1}{|j0 - p_1|} \tag{8-36}$$

The point j0 is plotted at the origin in Figure 8-7(c), and we interpret $j0 - p_1$ as a vector pointing from p_1 to j0, as illustrated in Figure 8-7(c). The magnitude of this vector is obvious from the figure: $|j0 - p_1| = 1/RC$. Therefore, we have

$$|\mathcal{H}(j0)| = \frac{1}{RC} \frac{1}{(1/RC)} = 1 \tag{8-37}$$

To evaluate the magnitude of the transfer function at

$$\omega = \frac{1}{RC} \tag{8-38}$$

we write

$$\mathcal{H}\left(j\frac{1}{RC}\right) = \frac{1}{RC} \frac{1}{|j(1/RC) - p_1|} \tag{8-39}$$

We interpret $j(1/RC) - p_1$ as a vector with head at $j(1/RC)$ and tail at p_1, as illustrated in Figure 8-7(d). A simple geometric computation tells us that the magnitude of the vector is $|j(1/RC) - p_1| = \sqrt{2}\,(1/RC)$. Hence,

$$\left|\mathcal{H}\left(j\frac{1}{RC}\right)\right| = \frac{1}{RC} \frac{1}{\sqrt{2}\,(1/RC)} = \frac{1}{\sqrt{2}} \approx 0.707 \tag{8-40}$$

The same procedure can be used to evaluate the magnitude of the transfer function for any ω. That is, to evaluate $|\mathcal{H}(j\omega)|$ we write

$$|\mathcal{H}(j\omega)| = \frac{1}{RC} \frac{1}{|j\omega - p_1|} \tag{8-41}$$

and construct the required vector on the pole-zero diagram, as illustrated in Figure 8(e). The magnitude of the vector is $\sqrt{[\omega^2 + (1/RC)^2]}$ so that

$$|\mathcal{H}(j\omega)| = \frac{1}{RC} \frac{1}{\sqrt{\omega^2 + (1/RC)^2}} = \frac{1}{\sqrt{R^2C^2\omega^2 + 1}} \tag{8-42}$$

Note that we can now visualize the behavior of $|\mathcal{H}(j\omega)|$, as ω varies from 0 to ∞, by simply observing the length of the vector while the head of the vector is moved from j0 up to the imaginary axis to $j\infty$. Since the network of Figure 8-7(a) is the same as that of Figure 8-5(a), the plot of $|\mathcal{H}(j\omega)|$ is given in Figure 8-5(b).

A plot of the angle of any transfer function versus frequency can be obtained directly from the pole-zero diagram by following a somewhat similar procedure. To show this, we return to the factored form for the transfer function

$$\mathcal{H}(p) = \frac{a_n}{b_m}\frac{(p - z_1)(p - z_2)\cdots(p - z_n)}{(p - p_1)(p - p_2)\cdots(p - p_m)} \tag{8-43}$$

and again interpret each of the bracketed quantities as a vector; for example, $(p - z_1)$ is interpreted as a vector with head at p and tail at z_1. The angle of $\mathcal{H}(p)$ is given by

$$\underline{/\mathcal{H}(p)} = \underline{/p - z_1} + \underline{/p - z_2} + \cdots + \underline{/p - z_n} - \underline{/p - p_1} - \underline{/p - p_2} - \cdots - \underline{/p - p_m} \tag{8-44}$$

That is, the angle of $\mathcal{H}(p)$ is interpreted to be the sum of the angles of the numerator vectors minus the angles of the denominator vectors. Hence, to evaluate $\underline{/\mathcal{H}(p = j\omega)}$, we go to the pole-zero diagram, draw the appropriate vectors, use geometry to determine the various angles, and add them according to Equation (8-44).

Example 8-9

The transfer function and its pole-zero diagram for the RC network of Figure 8-7(a) are presented in Figure 8-7(b). In this example we illustrate the method for obtaining a plot of the angle of $\mathcal{H}(j\omega)$ vs. ω directly from the pole-zero diagram.

We start by setting $p = j\omega$ in the transfer function equation,

$$\mathcal{H}(j\omega) = \frac{1}{RC}\frac{1}{(j\omega - p_1)} \tag{8-45}$$

and note that

$$\underline{/\mathcal{H}(j\omega)} = -\underline{/j\omega - p_1} \tag{8-46}$$

To evaluate the angle of $\mathcal{H}(j\omega)$ at $\omega = 0$, we write

$$\underline{/\mathcal{H}(j0)} = -\underline{/j0 - p_1} \tag{8-47}$$

Hence, $\underline{/\mathcal{H}(j0)}$ is the negative of the angle of the vector from the pole to the origin, as illustrated in Figure 8-7. Since the angle of this vector is zero, we conclude that $\underline{/\mathcal{H}(j0)} = 0$. To evaluate $\underline{/\mathcal{H}(j\omega)}$ (or any ω), we construct a vector from the pole p_1 to the point $j\omega$, as illustrated in Figure 8-7(e). From Figure 8-7(e) it is apparent that

$$\underline{/\mathcal{H}(j\omega)} = -\underline{/(j\omega - p_1)} = -\tan^{-1}\frac{\omega}{1/RC} = -\tan^{-1}\omega RC \tag{8-48}$$

Example 8-10

As another example, we consider the input admittance for the parallel RLC network of Figure 8-8(a). In Example 8-7 we constructed the pole-zero diagram for this transfer function and obtained the result presented in Figure 8-8(b). In this example we obtain plots of $|y_{in}(j\omega)|$ vs. ω and $\underline{/y_{in}(j\omega)}$ vs. ω from the pole-zero diagram.

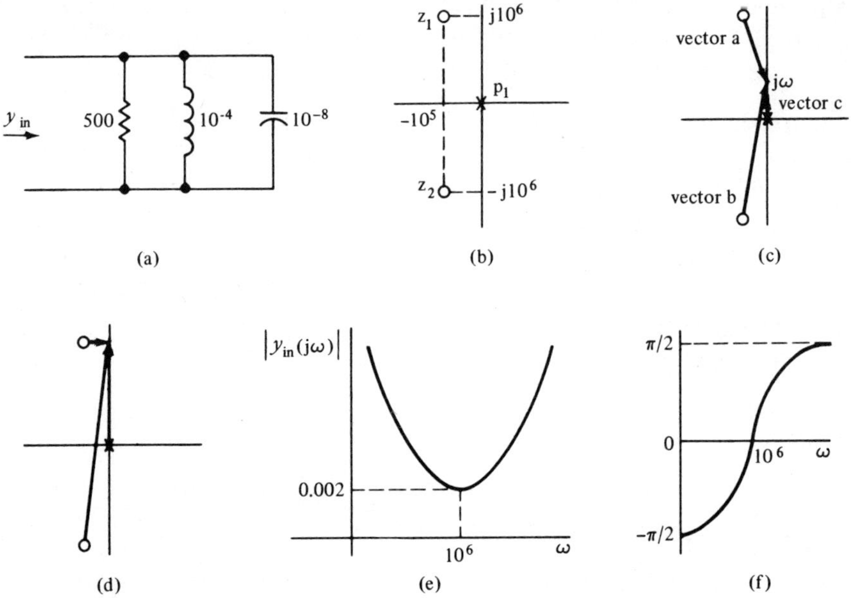

Figure 8-8 Example 8-10.

A plot of $|y_{in}(j\omega)|$ vs. ω can be obtained by noting that, for any ω, $|y_{in}(j\omega)|$ is given by 10^{-8} times the magnitude of the vector from z_1 to the point $j\omega$, times the magnitude of the vector from z_2 to the point $j\omega$, divided by the magnitude of the vector from p_1 to the point $j\omega$, as illustrated in Figure 8-8(c); that is,

$$|y_{in}(j\omega)| = 10^{-8}\,\frac{[\text{magnitude of vector a}][\text{magnitude of vector b}]}{[\text{magnitude of vector c}]} \tag{8-49}$$

We are interested in the value of this quantity as ω is varied from 0 to ∞. Observe that for $\omega = 0$, $|y_{in}(j\omega)| = \infty$, since the magnitude of vector c is 0, while the magnitudes of vectors a and b are finite. As ω increases from 0, all three vectors have finite magnitude; hence, $|y_{in}(j\omega)|$ is finite. For $\omega > 10^6$, the magnitude of vector b is approximately twice the magnitude of vector c. Hence, in Equation (8-49),

$$\frac{[\text{magnitude of vector b}]}{[\text{magnitude of vector c}]} \approx 2 \qquad \omega > 10^6$$

and

$$|y_{in}(j\omega)| \approx 2 \times 10^{-8}\,[\text{magnitude of vector a}] \qquad \omega > 10^6 \tag{8-50}$$

Hence, $|y_{in}(j\infty)| = \infty$. It is also apparent that $|y_{in}(j\omega)|$ goes through a minimum somewhere in the vicinity of $\omega = 10^6$, since the ratio of the magnitude of vector b to the magnitude of vector c does not change much as ω is varied in this vicinity, while the magnitude of vector a goes through a minimum at $\omega = 10^6$.

In summary, as ω is varied from 0 to ∞, $|y_{in}(j\omega)|$ behaves as indicated in Figure 8-8(e). Any point on the curve can be plotted by computing the magnitudes of the vectors on the pole-zero diagram for the corresponding value of ω. Important regions of such a plot can be visualized from the pole-zero diagram.

To plot $\underline{/y_{in}(j\omega)}$ vs. ω, we return to Figure 8-8(c) and note that

$$\underline{/y_{in}(j\omega)} = [\text{angle of vector a}] + [\text{angle of vector b}] - [\text{angle of vector c}] \tag{8-51}$$

At $\omega = 0$, the sum of the vector a and vector b angles is zero, while the vector c angle is $\pi/2$. Hence, $\underline{/y_{in}(j0)} = -\pi/2$. At $\omega = 10^6$, [angle of vector a] $= 0$, [angle of vector b] $\approx \pi/2$, [angle of vector c] $= \pi/2$, so that $\underline{/y_{in}(j10^6)} = 0$. For $\omega \gg 10^6$, all three vectors have angle $\pi/2$, so that Equation (8-51) yields $\underline{/y_{in}(j\infty)} = \pi/2$. The result is plotted in Figure 8-8(f).

Decibel plots. Another convenient method for graphically displaying the behavior of transfer functions as a function of the frequency is to make a log-log plot of $|\mathcal{H}(j\omega)|$ and a semilog plot of $\underline{/\mathcal{H}(j\omega)}$. For historical reasons, log $|\mathcal{H}(j\omega)|$ vs. log ω is expressed in decibels. A *decibel* (dB) is defined as $10 \log |\mathcal{H}(j\omega)|^2 = 20 \log |\mathcal{H}(j\omega)|$. Hence, to make a dB plot of a transfer function $\mathcal{H}(j\omega)$, we

1. plot $20 \log |\mathcal{H}(j\omega)|$ vs. $\log \omega$
2. plot $\underline{/\mathcal{H}(j\omega)}$ vs. $\log \omega$.

We have already shown that the magnitude of any transfer function can be written in the form [see Equation (8-32)]

$$|\mathcal{H}(j\omega)| = \frac{a_n}{b_m}\frac{|j\omega - z_1|\,|j\omega - z_2| \cdots |j\omega - z_n|}{|j\omega - p_1|\,|j\omega - p_2| \cdots |j\omega - p_m|} \tag{8-52}$$

Therefore, $20 \log |\mathcal{H}(j\omega)|$ is given by

$$20 \log |\mathcal{H}(j\omega)| = 20 \log \frac{a_n}{b_m} + 20 \log |j\omega - z_1| + \cdots + 20 \log |j\omega - z_n| - 20 \log |j\omega - p_1| - \cdots - 20 \log |j\omega - p_m| \tag{8-53}$$

Hence, a decibel plot of any transfer function can be obtained directly from the pole-zero diagram by a method similar to the one used to obtain a plot of $|\mathcal{H}(j\omega)|$ vs. ω. This time we draw the appropriate vectors on the pole-zero diagram, compute their magnitudes, take logarithms of each, and sum the individual contributions.

To plot $\underline{/\mathcal{H}(j\omega)}$ vs. $\log \omega$, we follow the same procedure used to plot $\underline{/\mathcal{H}(j\omega)}$ vs. ω except that the frequency variable is plotted on a logarithmic scale. This can be accomplished by plotting $\log \omega$ on a linear scale or by plotting ω on a logarithmic scale.

Example 8-11

In this example we make a decibel plot of the voltage gain of the RC network presented in Figure 8-9(a). The pole-zero diagram for this transfer function is presented in Figure 8-9(b).

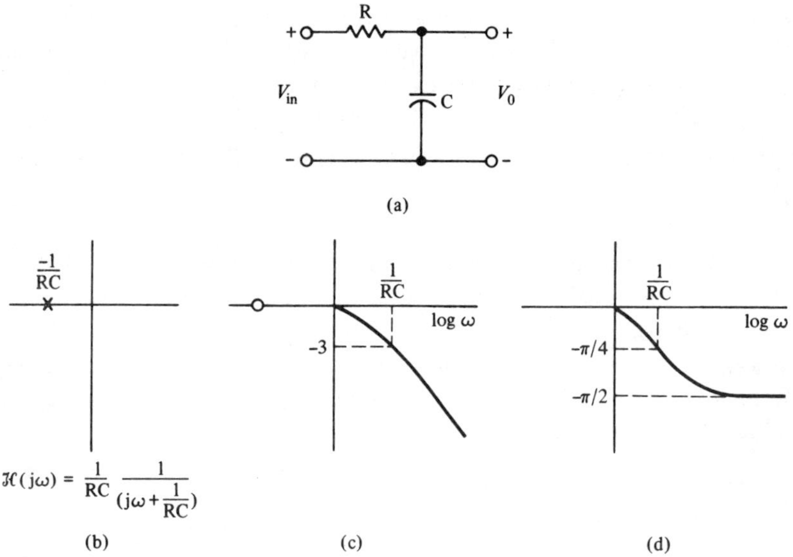

Figure 8-9 Example 8-11.

The magnitude of the transfer function is given by

$$|\mathcal{H}(j\omega)| = \frac{1}{RC} \frac{1}{|j\omega + (1/RC)|} \tag{8-54}$$

and the magnitude in decibels is given by

$$20 \log |\mathcal{H}(j\omega)| = 20 \log \frac{1}{RC} - 20 \log \left| j\omega + \frac{1}{RC} \right| \tag{8-55}$$

Referring to Figure 8-9(b), we observe that for $\omega = 0$ we have

$$20 \log |\mathcal{H}(j\omega)| = 20 \log \frac{1}{RC} - 20 \log \frac{1}{RC} = 0 \text{ dB} \tag{8-56}$$

For $\omega = 1/RC$ we have

$$\begin{aligned} 20 \log \left| \mathcal{H}\left(j \frac{1}{RC}\right) \right| &= 20 \log \frac{1}{RC} - 20 \log \frac{\sqrt{2}}{RC} \\ &= 20 \log \frac{1}{RC} - 20 \log \sqrt{2} - 20 \log \frac{1}{RC} \\ &= -20 \log \sqrt{2} \approx -3 \text{ dB} \end{aligned} \tag{8-57}$$

For $\omega = \infty$, we have

$$20 \log |\mathcal{H}(j\omega)| = 20 \log \frac{1}{RC} - 20 \log \infty = -\infty \text{ dB} \tag{8-58}$$

The complete plot of $20 \log |\mathcal{H}(j\omega)|$ vs. $\log \omega$ is presented in Figure 8-9(c).

The plot of $\underline{/\mathcal{H}(j\omega)}$ vs. $\log \omega$ is the same as the plot of $\underline{/\mathcal{H}(j\omega)}$ vs. ω obtained in Example 8-9, except that the abscissa is a logarithmic scale, as in Figure 8-9(d).

A considerable savings in the time required to make decibel plots can be achieved by first plotting the asymptotes for each of the terms in Equation (8-53). A typical term in Equation (8-53) is of the form $20 \log |j\omega - q|$, where q is a pole or zero. Consider first the case where q is real. Then, for small values of ω, the real part of the vector q is much larger than the imaginary part ω and

$$\begin{aligned} |j\omega - q| &\approx q \qquad \omega \ll q \\ 20 \log |j\omega - q| &\approx 20 \log q \qquad \omega \ll q \end{aligned} \tag{8-59a}$$

On the other hand, for large values of ω

$$\begin{aligned} |j\omega - q| &\approx \omega \qquad \omega \gg q \\ 20 \log |j\omega - q| &\approx 20 \log \omega \qquad \omega \gg q \end{aligned} \tag{8-59b}$$

The low-frequency asymptote given by Equation (8-59a) and the high-frequency asymptote given by Equation (8-59b) are plotted in Figure 8-10(a). The high-frequency asymptote is a plot of $20 \log \omega$ vs. $\log \omega$. This is a straight line with a slope of 20 dB/decade. A *decade* is a tenfold change in ω. When ω changes by a factor of 10, $\log \omega$ changes by a factor of 1, and $20 \log \omega$ changes by a factor of 20. (For example, if ω

changes from 100 to 1000, log ω changes from 2 to 3, and 20 log ω changes from 40 to 60.)

Although the low-frequency asymptote is valid only for small ω (solid line) we have extended this line to the right (dotted line). Also, even though the high-frequency asymptote is valid only for large ω (solid line), we have extended this line until it crosses the dotted port of the low-frequency asymptote line. Note that the point of intersection occurs at the point where 20 log q = 20 log ω or ω = q. The point of intersection is called the *break point*.

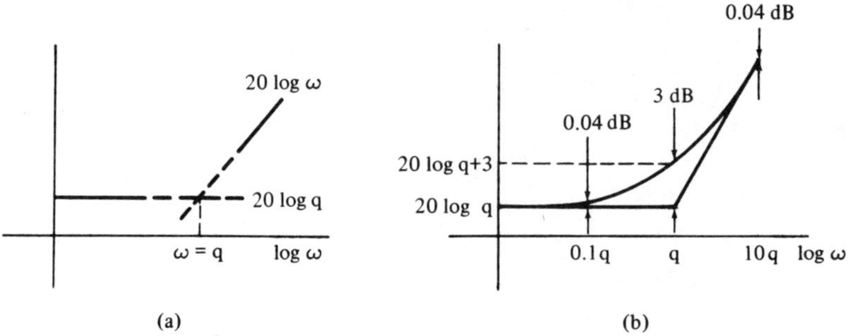

Figure 8-10 Asymptotes for dB plots.

In Figure 8-10(b) we have extended both the low- and high-frequency asymptotes until they meet. We have also plotted the original curve $20 \log |j\omega - q|$. Note that the maximum error occurs at $\omega = q$. At $\omega = q$ the asymptotic approximation gives 20 log q, while the actual value is $20 \log |jq - q| = 20 \log \sqrt{2}\, q = 20 \log q + 20 \log \sqrt{2} = 20 \log q + 3$. Hence, the maximum error is 3 dB. At $\omega = 0.1q$, the error is only 0.04 dB and at $\omega = 10q$ the error is also 0.04 dB. We conclude that a reasonably accurate sketch of $20 \log |j\omega - q|$ vs. log ω can be achieved by drawing the low- and high-frequency asymptotes, extending them until they meet, and then sketching in a curve as indicated in Figure 8-10(b).

Example 8-12

In this example we construct a decibel plot of the transfer function

$$\mathcal{H}(j\omega) = \frac{j\omega - 1}{j\omega(j\omega + 10)} \tag{8-59c}$$

The pole-zero diagram for this transfer function is presented in Figure 8-11(a).

We start by writing

$$20 \log |\mathcal{H}(j\omega)| = 20 \log |j\omega - 1| - 20 \log |j\omega| - 20 \log |j\omega + 10| \tag{8-60}$$

Our procedure will be first to sketch each of the three terms on the right-hand side of Equation (8-60) and then to add them up.

The first term 20 log $|j\omega - 1|$ is identical to the plot of Figure 8-10(b) with q = 1. Hence, the high- and low-frequency asymptotes and the plot of 20 log $|j\omega - 1|$ are easily constructed, as indicated in Figure 8-11(b).

We next consider the third term on the right-hand side of Equation (8-60). The high- and low-frequency asymptotes are given by

$$-20 \log |j\omega + 10| = \begin{cases} -20 \log 10 = -20 \text{ dB} & \omega \ll 10 \\ -20 \log \omega & \omega \gg 10 \end{cases} \tag{8-61}$$

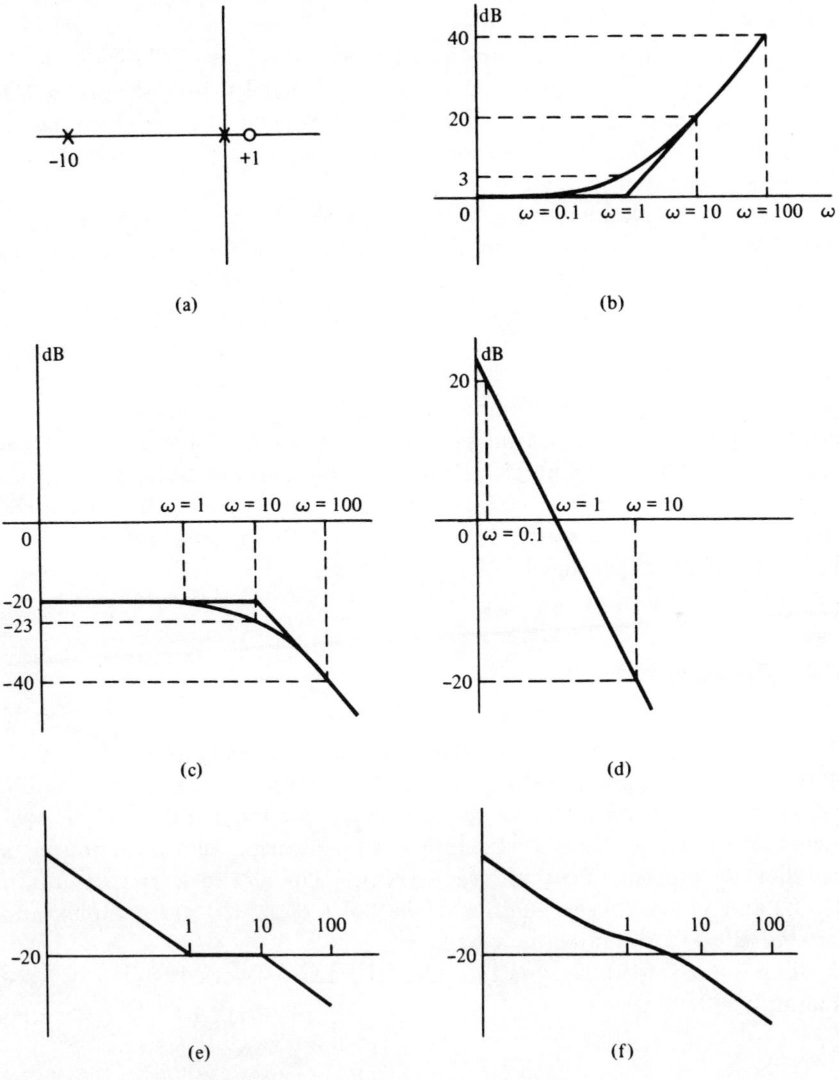

Figure 8-11 Example 8-12.

Both asymptotes and the plot of $-20 \log |j\omega + 10|$ are shown in Figure 8-11(c).

Finally, we treat the second term on the right-hand side of Equation (8-60). Since $|j\omega| = \omega$, we have

$$-20 \log |j\omega| = -20 \log \omega \qquad \text{all } \omega \tag{8-62}$$

This equation is valid for all ω. Plotted against $\log \omega$, it is a straight line with slope -20 dB/decade. It passes through 0 dB at $\omega = 1$, that is, $-20 \log 1 = 0$. A plot of $-20 \log |j\omega|$ vs. $\log \omega$ is presented in Figure 8-11(d).

We are now ready to sum up the three contributions of Figure 8-11(b), (c), and (d). This is most easily accomplished by first summing the asymptotic plots as illustrated in Figure 8-11(e). Then, keeping in mind the errors illustrated in the plot of Figure 8-10(b), we sketch in the curve as illustrated in Figure 8-11(f).

In this subsection we have only considered the case of real poles and zeros; that is, in the discussion at the beginning of the subsection q was assumed real, and in Example 8-12 only real poles and zeros were encountered. For the case of complex poles and zeros, a similar procedure can be followed. The most convenient approach is to consider the complex poles and zeros in pairs. This can always be done, since all complex poles and zeros appear in complex conjugate pairs. Therefore, if q is a complex pole or zero, the term $(j\omega - q)$ is always accompanied by a term $(j\omega - q^*)$. The procedure for handling these pairs of terms is not particularly difficult, but it is too far from our central theme to be considered here. The interested reader is invited to see References [14], [15], [23], and [32], all of which develop this procedure.

8-4 Resonant Networks

A *resonant network* is a network that is highly frequency-selective. Such networks have application in the high-frequency circuits used in radio, television, radar, and so on. A simple resonant network can be constructed by connecting a resistance, capacitance, and inductance in parallel, as illustrated in Figure 8-12(a). The network is resonant if R, L, and C are chosen such that the poles of $\mathfrak{z}_{in}(p)$ are complex and located close to the imaginary axis.

The input impedance for the parallel RLC network is given by (see Example 8-2)

$$\mathfrak{z}_{in}(p) = \frac{1}{C} \frac{p}{p^2 + (1/RC)p + (1/LC)} \tag{8-63}$$

Hence, $\mathfrak{z}_{in}$ has a zero at the origin and two poles at

$$\begin{Bmatrix} p_1 \\ p_2 \end{Bmatrix} = -\frac{1}{2RC} \pm \sqrt{\left(\frac{1}{2RC}\right)^2 - \frac{1}{LC}} \tag{8-64}$$

Suppose that we choose R, L, and C such that $1/LC \gg (1/2RC)^2$. Then the poles are complex and given by

$$\begin{Bmatrix} p_1 \\ p_2 \end{Bmatrix} \approx -\frac{1}{2RC} \pm j\sqrt{\frac{1}{LC}} \tag{8-65}$$

The pole-zero diagram for this case is presented in Figure 8-12(b). Note that the assumption $1/LC \gg (1/2RC)^2$ assures that the poles are much closer to the imaginary axis than they are to the real axis.

We next use the pole-zero diagram to obtain a plot of $|\mathfrak{z}_{in}(j\omega)|$ vs. ω and $\underline{/\mathfrak{z}_{in}(j\omega)}$ versus ω. Following the procedure of Section 8-3, we first write

$$|\mathfrak{z}_{in}(j\omega)| = \frac{1}{C}\frac{|j\omega - z_1|}{|j\omega - p_1|\,|j\omega - p_2|} \tag{8-66}$$

At $\omega = 0$, the numerator vector has magnitude zero because of the zero at the origin. Hence, $|\mathfrak{z}_{in}(j0)| = 0$. For large ω the magnitude of $\mathfrak{z}_{in}$ again approaches zero. Hence, $|\mathfrak{z}_{in}(j\infty)| = 0$.

The frequency range of interest for the parallel RLC resonant network is the range of frequencies in the vicinity of the pole p_1. Note in Figure 8-12(b) that the magnitude of the vector from p_2 is very nearly

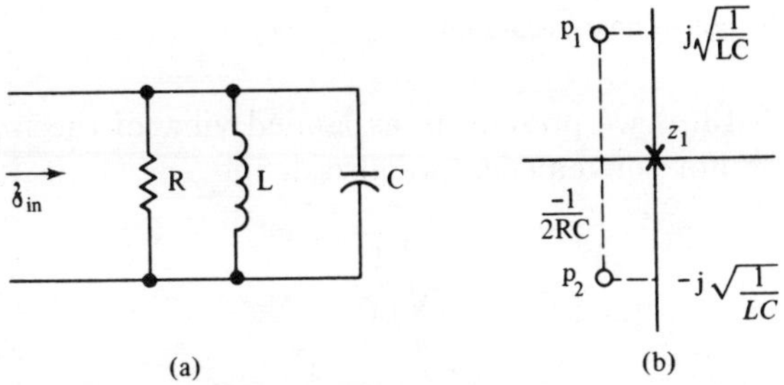

Figure 8-12 A resonant network and its pole-zero diagram.

twice the magnitude of the vector from z_1 for all ω in the frequency range of interest. Hence, we can write

$$\frac{|j\omega - z_1|}{|j\omega - p_2|} \approx \frac{1}{2} \qquad \omega \text{ near } p_1 \tag{8-67}$$

It follows that

$$|\mathfrak{z}_{in}(j\omega)| \approx \frac{1}{2C}\frac{1}{|j\omega - p_1|} \qquad \omega \text{ near } p_1 \tag{8-68}$$

We conclude that the frequency characteristics of the parallel RLC network for frequencies in the vicinity of p_1 are governed primarily by the pole at p_1. It follows that to plot $|\mathfrak{z}_{in}(j\omega)|$ for ω near p_1, we need only consider the magnitude of the vector from p_1.

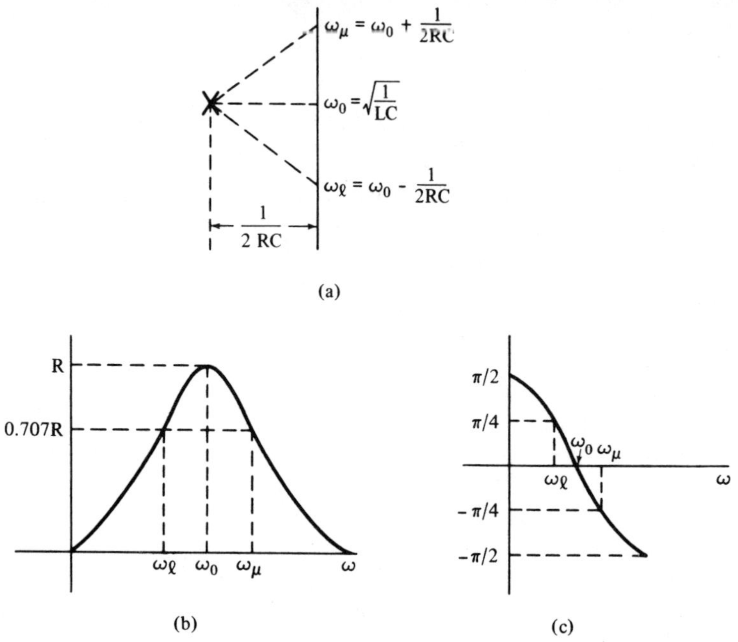

Figure 8-13 Response near resonance.

In Figure 8-13(a) we present an expanded view of the p plane in the vicinity of p_1. For convenience, we also define

$$\begin{aligned}\omega_0 &= \sqrt{\frac{1}{LC}}\\ \omega_\mu &= \omega_0 + \frac{1}{2RC}\\ \omega_\ell &= \omega_0 - \frac{1}{2RC}\end{aligned} \qquad \textbf{(8-69)}$$

At $\omega = \omega_0$ we observe that $|j\omega - p_1| = 1/(2RC)$ so that

$$|\mathfrak{z}_{in}(j\omega_0)| = \frac{1}{2C}\,\frac{1}{(1/(2RC))} = R \qquad \textbf{(8-70)}$$

As ω increases from $\omega = \omega_0$, the magnitude of the vector from p_1 increases and $|\mathfrak{z}_{in}(j\omega)|$ decreases. At $\omega = \omega_\mu$ the magnitude of the vector from p_1

has increased by a factor of $\sqrt{2}$ from its value at $\omega = \omega_0$. Hence,

$$|\mathfrak{z}_{in}(j\omega_\mu)| \approx 0.707R \tag{8-71a}$$

As ω increases from $\omega = \omega_\mu$, the magnitude of the vector from p_1 grows, and $|\mathfrak{z}(j\omega)|$ decreases further.

Next, we go back to $\omega = \omega_0$ and let ω decrease. As ω decreases from $\omega = \omega_0$, the magnitude of the vector from p_1 again increases, causing $|\mathfrak{z}_{in}(j\omega)|$ to decrease. At $\omega = \omega_\ell$, $|j\omega - p_1|$ has increased by a factor of $\sqrt{2}$ from its value at $\omega = \omega_0$. Hence,

$$|\mathfrak{z}_{in}(j\omega_\ell)| = 0.707R \tag{8-71b}$$

As ω decreases from $\omega = \omega_\ell$, $|j\omega - p_1|$ increases further, and $|\mathfrak{z}_{in}(j\omega)|$ continues to decrease. Hence, we arrive at the plot of $|\mathfrak{z}_{in}(j\omega)|$ vs. ω presented in Figure 8-13(b). ω_μ and ω_ℓ are called the upper and lower half-power points, since at these values of ω, $|\mathfrak{z}_{in}(j\omega)|^2$ is down by a factor of $\frac{1}{2}$ from its maximum value at ω_0.

To plot $\angle\mathfrak{z}_{in}(j\omega)$ vs. ω, for ω in the vicinity of p_1, we follow a similar procedure. For ω in the vicinity of p_1, we observe from Figure 8-13(a) that $\angle j\omega - z_1 \approx \angle j\omega - p_2$. Hence,

$$\angle\mathfrak{z}_{in}(j\omega) \approx -\angle j\omega - p_1 \tag{8-72}$$

Therefore, by observing the angle of the vector from p_1 to the point $j\omega$ as ω is varied about ω_0, we obtain the plot of Figure 8-13(c). Equation (8-72) is valid for all $\omega > \omega_0$. Hence, as $\omega \to \infty$, $\angle\mathfrak{z}_{in}(j\omega) \to -\pi/2$. For $\omega \ll \omega_0$, the angles of all three vectors must be considered. Note that for $\omega = 0^+$ the contributions from p_1 and p_2 cancel so that $\angle\mathfrak{z}_{in}(j0^+) = \angle j\omega - 0^+ = \pi/2$.

Networks having a pole or zero close enough to the imaginary axis to dominate the frequency characteristics for frequencies in that vicinity are called *resonant networks.* The *resonant frequency* is that value of ω for which the effect is most pronounced. *Resonance* is a description of the condition of a resonant network when it is excited by a signal at or near its resonant frequency. These definitions are not very precise, and they are not intended to be. The terms "resonance," "resonant network," and "resonant frequency" are used very loosely in the engineering literature. For example, some writers define the resonant frequency as the point on the imaginary axis closest to the pole or zero of interest; other writers define the resonant frequency as the frequency for which the magnitude of the transfer function achieves its maximum value; still other writers define the resonant frequency as the frequency at which the transfer function is real (its angle is zero). For the RLC network considered here, any of these definitions would be reasonable, since all three of these events occur at the same frequency, $\omega = \omega_0$. In general, however, these three events occur at three different frequencies. For example, if a small resistance is added in series with the inductance, as illustrated in Figure 8-14,[2] the resulting network still exhibits the same general frequency characteristics as before, but the three events mentioned occur at three slightly different frequencies.

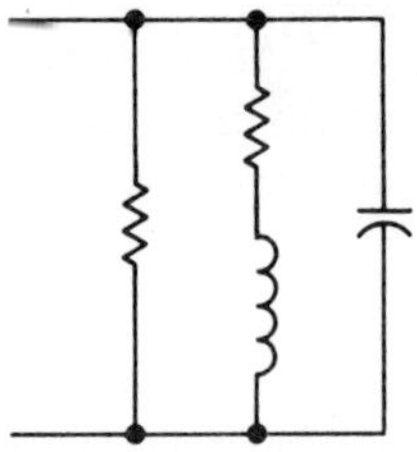

Figure 8-14 A resonant network using an inductor that is not lossless.

Further insight into the frequency behavior of the parallel RLC network can be gained by writing the input admittance in the form

$$\begin{aligned} y_{\text{in}}(j\omega) &= \frac{1}{R} + \frac{1}{j\omega L} + j\omega C \\ &= R^{-1} + j\left(\omega C - \frac{1}{\omega L}\right) \end{aligned} \tag{8-73}$$

Hence, the input admittance has real part R and imaginary part $\omega C - (1/\omega L)$. The magnitude of the admittance is the magnitude of the complex number given by Equation (8-73), that is, the square root of the sum of the squares of the real and imaginary parts. Since the real part does not depend on ω, the value of ω that minimizes the imaginary part also minimizes $|y_{\text{in}}(j\omega)|$. Hence, we set $\omega C - (1/\omega L) = 0$, and solve for ω. We find

$$\omega = \omega_0 = \frac{1}{\sqrt{LC}} \tag{8-74}$$

At this frequency the admittance of the parallel combination of the inductance and capacitance is zero (the impedance is infinite) so that the input admittance of the parallel RLC network at $\omega = \omega_0$ is R^{-1}. If ω is increased from ω_0, the admittance of the capacitance increases while the admittance of the inductance decreases. For sufficiently high frequencies, the capacitance behaves as a short circuit. On the other hand, if ω is decreased from ω_0, the admittance of the capacitance decreases, but the inductance admittance increases. For sufficiently low frequencies, the inductance behaves as a short circuit.

It is also instructive to consider resonant networks from an energy point of view. Consider the parallel RLC network of Figure 8-15(a).

[2] Since inductors are usually constructed by winding a length of wire onto a coil, and since all conductors have a finite resistance, a realistic model for an inductor would be an inductance in series with a resistance.

Suppose that the network is excited by a current source at its resonant frequency ω_0. At $\omega = \omega_0$, the parallel LC network has zero admittance so that the current into the LC combination is zero. Since the input admittance to the RLC network is R^{-1}, the voltage is

$$v(t) = R \cos \omega_0 t \tag{8-75}$$

Therefore, the three branch currents are given by

$$i_R(t) = \frac{v(t)}{R} = \cos \omega_0 t \tag{8-76}$$

$$i_L(t) = \frac{1}{L}\int v(t)\,dt = \frac{1}{L}\int R \cos \omega_0 t\,dt = \frac{R}{\omega_0 L} \sin \omega_0 t \tag{8-77}$$

$$i_C(t) = C\frac{d}{dt}[v(t)] = C\frac{d}{dt}[R \cos \omega_0 t] = -R\omega_0 C \sin \omega_0 t \tag{8-78}$$

These waveforms are sketched in Figure 8-15(b). Since $\omega_0 C = 1/\omega_0 L$, we have

$$i_L(t) + i_C(t) = 0 \tag{8-79}$$

as required. Hence, the inductance and capacitance currents are not zero, but one is the negative of the other (they have equal amplitudes but are π radians out of phase), so that the net current into the LC combination is zero.

We next compute the energy stored in the inductance, the energy stored in the capacitance, and the power dissipated in the resistance as functions of time. These quantities are given by

$$p_R(t) = Ri_R^2(t) = R \cos^2 \omega_0 t = \frac{R}{2} + \frac{R}{2}\cos 2\omega_0 t \tag{8-80}$$

$$\mathcal{E}_L(t) = \tfrac{1}{2}Li_L^2(t) = \frac{R^2}{2\omega_0^2} \sin^2 \omega_0 t = \frac{R^2 C}{2}\sin^2 \omega_0 t \tag{8-81}$$

$$\mathcal{E}_C(t) = \tfrac{1}{2}Cv^2(t) = \frac{R^2 C}{2}\cos^2 \omega_0 t \tag{8-82}$$

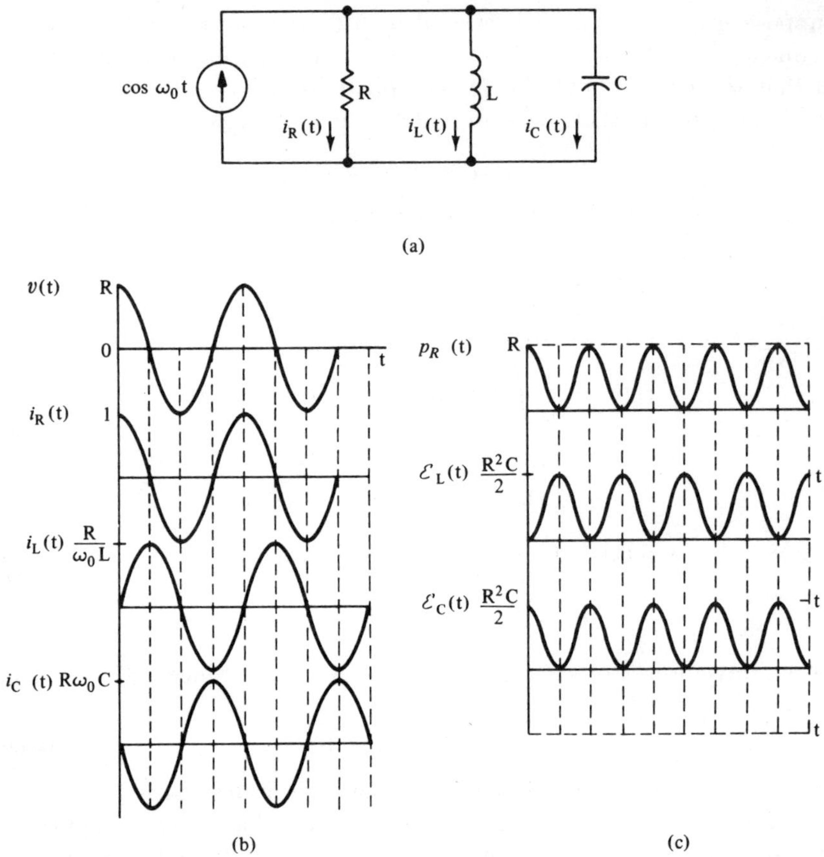

Figure 8-15 Waveforms for a resonant network.

These functions are sketched in Figure 8-15(c). The total energy stored in the LC combination is given by

$$\mathscr{E}_{\text{stored}}(t) = \mathscr{E}_C(t) + \mathscr{E}_L(t) = \frac{R^2C}{2}\sin^2\omega_0 t + \frac{R^2C}{2}\cos^2\omega_0 t$$
$$= \frac{R^2C}{2} \tag{8-83}$$

Since only the resistance dissipates power, we can also write

$$p_{\text{dissipated}}(t) = p_R(t) = \frac{R}{2} + \frac{R}{2}\cos^2\omega_0 t \tag{8-84}$$

We now define the Q of a resonant network to be 2π times the peak energy stored during one cycle to the total energy dissipated during one cycle when the network is excited at its resonant frequency. Equation (8-83) states that the total energy stored is constant at $R^2C/2$ joules.

The total energy dissipated per cycle is given by

$$\mathcal{E}_{\text{dissipated}} = \int_{\substack{\text{one}\\\text{cycle}}} p(t)\,dt = \int_0^{2\pi/\omega_0}\left(\frac{R}{2} + \frac{R}{2}\cos 2\omega_0 t\right)dt$$
$$= \frac{R\pi}{\omega_0} \tag{8-85}$$

Therefore, the Q of the parallel RLC network is given by

$$Q = 2\pi\,\frac{R^2C/2}{R\pi/\omega_0} = \omega_0 RC \tag{8-86}$$

The Q of a resonant network is a measure of the degree of the resonance phenomena. In high-Q networks, the resonance phenomena is quite pronounced; in low-Q networks it is not. Indeed, for the high Q parallel RLC network we note the following:

1. The peak energy stored in the LC networks is $Q/2\pi$ times the energy dissipated in the resistance during one cycle. (This is the definition of Q.)
2. The amplitudes of the inductance and capacitance currents are Q times the amplitude of the resistance current. (See Figure 8-15.)
3. The ratio of the resonant frequency to the "width" of the peak of $|\mathfrak{z}_{in}(j\omega)|$ is given by $\omega_0/(\omega_\mu - \omega_\ell) = Q$. [See Figure 8-13 (b).] Hence, in high-Q networks, the magnitude of the input impedance is very sharply peaked about the resonant frequency ω_0.

As a rule of thumb, networks having a Q greater than 10 are generally considered to be high-Q networks; networks having a Q of less than 10 are generally called low-Q.

8-5 Magnitude Scaling and Frequency Scaling

Introductory comments. One of the major uses of electronic networks is in the analysis and design of signal processors. The networks are models of the circuits used to process electrical signals. In many cases the desired signal-processing operation is an alteration of the magnitude and phase of an essentially steady-state signal as a function of frequency. The mathematical characterization of a network that models a circuit for performing this signal processing is the network transfer function. In many engineering problems the desired signal-processing operations are quite similar in some respects, although quite different in others. For example, most radio receivers could be characterized by the block diagram of Figure 1-2(b) whether they are for the AM broadcast band, any of the short-wave bands, the FM band, or any other frequency band.

Television and radar receivers also have most of these items in their block diagrams.

For many of the blocks, the pole-zero diagrams and the various magnitude or phase versus frequency plots described above look the same whether the block is from an AM broadcast receiver or a microwave radar set. The only difference is on the scales used on the axes of these graphical representation. The networks to realize the desired transfer functions do not have to be designed separately for each frequency range and for each desired gain and phase shift scale. In many applications, satisfactory designs for many situations can be obtained from one basic design by magnitude and frequency scaling. In this section we present the basic scaling concepts. Chapter 9 is a discussion of the application of these concepts to some practical problems.

Magnitude-scaling. A network is said to be *impedance-scaled* by a factor of k if each impedance in the network is scaled by a factor of k. Since the impedance of a resistance is R, impedance-scaling a resistance corresponds to scaling the resistance by a factor of k. The impedance for an inductance is $j\omega L$. Hence, impedance-scaling an L-henry inductance corresponds to scaling the inductance by a factor of k. On the other hand, the impedance of a C-farad capacitance is $1/j\omega C$. Hence, impedance-scaling a C-farad capacitance corresponds to scaling the capacitance by k^{-1}. These definitions are summarized in Table 8-1.

Table 8-1

ELEMENT VALUE	ELEMENT IMPEDANCE	ELEMENT IMPEDANCE AFTER IMPEDANCE-SCALING BY k	ELEMENT VALUE AFTER IMPEDANCE-SCALING BY k
R	R	$k(R) = kR$	kR
L	$j\omega L$	$k(j\omega L) = j\omega(kL)$	kL
C	$\dfrac{1}{j\omega C}$	$k\left(\dfrac{1}{j\omega C}\right) = \dfrac{1}{j\omega(C/k)}$	C/k

It can be shown that if a network is impedance-scaled by a factor of k, then the following is true:

1. All network parameters having units of impedance (all parameters that are ratios of a voltage to a current, for example, the 2-port z parameters $\mathfrak{z}_{11}$, $\mathfrak{z}_{12}$, $\mathfrak{z}_{21}$, $\mathfrak{z}_{22}$, the hybrid parameter $\mathfrak{h}_{11}$, input impedances, output impedances, and so on) are scaled by a factor of k.

2. All network parameters having units of admittance (all parameters that are ratios of a current to a voltage, for example, y parameters, IAM parameters, input admittances, and so on) are scaled by a factor of k^{-1}.

3. All network parameters that are ratios of a voltage to a voltage or a current to a current, for example, voltage gains and current gains, $\mathfrak{h}_{12}$, $\mathfrak{h}_{21}$, and so on, are not affected.

Finally, we point out that *admittance-scaling* by a factor of k is equivalent to impedance-scaling by a factor of k^{-1}. These results are summarized in Table 8-2.

Table 8-2

	TRANSFER FUNCTION	TRANSFER FUNCTION AFTER IMPEDANCE-SCALING BY k	TRANSFER FUNCTION AFTER ADMITTANCE-SCALING BY k
Input, output, and transfer impedances	$\mathcal{H}(j\omega)$	$k\mathcal{H}(j\omega)$	$k^{-1}\mathcal{H}(j\omega)$
Input, output, and transfer admittances	$\mathcal{H}(j\omega)$	$k^{-1}\mathcal{H}(j\omega)$	$k\mathcal{H}(j\omega)$
Voltage and current gains	$\mathcal{H}(j\omega)$	$\mathcal{H}(j\omega)$	$\mathcal{H}(j\omega)$

Frequency-scaling. A network is said to be *frequency-scaled* by a factor of k if the frequency-scaled network behaves at frequency $\omega' = k\omega$ the way the original network behaved at frequency ω. Therefore, frequency-scaling by a factor of k corresponds to making a change of the frequency variable in the transfer function. A frequency-scaled network will exhibit the same frequency characteristics at frequency $\omega' = k\omega$ that the original network exhibited at frequency ω if each element in the frequency-scaled network behaves at the new frequency $\omega' = k\omega$ the way the original element behaved at frequency ω. Therefore, to frequency-scale a network by a factor of k, we must frequency-scale each element in the network by a factor of k.

The complex impedance for an L-henry inductance at frequency ω is $j\omega L$. We wish to replace this element with another element that has this same impedance at frequency $\omega' = k\omega$. Therefore, if we replace the L-henry inductance with an L′-henry inductance with $L' = L/k$, the new inductance has impedance $j\omega L' = j\omega(L/k)$. Hence, at frequency $\omega' = k\omega$, the new inductance has impedance $j\omega' L' = j(k\omega)L' = j(k\omega)(L/k)$, which is identical to the impedance of the L-henry inductance of frequency ω.

The complex admittance for a C-farad capacitance is $j\omega C$. We wish to replace this element with another element that has the same admittance at frequency $\omega' = k\omega$ that the original element has at frequency ω. Clearly, if we assign the new capacitance, the value $C' = C/k$, then the new element behaves at frequency $\omega' = k\omega$ the way the original element behaved at frequency ω.

Since resistances are not frequency-dependent, they are not affected by frequency-scaling.

These results are summarized in Table 8-3.

Table 8-3

ELEMENT VALUE	ELEMENT IMPEDANCE OR ADMITTANCE AT FREQUENCY ω	IMPEDANCE OR ADMITTANCE AT FREQUENCY ω AFTER FREQUENCY-SCALING BY k	ELEMENT VALUE AFTER FREQUENCY-SCALING BY k
R	R	R	R
L	$j\omega L$	$j\omega L' = j\omega(L/k)$	L/k
C	$j\omega C$	$j\omega C' = j\omega(C/k)$	C/k

8-6 Computational Techniques for Computing Transfer Functions

Any transfer function for any linear network can be computed by the straightforward application of the theorems of Chapter 2 and the analysis techniques of Chapters 3 and 4. As the network analyst gains experience, he develops various shortcuts and tricks that allow him to carry out computations (such as computing transfer functions) more quickly and efficiently. The purpose of this text is to present general methods of network analysis rather than a list of such computational tricks that are useful only for special cases. Nevertheless, in this section we present an example of the application of the theorems of Chapter 2 and the analysis techniques of Chapters 3 and 4 to the computation of transfer functions. No new material on transfer functions is presented in this section; it can be omitted by readers requiring no further elaboration on computational techniques.

Suppose that we are required to compute the transfer function, V_0/V_s (voltage gain) for the network presented in Figure 8-16. We shall carry out this computation by a number of different methods.

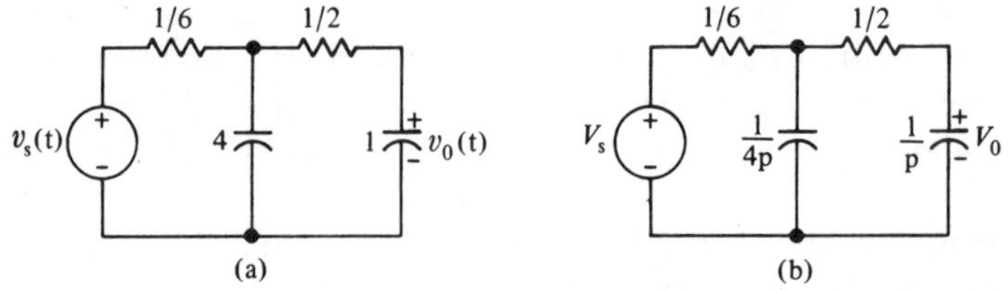

Figure 8-16 (a) A ladder network. (b) Its complex equivalent network.

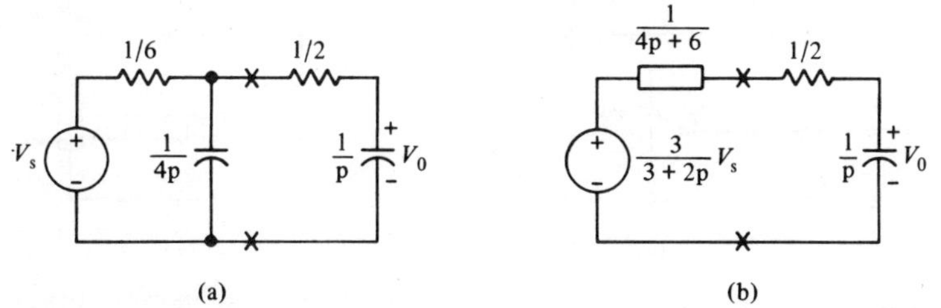

Figure 8-17 A network and its Thevenin equivalent.

Two procedures that involve only the straightforward application of the theorems presented in Chapter 2 are apparent. One is first to break the network to the right of the 4-farad capacitance and replace the left-hand network with its Norton equivalent network, as illustrated in Figure 8-17. Here we used the voltage-divider theorem to compute the Thevenin equivalent (open-circuit voltage), and we note that the Thevenin equivalent voltage is

$$V_T = \frac{1/4p}{(1/4p) + (1/6)} V_s = \frac{3}{2p + 3} V_s \tag{8-87}$$

We note that the Thevenin equivalent admittance is simply the sum of the two admittances 4p and 6. Hence, the Thevenin equivalent impedance is $1/(4p + 6)$. Another straightforward application of the voltage-divider principle to the network of Figure 8-17(b) yields

$$V_0 = \frac{1/p}{(1/p) + (1/2) + [1/(4p + 6)]} \left(\frac{3}{2p + 3} V_s\right) \tag{8-88}$$

or

$$\frac{V_0}{V_s} = \frac{3}{p^2 + 4p + 3} \tag{8-89}$$

Hence, we obtained the required transfer function by two applications of the voltage-divider theorem of Chapter 2.

The second procedure is to break the network to the left of the 4-farad capacitance and replace the right-hand network with its Thevenin equivalent impedance, as illustrated in Figure 8-18. Here we note that the admittance seen looking into the right-hand network is the admittance of the 4-farad capacitance plus the admittance of the series combination of the $\frac{1}{2}$-ohm resistance and 1-farad capacitance, that is, $4p + 1/[(1/p) + (1/2)] = (4p^2 + 10p)/(p + 2)$. Now referring to Figure 8-18(b) and using the voltage-divider technique, we compute the voltage V:

$$\begin{aligned} V &= \frac{(p + 2)/(4p^2 + 10p)}{(p + 2)/(4p^2 + 10p) + (1/6)} V_s \\ &= \frac{3(p + 2)}{2(p^2 + 4p + 3)} V_s \end{aligned} \tag{8-90}$$

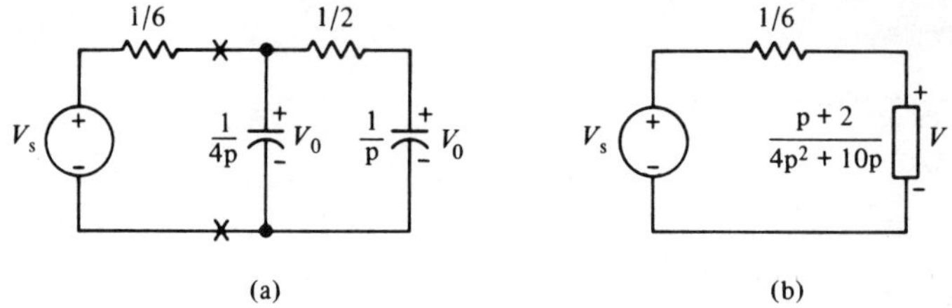

Figure 8-18 A network and its Thevenin equivalent.

Next we refer to Figure 8-18(a) and again use the voltage-divider theorem to note that V_0 and V are related by

$$\begin{aligned} V_0 &= \frac{1/\mathrm{p}}{(1/\mathrm{p}) + (1/2)} V \\ &= \frac{2}{\mathrm{p} + 2} V \end{aligned} \tag{8-91}$$

Equations (8-90) and (8-91) can be combined to give

$$\begin{aligned} \frac{V_0}{V_s} &= \left(\frac{V_0}{V}\right)\left(\frac{V}{V_s}\right) \\ &= \left(\frac{2}{\mathrm{p} + 2}\right)\left(\frac{3(\mathrm{p} + 2)}{2(\mathrm{p}^2 + 4\mathrm{p} + 3)}\right) \\ &= \frac{3}{\mathrm{p}^2 + 4\mathrm{p} + 3} \end{aligned} \tag{8-92}$$

Equations (8-89) and (8-92) were derived by using nothing more than two of the Chapter 2 theorems—Thevenin's theorem and the voltage-divider theorem. It is also possible to compute this transfer function by the method of Chapter 3. Here we start with the network of Figure 8-19(a) and construct a tree and cotree, and assign voltage synbols as illustrated in Figure 8-19(b). The tree branch voltage algorithm of

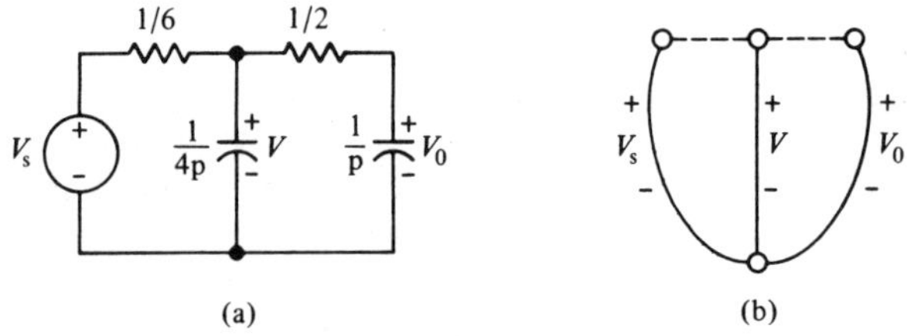

Figure 8-19 Application of the tree branch voltage algorithm.

Chapter 3 yields the two equations

$$\begin{aligned} 4pV &= 6(V_s - V) - 2(V - V_0) \\ pV_0 &= 2(V - V_0) \end{aligned} \tag{8-93}$$

or

$$\begin{aligned} 6V_s &= (4p + 8)V + (-2)V_0 \\ 0 &= (-2)V + (p + 2)V_0 \end{aligned} \tag{8-94}$$

Solving for V_0, we obtain

$$\begin{aligned} V_0 &= \frac{\begin{vmatrix} 4p + 8 & 6V_s \\ -2 & 0 \end{vmatrix}}{\begin{vmatrix} 4p + 8 & -2 \\ -2 & p + 2 \end{vmatrix}} = \frac{12V_s}{(4p + 8)(p + 2) - 4} \\ &= \frac{3}{p^2 + 4p + 3} V_s \end{aligned} \tag{8-95}$$

or

$$\frac{V_0}{V_s} = \frac{3}{p^2 + 4p + 3} \tag{8-96}$$

Another method for computing transfer functions is to start at the output and write a sequence of transfer functions, gradually working

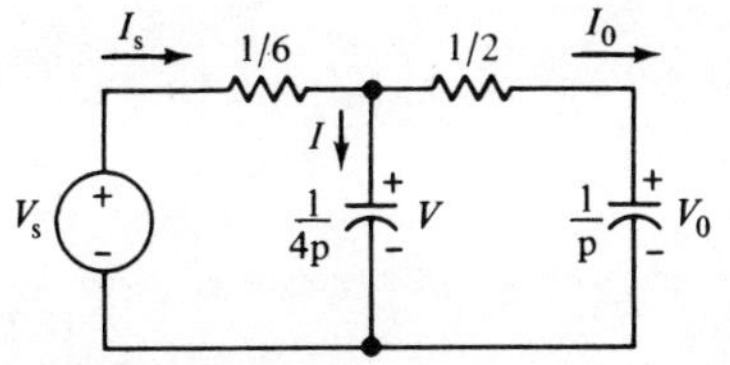

Figure 8-20 Ladder network calculation method.

back to the input. To illustrate this technique, we refer to Figure 8-20 and note that

$$I_0 = pV_0 \tag{8-97}$$

Next, we use KVL to write

$$\begin{aligned} V &= \left(\frac{1}{2} + \frac{1}{p}\right) I_0 \\ &= \frac{p + 2}{2p} I_0 \end{aligned} \tag{8-98}$$

Now substituting Equation (8-97) into Equation (8-98), we have

$$\begin{aligned} V &= \frac{p + 2}{2p} pV_0 \\ &= \frac{p + 2}{2} V_0 \end{aligned} \tag{8-99}$$

We next compute the current down through the 4-farad capacitance; that is,

$$I = 4pV \tag{8-100}$$

and substituting Equation (8-99) into (8-100), we have

$$\begin{aligned} I &= 4p\frac{p+2}{2}V_0 \\ &= (2p^2+4p)V_0 \end{aligned} \tag{8-101}$$

We next use KCL to write

$$I_s = I + I_0 \tag{8-102}$$

Substituting Equations (8-97) and (8-101) into (8-102), we have

$$\begin{aligned} I_s &= (2p^2+4p)V_0 + pV_0 \\ &= (2p^2+5p)V_0 \end{aligned} \tag{8-103}$$

Again using KVL, we have

$$V_s = \tfrac{1}{6}I_s + V \tag{8-104}$$

But I_s and V are given by Equations (8-99) and (8-103), so that

$$\begin{aligned} V_s &= \tfrac{1}{6}(2p^2+5p)V_0 + \frac{p+2}{2}V_0 \\ &= \frac{p^2+4p+3}{3}V_0 \end{aligned} \tag{8-105}$$

Hence, the final result is

$$\frac{V_0}{V_s} = \frac{3}{p^2+4p+3} \tag{8-106}$$

Hence, this method requires only KVL, KCL, and the element equations.

We have computed the transfer function for the network of Figure 8-16 by four different methods. All four methods require roughly the same amount of computation.

PROBLEMS

8-1 Compute the indicated transfer functions for each of the networks shown in Figure P8-1.

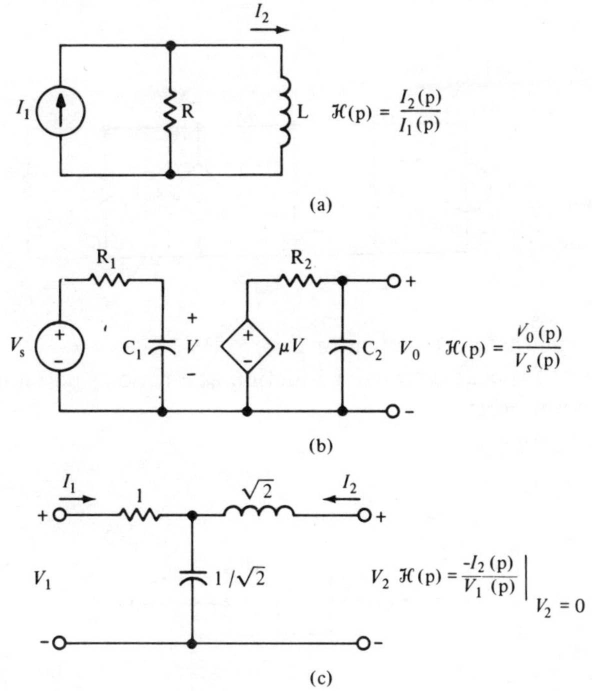

Figure P8-1

8-2 For the transfer function obtained in Problem 8-1 (c), verify the three properties of transfer functions listed in the text.

8-3 For the network shown in Figure P8-3

a. Write the 2-port y parameter y_{21} as a ratio of polynomials in p.

b. Compute the roots of the numerator and denominator polynomials.

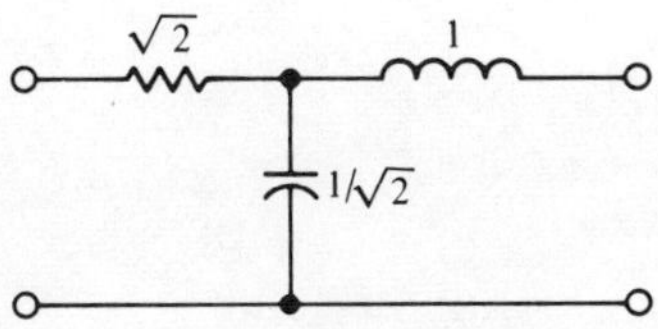

Figure P8-3

8-4 Compute the transfer function $\mathcal{H}(p) = V_0/I_s$ for the network shown in Figure P8-4. Write $\mathcal{H}(p)$ as a ratio of polynomials in p. Compute the roots of the polynomials. Use the numbers used in the FET example of Section 7-4.

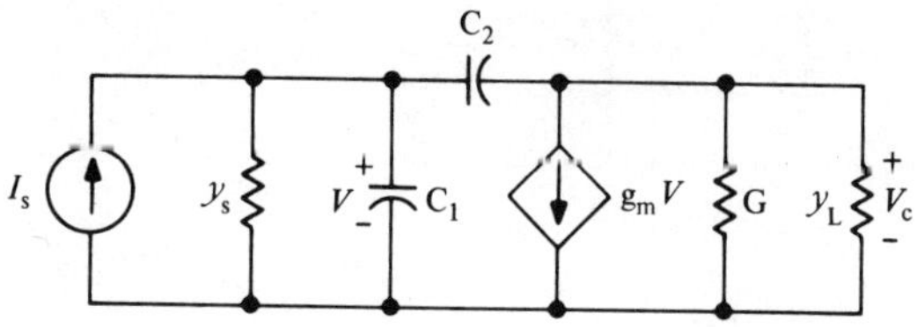

Figure P8-4

8-5 For each network shown in Figure P8-5

a. Write the specified transfer function as a ratio of polynomials in p.
b. Verify property 2.
c. Verify property 3.

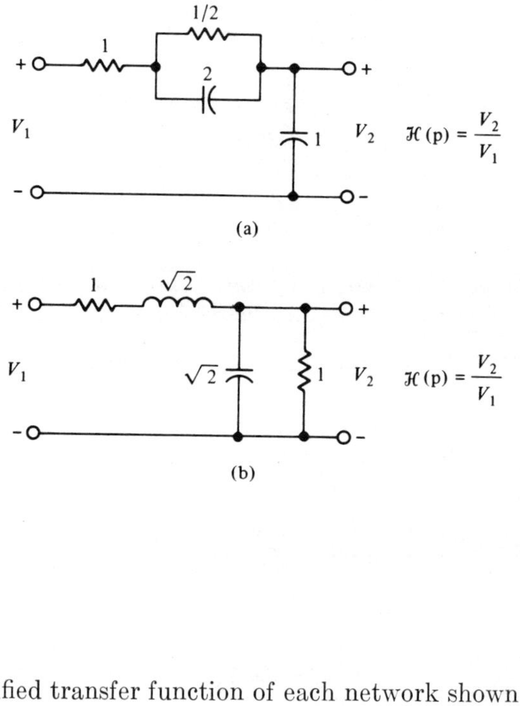

Figure P8-5

8-6 For the specified transfer function of each network shown in Figure P8-6

a. Make a pole-zero diagram.
b. Make a magnitude and angle versus frequency plot (first two networks only).
c. Make a decibel plot (first two networks only).

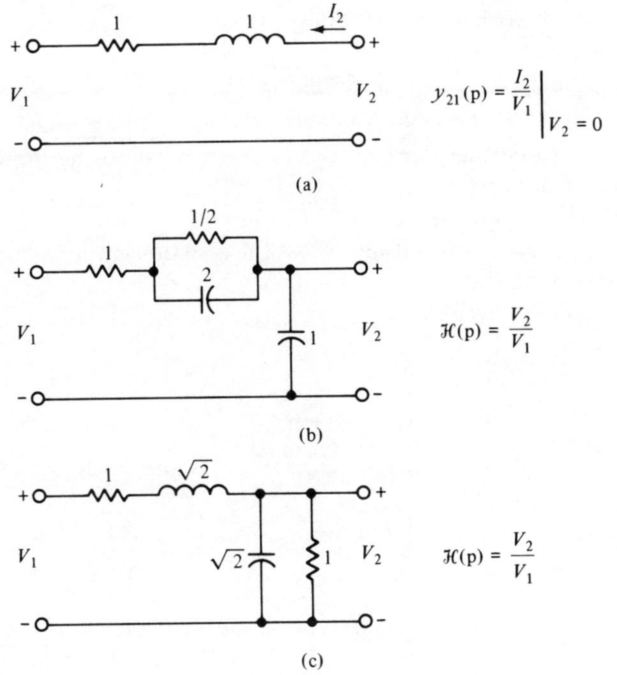

Figure P8-6

8-7 Is the second network of Problem 8-5 a "resonant network"? Why?

8-8 For the network shown in Figure P8-8

a. Compute the specified transfer function.

b. Make a pole-zero diagram.

c. Sketch the magnitude and angle of the transfer function versus the frequency ω.

d. What is the center frequency, bandwidth, and Q?

e. What is the voltage gain at the center frequency?

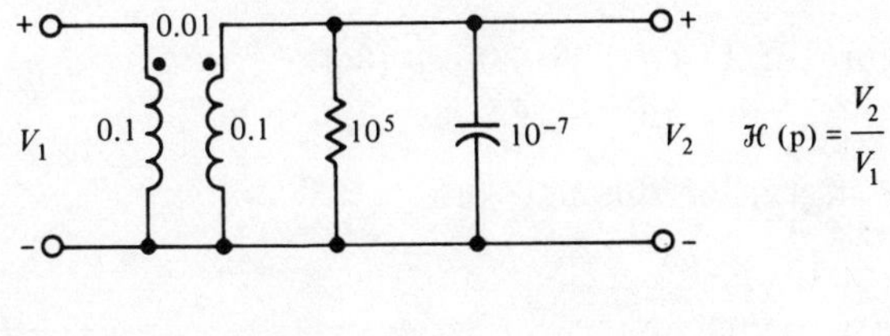

Figure P8-8

8-9 For the network shown in Figure P8-9

a. Compute $\mathcal{H}(p) = \mathfrak{z}_{in}(p)$.

b. Make a pole-zero diagram for $\mathcal{H}(p)$.

c. From the pole-zero diagram sketch $|\mathcal{H}(j\omega)|$ and $\underline{/\mathcal{H}(j\omega)}$ vs. ω.

d. Define ω_0 to be the point on the $j\omega$ axis closest to the upper left-half plane pole. Compute ω_ℓ, ω_μ.

e. What is the bandwidth? What is the Q?

f. What is the complex voltage across the resistance, the inductance, and the capacitance at the resonant frequency?

g. Sketch to scale $i_L(t)$, $i_C(t)$, $v_C(t)$, $v_L(t)$, and $v_R(t)$ as functions of time.

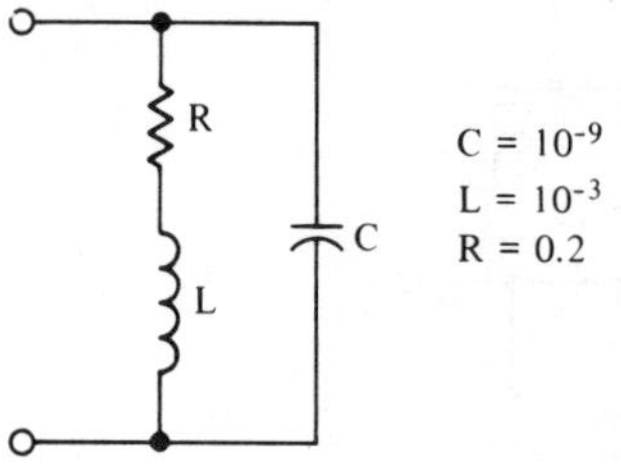

Figure P8-9

8-10 Design a frequency-selective network for a local AM radio station, to be inserted between the antenna and RF amplifier of a standard broadcast radio receiver as shown in Figure P8-10. What is the Q? (Hint: Decide on the center frequency and bandwidth and work from the pole-zero diagram.)

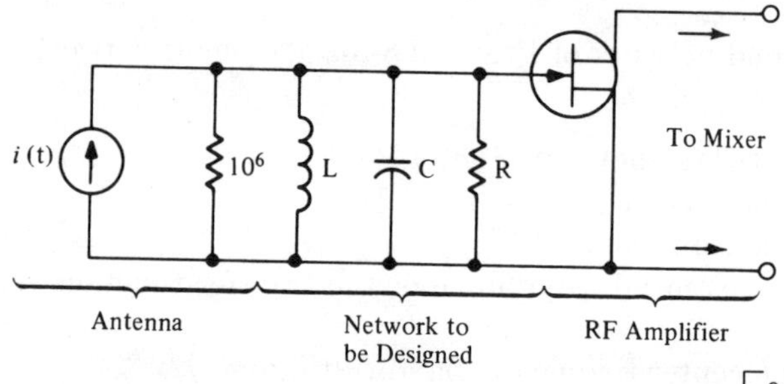

$$\mathbf{Y}(\text{FET}) = \begin{bmatrix} 0 & 0 \\ 10^{-2} & 10^{-4} \end{bmatrix}$$

Figure P8-10

8-11 In Figure P8-11 $v_1(t) = 5.0 \cos(3t)$

$$v_2(t) = 4.0 \cos(10t + 90°)$$

$$\frac{V_{out}}{V_{in}} = \mathcal{H}(j\omega) \text{ for this network}$$

$$\mathcal{H}(j3) = 1.0\underline{/-5°}$$

$$\mathcal{H}(j10) = 0.707\underline{/-45°}$$

Compute $v(t)$.

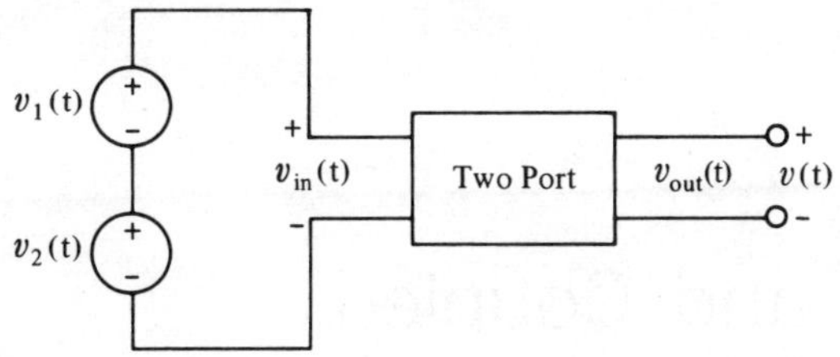

Figure P8-11

8-12 Plot the magnitude in decibels versus log ω (magnitude decibel plots) for the three networks of Problem 8-6 by first determining the asymptotes. Compute only the asymptotes and the 3-dB points. (Note the third network has complex poles. There is only one 3-dB point, and it is at 1 rad/sec.)

8-13 For the network shown in Figure P8-13

a. Magnitude-scale by 10^3.

b. Frequency-scale the result of part a by 10^6.

c. For the resulting network (after magnitude-scaling and frequency-scaling)

1. Compute the transfer function

$$y_{21}(p) = \left.\frac{I_2}{V_1}\right|_{V_2 = 0}$$

2. Make a pole-zero diagram.
3. Sketch $|y_{21}|$ vs. ω.

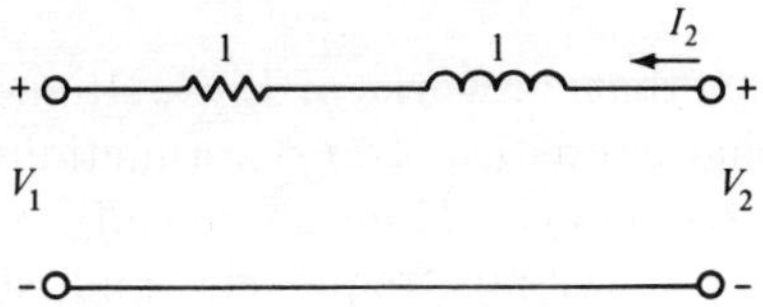

Figure P8-13

CHAPTER

9

Filters and Coupled Amplifiers

9-1 Introduction

In the previous chapter we saw that networks of L, C, R, and controlled sources have input-output characteristics (transfer functions) that are, in general, functions of the frequency. More specifically, we saw that resonant circuits select signals at certain frequencies and reject others. Such networks that select and reject signals on a frequency basis are called *filters.*

The design of filters is one important application of the theory of electronic networks. With the background material of the previous chapters and some additional elementary mathematical techniques, we are now in a position to investigate some practical methods of filter design. The material presented here is a very small part of what can be done with network theory in the design of practical filters.

To demonstrate the practicality of filters as signal processors requires more signal theory than most students have learned at this stage of their course work.[1] Nevertheless, everyone is familiar with radio and tele-

[1] See, for example, Reference [14].

vision receivers. From an elementary knowledge of the operation of these familiar devices, we can get some knowledge of the practical application of filters.

One application of filters is to provide selectivity in radio receivers. Consider, for example, the frequencies of the signals present in the AM broadcast band. Each radio station transmits a signal that occupies a band of frequencies, as shown in Figure 9-1(a). All of these signals

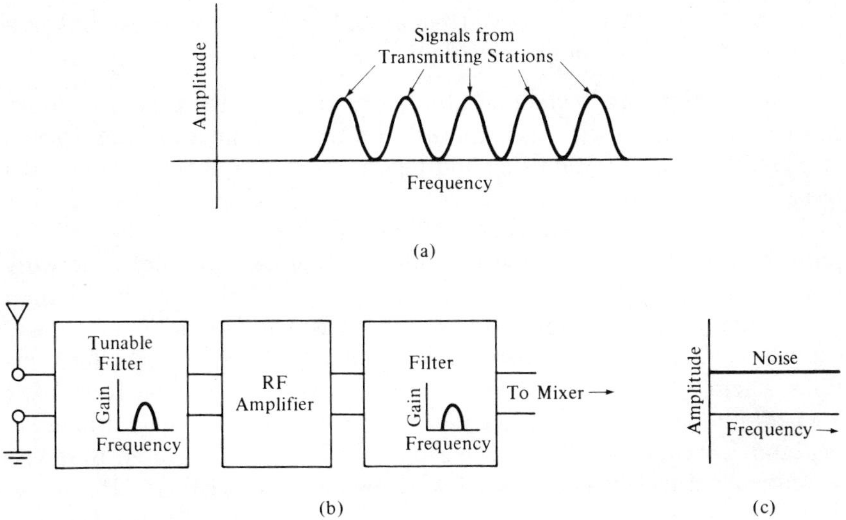

Figure 9-1 Radio receiver problem.

appear at the receiver antenna terminals. If radio receivers were not frequency-selective, the listener would hear a number of stations superimposed on each other. Hence, a filter is usually inserted between the antenna terminals and the RF amplifier, as illustrated in Figure 9-1(b). When the listener "tunes" the radio, he moves the filter position up or down the frequency band in order to select the station he wants and to filter out those that he does not want. Hence, even though all the signals appear at the filter input, only the selected signal is passed to the RF amplifier.

The filter following the RF amplifier illustrates a second application of filters—to eliminate noise. All electronic circuits generate noise. (The reader can verify this fact by short-circuiting the antenna terminals of his favorite radio or TV receiver, turning the volume up, and listening to the "static.") It is well known that this noise appears at all frequencies, as illustrated in Figure 9-1(c). Hence, a filter that attenuates all frequencies other than those of the desired signal is always placed after the RF amplifier in order to eliminate as much of this noise as possible.

Filters are usually classified in terms of the frequencies they pass or reject. The most common filter types are the following:

1. *lowpass filters* that pass low frequencies, but attenuate high frequencies
2. *highpass filters* that pass high frequencies, but attenuate low frequencies
3. *bandpass filters* that pass a band of frequencies, but attenuate all frequencies outside the band
4. *bandreject filters* that attenuate a band of frequencies, but pass all frequencies outside the band.

There are many techniques for designing circuits that are filters. Most of these begin with the selection of a transfer function whose graph of magnitude versus frequency and phase versus frequency has the right basic shape. Next, a basic network is chosen. This network is known to be one whose transfer function has the desired basic shape. The problem now becomes a question of selecting the particular network element values so that the precise shape is obtained. In this chapter we consider only one simple network type—the ladder of inductances and capacitances. Furthermore, we discuss only that class of transfer functions where all of the types of filters listed can be obtained from basic lowpass filter prototypes by frequency transformations, magnitude-scaling, and frequency-scaling. The same techniques can be used to design a frequency-selective coupling network between transistors. Hence, we first consider a known class of lowpass ladder filters in some detail. This is not the only class available, but it is adequate to illustrate the techniques.

9-2 Lowpass Filters

Lowpass filters pass low frequencies and attenuate high frequencies. In this section we first define an ideal lowpass-filter response, and then discuss some practical lowpass filters that approximate the ideal filter.

The ideal lowpass filter. The ideal lowpass filter has a transfer function having the gain-versus-frequency and phase-versus-frequency characteristics shown in Figure 9-2. The ideal filter passes all frequencies below 1 rad/sec with unit gain, and attenuates all frequencies higher than 1 rad/sec with infinite attenuation. The phase angle of the ideal lowpass filter is equal to the frequency. The "idealness" of the gain-versus-frequency response is obvious: For a signal consisting of a number of frequencies (such as a voice waveform), all frequencies are either passed unaltered or not passed at all. If the gain in the pass band (0–1 rad/sec) were not flat, some frequencies would be attenuated more than others;

this results in amplitude distortion of the signal. The "idealness" of the phase-versus-frequency response is best illustrated by noting that if the filter input is cos $(\omega t + \theta)$, $\omega < 1$, then the filter output is cos $(\omega t + \theta - \omega)$ $= \cos[\omega(t - 1) + \theta]$. Hence, the linear phase response of the filter has caused the output signal to be delayed by one second. Since the delay is not a function of frequency, all frequencies are delayed the same amount. Hence, the phase response of the ideal lowpass filter is "ideal" because it does not introduce any delay distortion in the signals the filter passes.

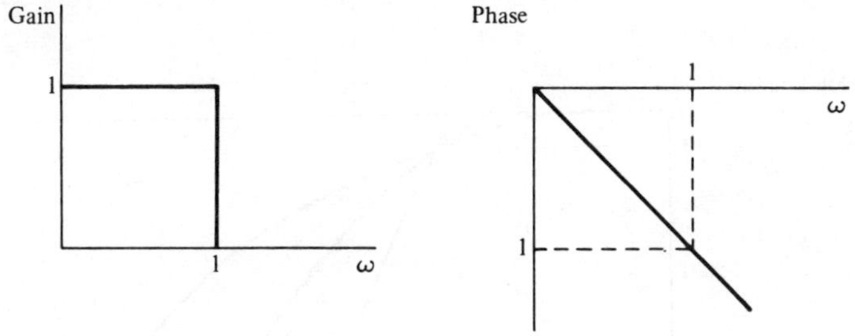

(a) (b)

Figure 9-2 Ideal lowpass-filter response. (a) Gain versus frequency. (b) Phase versus frequency.

Unfortunately, it is not possible to construct a filter having ideal response from a finite number of the network elements described in Chapter 1. All practical filters introduce both amplitude and delay distortion. The major problem of filter design is to find reasonably simple networks whose gain-versus-frequency and/or phase-versus-frequency response approximates the response of the mythical ideal lowpass filter. In most applications, filters are designed for either minimum amplitude distortion or minimum delay distortion.

The Butterworth function. One approximation to the amplitude response of the ideal lowpass filter is the nth-order Butterworth function defined by

$$|\mathcal{H}(j\omega)|^2 = \frac{1}{1 + \omega^{2n}} \qquad \textbf{(9-1)}$$

The Butterworth function is sketched in Figure 9-3(a) as a function of ω for several values of n. As n increases, the Butterworth function becomes a better approximation to the ideal amplitude response. Note that $|\mathcal{H}(j0)|^2 = 1$, $|\mathcal{H}(j1)|^2 = \frac{1}{2}$, and $|\mathcal{H}(j\,\infty)|^2 = 0$ for all n.

To make a decibel plot of the nth-order Butterworth function, we plot $10 \log |\mathcal{H}(j\omega)|^2$ vs. $\log \omega$. We have

$$10 \log |\mathcal{H}(j\omega)|^2 = 10 \log \frac{1}{1 + \omega^{2n}} = -10 \log (1 + \omega^{2n}) \qquad \textbf{(9-2)}$$

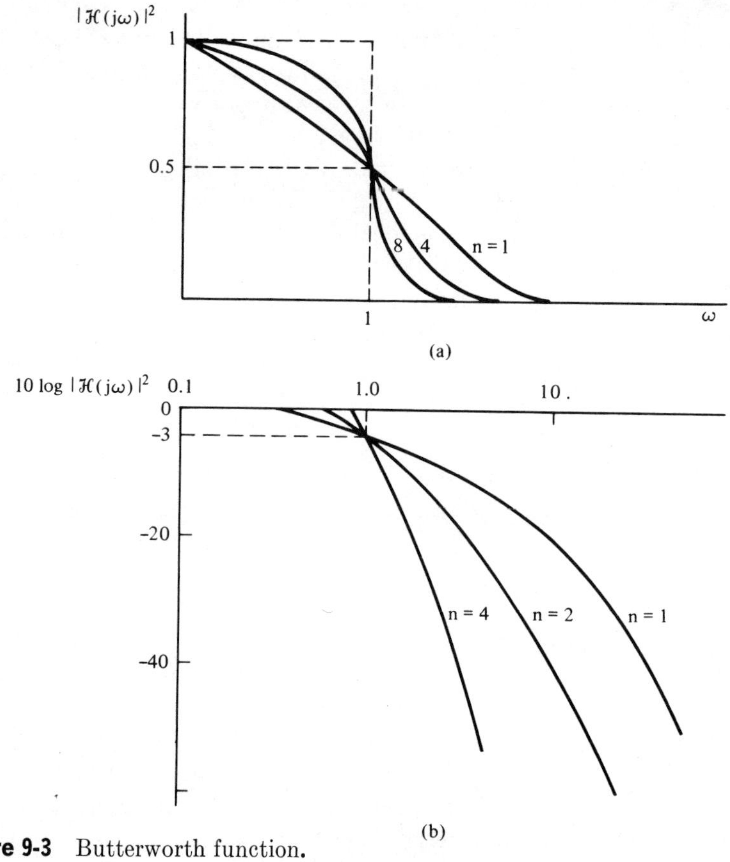

Figure 9-3 Butterworth function.

The low-frequency ($\omega \ll 1$) asymptote is given by

$$10 \log |\mathcal{H}(j\omega)|^2 \approx -10 \log 1 = 0, \qquad \omega \ll 1 \tag{9-3}$$

and the high-frequency asymptote ($\omega \gg 1$) is given by

$$10 \log |\mathcal{H}(j\omega)|^2 \approx -10 \log \omega^{2n} = -20n \log \omega \tag{9-4}$$

Hence, the high-frequency rolloff is $-20n$ dB/decade or $-6n$ dB/octave,[2] as shown in Figure 9-3(b). At $\omega = 1$ we have, for all n,

$$10 \log |\mathcal{H}(j1)|^2 = -10 \log 2 \approx -3 \text{ dB} \tag{9-5}$$

Butterworth filters. Filters having Butterworth transfer functions are called Butterworth filters. Three sets of Butterworth filters are presented in Figure 9-4. The set in Figure 9-4(a) is called Butterworth

[2] An octave is a factor of 2 in frequency. In music, the eight note natural and tempered scales both cover a 2:1 frequency spread. For example, in concert pitch, the note A is 110 Hz, 220 Hz, 440 Hz, 880 Hz, and so on.

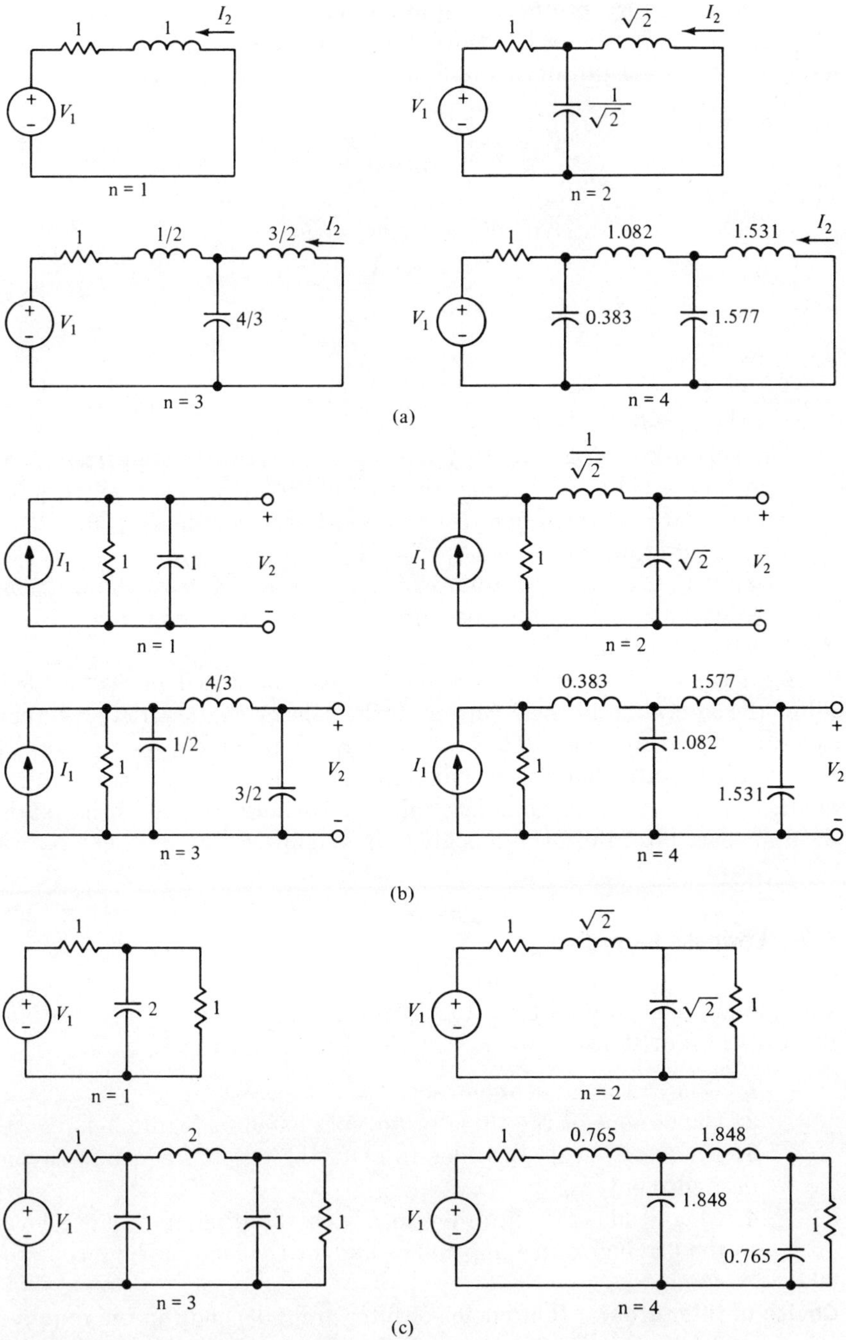

Figure 9-4 Butterworth prototype filters. (a) Transfer-admittance prototypes. (b) Transfer-impedance prototypes. (c) Insertion-power-gain prototypes.

transfer-admittance prototype filters because the magnitude squared of their transfer admittance is a Butterworth function. Note, for example, that for the n = 1 prototype filter we have

$$y_{21} = I_2 \Big|_{V_1=1, V_2=0} = \frac{-1}{1 + j\omega} \tag{9-6}$$

so that

$$\begin{aligned} |y_{21}(j\omega)|^2 &= y_{21}(j\omega) y_{21}^*(j\omega) \\ &= \frac{-1}{1 + j\omega} \cdot \frac{-1}{1 - j\omega} \\ &= \frac{1}{1 + \omega^2} \end{aligned} \tag{9-7}$$

This is the first-order Butterworth function as required.

The networks in Figure 9-4(b) are called Butterworth transfer-impedance prototype networks because the magnitude squared of their transfer impedances are Butterworth functions. For example, for n = 2, a simple computation shows that $|\mathfrak{z}_{21}|^2 = 1/(1 + \omega^4)$.

Finally, the set in Figure 9-4(c) is called the Butterworth insertion-power-gain prototype filters because their insertion power gain is a Butterworth function.

As a consequence of the reciprocity theorem stated in Section 5-3, all of the filters retain their Butterworth transfer characteristics if they are turned end for end. That is, since $y_{21} = y_{12}$ and $\mathfrak{z}_{21} = \mathfrak{z}_{12}$, ports 1 and 2 can be interchanged without altering their Butterworth transfer characteristic. A more complete table of Butterworth and other standard lowpass filter prototypes is given in Weinberg [33].

9-3 Lowpass Filter Design

The design of a lowpass filter to meet a particular set of specifications proceeds in four steps:

1. Choose the filter order.
2. Choose one of the three prototypes listed in Figure 9-4.
3. Frequency-scale in order to shift the cutoff frequency to the desired point.
4. Magnitude-scale in order to match the filter impedance level to the driving-source impedance and/or the load impedance level.

Choice of filter order. The choice of filter order depends on the required sharpness of the high frequency rolloff. Since the filter complexity is proportional to n, the filter order is chosen according to the smallest value of n that meets the required specifications. Hence, if a minimum rolloff of 10 dB/octave is specified, a second-order filter will suffice. If only 6 dB/octave were required, a first-order filter would do.

Choice of prototype. Filters are usually inserted between a "driving source" and a "load," as illustrated in Figure 9-5(a). That is, rather

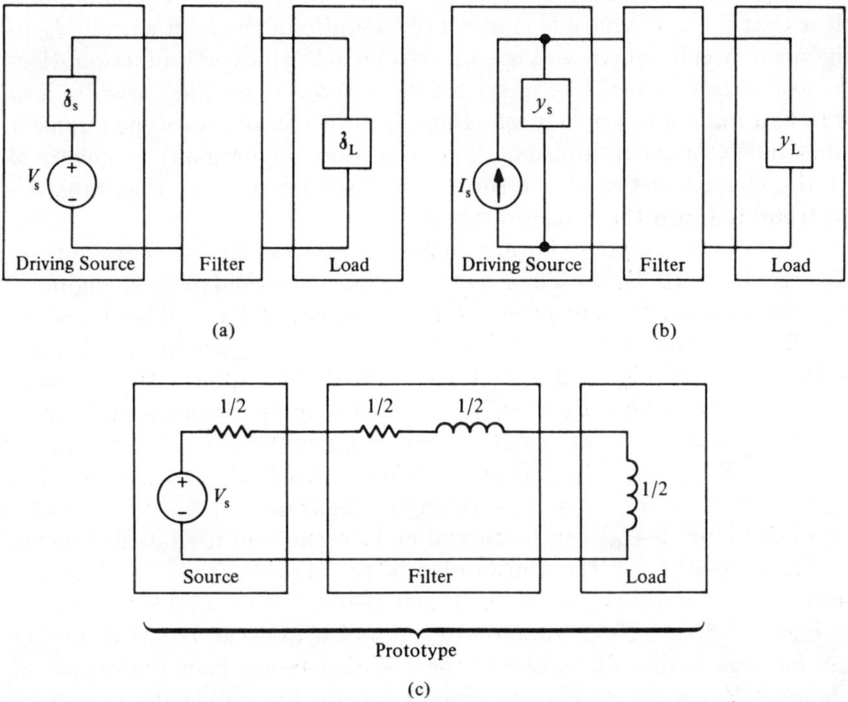

Figure 9-5 Incorporation of a filter with the source and load.

than connect the load impedance directly to the driving source, the circuit designer inserts a filter in order to obtain the frequency-selective properties desired. According to Section 5-2, the behavior of the driving source relative to its terminals can be characterized by its open-circuit voltage V_s and its Thevenin equivalent impedance $\mathfrak{z}_s$, as illustrated in Figure 9-5(a), or its short-circuit current I_s and its Norton equivalent admittance y_s, as illustrated in Figure 9-5(b). Furthermore, assuming that the load contains no independent sources, it can be represented by either its Thevenin equivalent impedance $\mathfrak{z}_L$, as illustrated in Figure 9-5(a), or its Norton equivalent admittance y_L, as illustrated in Figure 9-5(b).

The frequency characteristics of the signal delivered to the load depend on $\mathfrak{z}_s$ and $\mathfrak{z}_L$ as well as the elements in the network labeled FILTER. For example, suppose that in Figure 9-5(a) $\mathfrak{z}_s = \frac{1}{2}$, the network labeled FILTER consists of a $\frac{1}{2}$-ohm resistance in series with a $\frac{1}{2}$-henry inductance, and the LOAD is a $\frac{1}{2}$-henry inductance, as illustrated in Figure 9-5(c). Then the total network, consisting of driving source, filter, and load, is equivalent to the first-order transfer-admittance prototype [see Figure 9-4(a)], and the magnitude squared of the ratio of the load current to source voltage is a first-order Butterworth function. If the load is changed, or if $\mathfrak{z}_s \neq \frac{1}{2}$ (say, $\mathfrak{z}_s = 2 - j2$), then the frequency characteristics of the load current change. Therefore, it is clear that if the magnitude squared of the ratio of the load current I_L to the open circuit source voltage V_s is to be a Butterworth function, then *the total network consisting of the source impedance, the filter, and the load must be a Butterworth prototype.* Indeed, the choice of prototype (transfer admittance, transfer impedance, or insertion power gain) is governed by the characteristics of the source and load impedances that must be incorporated into the filter prototype.

For example, suppose both $\mathfrak{z}_s$ and $\mathfrak{z}_L$ are real and that $\mathfrak{z}_s = R_s \gg \mathfrak{z}_L = R_L$ (say, $R_s > 100R_L$). Then, to the accuracy required in most engineering applications, we can often assume that either $R_L = 0$ or $R_s = \infty$. For $R_L = 0$, any of the transfer-admittance prototypes are applicable, as illustrated in Figure 9-6(a), for the third-order filter. If it is more reasonable to assume $R_s = \infty$ ($y_s = 0$), any of the transfer-impedance prototypes can be chosen, as illustrated in Figure 9-6(d). Next, suppose that $\mathfrak{z}_s = R_s \ll \mathfrak{z}_L = R_L$. Then, it is often reasonable to assume that $R_s = 0$ or $R_L = \infty$. For $R_s = 0$, any of the transfer-admittance prototypes of Figure 9-4(a) can be turned end for end and used as illustrated in Figure 9-6(b) for the third-order filter. For $R_L = \infty$, any of the transfer-impedance prototypes of Figure 9-4(b) can be used as illustrated in Figure 9-6(c). Next, suppose that $\mathfrak{z}_s$ and $\mathfrak{z}_L$ are real, but that neither can be neglected. Then, one of the insertion-power-gain prototypes of Figure 9-4(c) must be chosen, since they are the only ones that have resistances at both ends. See Figures 9-6(e) and 9-6(f). Finally, we point out that the Butterworth prototype filters can be used with complex source and/or complex load impedances, provided that their form allows them to be incorporated into one of the prototype filters. For example, the third-order insertion-power-gain prototype can be used with a driving source containing a shunt capacitance, as well as a shunt resistance, and a load consisting of a shunt resistance and shunt capacitance, as illustrated in Figure 9-6(g).

Frequency-scaling. All of the Butterworth prototype filters cut off at $\omega = 1$ rad/sec. The cutoff frequency can be shifted from $\omega = 1$ to $\omega = \omega_c$ rad/sec by frequency-scaling by ω_c. Frequency-scaling any of the Butterworth prototypes is equivalent to replacing ω by ω/ω_c in the Butterworth function, as illustrated in Figure 9-7. (See Section 8-5.)

Magnitude-scaling. After a particular prototype filter has been chosen and frequency-scaled to obtain the required cutoff frequency, the imped-

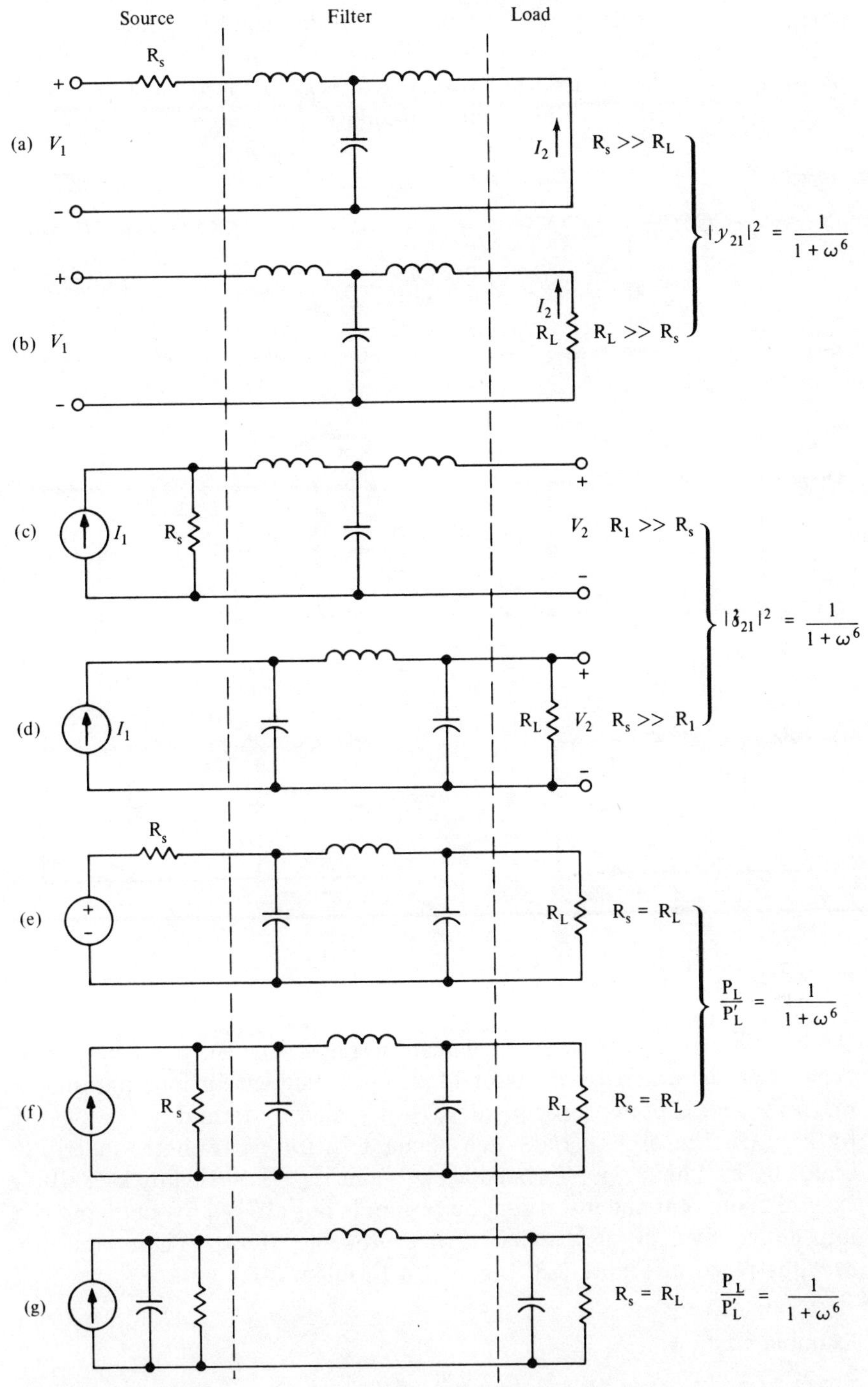

Figure 9-6 Some source, coupling filter, and load configurations.

ance level of the prototype must be magnitude-scaled in order to match it to the impedance level of the source and load impedances. For example, if $\mathfrak{z}_L = 0$ and $\mathfrak{z}_s = 10^3$, the prototype of Figure 9-7(a) must be impedance-scaled by 10^3 in order to accommodate the source impedance. If

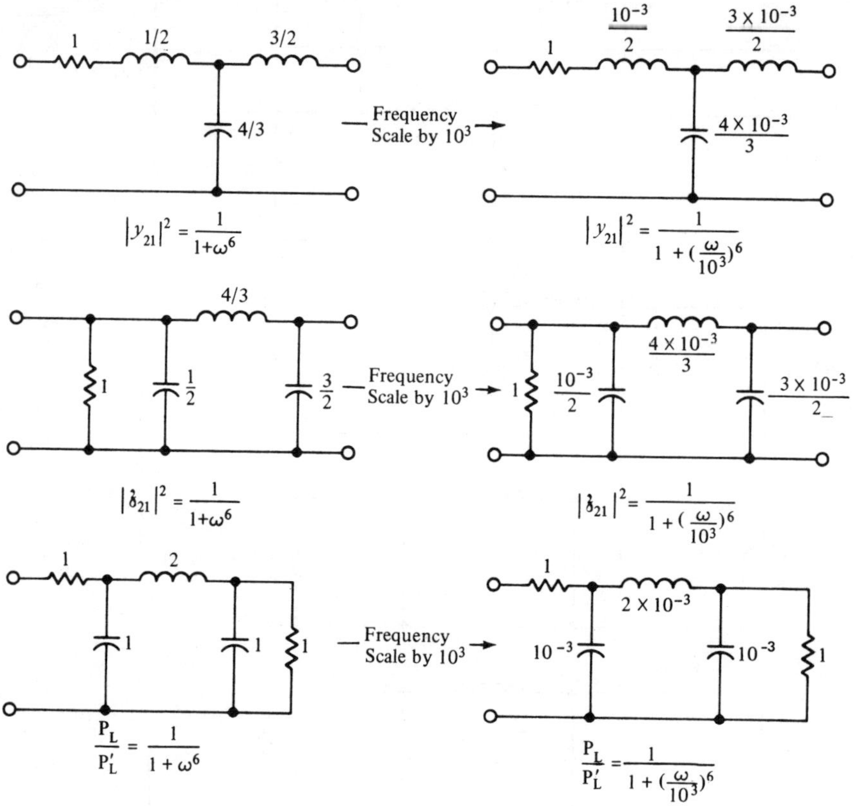

Figure 9-7 Frequency-scaling.

the impedance of each network element in any of the Butterworth prototype filters is scaled by a factor of k, each element in its z parameter matrix is scaled by k; consequently, $\mathfrak{z}_{21}$ is scaled by k, and $|\mathfrak{z}_{21}|^2$ is scaled by k^2. On the other hand, each element in its y parameter matrix is scaled by k^{-1}; hence, y_{21} is scaled by k^{-1}, and $|y_{21}|^2$ is scaled by k^{-2}. It is easy to show that the insertion power gain is not altered by changing the impedance level of the Butterworth prototype filters. These concepts are illustrated in Figure 9-8. (See also Problem 9-7.)

Example 9-1

In this example we design a filter for use with the driving source and load presented in Figure 9-9. The filter is to exhibit the lowpass Butterworth

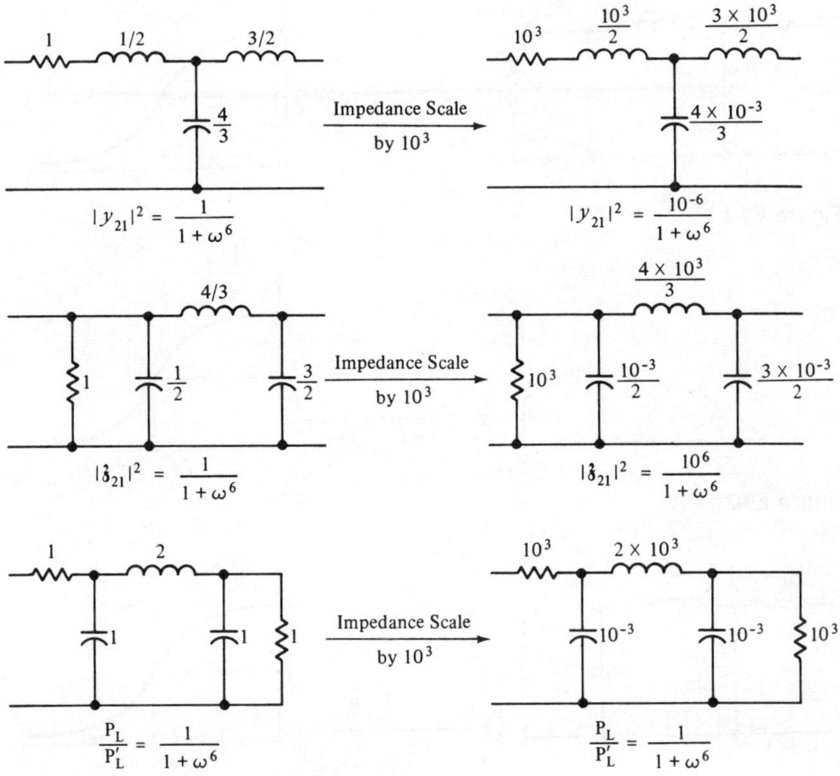

Figure 9-8 Impedance-scaling.

response with a frequency cutoff of 100 kHz. The high-frequency rolloff is to be 18 dB/octave.

Following the four steps listed at the beginning of Section 9-3, we first note that a third-order filter results in a high-frequency rolloff of $6n = 18$ dB/octave. Second, to accommodate both a source resistance and a load resistance, we must choose an insertion power-gain prototype. The third-order insertion-power-gain prototype and its transfer function are shown in Figure E9-1. To change the frequency cutoff from $\omega = 1$ to $\omega = 2\pi \times 10^5$, we next frequency-scale by $2\pi \times 10^5$. The resulting network and its transfer function are shown in Figure E9-2. Finally, we magnitude-scale by 10^3 to change the 1-ohm resistances to 1000-ohm resis-

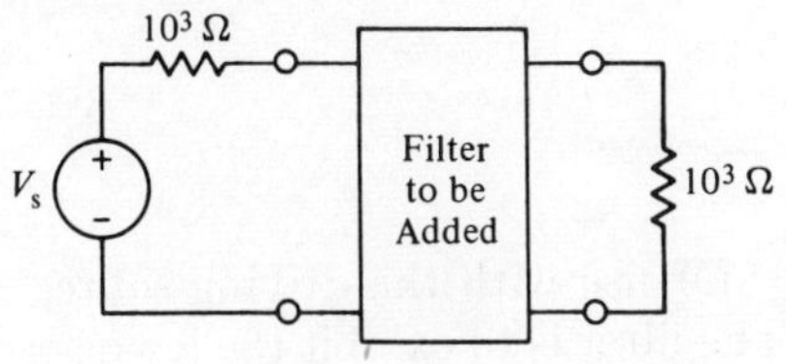

Figure 9-9 Problem of Example 9-1.

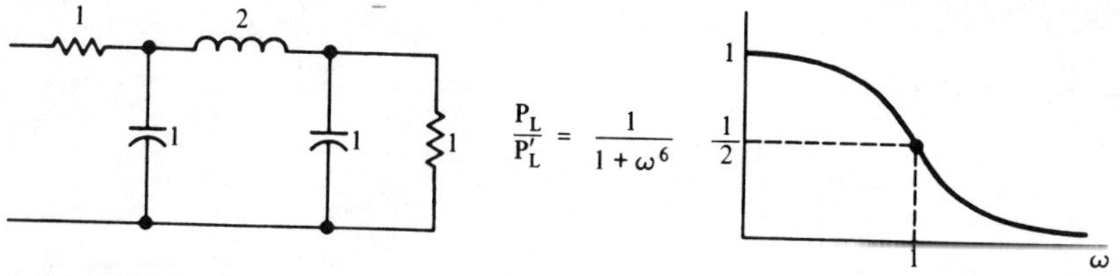

Figure E9-1

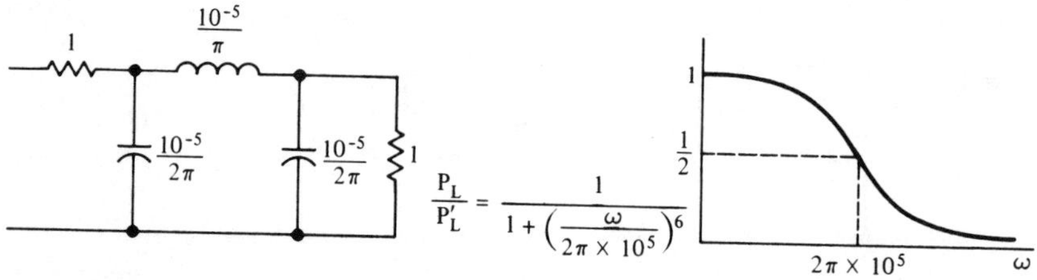

Figure E9-2

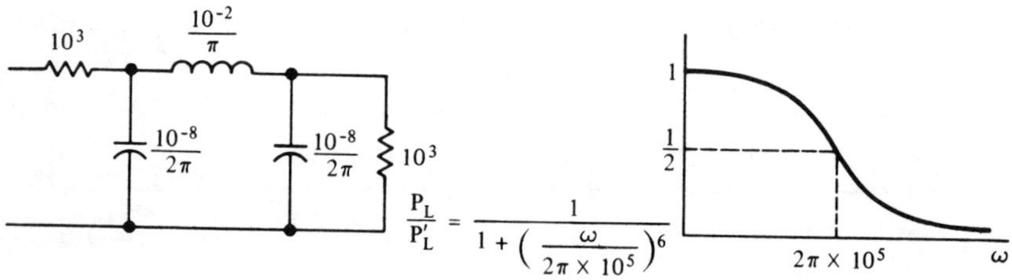

Figure E9-3

tances. The resulting network and its insertion power gain are given in Figure E9-3. As illustrated in Figure 9-10, an inductance and two capacitances must be added between the driving source and the load to obtain the required performance.

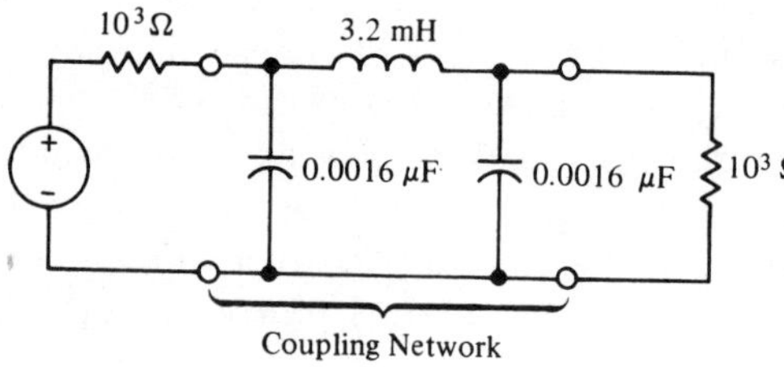

Figure 9-10 Solution to Example 9-1.

Example 9-2

For a second example, we design a filter for use with the driving source and load presented in Figure 9-11(a). The filter is to exhibit the lowpass

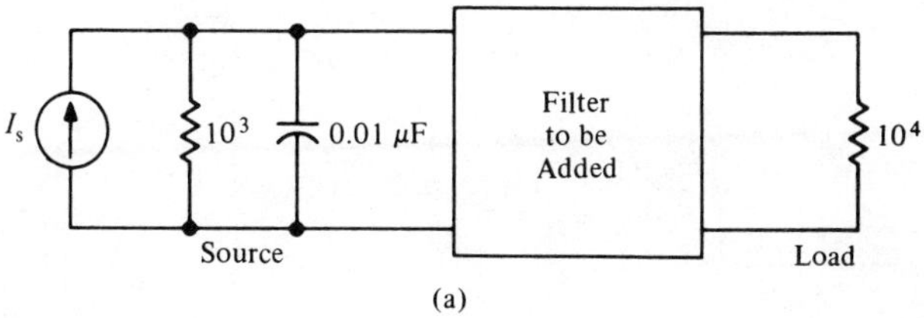

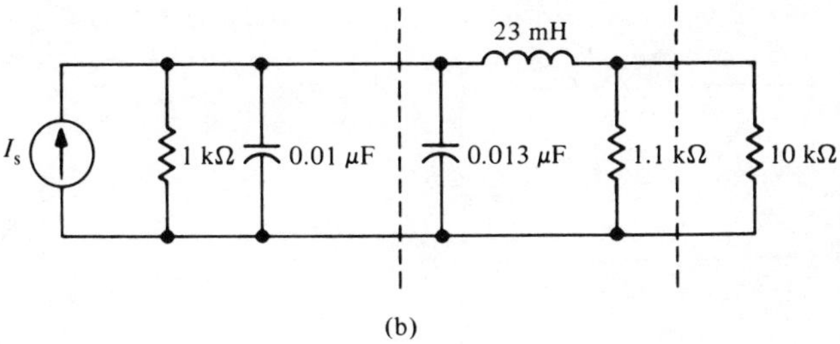

Figure 9-11 Example 9-2.

Butterworth response with a frequency cutoff of 10 kHz. The high-frequency rolloff is to be at least 10 dB/octave.

Following the prescribed four steps, we first choose a second-order filter, which gives a high-frequency rolloff of 12 dB/octave; this is the lowest-order filter that satisfies the 10 dB/octave specification. Next, we choose the insertion-power-gain prototype because both a source resistance and a load resistance must be accommodated. To acommodate the source capacitance, the prototype must have a capacitance shunting its input. Hence, we try the second-order insertion power-gain prototype of Figure 9-4(c) turned end for end. This network and its insertion power gain are given in Figure E9-4. We next frequency-scale

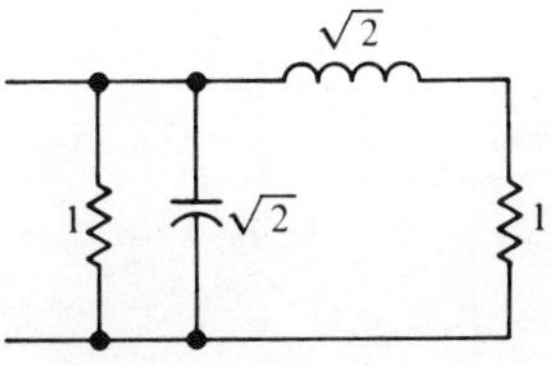

$$\frac{P_L}{P_L'} = \frac{1}{1 + \omega^4}$$

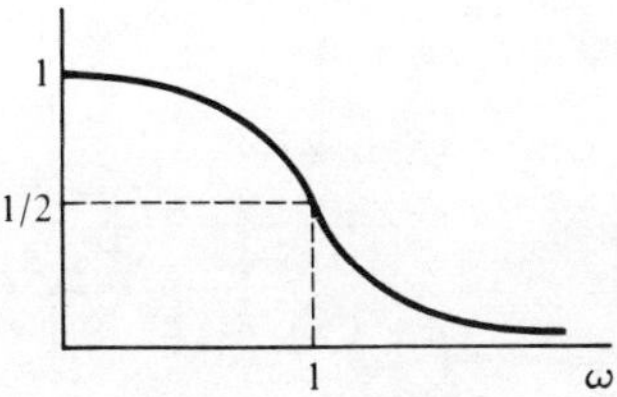

Figure E9-4

by the desired cutoff frequency $2\pi \times 10^4$ to obtain Figure E9-5. Next, we magnitude-scale by the source resistance 10^3 to obtain Figure E9-6. This network has the desired characteristics, but the capacitance did not turn out to be the source capacitance of 0.01 μF, nor did the right-hand resistance end up 10^4 ohms. We can, however, solve the problem by

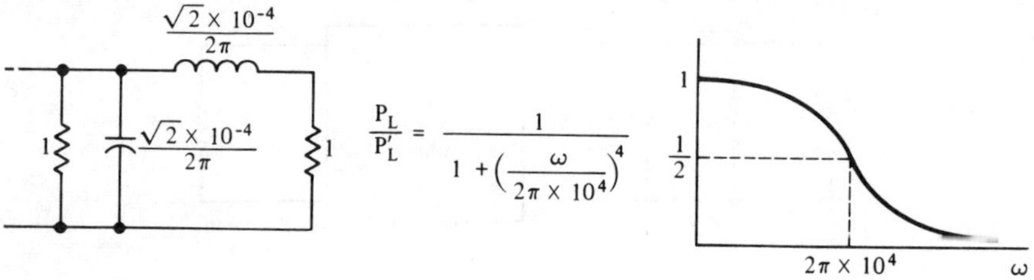

Figure E9-5

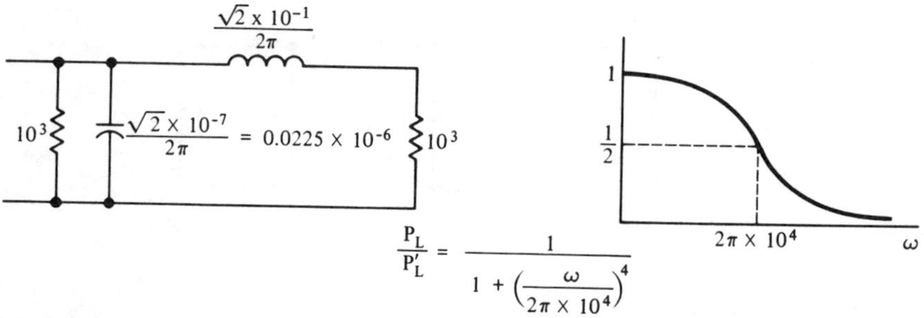

Figure E9-6

adding 0.0125 μF across the source terminals to give the required total of 0.0225 μF. Similarly, we can add a 1111-Ω resistance across the load terminals so that the parallel combination of this and the 4000-Ω load is equivalent to 1000 ohms. The result is illustrated in Figure 9-11(b).

Example 9-3

In this example we design a lowpass filter to be used as an interstage coupling in the two-transistor amplifier presented in Figure 9-12. The

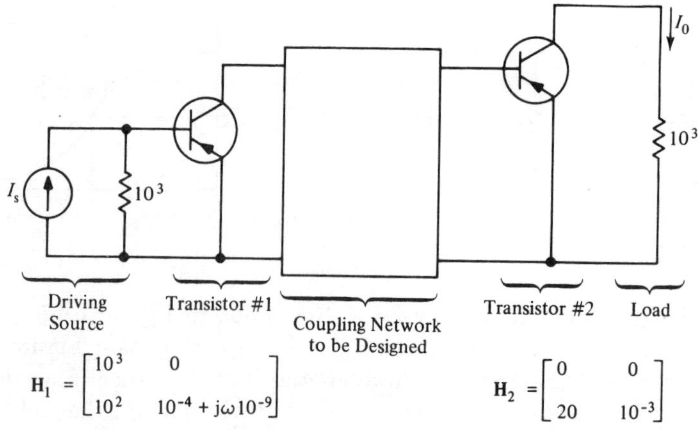

Figure 9-12 A frequency-selective two-stage transistor amplifier.

h parameters for both transistors are given in the figure. The magnitude squared of the current gain $|I_0/I_s|^2$ is to be a second-order Butterworth function, with a frequency cutoff of 10 kHz. The dc gain is to be as high as possible.

We start by replacing both transistor symbols with their h parameter models, as shown in Figure 9-13. Next, we note that the transfer

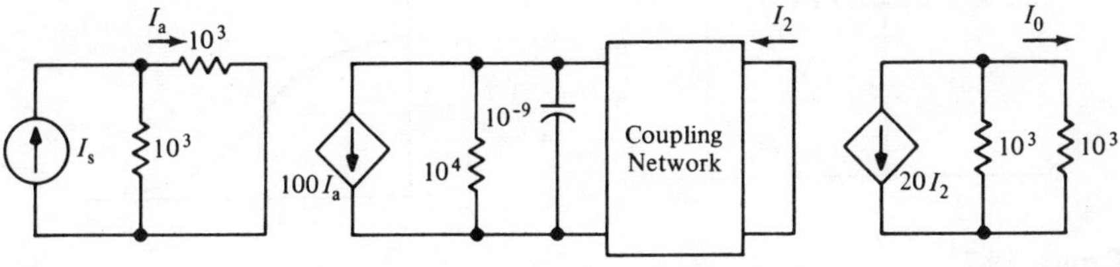

Figure 9-13 Network model for the circuit of Figure 9-12.

function from I_2 to I_0 is independent of frequency; that is,

$$\frac{I_0}{I_2} = -10 \tag{9-8}$$

Starting at the input to the two-transistor amplifier, we note that the first frequency-dependent element is the shunt capacitance associated with the output of the first transistor. Hence, we replace everything from the amplifier input to the first frequency-dependent element with its Thevenin equivalent network, as shown in Figure 9-14. Now, $|I_0/I_2|^2 =$

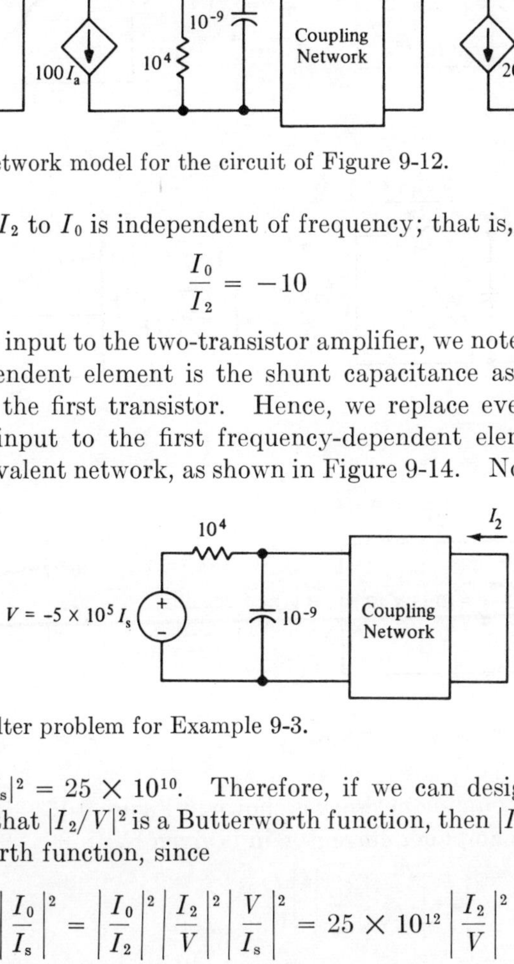

Figure 9-14 Filter problem for Example 9-3.

100, and $|V/I_s|^2 = 25 \times 10^{10}$. Therefore, if we can design a coupling network such that $|I_2/V|^2$ is a Butterworth function, then $|I_0/I_s|^2$ will also be a Butterworth function, since

$$\left|\frac{I_0}{I_s}\right|^2 = \left|\frac{I_0}{I_2}\right|^2 \left|\frac{I_2}{V}\right|^2 \left|\frac{V}{I_s}\right|^2 = 25 \times 10^{12} \left|\frac{I_2}{V}\right|^2 \tag{9-9}$$

Referring to Figure 9-14, we note that I_2/V is the transfer admittance for the total network shown (coupling network plus the RC network), since V is the input voltage and I_2 is the output short-circuit current. Hence, $|I_2/V|^2$ will be a second-order Butterworth function if the total network of Figure 9-14 is a second-order Butterworth transfer-admittance prototype filter. From Figure 9-4(a) we note that the second-order Butterworth transfer-admittance prototype provides the proper match with the source resistance and capacitance. Therefore, we start with this prototype (see Figure E9-7) and proceed as follows. Frequency-scale by $2\pi \times 10^4$ (see Figure E9-8); then impedance-scale by 10^4 (see Figure E9-9). The resistance and 10^{-9} farads are provided by the source. Hence, we must add 0.13×10^{-9} farads and a 0.225-henry inductance.

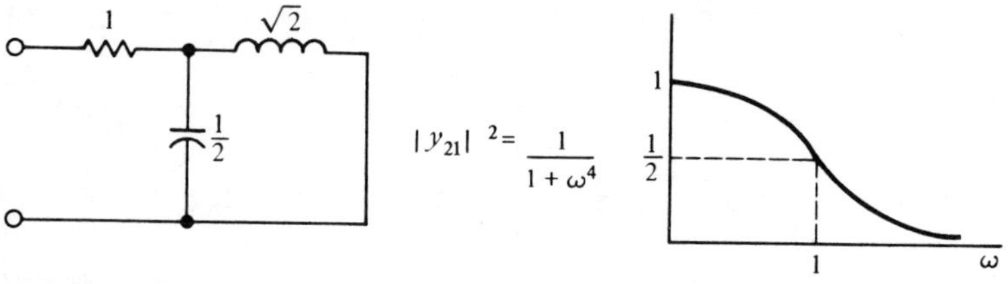

Figure E9-7

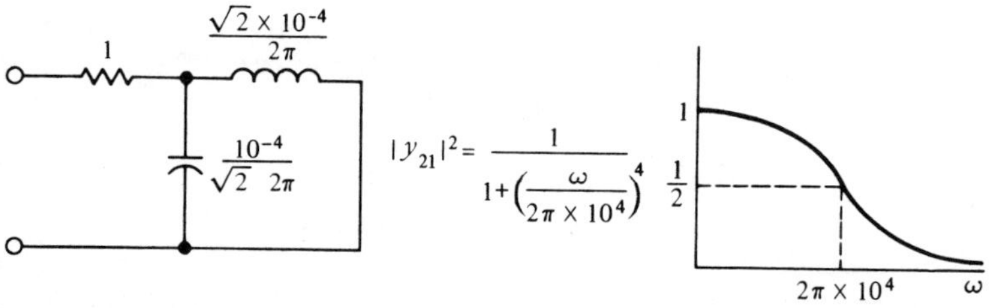

Figure E9-8

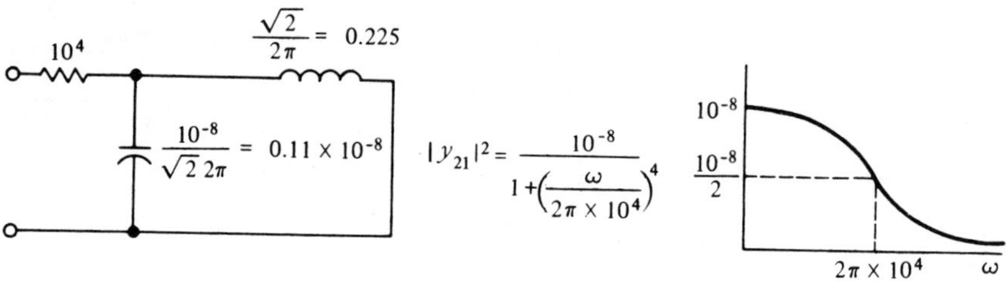

Figure E9-9

The resulting coupling network is shown in Figure 9-15. The magnitude squared of the amplifier current gain is given by

$$\left|\frac{I_0}{I_s}\right|^2 = \left|\frac{V_1}{I_s}\right|^2 \left|\frac{I_2}{V_1}\right|^2 \left|\frac{I_0}{I_2}\right|^2$$

$$= |25 \times 10^{12}| \frac{10^{-8}}{1 + \left(\dfrac{\omega}{2\pi \times 10^4}\right)^4} \tag{9-10}$$

$$= \frac{25 \times 10^4}{1 + \left(\dfrac{\omega}{2\pi \times 10^4}\right)^4}$$

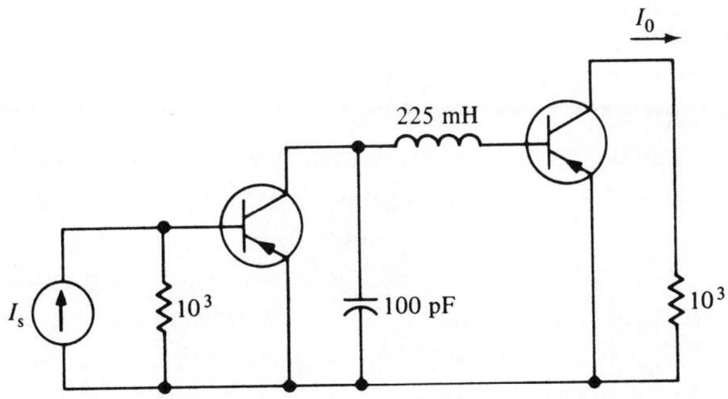

Figure 9-15 Final amplifier circuit for Example 9-3.

The magnitude squared of the current gain is a second-order Butterworth function, with a cutoff frequency of $2\pi \times 10^4$ rad/sec. The dc current gain is 500.

9-4 Highpass Filters

A highpass filter passes high frequencies and attenuates low frequencies. Highpass filters can be designed by starting with one of the lowpass Butterworth prototype filters and making a lowpass-to-highpass transformation.

Lowpass-to-highpass transformation. The nth-order lowpass Butterworth function defined by Equation (9-1) can be transformed into an nth-order highpass Butterworth function by making the change in frequency variable $j\omega' = 1/j\omega$, which implies $\omega' = -1/\omega$ and $\omega'^2 = 1/\omega^2$. This transformation results in the highpass Butterworth function

$$|\mathcal{H}(j\omega)|^2 = \frac{1}{1 + (1/\omega)^{2n}} \tag{9-11}$$

Equation (9-11) is sketched in Figure 9-16 for various values of n. The

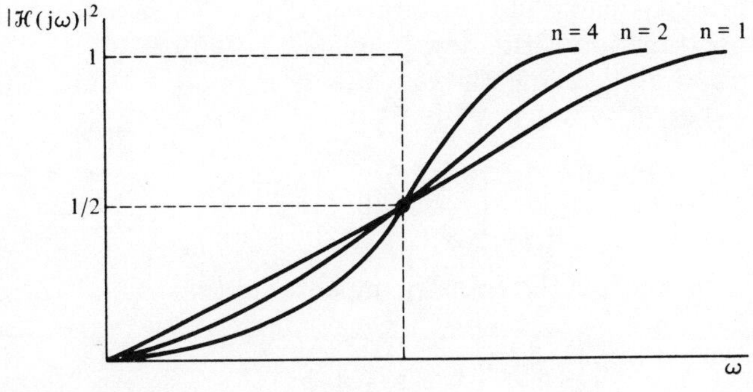

Figure 9-16 Highpass Butterworth function.

highpass Butterworth function has the same value at frequency $1/\omega$ that the lowpass Butterworth function has at frequency ω.

Any of the lowpass Butterworth prototype filters can be transformed into highpass Butterworth filters exhibiting the highpass Butterworth characteristic if we replace each element in the lowpass prototype with a new element that behaves at frequency $1/j\omega$ the same way the original element behaved at frequency $j\omega$. Hence, each capacitance having admittance $j\omega C$ must be replaced with an element having admittance $C/j\omega$. But a C^{-1}-henry inductance has admittance $1/j\omega C^{-1} = C/j\omega$. Hence, each C-farad capacitance must be replaced with a C^{-1}-henry inductance. Furthermore, each inductance having admittance $1/j\omega L$ in the lowpass prototype must be replaced with an element having admittance $j\omega/L$ in the highpass prototype. Since an L^{-1}-farad capacitance has admittance $j\omega L^{-1}$, we replace all L-henry inductances in the lowpass prototype with L^{-1}-farad capacitances. Since resistances are not frequency-dependent, they are left unchanged.

Highpass filter design. The design of a Butterworth highpass filter to meet a particular set of specifications proceeds in five steps:

1. Choose the filter order.
2. Choose one of the prototypes listed in Figure 9-4.
3. Change the lowpass prototype to a highpass prototype by replacing all C-farad capacitances with C^{-1}-henry inductances and replacing all L-henry inductances with L^{-1}-farad capacitances; leave all resistances unchanged.
4. Frequency-scale.
5. Magnitude-scale.

Example 9-4

In this example we design a coupling network to be used between two field-effect transistors (FET's) in the amplifier shown in Figure 9-17(a). Both field-effect transistors have the y parameters shown in the figure. The magnitude squared of the voltage gain $|V_0/V_i|^2$ is to be a third-order Butterworth function with a low-frequency cutoff of 70 kHz.

We start by sketching the equivalent network as shown in Figure 9-17(b). Since $I_1 = -0.01V_i$ and $V_0 = -500V_2$, we have

$$\left|\frac{V_0}{V_i}\right|^2 = \left|\frac{V_0}{V_2}\right|^2 \left|\frac{V_2}{I_1}\right|^2 \left|\frac{I_1}{V_i}\right|^2 = 10^{-4}\left|\frac{V_2}{I_1}\right|^2 \times (500)^2 = 25\left|\frac{V_2}{I_1}\right|^2$$

Since $I_2 = 0$, V_2/I_1 is the transfer impedance; that is,

$$\mathfrak{z}_{21} = \left.\frac{V_2}{I_1}\right|_{I_2=0}$$

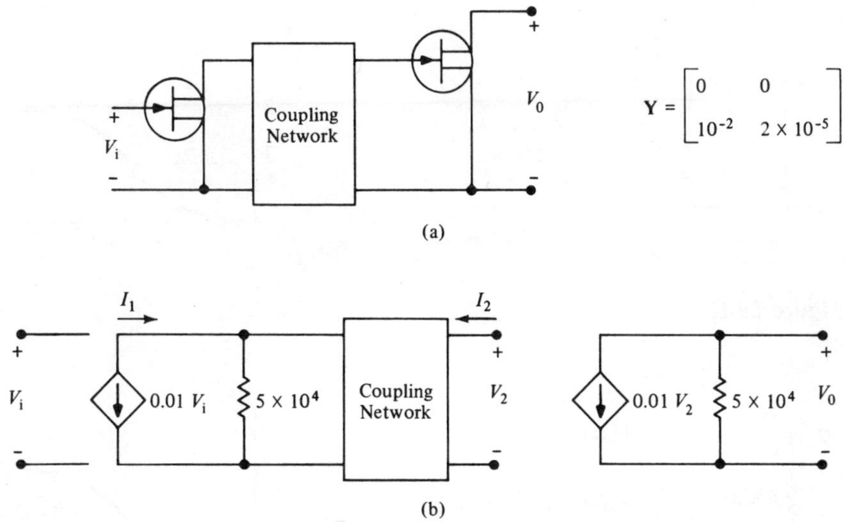

Figure 9-17 (a) A two-stage FET amplifier. (b) Network model for the FET amplifier.

Thus we choose the third-order transfer-impedance prototype filter of Figure 9-4(b). We start with this prototype (see Figure E9-10) and proceed as follows. First transform from lowpass to highpass (see Figure E9-11). Next frequency-scale by $14\pi \times 10^4$ (see Figure E9-12), then magnitude-scale by 5×10^4 (see Figure E9-13). The required coupling network is shown in Figure E9-14. The magnitude squared of the volt-

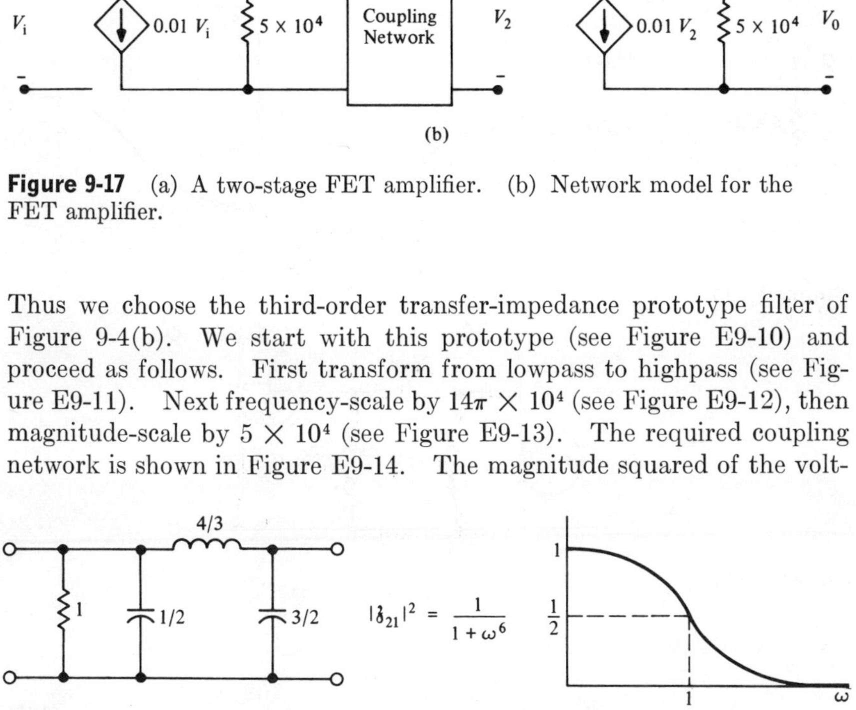

Figure E9-10

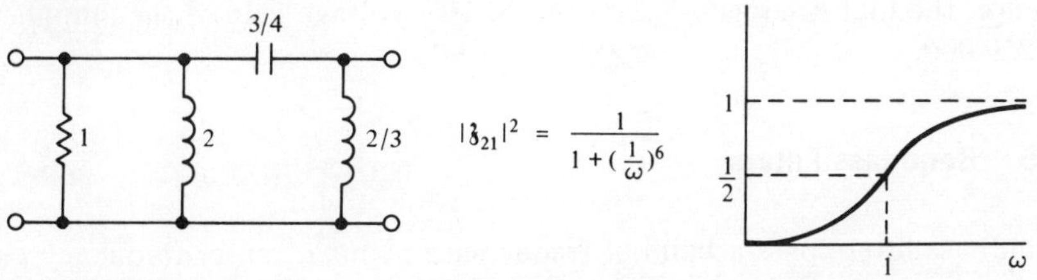

Figure E9-11

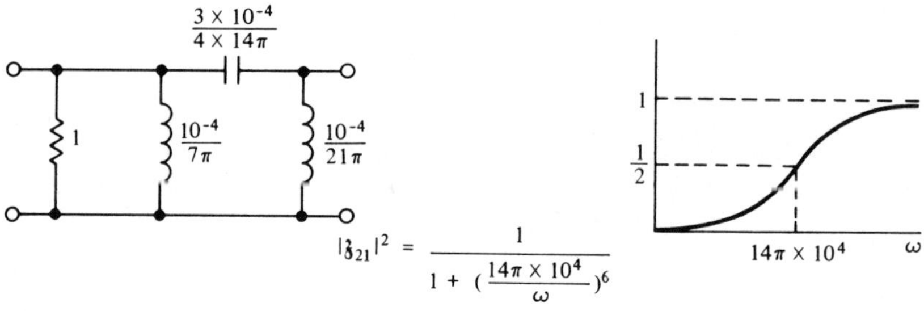

Figure E9-12

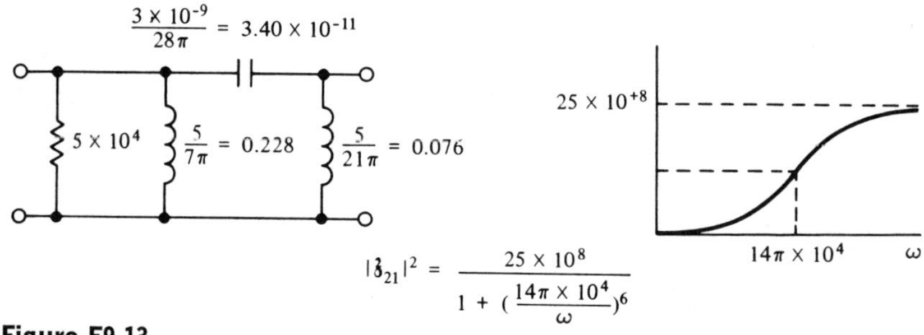

Figure E9-13

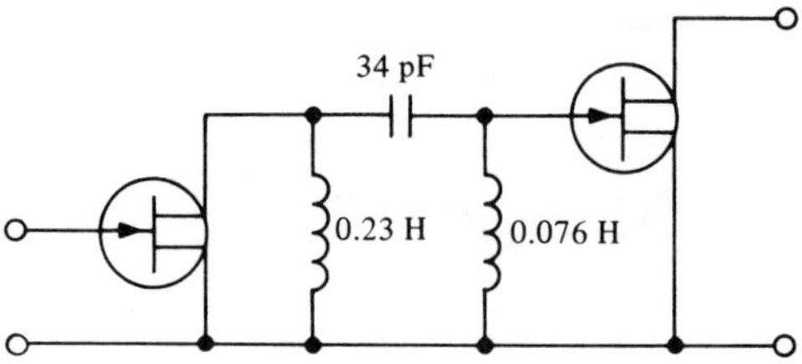

Figure E9-14

age gain is given by

$$25\left|\frac{V_2}{I_1}\right|^2 = 25\,\frac{25 \times 10^8}{1+\left(\dfrac{14\pi \times 10^4}{\omega}\right)^6} = \frac{625 \times 10^8}{1+\left(\dfrac{14\pi \times 10^4}{\omega}\right)^6} \tag{9-12}$$

Hence, the high-frequency ($\omega \gg 14\pi \times 10^4$) voltage gain of the amplifier is 250,000.

9-5 Bandpass Filters

Bandpass filters pass a band of frequencies about a center frequency ω_0, and attenuate all frequencies above and below this band. Bandpass filters can be designed by starting with one of the lowpass prototype filters in Figure 9-4 and making a lowpass-to-bandpass transformation.

Lowpass-to-bandpass transformation. The nth-order lowpass Butterworth function defined by Equation (9-1) can be transformed into an nth-order bandpass Butterworth function centered at frequency ω_0 and having bandwidth W by making the change in frequency variable $j\omega' = [(j\omega)^2 + \omega_0^2]/j\omega W$ which implies $\omega' = (\omega^2 - \omega_0^2)/\omega W$. Applying this transformation to Equation (9-1), we obtain the nth-order bandpass Butterworth function

$$|\mathcal{H}(j\omega)|^2 = \frac{1}{1 + \left(\dfrac{\omega^2 - \omega_0^2}{\omega W}\right)^{2n}} \tag{9-13}$$

It is instructive to study Equation (9-13) in some detail to see that the resultant filter is as stated. First note that $|\mathcal{H}(j0)|^2 = 0$, $|\mathcal{H}(j\omega_0)|^2 = 1$, and $|\mathcal{H}(j\infty)|^2 = 0$. Furthermore, $|\mathcal{H}(j\omega)|^2$ is a monotonically increasing function for $0 < \omega < \omega_0$ and a monotonically decreasing function for $\omega > \omega_0$. Hence, for positive frequencies, $|\mathcal{H}(j\omega)|^2$ has one smooth hump at $\omega = \omega_0$, as illustrated in Figure 9-18 for two values of n.

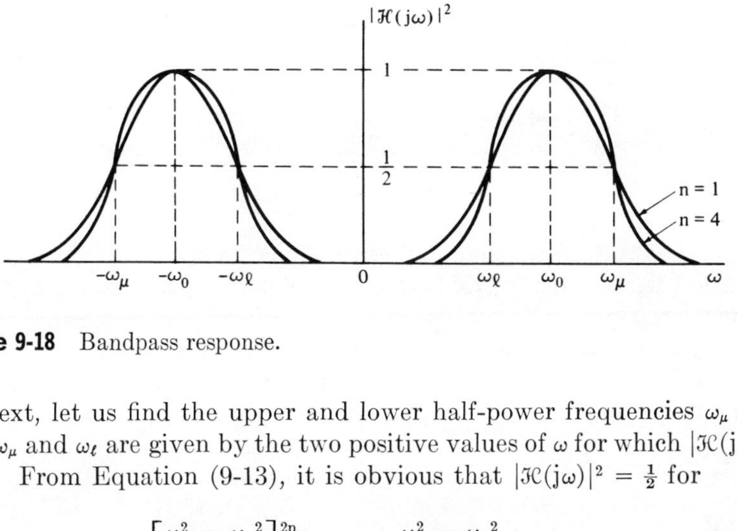

Figure 9-18 Bandpass response.

Next, let us find the upper and lower half-power frequencies ω_μ and ω_ℓ. ω_μ and ω_ℓ are given by the two positive values of ω for which $|\mathcal{H}(j\omega)|^2 = \frac{1}{2}$. From Equation (9-13), it is obvious that $|\mathcal{H}(j\omega)|^2 = \frac{1}{2}$ for

$$\left[\frac{\omega^2 - \omega_0^2}{\omega W}\right]^{2n} = 1 \text{ or } \frac{\omega^2 - \omega_0^2}{\omega W} = \pm 1 \tag{9-14}$$

Equation (9-14) has four solutions

$$\omega = \pm \frac{W}{2} \pm \sqrt{\left(\frac{W}{2}\right)^2 + \omega_0^2} \tag{9-15}$$

but, since

$$\frac{W}{2} \leq \sqrt{\left(\frac{W}{2}\right)^2 + \omega_0^2}$$

only the two solutions given by

$$\omega = \pm \frac{W}{2} + \sqrt{\left(\frac{W}{2}\right)^2 + \omega_0^2} \tag{9-16}$$

are positive. Hence,

$$\omega_\mu = \frac{W}{2} + \sqrt{\left(\frac{W}{2}\right)^2 + \omega_0^2}, \qquad \omega_\ell = -\frac{W}{2} + \sqrt{\left(\frac{W}{2}\right)^2 + \omega_0^2} \tag{9-17}$$

Note also that

$$\omega_\mu - \omega_\ell = W \tag{9-18}$$

Therefore, if we define the bandwidth as the distance (in radians per second) between the upper and lower half-power frequencies, we conclude that the bandwidth is given by the parameter W.

Finally, we note from Equations (9-17) that

$$\omega_\mu \omega_\ell = \omega_0^2 \tag{9-19}$$

which indicates that the center frequency ω_0 is the geometric mean of the two half-power frequencies. Therefore, when plotted on a linear frequency scale, as in Figure 9-18, $|\mathcal{H}(j\omega)|^2$, is not symmetrical about ω_0. Rather, it is somewhat skewed. The amount of skewness depends on the ratio ω_0/W. For reasonable values of ω_0/W, say $\omega_0/W > 10$, the amount of skewness is very slight. From Equation (9-19) we note that

$$\log \omega_0 = \tfrac{1}{2}[\log \omega_\mu + \log \omega_\ell] \tag{9-20}$$

so that if $|\mathcal{H}(j\omega)|^2$ is plotted on a logarithmic frequency scale, the center frequency is the arithmetic mean of the upper and lower half-power frequencies.

Any of the lowpass prototype filters can be transformed into a bandpass filter with center frequency ω_0 and bandwidth W if we replace each element in the lowpass prototype with a network that behaves at frequency $\omega' = \omega/W - \omega_0^2/(\omega W)$ the same way the original element behaved at frequency ω. For example, each inductance in the prototype filter must be replaced with a 2-terminal network that, at frequency ω', has the same impedance that the inductor has at frequency ω. Since, at frequency ω, an L-henry inductance has impedance $j\omega L$, each L-henry inductance must be replaced with a network having impedance

$$j\left(\frac{\omega}{W} - \frac{\omega_0^2}{\omega W}\right)L = j\omega\left(\frac{L}{W}\right) + \frac{1}{j\omega\left(\frac{W}{\omega_0^2 L}\right)}$$

Such an impedance can be realized with an L/W-henry inductance in series with a $W/(\omega_0^2 L)$-farad capacitance. Similarly, each C-farad capacitance in the lowpass prototype network must be replaced with a network having admittance given by

$$j\left(\frac{\omega}{W} - \frac{\omega_0^2}{\omega W}\right)C = j\omega\left(\frac{C}{W}\right) + \frac{1}{j\omega\left(\frac{W}{\omega_0^2 C}\right)}$$

Such an admittance can be synthesized with a C/W-farad capacitance in parallel with an $W/(\omega_0^2 C)$-henry inductance. Since the impedance associated with a resistance is not frequency-dependent, resistances are not changed.

Note that at frequency ω_0, the series LC networks have impedance

$$j\omega_0\left(\frac{L}{W}\right) + \frac{1}{j\omega_0\left(\frac{W}{\omega_0^2 L}\right)} = 0$$

and the parallel LC networks have admittance

$$j\omega_0\left(\frac{C}{W}\right) + \frac{1}{j\omega_0\left(\frac{W}{\omega_0^2 C}\right)} = 0$$

Therefore, at $\omega = \omega_0$, the series LC networks are short circuits, and the parallel LC networks are open circuits.

Bandpass filter design. The design of a bandpass filter from the lowpass Butterworth prototype filter is easily accomplished in four steps:

1. Choose the filter order.
2. Choose one of the prototype filters listed in Figure 9-4.
3. Transform the lowpass prototype filter to a bandpass filter of center frequency ω_0 and bandwidth W by replacing each L-henry inductance with an L/W-henry inductance in series with an $W/(\omega_0^2 L)$-farad capacitance; replace each C-farad capacitance with a C/W-farad capacitance in parallel with an $W/(\omega_0^2 C)$-henry inductance; leave each resistance unchanged.
4. Magnitude-scale.

Example 9-5

In this example we design a bandpass coupling network for the two-stage transistor amplifier presented in Figure 9-19(a). The h parameters for both transistors are given in the figure. The voltage gain $|V_0/V_i|^2$ is to be a second-order Butterworth function, the center frequency is to be 100 kHz, and the bandwidth, 10 kHz.

We start by sketching the complete equivalent network for the ampli-

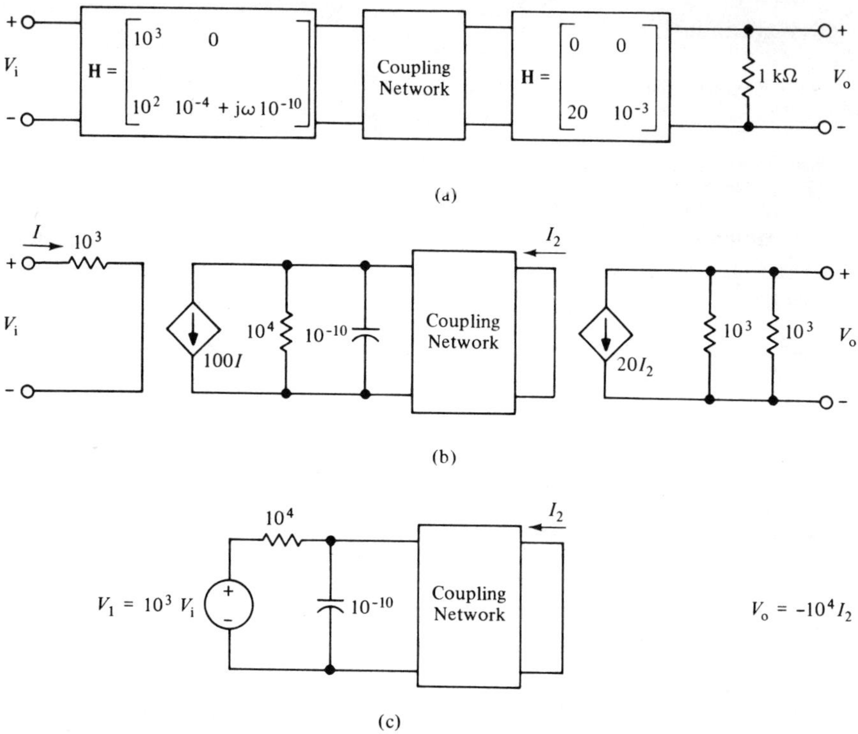

Figure 9-19 Bandpass amplifier.

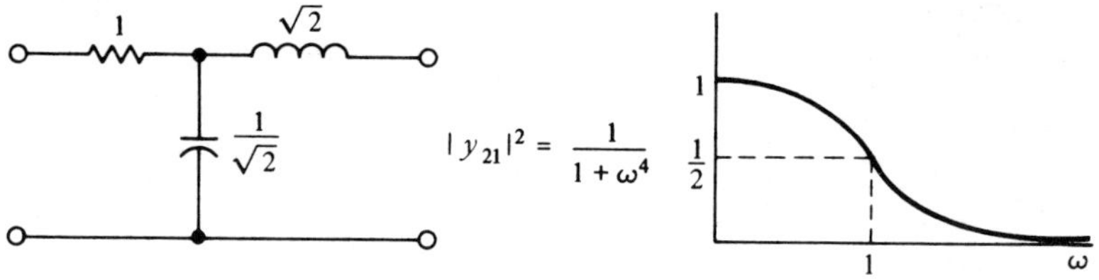

Figure E9-15

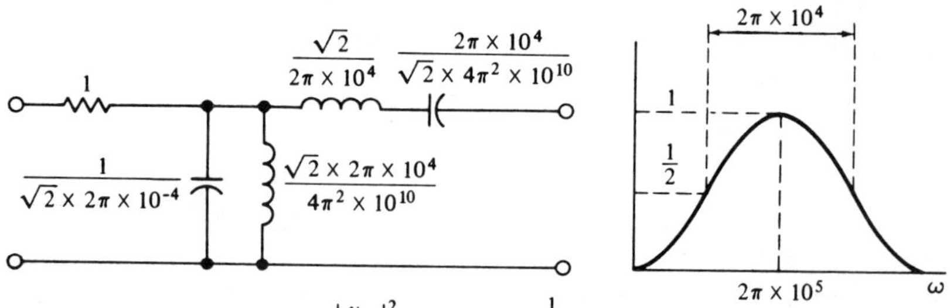

$$|y_{21}|^2 = \frac{1}{1 + \left(\frac{\omega^2 - 4\pi^2 \times 10^{10}}{2\pi \times 10^4 \omega}\right)^4}$$

Figure E9-16

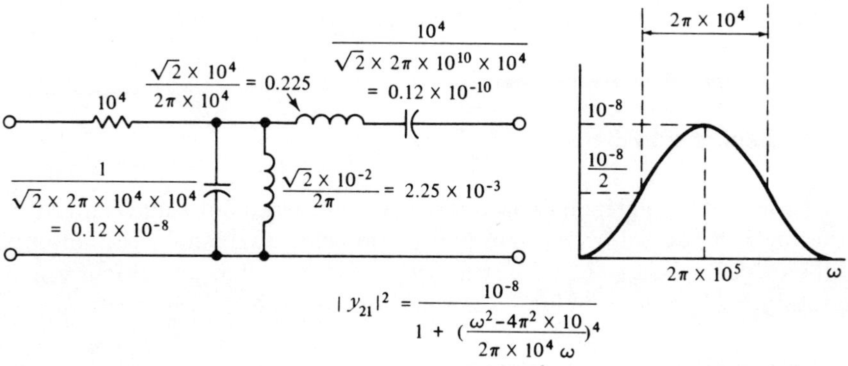

Figure E9-17

fier, as illustrated in Figure 9-19(b). Next, we isolate the frequency-dependent part of the network by replacing everything before the capacitance with its Thevenin equivalent network, as illustrated in Figure 9-19(c). On the right $V_0 = -0.10^4 I_2$. The network having input V_1 and output I_2 must be designed to give the second-order bandpass Butterworth response. Since the load impedance seen by this network is zero, we choose the second-order transfer-admittance prototype filter (see Figure E9-15) and proceed as follows. Transform from lowpass to bandpass (see Figure E9-16), then impedance-scale by 10^4 (see Figure (E9-17). The required coupling network is shown in Figure 9-20. The magnitude squared of the voltage gain is given by

$$\left|\frac{V_0}{V_i}\right|^2 = \left|\frac{V_1}{V_i}\right|^2 |y_{21}|^2 \left|\frac{V_0}{I_2}\right|^2$$

$$= |-10^3|^2 \frac{10^{-8}}{1 + \left(\dfrac{\omega^2 - 4\pi^2 \times 10^{10}}{2\pi \times 10^4 \omega}\right)^4} |10^4|^2 \tag{9-21}$$

$$= \frac{10^6}{1 + \left(\dfrac{\omega^2 - 4\pi^2 \times 10^{10}}{2\pi \times 10^4 \omega}\right)^4}$$

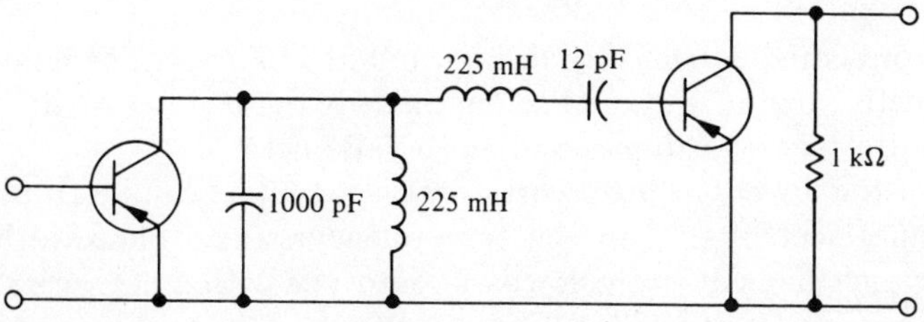

Figure 9-20 Final design of the bandpass amplifier.

Hence, the center-frequency voltage gain is 1000. (See also Example 9-7.)

9-6 Bandreject Filters

Bandreject filters attenuate a band of frequencies about some center frequency and pass all frequencies above and below this band. Bandreject filters can be designed by starting with one of the lowpass Butterworth prototype filters and making a lowpass-to-bandreject transformation.

Lowpass-to-bandreject transformation. The nth-order lowpass Butterworth function defined by Equation (9-1) can be transformed into an nth-order bandreject Butterworth function centered at frequency ω_0 and having bandwidth W by making the change in frequency variable $\omega' = \omega W/(\omega^2 - \omega_0^2)$. With this change in variable, Equation (9-1) becomes

$$|\mathcal{H}(j\omega)|^2 = \frac{1}{1 + \left(\dfrac{\omega W}{\omega^2 - \omega_0^2}\right)^{2n}} \tag{9-22}$$

The same result can be obtained by making a sequence of the two transformations already discussed. That is, the lowpass-to-bandreject transformation is equivalent to making first a lowpass-to-highpass transformation, and then a lowpass-to-bandpass transformation. Equation

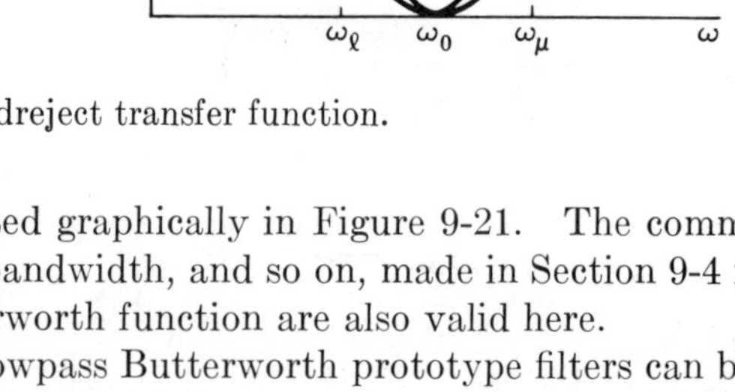

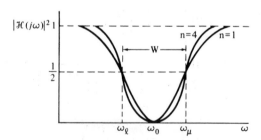

Figure 9-21 Bandreject transfer function.

(9-22) is presented graphically in Figure 9-21. The comments concerning symmetry, bandwidth, and so on, made in Section 9-4 relative to the bandpass Butterworth function are also valid here.

Any of the lowpass Butterworth prototype filters can be transformed into a bandreject filter with center frequency ω_0 and bandwidth W by replacing each element in the lowpass prototype with a network that has the same impedance at frequency $\omega' = \omega W/(-\omega^2 + \omega_0^2)$ that the original element had at frequency ω. Hence, each inductance that originally had impedance $j\omega L$ must be replaced by a network having impedance

$$j\left(\frac{\omega W}{-\omega^2 + \omega_0^2}\right)L = \frac{1}{j\omega\left(\frac{1}{WL}\right) + \frac{1}{j\omega\left(\frac{WL}{\omega_0^2}\right)}}$$

Such a network has an admittance given by $j\omega(1/WL) + 1/[j\omega(WL/\omega_0^2)]$, and can be synthesized as an $1/WL$-farad capacitance in parallel with an WL/ω_0^2-henry inductance. Proceeding in a similar fashion, we find that each C-farad capacitance in the lowpass prototype must be replaced with a $1/WC$-henry inductance in series with a CW/ω_0^2-farad capacitance. Resistances are left unchanged.

Note that at the center frequency ω_0, the parallel LC networks have admittance given by

$$j\omega_0\left(\frac{1}{WL}\right) + \frac{1}{j\omega_0\left(\frac{WL}{\omega_0^2}\right)} = 0$$

while the series LC networks have impedance given by

$$j\omega_0\left(\frac{1}{WL}\right) + \frac{1}{j\omega_0\left(\frac{WC}{\omega_0^2}\right)} = 0$$

Hence, at $\omega = \omega_0$, the series LC networks are short circuits, while the parallel LC networks are open circuits.

Bandreject filter design. The design of a bandreject filter with center frequency ω_0 and bandwidth W can be achieved in four steps.

1. Choose the filter order.
2. Choose one of the prototype filters listed in Figure 9-4.
3. Transform the prototype filter to a bandreject filter with center frequency ω_0 and bandwidth W by replacing each L-henry inductance with a $1/WL$-farad capacitance in parallel with a WL/ω_0^2-henry inductance; replace each C-farad capacitance with a $1/WC$-henry inductance in series with a WC/ω_0^2-farad capacitance; leave resistances unchanged.
4. Magnitude-scale.

9-7 Impedance-Transforming Filters

All of the filter design techniques discussed in the preceeding section allowed only one degree of freedom so far as impedance level is concerned. That is, when a filter was magnitude-scaled, every element in the filter was impedance-scaled by the same amount. Hence, if the filter must be matched to both a driving source resistance and load resistance when the two are not equal, some resistance must be added. (See Example 9-2.) Adding resistances, however, implies power dissipation in other than the driving source and load, with a resulting loss of gain.

In this section we introduce a technique that allows one to magnitude-scale the filter input and output impedances by different amounts, and thus allows matching the filter to unequal source and load impedances without the addition of any resistances. This technique is applicable only to the case of bandpass and bandreject filters.

Before attempting a discussion of impedance-transforming filters, we

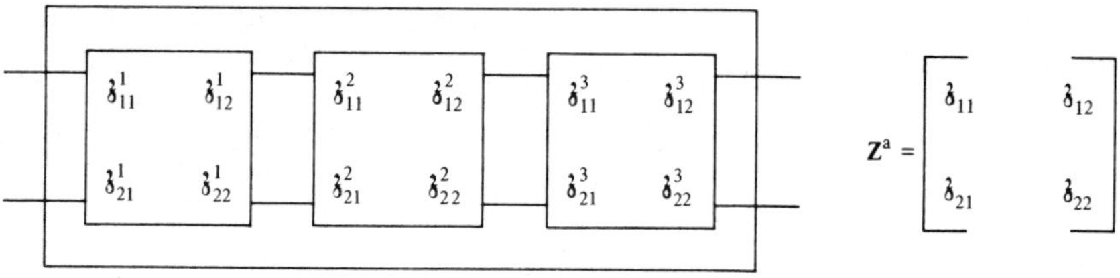

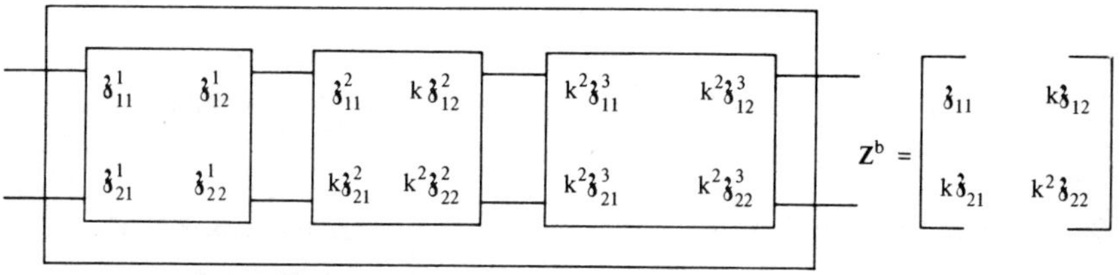

Figure 9-22 Nonuniform impedance-scaling.

must establish the result illustrated in Figure 9-22. Figure 9-22(a) shows a network consisting of three component networks in cascade. Each of the three cascaded networks is characterized by the z parameters shown inside the boxes representing the networks. The z parameters for the total network are given to the right of Figure 9-22(a). The network of Figure 9-22(b) is obtained by making the first of the three component networks identical to the first component network of Figure 9-22(a); the second component network of Figure 9-22(b) is obtained by starting with the second component network of Figure 9-22(a) and retaining the same open-circuit input impedance (z_{11}^2), magnitude-scaling both open-circuit transfer impedances (z_{12}^2 and z_{21}^2) by k, and magnitude-scaling the open-circuit output impedance (z_{22}^2) by k^2; the third component network of Figure 9-22(b) is obtained by magnitude-scaling each

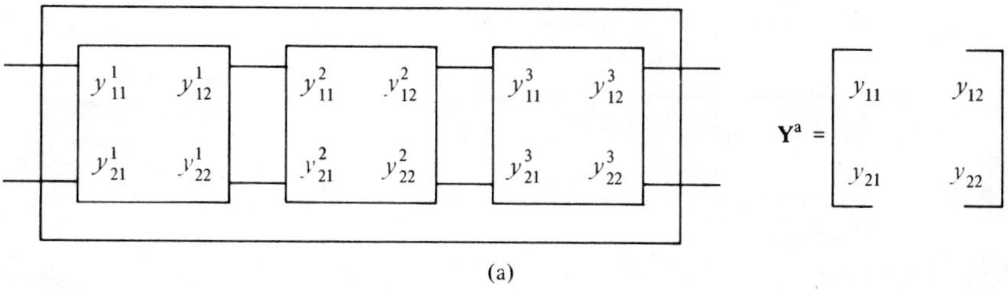

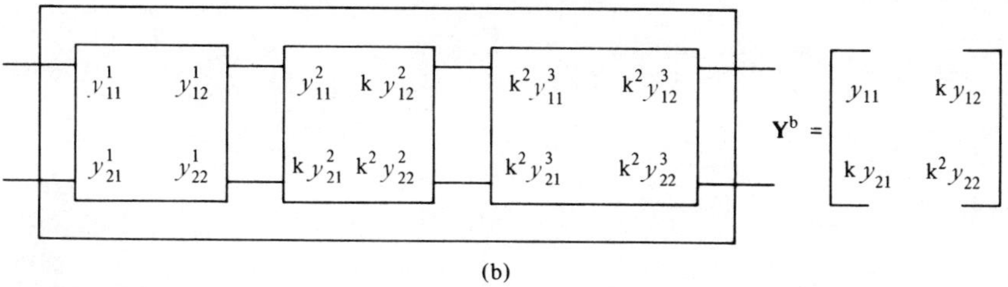

Figure 9-23 Nonuniform admittance-scaling.

impedance of the corresponding network in Figure 9-22(a) by k^2. The z parameters for the total network of Figure 9-22(b) are given to the right of Figure 9-22(b). Note that the network of Figure 4-22(b) has the same input impedance as the network of Figure 9-22(a), but that the output impedance is different by a factor of k^2, and the transfer impedance has been scaled by k. The utility of this method is due to this very interesting result. Similar results can be established for the short-circuit admittance parameters, as illustrated in Figure 9-23.

One point remains to be settled. Given the three component networks of Figure 9-22(a) [or Figure 9-23(a)], how does one synthesize the three component networks of Figure 9-22(b) [or Figure 9-23(b)]? The first and third networks are easily obtained. They require only magnitude-scaling by 1 and k^2, respectively. The second network requires more study. Suppose the second network is a Tee of inductance; that is, we start with the network of Figure 9-A. Let us consider replacing this Tee of inductances with another Tee of inductances (see Figure 9-B). If these two inductance Tee's are second networks of Figures 9-23(a) and

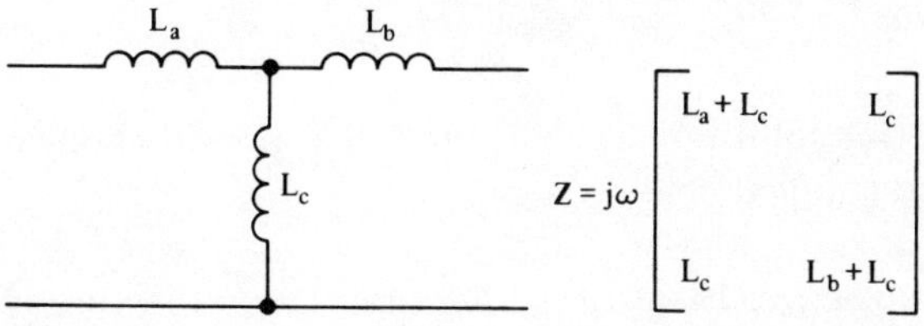

Figure 9-A

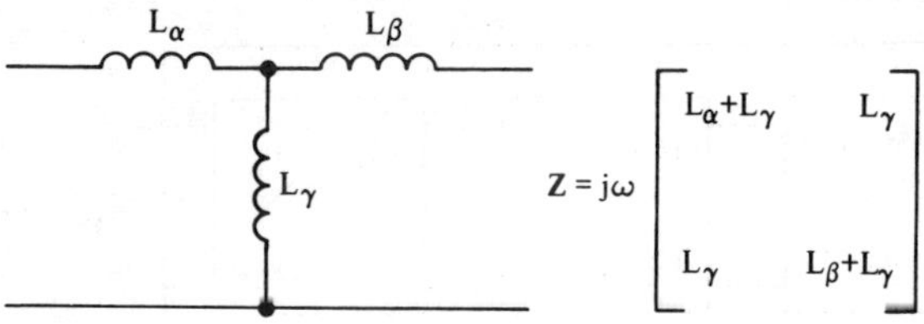

Figure 9-B

9-23(b), respectively, then their impedance matrices are related by

$$\begin{bmatrix} L_\alpha + L_\gamma & L_\gamma \\ L_\gamma & L_\beta + L_\gamma \end{bmatrix} = \begin{bmatrix} L_a + L_c & kL_c \\ kL_c & k^2(L_b + L_c) \end{bmatrix} \tag{9-23}$$

This implies

$$L_\alpha = L_a + (1 - k)L_c \tag{9-24}$$
$$L_\beta = k^2L_b + k(k - 1)L_c \tag{9-25}$$
$$L_\gamma = kL_c \tag{9-26}$$

This result has utility for only a limited range of values for k. For example, if $L_a = 0$, it is obvious from Equation (9-24) that $k > 1$ results in a negative L_α. On the other hand, if $L_b = 0$, Equation (9-25) reveals that L_β is negative for $k < 1$. Hence, the usable range of values for k depends on L_a and L_b.

Next, let us replace the Tee of inductances with a pair of coupled coils rather than another Tee of inductances (see Figure 9-C). For this

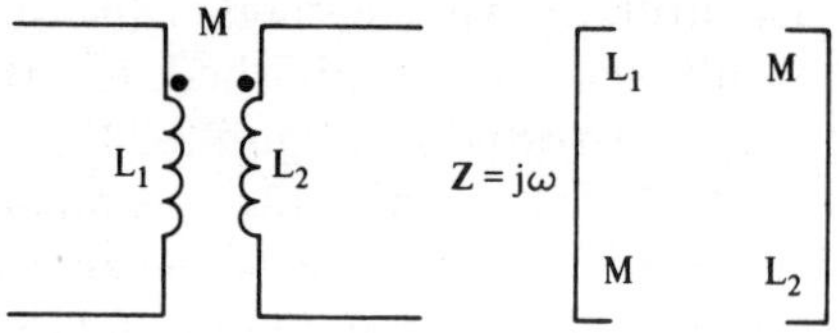

Figure 9-C

replacement we have

$$\begin{bmatrix} L_1 & M \\ M & L_2 \end{bmatrix} = \begin{bmatrix} L_a + L_c & kL_c \\ kL_c & k^2(L_b + L_c) \end{bmatrix} \tag{9-27}$$

Equation (9-27) requires

$$\begin{aligned} L_1 &= L_a + L_c \\ L_2 &= k^2(L_b + L_c) \\ M &= kL_c \end{aligned} \tag{9-28}$$

Apparently, this solution has utility for all k, provided one can physically construct the coupled coils.[3]

[3] This may not always be possible. For example, if $L_a = L_b = L_c = 0.001$ and $k = 100$, the $L_1 = 0.002$, $M = 0.1$, and $L_2 = 20$, the large ratio of L_2 to L_1 poses severe construction problems.

The same technique can be applied to a pi of inductances, a Tee of capacitances, and a pi of capacitances, except that y parameters are more convenient than z parameters for characterizing the pi networks.

The application of these techniques to the design of bandpass filters that must be matched to unequal driving source and load impedances is best illustrated by example.

Example 9-6

In this example we design a second-order bandpass filter for use between the driving source and load of Figure 9-24. The filter is to have a center frequency of 100 kHz, and a bandwidth of 1 kHz.

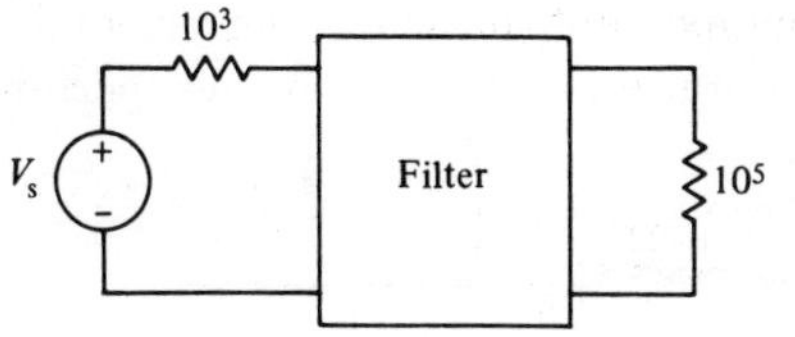

Figure 9-24 Example 9-6.

To accommodate both the driving source and load impedances, we choose the (second-order) insertion-power-gain prototype (see Figure E9-18) and proceed as follows. First, transform from lowpass to bandpass (see Figure E9-19), then impedance-scale by 10^3 (see Figure E9-20).

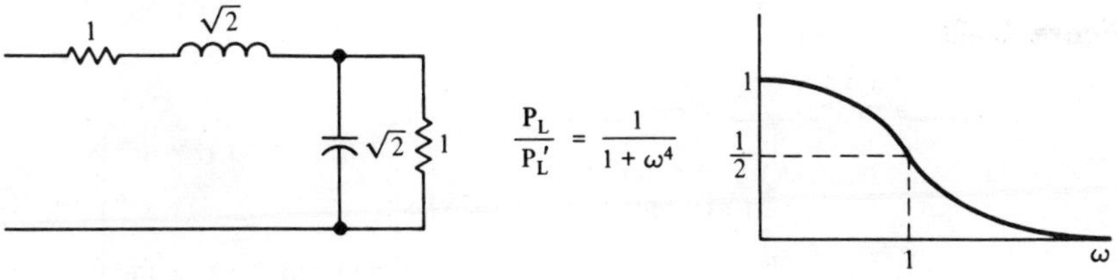

Figure E9-18

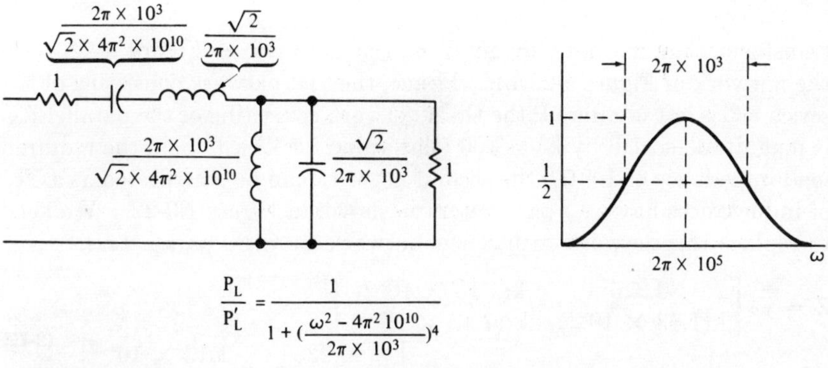

Figure E9-19

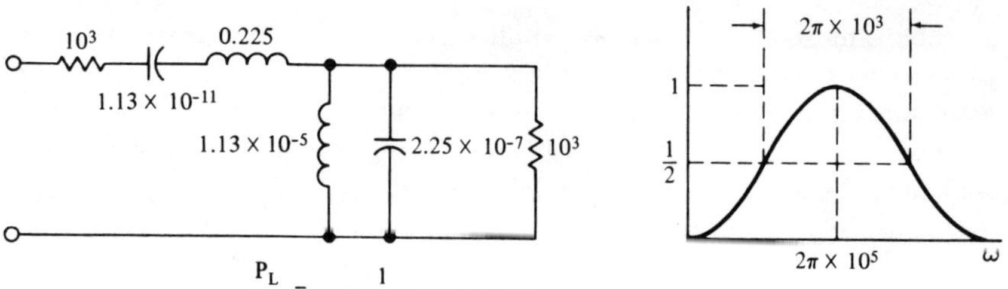

$$\frac{P_L}{P'_L} = \frac{1}{1 + \left(\frac{\omega^2 - 4\pi^2 \times 10^{10}}{2\pi \times 10^3\, \omega}\right)^4}$$

Figure E9-20

At this point we divide the network into three component networks (see Figure E9-21). These three component networks correspond to the three component networks of Figure 9-22(a). Now we perform the

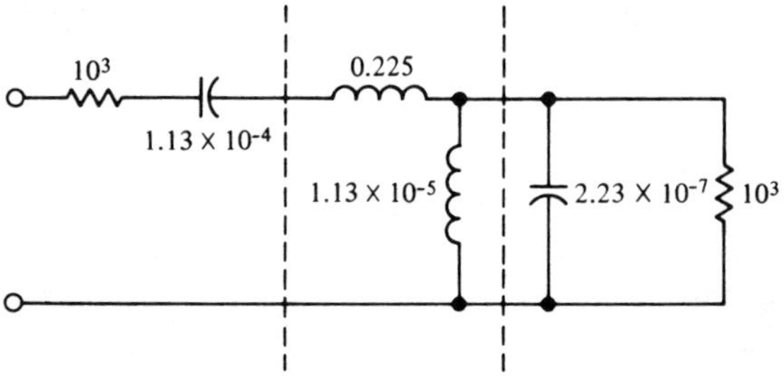

Figure E9-21

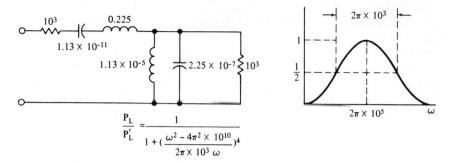

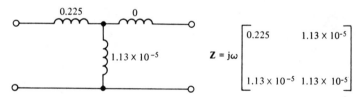

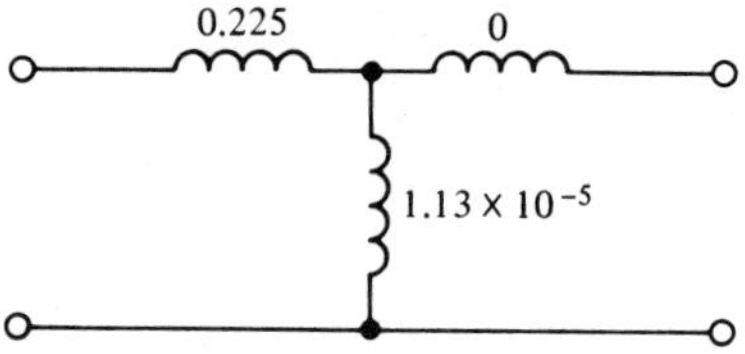

$$Z = j\omega \begin{bmatrix} 0.225 & 1.13 \times 10^{-5} \\ 1.13 \times 10^{-5} & 1.13 \times 10^{-5} \end{bmatrix}$$

Figure E9-22

transformation we used to go from the network of Figure 9-22(a) to the network of Figure 9-22(b). Hence, the first network consisting of the series RC is left unaltered; the third network consisting of the parallel RC is magnitude-scaled by $k^2 = 100$ (this value for k^2 will yield the required load impedance of $10^5\ \Omega$); the second network can be interpreted as a Tee of inductances having z parameters as shown in Figure E9-22. We need to replace this network with a new network having z parameters:

$$Z = j\omega \begin{bmatrix} 0.225 & k(1.13 \times 10^{-5}) \\ k(1.13 \times 10^{-5}) & k^2(1.13 \times 10^{-5}) \end{bmatrix} = j\omega \begin{bmatrix} 0.225 & 1.13 \times 10^{-4} \\ 1.13 \times 10^{-4} & 1.13 \times 10^{-3} \end{bmatrix} \tag{9-29}$$

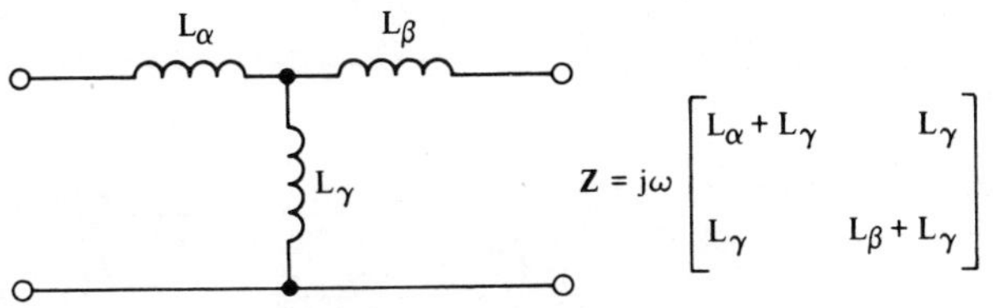

Figure E9-23

We try another Tee of inductances (see Figure E9-23), which appears to work with

$$\begin{aligned} L_\alpha &= 0.225 \text{ H} \\ L_\beta &= 1.02 \text{ mH} \\ L_\gamma &= 0.113 \text{ mH} \end{aligned} \tag{9-30}$$

The resulting network is presented in Figure 9-25. In Problem 9-21 the

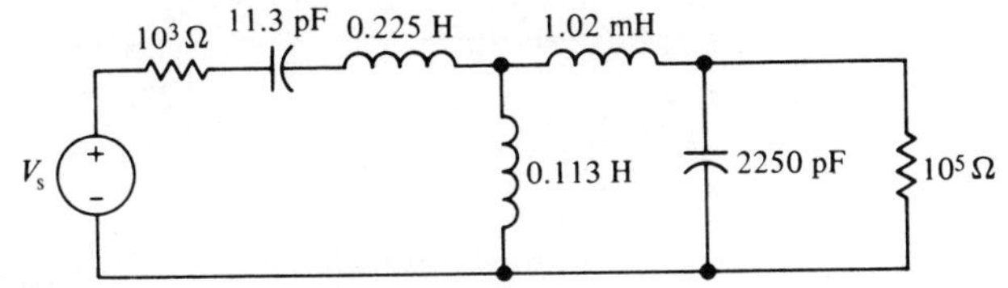

Figure 9-25 Final design for Example 9-6.

reader is asked to compare this result to the filter obtained by stopping after the magnitude-scale-by-10^3 step and adding a resistance in parallel with the load such that the equivalent resistance of the parallel combination is 10^3 ohms.

Example 9-7

As a second example we consider the problem discussed in Example 9-5 and show how the overall amplifier voltage gain can be increased by using the impedance-transformation technique developed in this section. At the end of Example 9-5, we ended up with the network presented in Figure 9-26 where the 10^4-ohm resistance and the 100-pF capacitance are

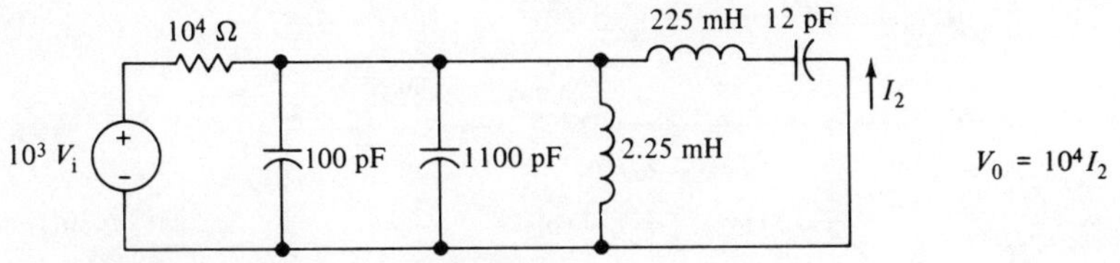

Figure 9-26 Example 9-7.

part of the first transistor. The magnitude squared of the voltage gain was [see Equation (9-20)]

$$\left|\frac{V_0}{V_i}\right|^2 = \frac{10^6}{1 + \left(\dfrac{\omega^2 - 4\pi^2 \times 10^{10}}{2\pi \times 10^4\omega}\right)^4} \tag{9-31}$$

There are two ways in which we can use the impedance-transformation technique to increase the voltage gain.

The first method is to partition the network of Figure 9-26 as shown in Figure E9-24. The three component networks correspond to the

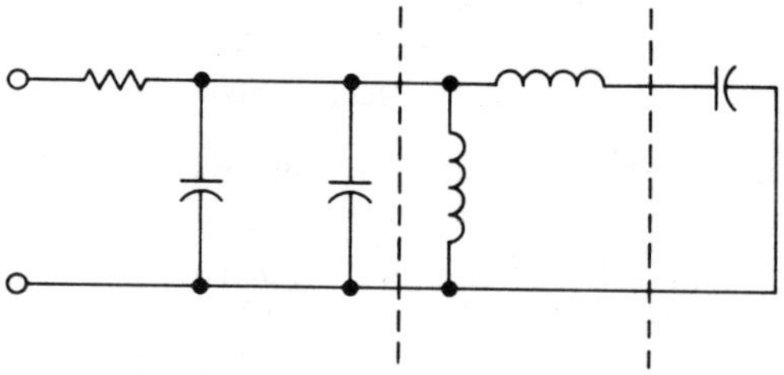

Figure E9-24

three component networks of Figure 9-22(a). Now, the 12-pF capacitor that corresponds to network 3 can be scaled to any practical value, since it is not a part of either transistor. Let us impedance-scale this capacitance by k^{-2}. Then we must replace the network of inductances that correspond to network 2 of Figure 9-22(a) with a new network having the z parameter matrix

$$\mathbf{Z}_2 = j\omega \begin{bmatrix} 2.25 & 2.25^{-}k^{1} \\ 2.25k^{-1} & 227.25k^{-2} \end{bmatrix} \tag{9-32}$$

The maximum value of k that can be used and still allow this z parameter matrix to be synthesized without transformers is given by the solution to the equation (this is the maximum value of k for which L_α, L_β, and L_γ are nonnegative and which occurs for $L_\beta = 0$)

$$2.25k^{-1} = 227.25k^{-2} \tag{9-33}$$

or

$$k = 101 \tag{9-34}$$

The resultant network is shown in Figure 9-27. The magnitude

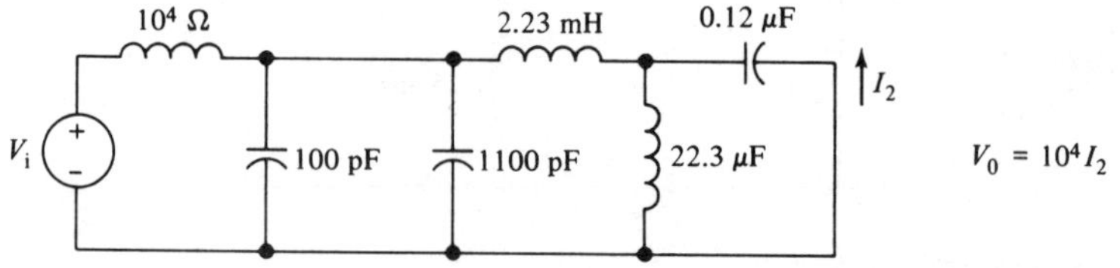

Figure 9-27 Final design for Example 9-7.

squared of the overall amplifier voltage gain is now given by

$$\left|\frac{V_0}{V_i}\right|^2 = \frac{(101)^2 10^6}{1 + \left(\dfrac{\omega^2 - 4\pi^2 \times 10^{10}}{2\pi \times 10^4 \omega}\right)^4} \tag{9-35}$$

Hence, we have increased the amplifier voltage gain by a factor of 101. By using coupled coils we could do even better, at least on paper. However, any further improvement through the use of coupled coils would require the secondary inductance to be more than 101 times greater than the primary inductance. This would be impractical because of construction problems.

The second method requires that we redraw the network of Figure 9-26, as shown in Figure 9-28. This time we partition the network such

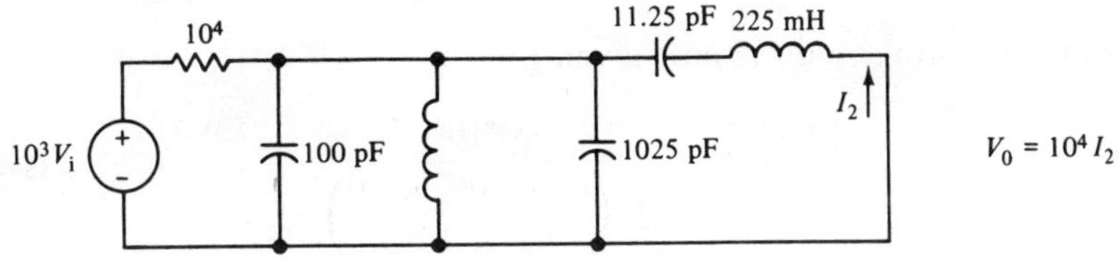

Figure 9-28 Alternative procedure for Example 9-7.

that the capacitances correspond to network 2 of Figure 9-23(a) (see Figure E9-25). First we admittance-scale the 225-mH inductance that corresponds to network 3 of Figure 9-23(a) by k^2. Then the short-circuit admittance-parameter matrix for the capacitances corresponding to network 2

$$\mathbf{Y}_2 = j\omega \begin{bmatrix} 1112 & -12 \\ -12 & -12 \end{bmatrix} \tag{9-36}$$

must be replaced by a new network having the y parameter matrix

$$\mathbf{Y}_2' = j\omega \begin{bmatrix} 1112 & -12k \\ -12k & -12k^2 \end{bmatrix} \tag{9-37}$$

The maximum value of k that can be used under the constraint that this y parameter matrix be realized by a Tee of capacitances is given by the

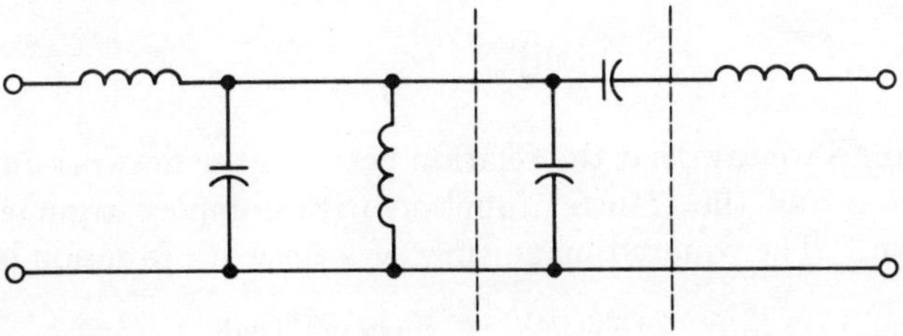

Figure E9-25

solution to the equation

$$1112 = 12k \tag{9-38}$$

or

$$k = 93 \tag{9-39}$$

This value for k gives the network of Figure 9-29. The squared magnitude of the voltage gain is now given by

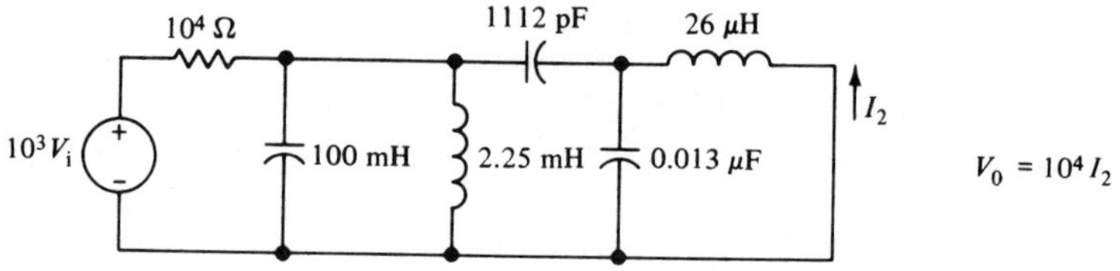

Figure 9-29 Results of capacitance scaling for Example 9-7.

$$\left|\frac{V_0}{V_i}\right|^2 = \frac{(93)^2 10^6}{1 + \left(\dfrac{\omega^2 - 4\pi^2 \times 10}{2\pi \times 10^4 \omega}\right)^4} \tag{9-40}$$

Hence, this method increases the voltage gain by a factor of 93.

9-8 Poles and Zeros of Filter Transfer Functions

In the preceding sections we have discussed one approximation to the ideal filter (the Butterworth approximation) and one form of realization (the LC ladder). The ideas presented in those sections apply to frequency-selective amplifiers in general. Many of the generalizations can be deduced by considering the poles and zeros of the Butterworth filter transfer functions. From these pole-zero diagrams we can get a better understanding of the Butterworth approximation and other functions that are in some ways better approximations to the ideal filter. We can also consider other network configurations that realize these pole-zero diagrams.

The general nth-order Butterworth network function has a squared magnitude:

$$|\mathcal{H}(j\omega)|^2 = \frac{1}{1 + \omega^{2n}} \tag{9-41}$$

In Chapter 8 we saw that the relation between the network function with argument ω and the transfer function with complex argument p was a trivial one. The squared magnitude of a network function is

$$|\mathcal{H}(j\omega)|^2 = \mathcal{H}(j\omega)\mathcal{H}^*(j\omega) \tag{9-42}$$

Since the coefficients of the numerator and denominator polynomials of network and transfer functions are real, $\mathcal{H}^*(j\omega)$ is readily constructed from $\mathcal{H}(j\omega)$ by changing $+j\omega$ to $-j\omega$. Thus,

$$\mathcal{H}^*(j\omega) = \mathcal{H}(-j\omega) \tag{9-43}$$

Consequently,

$$|\mathcal{H}(j\omega)|^2 = [\mathcal{H}(p)\mathcal{H}(-p)]_{p=j\omega} \tag{9-44}$$

In Chapter 8 we stated that the poles and zeros of a transfer function are either real or they occur in conjugate pairs. Thus for each pole above the real axis of a pole-zero diagram, there is another, equally spaced below. When the product $\mathcal{H}(p)\mathcal{H}(-p)$ is formed, the result has a pole-zero diagram with real poles paired on the left and right of the origin, and complex poles occurring in quadruplets, equally spaced from the origin in the four quadrants of the p plane. As an example, let us consider the second-order Butterworth network function. It is

$$|\mathcal{H}(j\omega)|^2 = \frac{1}{1 + \omega^4} \tag{9-45}$$

Then

$$\mathcal{H}(p)\mathcal{H}(-p) = \frac{1}{p^4 + 1}$$

$$= \frac{1}{\left(p + \frac{1}{\sqrt{2}} + j\frac{1}{\sqrt{2}}\right)\left(p + \frac{1}{\sqrt{2}} - j\frac{1}{\sqrt{2}}\right)\left(p - \frac{1}{\sqrt{2}} + j\frac{1}{\sqrt{2}}\right)\left(p - \frac{1}{\sqrt{2}} - j\frac{1}{\sqrt{2}}\right)}$$

The right side of Equation (9-46) has four poles and no zeros. The pole-zero diagram is shown in Figure 9-30. This diagram has the required

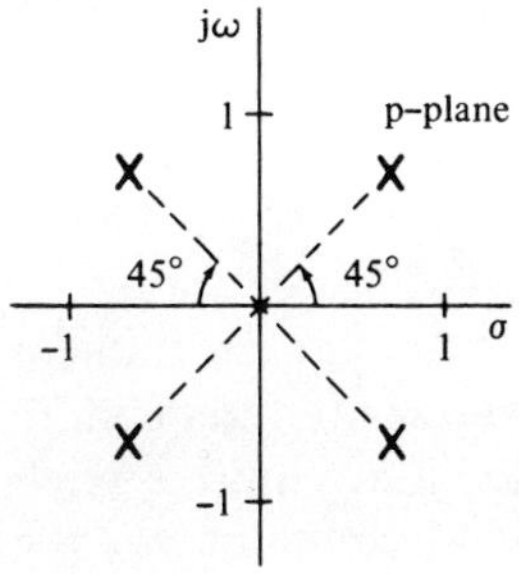

Figure 9-30 Pole-zero diagram for a second-order Butterworth function.

symmetry. The four poles, whose locations are the fourth roots of -1, all lie on a circle of radius 1.

The pole-zero diagram of Figure 9-30 has four poles. Two belong to $\mathcal{H}(p)$ and two to $\mathcal{H}(-p)$. In Chapters 10 and 11 we shall see that a network is unstable if any of its poles lie in the right half of the p plane. Consequently, for a stable filter, the poles of $\mathcal{H}(p)$ must be the two on the left. We conclude that

$$\mathcal{H}(p) = \frac{1}{\left(p + \frac{1}{\sqrt{2}} + j\frac{1}{\sqrt{2}}\right)\left(p + \frac{1}{\sqrt{2}} - j\frac{1}{\sqrt{2}}\right)} \tag{9-47}$$

This is the transfer function for each of the second-order prototypes of Figure 9-4 above. The result can be verified by analysis of the networks in that figure.

In the nth-order case, the transfer function of the Butterworth prototype has n poles on the left half of the unit circle. If n is odd, there is one pole on the negative real axis. The poles divide the unit circles with equal arcs between poles and half-arcs between the $j\omega$ axis and the right-most poles of the group. When the frequency of a lowpass filter is scaled so that the cutoff is at W radians instead of one radian, the poles move to corresponding points on a circle of radius W. Thus, the pole-zero diagram remains the same under frequency-scaling except for a change of scale.

When the lowpass-to-bandpass transformation of Section 9-5 above is applied to a Butterworth prototype, the resulting pole-zero diagram has two clusters of poles. One cluster is to the left of the point $j\omega_0$ and the

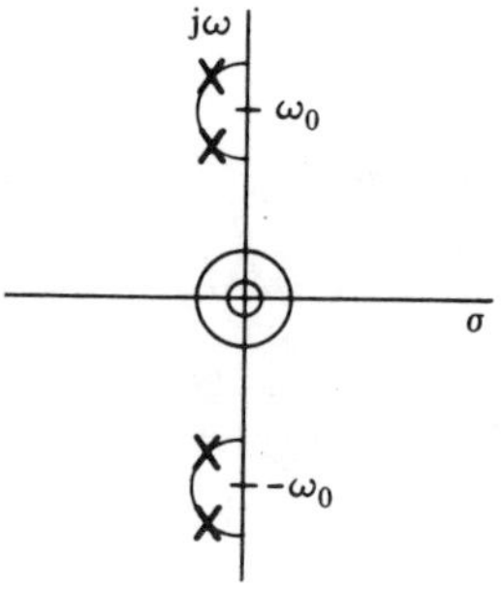

Figure 9-31 Pole-zero diagram—bandpass case.

other to the left of $-j\omega_0$, as shown in Figure 9-31 for $n = 2$. The transfer function also has an nth-order zero at the origin, as shown in the diagram. In Section 9-5 we found that the half-power frequencies of a bandpass filter designed through a lowpass-to-bandpass transformation were such that their geometric mean was the center frequency. Thus the line on which the poles of each cluster lie is not quite a circle in the bandpass case. When the bandwidth is small compared to the center frequency, the poles are very near to a circle centered at ω_0 with diameter equal to the bandwidth. Thus the circles of the bandpass filter poles are each one-half the radius of the circle of the lowpass filter with the same bandwidth.

The pole-zero diagrams for the highpass filter and the bandreject filter are readily derived from the diagrams for lowpass and bandpass, respectively. One merely interchanges the role of poles and zeros.

Improved filters through pole-zero modification. The design engineer can use the pole-zero diagram as a guide to better filters than the precise results of the manipulations of the Butterworth prototypes outlined in the previous sections. By moving the poles slightly, he can get a better approximation to the ideal lowpass filter for certain applications. Also, for practical reasons of constructing a circuit of real devices, he may have to let the poles and zeros move slightly. Finally, as discussed in the next section, he can use other networks to realize a desired pole-zero pattern.

The Butterworth approximation to the ideal lowpass filter is perfect at dc but not very good at the cutoff frequency. If the poles of the second-order filter, as shown in Figure 9-30, were moved to the right a little, they would increase the value of the network function $|\mathcal{H}(j\omega)|$ near the middle of the band more than near dc. By proper choice of pole location, the response of Figure 9-32 could be obtained. At the origin

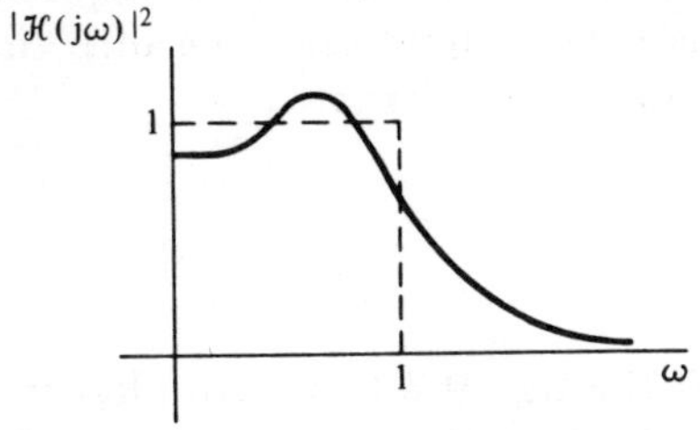

Figure 9-32 Chebyshev-filter response.

(direct current) and at the cutoff frequency, the response is a little low, whereas in midband, the response is equally high. Such equiripple response is a property of a Chebyshev filter. Ladder filters, similar to the Butterworth prototypes of Figure 9-4, which give such a Chebyshev response, have the same configurations as the Butterworth prototypes but with slightly different element values. A tabulation of these element values and a discussion of the poles-zero diagrams for Chebyshev filters are given in Reference [33].

Both the Butterworth and Chebyshev filters are monotonic in the stop band. For some applications it is important that the transition from gain to high attenuation be very abrupt. If a filter with zeros along the $j\omega$ axis, such as that of Figure 9-33(a), can be designed, the response will be that of Figure 9-33(b). One class of these, called elliptic-function filters because elliptic functions are used in deriving the response, are discussed in References [23] and [33].

The LC ladder filters of Figure 9-4 and all those of Reference [33] are amenable to the scaling and transformation of Sections 9-3 through 9-7.

In addition to designs that start from lowpass prototypes, there are other filters that are designed directly as bandpass filters. In these direct designs, a pole-zero pattern is constructed exactly or approximately, and then the transfer function is realized. With modern large-scale com-

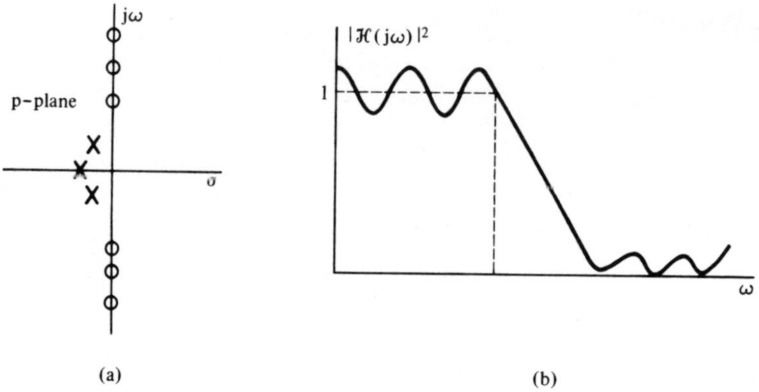

Figure 9-33 Elliptic-function filter response.

puters available, it is often more convenient to make an approximate design using poles and zeros. Then the computer can perturb the element values systematically and optimize the design.

9-9 Other Filter Networks

The networks used in Sections 9-3 through 9-7 to realize filter transfer functions are one class that is very easy to understand. Other configurations, many of which have practical advantages over the LC ladders, are available for use by design engineers. In this section we discuss two such filters very briefly.

One way of constructing a transfer function with a number of poles is to make a cascade consisting of individual resonant circuits separated by isolating amplifiers, as shown in Figure 9-34. The values of the elements in the resonant circuits are chosen to give the correct location and

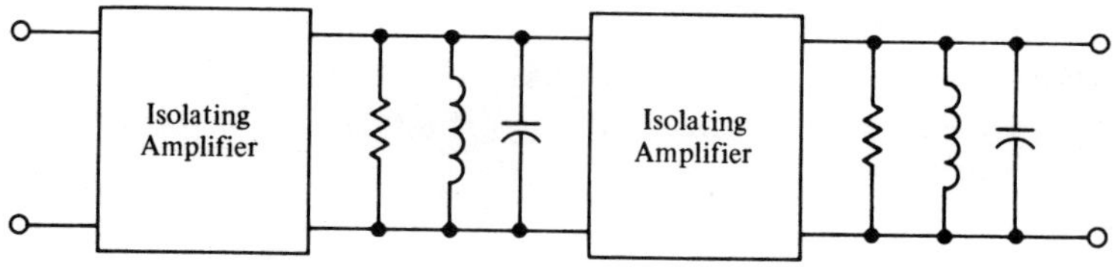

Figure 9-34 Stagger-tuned stages.

multiplier constant for each filter pole separately. The overall transfer function has all the correct poles and zeros. This technique of using individual circuits with different resonant frequencies to get the individual poles of a filter function is known as *stagger-tuning.*

For very low frequency filters, the inductance and capacitance values necessary to realize the desired transfer functions are quite large. Inductors with large values of inductance are comparatively large and are not well represented by an inductance model. They have considerable resistance as well. Better large inductance, high-Q inductors are being built with modern ferrite core material. Even so, there are circumstances where it is desirable to build filters without inductors. Reference [25] is devoted entirely to the problem of filter design without inductance.

A network consisting of only resistances and capacitances has all its poles on the real axis in the p plane. By judicious use of controlled sources along with R's and C's a network with poles anywhere in the p plane can be constructed. In the early days of RC active filter design, when amplifiers (tubes) were more expensive than R's and C's, engineers developed techniques for multiple pole filters, using only one controlled source along with many R's and C's. With modern integrated circuit techniques, transistors are cheaper than R's and C's.

For fabrication considerations, a good design is one where the elements that realize each individual pole can be built and aligned separately. Thus a common design technique is to take a basic circuit that realizes one pole pair and substitute it for the RLC circuits of Figure 9-34.

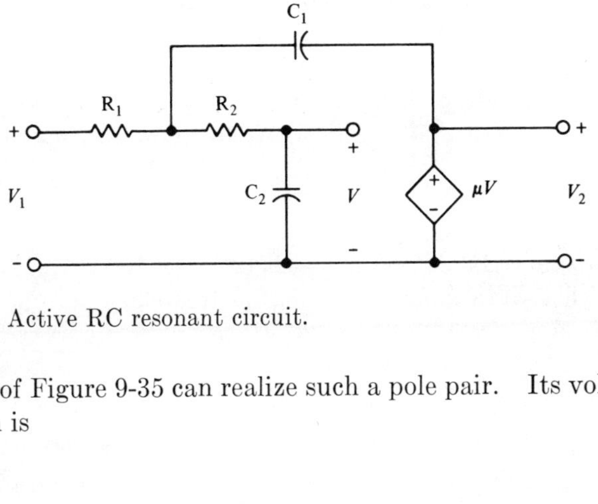

Figure 9-35 Active RC resonant circuit.

The circuit of Figure 9-35 can realize such a pole pair. Its voltage transfer function is

$$\mathcal{H}(p) = \frac{V_2}{V_1}$$

$$= \frac{\mu}{R_1R_2C_1C_2} \frac{1}{p^2 + \left[\frac{(1-\mu)}{R_2C_2} + \frac{1}{R_1C_1} + \frac{1}{R_2C_1}\right]p + \frac{1}{R_1R_2C_1C_2}} \tag{9-48}$$

One can put the denominator roots anywhere in the complex p plane by proper choice of the element values.

The above discussion includes only a few of the basic aspects of filter and coupled amplifier design with regard to the circuits. References [22], [23], [25]–[27], [31], and [33] are a few of the many texts dealing with this subject. The engineer who designs filters will need more study.

The engineer who uses filters can discuss the problem with the background presented here. Furthermore, the frequency- and magnitude-scaling and transforming techniques discussed above in connection with filter design have much wider application in engineering practice.

PROBLEMS

9-1 Compute $\mathfrak{z}_{21}$ and $|\mathfrak{z}_{21}|^2$ for the network shown in Figure P9-1.

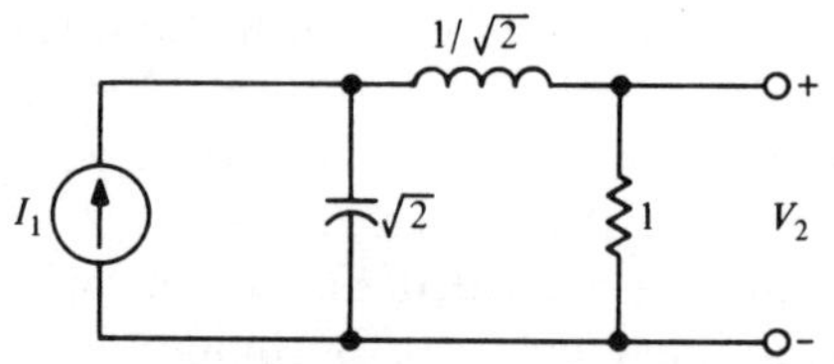

Figure P9-1

9-2 Compute the insertion power gain for the network shown in Figure P9-2. Assume 1-Ω source and load impedances.

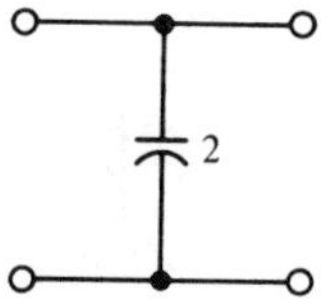

Figure P9-2

9-3 Design a lowpass filter for use between a 100-Ω source impedance and a 100-Ω load impedance. The cutoff frequency is to be 20 kHz and the high-frequency rolloff at least 30 dB per decade.

9-4 Design a lowpass filter to be used as an interstage coupling between the two transistors shown in Figure P9-4. The voltage gain is

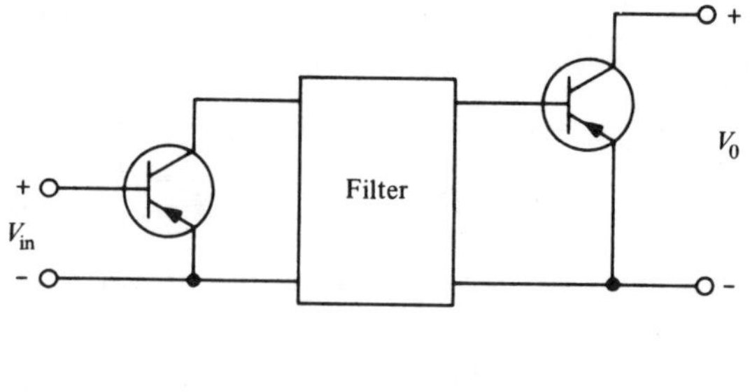

Figure P9-4

to be as high as possible and the high-frequency rolloff must be at least 20 dB/decade.

a. Make the cutoff frequency 1 kHz. Compute and sketch $|V_0/V_{in}|^2$ vs. ω.
b. Make the cutoff frequency 10 MHz. Compute and sketch $|V_0/V_{in}|^2$ vs. ω. Hint: The input admittance to the second transistor is 100 times greater than the output admittance of the first transistor. Therefore, consider either the input admittance to the second transistor to be infinite, or consider the output admittance of the first transistor to be zero.

$$\mathbf{Y}^{(\text{first transistor})} = \begin{bmatrix} 10^{-1} & 0 \\ 50 & 10^{-3} + j\omega 10^{-9} \end{bmatrix}$$

$$\mathbf{Y}^{(\text{second transistor})} = \begin{bmatrix} 10^{-1} & 0 \\ 50 & 10^{-3} \end{bmatrix}$$

9-5 An audio interstage filter is needed for an audio preamplifier. The following specifications are given:

1. Bandwidth: dc to $\sqrt{2} \times 20$ kHz $\simeq 28.3$ kHz.
2. Attenuation should exceed 30 dB per decade outside the passband.
3. The first stage output admittance is 10^{-3} mho $+ j\omega(1/4\pi \times 10^{-8})$ mho; that is, it consists of a 1-kΩ resistance in parallel with a $1/4\pi \times 10^{-6}$F capacitance.
4. The second-stage input impedance is 1 kΩ.

a. Design a filter to meet the above specifications. Do not neglect the $1/4\pi \times 10^{-8}$F capacitor.
b. State qualitatively how you would have designed the above filter if the second-stage input impedance was
(i) 500 ohms.
(ii) 10 ohms.
c. Suppose the second-stage input impedance included a series inductance. Could this inductance value be included in your filter design? How? If the order of the filter were one order higher, could this inductance value be included in the filter design?

9-6 A lowpass filter is to be used as an interstage coupling between two FET's, as illustrated in Figure P9-6. Both FET's have the model shown. The magnitude squared of the voltage gain is to be a first-order Butterworth function with a cutoff frequency of 20 kHz.
a. Neglect the capacitance in the second FET and design the filter. (Take into account the capacitance of the first FET.) Write the equation for $|V_0/V_{in}|^2$ and sketch as a function of ω.
b. Now put the 100-pF capacitance back in the second FET model. Write the new equation for $|V_0/V_{in}|^2$ and sketch as a function of frequency.

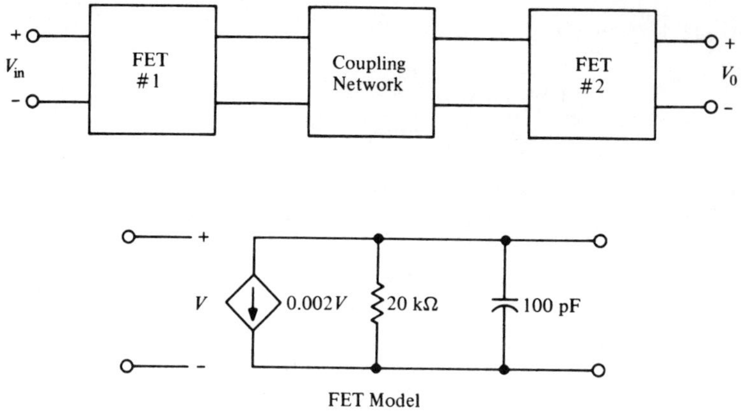

Figure P9-6

9-7 Compute the insertion power gain at $\omega = 0$ for the network designed in Example 9-2. Be careful—the 4000-Ω resistor is the load and the 1111-Ω is part of the filter, as in Figure 9-11(b). Sketch P_L/P'_L for all ω.

9-8 Try a different approach to the problem of Example 9-2. Magnitude-scale by 10^4 rather than 10^3 and solve by adding some resistance at the source end. (Do not be afraid to use Thevenin and Norton's theorems.) Compute the insertion power gain at $\omega = 0$ and compare to the result of Problem 9-7. Sketch P_L/P'_L for all ω.

9-9 Design a highpass filter for use between a 1000-Ω source and a 1000-Ω load. The half-power point is to be at 3 kHz and the low-frequency rolloff at least 18 dB/octave.

9-10 Transistorized audio power amplifiers generally have very low output impedances. Thus, these amplifiers can be represented quite accurately by an ideal voltage source.

Suppose we have an amplifier as described above, which is to drive two speakers. One speaker, a woofer, has a good low-frequency response and the other speaker, a tweeter, has a good high-frequency response.

Your task is to design an electronic crossover network using a pair of third-order Butterworth filters such that the amplifier delivers the majority of the low-frequency energy to the woofer and the majority of the high-frequency energy to the tweeter. The crossover frequency (that is, the frequency at which half of the energy is delivered to the woofer and half to the tweeter) is to be $80/2\pi$ kHz. The woofer and tweeter impedances are shown in Figure P9-10. Hint: Look at Figure 9-4(a) and Figure 9-6(b).

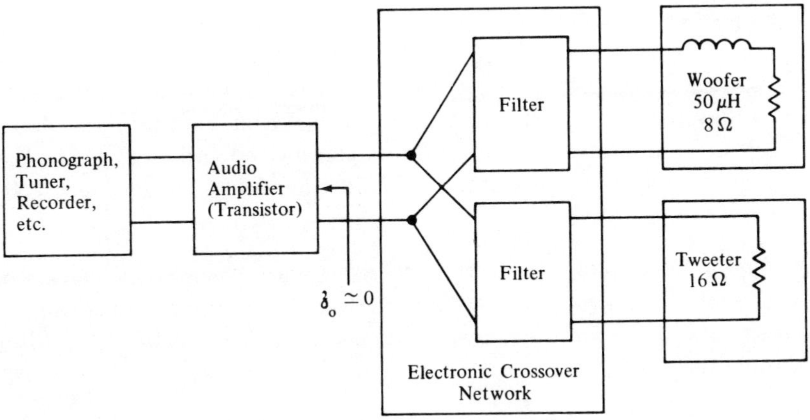

Figure P9-10

9-11 Design a two-stage transistor amplifier using transistors defined by Equation (7-26). The source admittance is 1 ohm and the load admittance is 100 ohms as shown in Figure P9-11. The magnitude squared of the voltage gain is to be a third-order, highpass Butterworth function with a cutoff of 1 kHz. Make the high-frequency voltage gain as high as possible. Hint: Use one of the approximations discussed in connection with Figure 9-6 above.

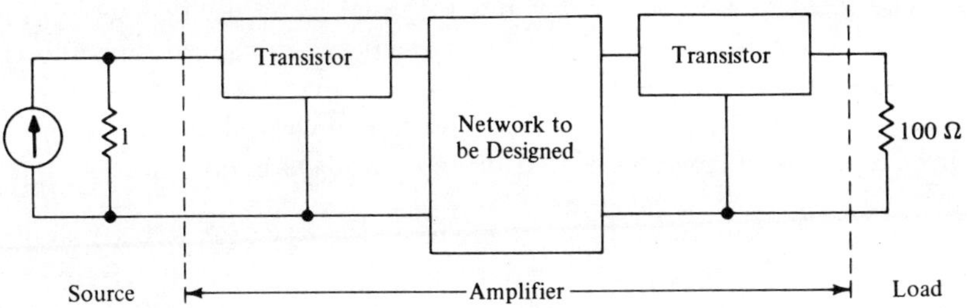

Figure P9-11

9-12 Design a two-stage FET amplifier using 2N3819 FET's. The source and load impedances are given in Figure P9-12. The magnitude

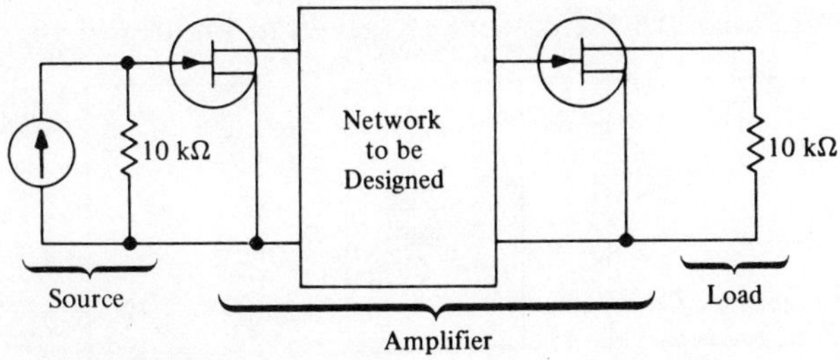

Figure P9-12

squared of the voltage gain is to be a third-order highpass Butterworth function with break point at 10^5 Hz. Write an equation for and sketch the magnitude squared of the voltage gain as a function of frequency. Use the low-frequency FET model of Figure 7-6(a) in your design. After completing your design, consider the effects of the capacitances in the actual FET model. (See Figure 7-5 and the discussion on pages 230, 231.)

9-13 Design a second-order bandpass filter with a center frequency of 45 kHz and a bandwidth of 20 kHz. Both the source and load impedances are 890 Ω. Carefully sketch the insertion power gain of the filter as a function of frequency.

9-14 Design a second-order bandreject filter with a center frequency of 45 kHz and a bandwidth of 20 kHz. Both the source and load impedances are 890 Ω. Carefully sketch the insertion power gain of the filter as a function of frequency.

9-15 Design a second-order bandreject filter for use with a 1-kΩ source. The load impedance can be assumed infinite. Make the center frequency 50.3 kHz and the bandwidth 17.5 kHz. Write an equation for the resulting transfer impedance and sketch it as a function of ω.

9-16 The local FM station wipes out Channel 13 on my TV set. (The antenna for the FM station is in a direct line between my antenna and the TV station antenna.) Design something to correct the situation. Be complete. Give a network with all values shown and tell us where to connect it. Use only values listed in any standard catalogue.

9-17 Design a second-order bandpass filter for use with a 1-kΩ source. The load impedance can be assumed infinite. Make the center frequency 50.3 kHz and the bandwidth 17.5 kHz. Write an equation for the resulting transfer impedance and sketch it as a function of ω.

9-18 Design the coupling network for the two-stage transistor amplifier so that the voltage gain is maximum, the center frequency is 100 kHz, the bandwidth is 10 kHz, and the rolloff is at least 12 dB/octave. The h parameters for both transistors are given in Figure P9-18.

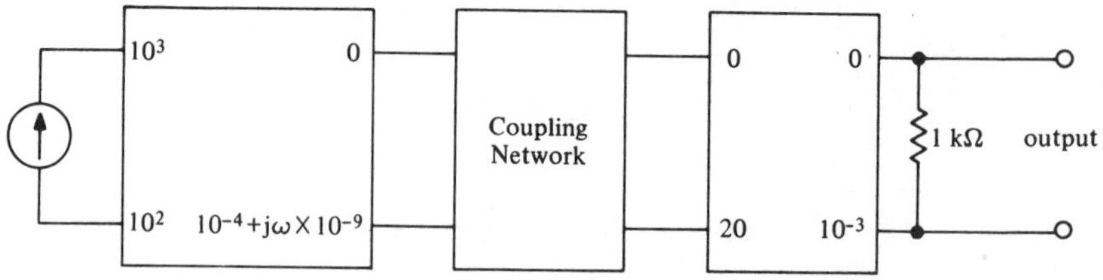

Figure P9-18

9-19 A bandpass filter is to be designed for an interstate coupling for a 2-FET amplifier, as illustrated in Figure P9-19. Both FET's have the model shown. The magnitude squared of the voltage gain is to be a first-order bandpass Butterworth function with bandwidth 10^3 Hz and center frequency 10^5 Hz.

a. Neglect the capacitance of the second FET and design the filter. (Take into account the capacitance of the first FET.) Write the equation for $|V_0/V_{in}|^2$ and sketch it as a function of the frequency.

b. Put the 100-pF capacitance back in the second FET model. Write the new equation for $|V_0/V_{in}|^2$ and sketch it as a function of the frequency.

c. Negate the effect of the capacitance in the second FET by adding a resistance across the output. How does this affect the voltage gain? Write the new equation for $|V_0/V_{in}|^2$ and sketch it as a function of the frequency.

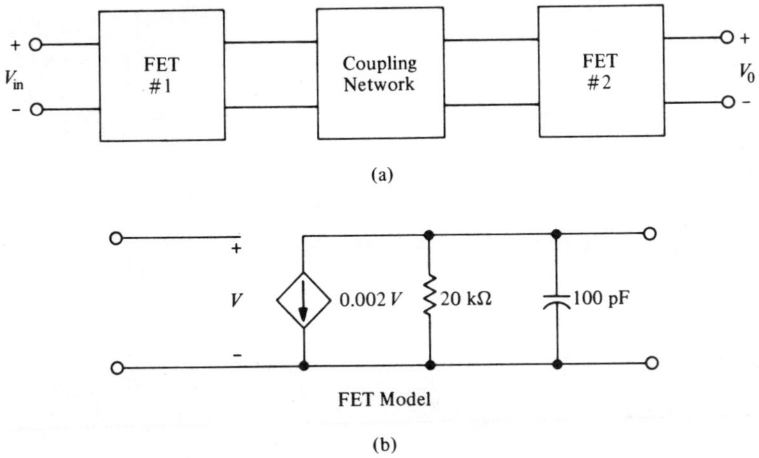

Figure P9-19

9-20 Design a two-stage transistor amplifier using transistors defined by Equation (7-26). The source admittance is 1 ohm and the load admittance is 100 ohms. The magnitude squared of the voltage gain is a third-order bandreject Butterworth function with a 10-kHz center frequency and a 1-kHz bandwidth. Use both transistors in the grounded-source configuration. Compute and sketch the transfer function as a function of ω. Make the high-frequency voltage gain as high as possible. Hint: Use one of the approximations discussed in connection with Figure 9-6 above.

9-21 Consider the filter of example 9-6. Design a filter with the same source and load impedance by separating the 10^3 Ω load of Figure E9-21 into two parallel resistances. One of these resistances is the 10^5 Ω load of the filter. Consider the other resistor internal to the filter. Compare the insertion power gains of this filter to that of Figure 9-25.

9-22 Choose a set of numbers for the network of Figure 9-35 so that a second-order lowpass Butterworth filter of bandwidth 1 MHz results.

9-23 If two stages of Figure 9-35 are cascaded, a four-pole filter results. Design a second-order filter whose poles are those of a second-order Butterworth filter of bandwidth 1 kHz and center frequency 10 kHz.

9-24 Show how two blocking (series) capacitors can be added to the two-stage filter of Problem 9-22 to insert the second-order zero at the origin as required by a second-order Butterworth filter derived by lowpass-to-bandpass transformation. Prove the validity of your network by writing the transfer function.

CHAPTER

10

General Solution of Linear Network Equations—First- and Second-Order Systems

10-1 Introduction

In Chapter 3 we presented two algorithms for writing a set of linear differential equations that describe a network. For networks containing no frequency-dependent elements (no inductances or capacitances), the equations are linear algebraic equations and can be solved by any of the methods of linear algebra. For networks containing frequency-dependent elements and sinusoidal sources, the steady-state response can be obtained by reducing the set of linear differential equations to a set of linear algebraic equations by the method discussed in Chapter 4 and then solving the algebraic equations. Except for these two special cases, the network equations do not reduce to algebraic equations. Therefore, in order to handle the more general case, the network analyst must be capable of solving sets of linear differential equations.

In Chapters 10 and 11 we introduce a method for solving a set of linear differential equations. This method, called the *state-variable* method, is completely general in that it can be used to solve any set of network equations. According to this method, we first write the set of linear differential equations in a *normal form,* and then we solve them by the method of variation of parameters.

In Section 10-2 we show how to write network equations in the normal form. In Section 10-3 we carry out the solution for the first-order case (networks containing one energy storage element) and apply it to some network problems. In Section 10-4 we carry out the solution for the second-order case (networks containing two energy storage elements) and apply this result to a number of network problems. In Chapter 11 we derive the solution for the general nth-order case (networks containing n energy storage elements).

10-2 Normal-Form Equations

In this section we first define a *set of normal-form equations.* Then we use the method of Chapter 3 to construct an algorithm for writing a set of normal-form equations for most networks.

The normal form for a set of n linear first-order differential equations in n unknowns is given by ($\dot{x}_i = dx_i/dt$)

$$\begin{aligned}
\dot{x}_1 &= a_{11}x_1 + a_{12}x_2 + \cdots + a_{1n}x_n + u_1\\
\dot{x}_2 &= a_{21}x_1 + a_{22}x_2 + \cdots + a_{2n}x_n + u_2\\
&\quad\vdots\\
\dot{x}_n &= a_{n1}x_1 + a_{n2}x_2 + \cdots + a_{nn}x_n + u_n
\end{aligned} \tag{10-1}$$

Equation (10-1) can also be written in matrix notation:

$$\begin{bmatrix}\dot{x}_1\\ \dot{x}_2\\ \cdot\\ \cdot\\ \cdot\\ \dot{x}_n\end{bmatrix} = \begin{bmatrix}a_{11} & a_{12} & \cdots & a_{1n}\\ a_{21} & a_{22} & \cdots & a_{2n}\\ \cdot & \cdot & \cdot & \cdot\\ \cdot & \cdot & & \cdot\\ \cdot & \cdot & \cdot & \cdot\\ a_{n1} & a_{n2} & \cdots & a_{nn}\end{bmatrix}\begin{bmatrix}x_1\\ x_2\\ \cdot\\ \cdot\\ \cdot\\ x_n\end{bmatrix} + \begin{bmatrix}u_1\\ u_2\\ \cdot\\ \cdot\\ \cdot\\ u_n\end{bmatrix} \tag{10-2}$$

or, more compactly, as

$$\dot{\boldsymbol{x}} = \mathbf{a}\boldsymbol{x} + \boldsymbol{u} \tag{10-3}$$

Here $\boldsymbol{x}$ is an n × 1 matrix of unknown time functions, sometimes called the *state vector;* $\dot{\boldsymbol{x}}$ is an n × 1 matrix whose elements are the time derivatives of the elements of $\boldsymbol{x}$; $\mathbf{a}$ is an n × n matrix of known constants, sometimes called the *network matrix* or *system matrix;* $\boldsymbol{u}$ is an n × 1 matrix of known time functions or *forcing functions.*

In network analysis problems, the x_j (j = 1, 2, · · · , n) are the n network voltages and currents (time functions) for which we must solve; the a_{ij} (i, j = 1, 2, · · · , n) are numbers that depend on the network elements—resistances, capacitances, self- and mutual inductances, and controlled source parameters; the u_j (j = 1, 2, · · · , n) are time functions due to the independent sources.

In the normal-form equations, Equations (10-1, 2, 3), the derivatives of electrical signals occurring in the equations are singled out. Such derivatives occur because of the inductances and capacitances in the networks. These normal-form equations can be constructed directly from a tree-cotree partition of a network, as discussed in Chapter 3. To this end, the tree and cotree must be selected so that the inductances and capacitances are appropriately located. Then Kirchhoff's laws yield normal-form equations directly. The following algorithm accomplishes the desired results for almost all networks of elements defined in Chapter 1.

1. Select a *normal tree:*
a. Put all branches corresponding to voltage sources and capacitances in the tree. This is possible unless there are loops containing capacitances and voltage sources only.
b. Put all branches corresponding to current sources and inductances in the cotree. This is possible unless there are cut sets containing current sources and inductances only.
c. Put resistances in either tree or cotree.
2. Assign a symbol (including polarity) to each branch:
a. For each branch corresponding to an independent source use the source symbol.
b. For each tree branch corresponding to a capacitance or resistance assign a voltage symbol.
c. For each link corresponding to an inductance or resistance assign a current symbol.
d. For each branch corresponding to a controlled source use the control equation to label the branch in terms of symbols already defined. (Use only the symbols defined in a, b, and c above—not their derivatives or integrals.)
3. For each capacitance use KCL to write the capacitance current as a sum of link currents.

$$C\dot{v}_C = \Sigma \textit{ link currents}$$

For each inductance use KVL to write the inductance voltage as a sum of tree branch voltages:

$$L\dot{i}_L = \Sigma \textit{ tree branch voltages}$$

4. For each tree branch corresponding to a resistance, use KCL to write the resistance current as a sum of link currents:

$$\frac{v_R}{R} = \Sigma \textit{ link currents}$$

For each link corresponding to a resistance, use KVL to write the resistance voltage as a sum of tree branch voltages:

$$Ri_R = \Sigma \textit{ tree branch voltages}$$

5. Solve the equations obtained in step 4 for the resistance voltages and currents (v_R's and i_R's) in terms of inductance currents and capacitance voltages.

6. Use the step-5 equations to eliminate all resistance voltages and resistance currents from the step-3 equations.

For most networks containing a total of n inductances and capacitances (n is the number of inductances plus the number of capacitances), this algorithm yields n normal-form equations in which the n unknowns are the inductance currents and capacitance voltages.

When there is a loop containing only capacitances and voltage sources or a cut set of inductances and current sources, step 1 fails. The reason is that the capacitance voltages in such a loop or inductance currents in such a cut set are not independent. By careful application of Kirchhoff's laws, the extra variable(s) can be eliminated. Then the number of equations is reduced by one for each such loop or cut set. A detailed set of steps that handles all cases is very cumbersome. Hence, we suggest the application of the algorithm when it applies, and the application of common sense when it does not. An illustration of the latter is presented in Example 10-2.

Example 10-1

To illustrate the use of the algorithm, we formulate the normal-form equations for the network of Figure 10-1(a). Following step 1, we first construct a normal tree, as shown in Figure 10-1(b). Note that other normal trees also exist. Next, we assign a voltage symbol to each tree branch and a current symbol to each link, as illustrated in Figure 10-1(c). Then, according to step 3, we write for the capacitance

$$C\dot{v} = i + i_2 - \alpha i_2 \tag{10-4}$$

and for the inductance we write

$$L\dot{i} = -v_1 + v_s - v_3 - v \tag{10-5}$$

Proceeding according to step 4, we write for the two tree branches corresponding to resistances

$$\frac{v_1}{R_1} = i \tag{10-6}$$

$$\frac{v_3}{R_3} = i + i_2 \tag{10-7}$$

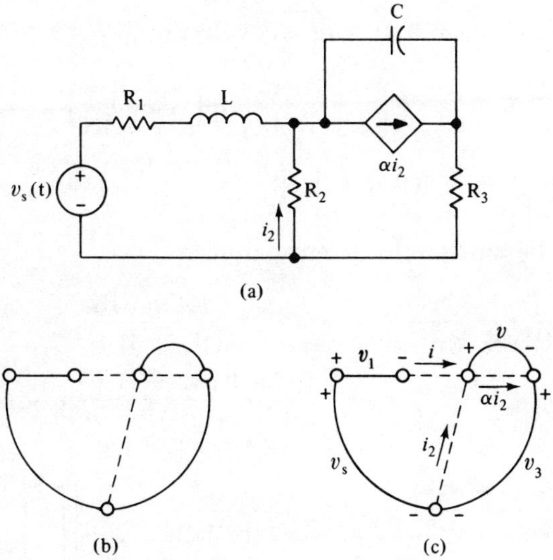

Figure 10-1 Network for Example 10-1.

and for the link corresponding to the resistance, we write

$$i_2 R_2 = -v - v_3 \tag{10-8}$$

As directed by step 5, we now solve Equations (10-6), (10-7), and (10-8) for the three unknown branch voltages and currents that correspond to resistances, that is, v_1, v_3, and i_2. Equation (10-6) gives

$$v_1 = R_1 i \tag{10-9}$$

To solve Equations (10-7) and (10-8) for v_3 and i_2, we write them in the form

$$\begin{aligned} v_3 - R_3 i_2 &= R_3 i \\ v_3 + R_2 i_2 &= -v \end{aligned} \tag{10-10}$$

and use Cramer's rule

$$v_3 = \frac{\begin{vmatrix} R_3 i & -R_3 \\ -v & R_2 \end{vmatrix}}{\begin{vmatrix} 1 & -R_3 \\ 1 & R_2 \end{vmatrix}} = \frac{R_2 R_3 i - R_3 v}{R_2 + R_3} \tag{10-11}$$

$$i_2 = \frac{\begin{vmatrix} 1 & R_3 i \\ 1 & -v \end{vmatrix}}{\begin{vmatrix} 1 & -R_3 \\ 1 & R_2 \end{vmatrix}} = \frac{-v - R_3 i}{R_2 + R_3} \tag{10-12}$$

Finally, according to step 6, we substitute Equations (10-9), (10-11), and

(10-12) into Equations (10-4), and (10-5) to obtain

$$
\begin{aligned}
C\dot{v} &= i - \left(\frac{v + R_3 i}{R_2 + R_3}\right) + \alpha\left(\frac{v + R_3 i}{R_2 + R_3}\right) \\
L\dot{i} &= -R_1 i + v_s - \left(\frac{R_2 R_3 i - R_3 v}{R_2 + R_3}\right) - v
\end{aligned}
\qquad \textbf{(10-13)}
$$

which can also be written in the normal form

$$
\begin{aligned}
\dot{v} &= \frac{-1 + \alpha}{(R_2 + R_3)C} v + \frac{R_2 + \alpha R_3}{(R_2 + R_3)C} i \\
\dot{i} &= \frac{-R_3}{(R_2 + R_3)L} v - \frac{R_1R_2 + R_1R_3 + R_2R_3}{(R_2 + R_3)L} i + \frac{1}{L} v_s
\end{aligned}
$$

or

$$
\begin{bmatrix} \dot{v} \\ \dot{i} \end{bmatrix} = \begin{bmatrix} \dfrac{-1 + \alpha}{(R_2 + R_3)C} & \dfrac{R_2 + \alpha R_3}{(R_2 + R_3)C} \\ \dfrac{-R_3}{(R_2 + R_3)L} & -\dfrac{R_1R_2 + R_1R_3 + R_2R_3}{(R_2 + R_3)L} \end{bmatrix} \begin{bmatrix} v \\ i \end{bmatrix} + \begin{bmatrix} 0 \\ \dfrac{1}{L} v_s \end{bmatrix}
$$

or

$$
\dot{\boldsymbol{x}} = \mathbf{a}\boldsymbol{x} + \boldsymbol{u} \qquad \textbf{(10-14)}
$$

where

$$
\boldsymbol{x} = \begin{bmatrix} v \\ i \end{bmatrix}, \quad \mathbf{a} = \begin{bmatrix} \dfrac{-1 + \alpha}{(R_2 + R_3)C} & \dfrac{R_2 + \alpha R_3}{(R_2 + R_3)C} \\ \dfrac{-R_2}{(R_2 + R_3)L} & -\dfrac{R_1R_2 + R_1R_3 + R_2R_3}{(R_2 + R_3)L} \end{bmatrix}, \quad \boldsymbol{u} = \begin{bmatrix} 0 \\ \dfrac{1}{L} v_s \end{bmatrix} \qquad \textbf{(10-15)}
$$

Example 10-2

To see how to handle the degenerate cases of capacitance and voltage source loops or inductance and current source cut sets, we consider the network of Figure 10-2(a).

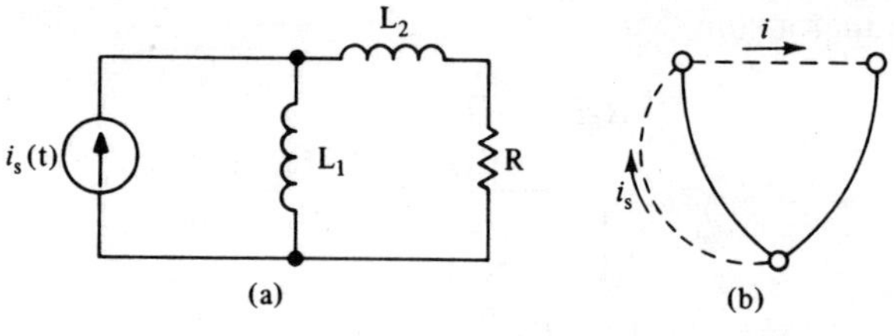

Figure 10-2 Network for Example 10-2.

For this network we must include one of the inductances or the current source in the tree. Suppose we select the tree and cotree illustrated in Figure 10-2(b). According to the link-current algorithm of Section

3-5, we should be able to find one equation in the link current i. Setting the link voltage equal to the sum of the two tree branch voltages, we write

$$L_2 \dot{i} = L_1(\dot{i}_s - \dot{i}) - iR \tag{10-16}$$

This equation can now be rearranged into the normal form

$$\dot{i} = \frac{-R}{L_1 + L_2} i + \frac{L_1}{L_1 + L_2} \dot{i}_s \tag{10-17}$$

The fact that the forcing function turned out to be a function of $\dot{i}_s$ rather than i_s is of no particular significance. When i_s is a known time function, then $\dot{i}_s$ is also a known time function.

10-3 First-Order Systems

A first-order system is a system that can be characterized by one normal-form differential equation in one unknown; that is,

$$\dot{x} = ax + u \tag{10-18}$$

Hence, networks containing one energy storage element are called first-order networks. Any first-order network can be described by one equation of the form of $\dot{x} = ax + u$ by applying the algorithm of Section 10-2. If the energy storage element is a capacitance, x is the capacitance voltage; if the energy storage element is an inductance, x is the inductance current. a is a constant depending only on the system or network, and u is a time function due to the input and is called the forcing function. The equation $\dot{x} = ax + u$ can be solved by following a two-step procedure. First, we find a solution to the homogeneous equation

$$\dot{y} = ay \tag{10-19}$$

and then we use this result to obtain the solution to the complete equation $\dot{x} = ax + u$.

Solution to the homogeneous equation. We seek the time function $y = y(t)$ that satisfies $\dot{y} = ay$. Recall that a is a known constant. Hence, we seek a function whose derivative is the same as the original function except for a constant factor. Since exponential functions have this property, that is, the derivative of an exponential is an exponential, we try an exponential function. The most general exponential function is given by

$$y = Be^{pt} \tag{10-20}$$

where B and p are constants. Next, we determine the constants B and p, if any, for which $y = Be^{pt}$ satisfies $\dot{y} = ay$. The derivative of $y = Be^{pt}$

is given by

$$\dot{y} = \mathrm{pBe^{pt}} \tag{10-21}$$

Substituting Equations (10-20) and (10-21) into Equation (10-19) yields

$$\mathrm{pBe^{pt}} = \mathrm{aBe^{pt}} \tag{10-22}$$

Hence, $y = \mathrm{Be^{pt}}$ satisfies $\dot{y} = \mathrm{a}y$ for any B (it cancels) if we set $\mathrm{p} = \mathrm{a}$. Hence, we conclude that

$$y = \mathrm{Be^{at}} \tag{10-23}$$

is a solution to Equation (10-19) for any B.

This is as far as we can go with the homogeneous equation, and we now proceed to the solution to the complete equation.

Solution to the complete equation. We now seek the solution $x = x(\mathrm{t})$ to the complete equation

$$\dot{x} = \mathrm{a}x + u \tag{10-24}$$

where a is a known constant, and $u = u(\mathrm{t})$ is a known time function. The procedure is very similar to the procedure we used to solve the homogeneous equation. Following the method of variation of parameters, we start by assuming a solution of the form

$$x = b\mathrm{e^{at}}, \qquad b = b(\mathrm{t}) \tag{10-25}$$

(This is identical to the solution to the homogeneous equation given by Equation (10-23), except that we have replaced the constant B with the time function b.) Next, we substitute $x = b\mathrm{e^{at}}$ into $\dot{x} = \mathrm{a}x + u$ and attempt to find a time function b for which $x = b\mathrm{e^{at}}$ satisfies $\dot{x} = \mathrm{a}x + u$.

Since b is a time function, the time derivative of x is given by

$$\dot{x} = \dot{b}\mathrm{e^{at}} + \mathrm{a}b\mathrm{e^{at}} \tag{10-26}$$

Substituting Equations (10-25) and (10-26) into Equation (10-24) yields

$$\dot{b}\mathrm{e^{at}} + \mathrm{a}b\mathrm{e^{at}} = \mathrm{a}b\mathrm{e^{at}} + u$$

or

$$\dot{b} = u\mathrm{e^{-at}} \tag{10-27}$$

The unknown function $b = b(\mathrm{t})$ can be found to within a constant by integrating Equation (10-27). Later, we will see how to relate the constant of integration to energy stored in the inductances and capacitances in the network. A more physical way to establish the constant of integration is to consider all network inputs back to minus infinity. For the networks discussed in this text, the improper integral is well defined. Therefore, $x = b\mathrm{e^{at}}$ is a solution to $\dot{x} = \mathrm{a}x + u$, provided we set

$$b = \int_{-\infty}^{\mathrm{t}} u(\lambda)\mathrm{e^{-a\lambda}}\, d\lambda \tag{10-28}$$

Substituting Equation (10-28) into Equation (10-25) gives the solution to the first-order differential equation

$$x(t) = e^{at} \int_{-\infty}^{t} u(\lambda)e^{-a\lambda}\, d\lambda \tag{10-29}$$

Note that $x(t)$ (the system response at time t) depends on the total past history of the input, but does not depend on any future inputs; that is, $x(t)$ depends on $u(\lambda)$ for all $\lambda \leq t$, but not on $u(\lambda)$ for $\lambda > t$. Because we often do not know the total past history of the input, it is usually more convenient to write the integral over $(-\infty, t)$ in Equation (10-29) as the sum of the integrals over $(-\infty, 0)$ and $(0, t)$; that is, we write

$$x(t) = e^{at} \left\{ \int_{-\infty}^{0} u(\lambda)e^{-a\lambda}\, d\lambda + \int_{0}^{t} u(\lambda)e^{-a\lambda}\, d\lambda \right\} \tag{10-30}$$

Evaluating Equation (10-30) at $t = 0$, we find

$$x(0) = \int_{-\infty}^{0} u(\lambda)e^{-a\lambda}\, d\lambda \tag{10-31}$$

so that we can write

$$x(t) = x(0)e^{at} + e^{at} \int_{0}^{t} u(\lambda)e^{-a\lambda}\, d\lambda \tag{10-32}$$

Equation (10-32) gives the response at time t, $t \geq 0$ in terms of $x(0)$ (*the state of the system at* $t = 0$), and $u(\lambda)$, $0 \leq \lambda \leq t$ [the input during the time interval $(0, t)$]. Therefore, if we are given $x(0)$ and $u(\lambda)$, $0 \leq \lambda \leq t$, we can use Equation (10-32) to compute the response for all $t \geq 0$. Suppose that the energy storage element in the first-order network is a capacitance. Then $x(t)$ is the capacitance voltage at time t, and $x(0)$ is the capacitance voltage at $t = 0$. The energy stored in the capacitance at $t = 0$ is $\frac{1}{2}Cx^2(0)$, so that knowledge of the state of the network at $t = 0$ is equivalent to knowing the energy stored in the network at $t = 0$. If the energy storage element is an inductance, $x(0)$ is the inductance current at $t = 0$, and the energy stored in the inductance at $t = 0$ is $\frac{1}{2}Lx^2(0)$. Again, knowledge of the state of the network at $t = 0$ is equivalent to knowing the energy stored at $t = 0$. Hence, the first term in Equation (10-32) is due to energy stored in the system at $t = 0$ (if no energy is stored in the network at $t = 0$, then $x(0) = 0$ and this term is zero), and is sometimes called the *energy storage term*. On the other hand, the second term in Equation (10-32) depends on the input (or forcing function) and is called the *forced response*. Because the network is linear, the principle of superposition applies, and these two terms are additive. (If we set the independent source $u(\lambda)$ to zero and find the response due to the stored energy, we would find $x(t) = x(0)e^{at}$; if we then assumed zero energy storage and found the response due only to the independent source, we would obtain

$$x(t) = e^{at} \int_0^t u(\lambda)e^{-a\lambda}$$

The total response is the sum of these two.)

Finally, we note that Equation (10-29) can be written in the form

$$x(t) = \int_{-\infty}^{t} u(\lambda)e^{a(t-\lambda)}\, d\lambda \tag{10-33}$$

which is an integral of the form

$$x(t) = \int_{-\infty}^{t} g(\lambda)h(t - \lambda)\, d\lambda \tag{10-34}$$

This integral occurs so often in applied mathematics that it has been given a name—the *convolution integral.* We say that $x(t)$ is the convolution of $g(\lambda)$ and $h(\lambda)$, or $x(t)$ is $g(\lambda)$ convolved with $h(\lambda)$. Hence, in this terminology, the system response $x(t)$ is the convolution of the forcing function $u(\lambda)$ with the function $e^{a\lambda}$.

We also point out that the constant a in these equations is called the *natural frequency* of the network. For the first-order equation, the natural frequency is the network parameter a, that is, the parameter in the normal-form equation $\dot{x} = ax + u$.

Example 10-3

In this example we use the result derived in the preceding pages to find the capacitance voltage as a function of time for the network presented in Figure 10-3(a).

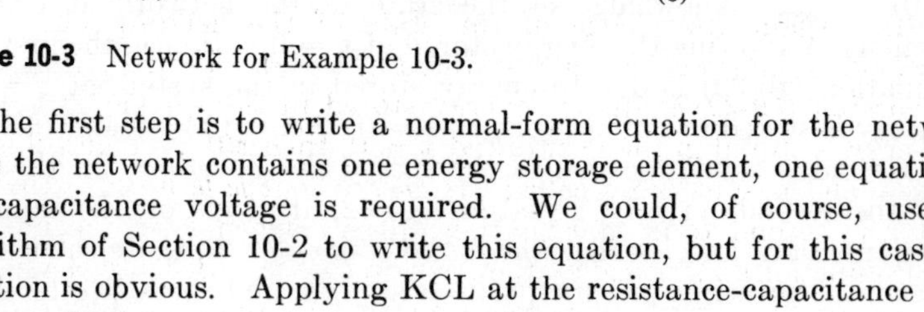

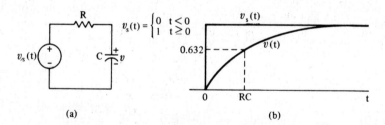

Figure 10-3 Network for Example 10-3.

The first step is to write a normal-form equation for the network. Since the network contains one energy storage element, one equation in the capacitance voltage is required. We could, of course, use the algorithm of Section 10-2 to write this equation, but for this case the equation is obvious. Applying KCL at the resistance-capacitance junction, we obtain

$$C\dot{v} = \frac{v_s - v}{R} \tag{10-35}$$

or

$$\dot{v} = -\frac{1}{RC}v + \frac{1}{RC}v_s \tag{10-36}$$

which is of the form

$$\dot{x} = ax + u \tag{10-37}$$

with

$$x = v, \qquad a = -\frac{1}{RC}, \qquad u = \frac{1}{RC}v_s = \begin{cases} 0 & t < 0 \\ \dfrac{1}{RC} & t \geq 0 \end{cases} \tag{10-38}$$

Using these quantities in Equation (10-29), we obtain

$$\begin{aligned} v(t) &= e^{-t/RC}\int_0^t \frac{1}{RC} e^{\lambda/RC}\, d\lambda \\ &= e^{-t/RC}[e^{t/RC} - 1] \\ &= 1 - e^{-t/RC} \qquad t \geq 0 \end{aligned} \tag{10-39}$$

The capacitance voltage is the sum of two terms. The first term, which is of the same functional form as the input, is called the *steady-state response*. That is, for $t > 0$, the input is constant at 1 volt. Hence, the capacitance eventually charges to its steady-state value of 1 volt.

The second term, which goes to zero for large t, is called the *transient response*. The transient term describes the transition period between the initial state of the system and the steady state. The sum of these two terms gives the forced response; there is no energy storage term because no energy is stored in the capacitance at $t = 0$. If there had been energy stored in the capacitance at $t = 0$, then $v(t)$ would contain a third term.

In Figure 10-3(b) we sketch the capacitance voltage v along with the input v_s for all t. The capacitance voltage starts at zero and rises exponentially to its final value of 1 volt. Note that at $t = RC$ we have $v(RC) = 1 - e^{-1} = 0.632$. The time required for the capacitance voltage to reach 63 percent of its final value is called the *time constant* of the network. Hence, this network has a time constant of RC seconds. Note that the time constant is the negative of the reciprocal of the natural frequency.

Example 10-4

In this example we compute the voltage across the capacitance as a function of time for all $t \geq 0$ for the network presented in Figure 10-4. This is the same as example 10-3, except that the input voltage is a square pulse rather than a step function.

The differential equation for this network was derived in Example

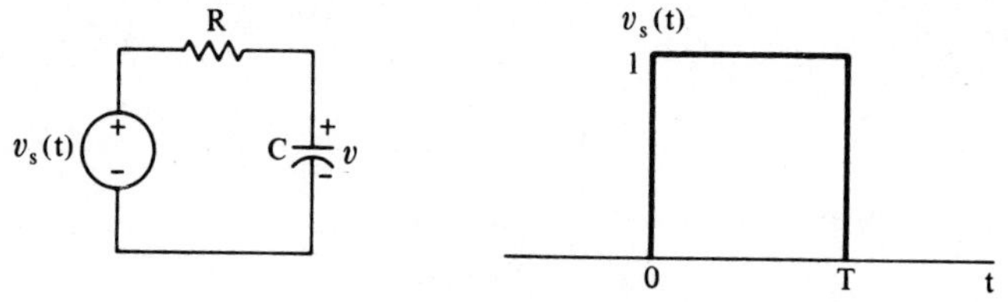

Figure 10-4 Network for Example 10-4.

10-3, where we found

$$\dot{v} = -\frac{1}{RC}v + \frac{1}{RC}v_s \tag{(10-40)}$$

This equation is of the form $\dot{x} = ax + u$ if we set

$$x = v, \qquad a = -\frac{1}{RC}, \qquad u = \frac{1}{RC}, \qquad v_s = \begin{cases} 0 & t < 0 \\ \dfrac{1}{RC} & 0 \le t \le T \\ 0 & t > T \end{cases} \tag{10-41}$$

Therefore, the solution is given by Equation (10-29):

$$v(t) = e^{-t/RC} \int_{-\infty}^{t} \frac{1}{RC} v_s(\lambda) e^{-\lambda/RC}\, d\lambda$$

$$= \begin{cases} 0 & t < 0 \\ e^{-t/RC} \displaystyle\int_0^t \frac{1}{RC} e^{\lambda/RC}\, d\lambda & 0 \le t \le T \\ e^{-t/RC} \displaystyle\int_0^T \frac{1}{RC} e^{\lambda/RC}\, d\lambda & t > T \end{cases}$$

$$= \begin{cases} 0 & t < 0 \\ 1 - e^{-t/RC} & 0 \le t \le T \\ (e^{T/RC} - 1)e^{-t/RC} & t > T \end{cases} \tag{10-42}$$

Equation (10-42) is sketched in Figure 10-5, along with a sketch of the input voltage for various relationships between the pulse duration T and the network time constant RC. For RC $\ll$ T, the capacitance voltage is nearly identical to the input voltage, whereas, for RC > T, the input pulse is severely distorted. Hence, pulse circuits (circuits for transmitting or processing pulses) must have time constants that are small compared to the pulse duration if the pulse shape is to be preserved.

This problem can also be solved in a slightly different manner. Let us first assume a unit step function input as in Example 10-3. Then the capacitance voltage as a function of time as given by Equation (10-39); that is,

$$v(t) = 1 - e^{-t/RC} \tag{10-43}$$

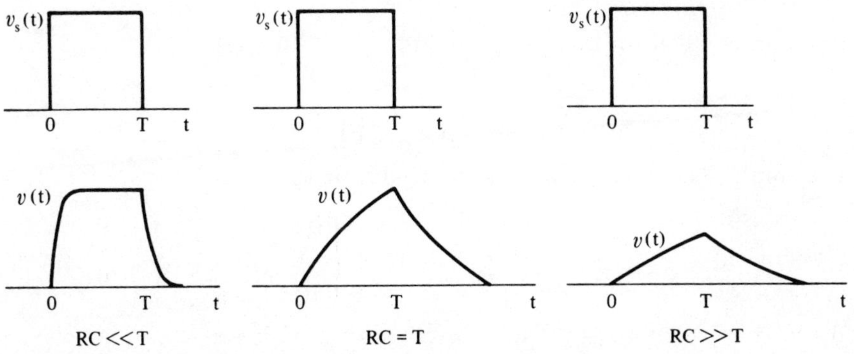

Figure 10-5 Input and output of an RC network for RC $\ll$ T, RC = T, RC $\gg$ T.

In the present problem, however, this equation is valid only on (0, T), since for t > T the pulse input is no longer identical to the unit step input. To compute the capacitance voltage for t > T, we first note that at t = T, the capacitance voltage is given by

$$v(T) = 1 - e^{-T/RC} \tag{10-44}$$

and that the input is zero for t > *T*. Therefore, we can use Equation (10-32), in which the forced response is now zero so that the capacitance voltage is due only to the energy storage term. Since the transition takes place at t = T rather than at t = 0, we must shift the time variable by T seconds; that is, replace t by t − T, and write

$$\begin{aligned} v(t) &= v(T)e^{-(t-T)/RC} \\ &= (1 - e^{-T/RC})e^{-(t-T)/RC} \\ &= (e^{T/RC} - 1)e^{-t/RC} \qquad t \geq T \end{aligned} \tag{10-45}$$

which gives the same result obtained previously.

Example 10-5

In this example we consider the network presented in Figure 10-6 and compute the capacitance voltage as a function of time when a sine wave is applied at t = 0. We further assume that at t = 0^- the capacitance voltage is $v(0^-) = v_0$ volts.

The differential equation for this network was derived in Example

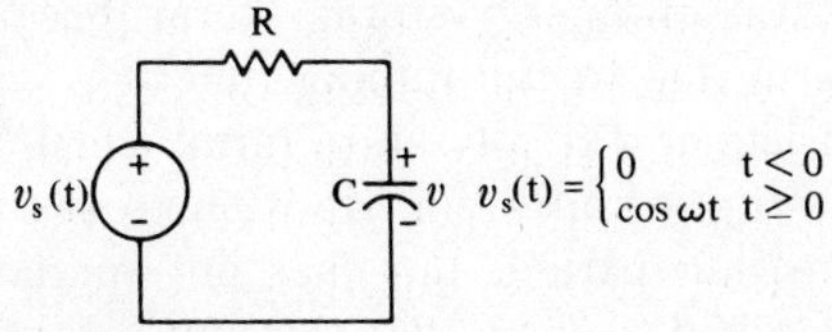

Figure 10-6 Network for Example 10-5.

10-3 and is given by Equation (10-36):

$$\dot{v} = -\frac{1}{RC}v + \frac{1}{RC}v_s \tag{10-46}$$

The solution is given by Equation (10-32), with

$$x = v, \qquad a = -\frac{1}{RC}, \qquad u = \frac{1}{RC}, \qquad v_s = \begin{cases} 0 & t < 0 \\ \dfrac{1}{RC}\cos\omega t & t \geq 0 \end{cases} \tag{10-47}$$

We have

$$\begin{aligned} v(t) &= v_0 e^{-t/RC} + e^{-t/RC}\int_0^t \frac{1}{RC}\cos\omega\lambda\, e^{\lambda/RC}\, d\lambda \qquad t \geq 0 \\ &= v_0 e^{-t/RC} + \frac{e^{-t/RC}}{RC}\int_0^t e^{\lambda/RC}\cos\omega\lambda\, d\lambda \qquad t \geq 0 \end{aligned} \tag{10-48}$$

From an integral table, or by integrating by parts, we note that the integral in Equation (10-48) is of the form

$$\begin{aligned} \int e^{\alpha t}\cos\omega t\, dt &= \frac{e^{\alpha t}[\alpha\cos\omega t + \omega\sin\omega t]}{\alpha^2 + \omega^2} \\ &= \frac{e^{\alpha t}\cos(\omega t - \tan^{-1}\omega/\alpha)}{\sqrt{\alpha^2 + \omega^2}} \end{aligned} \tag{10-49}$$

If we use Equation (10-49), the integral in Equation (10-48) can be evaluated, and we have

$$\begin{aligned} v(t) &= v_0 e^{-t/RC} \\ &\quad + \frac{e^{-t/RC}}{RC}\left[\frac{e^{t/RC}\cos(\omega t - \tan^{-1}\omega RC)}{\sqrt{\left(\frac{1}{RC}\right)^2 + \omega^2}} - \frac{\cos(-\tan^{-1}\omega RC)}{\sqrt{\left(\frac{1}{RC}\right)^2 + \omega^2}}\right] \\ &= \underbrace{v_0 e^{-t/RC}}_{\text{energy-storage response}} \underbrace{\underbrace{- \frac{1}{\sqrt{1 + (\omega RC)^2}} e^{-t/RC}}_{\text{transient response}} + \underbrace{\frac{\cos(\omega t - \tan^{-1}\omega RC)}{\sqrt{1 + (\omega RC)^2}}}_{\text{steady-state response}}}_{\text{forced response}} \qquad t \geq 0 \end{aligned} \tag{10-50}$$

The solution contains an energy-storage term [because $v(0) \neq 0$] and a forced-response term due to the forcing function. The forced-response term, in turn, consists of a steady-state term, which is of the same functional form as the input and a contribution to the transient term that describes the transition period, but dies out for large t. In Problem 10-11 the reader is asked to show that the sinusoidal steady-state method described in Chapter 4 gives the steady-state term.

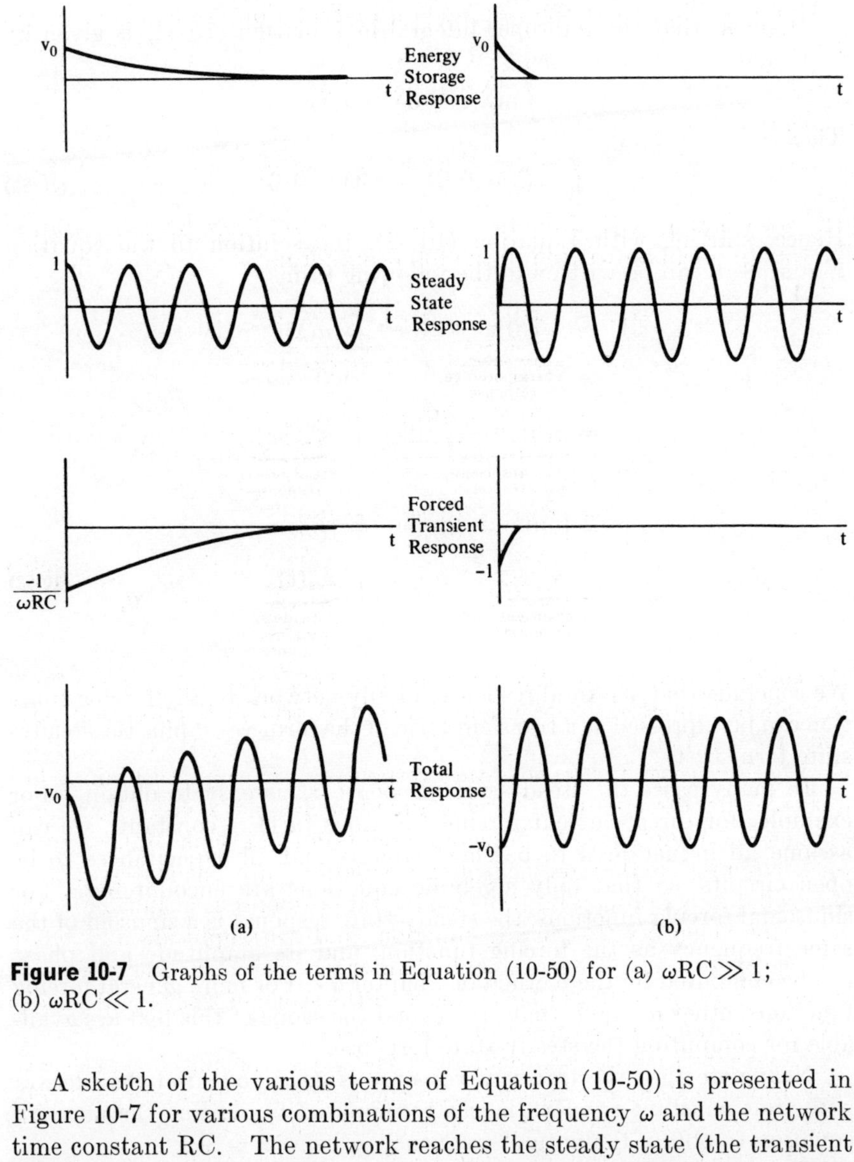

Figure 10-7 Graphs of the terms in Equation (10-50) for (a) $\omega RC \gg 1$; (b) $\omega RC \ll 1$.

A sketch of the various terms of Equation (10-50) is presented in Figure 10-7 for various combinations of the frequency ω and the network time constant RC. The network reaches the steady state (the transient terms die out) after approximately three time constants, that is, for $t > 3RC$ seconds.

The time-constant concept. So far in Section 10-3, we have obtained two forms for the solution to the first-order equation $\dot{x} = ax + u$ [see Equations (10-29) and (10-32)], and we have considered several examples. In this section we start with the form of the solution given by Equation (10-32) and write it in a somewhat different form that is sometimes easier to evaluate.

Suppose that the improper integral in Equation (10-32) is given by $f(\lambda)$; that is,

$$\int u(\lambda)e^{-a\lambda}\,d\lambda = f(\lambda)$$

Then

$$\int_0^t u(\lambda)e^{-a\lambda}\,d\lambda = f(t) - f(0) \tag{10-51}$$

Hence, starting with Equation (10-32), the solution to the equation $\dot{x} = ax + u$ can be written in the following forms:

$$\begin{aligned} x(t) &= \underbrace{x(0)e^{at}}_{\substack{\text{energy-storage}\\ \text{response}}} + \underbrace{e^{at}\int_0^t u(\lambda)e^{-a\lambda}\,d\lambda}_{\text{forced response}} \\ &= \underbrace{x(0)e^{at} - f(0)e^{at}}_{\substack{\text{transient}\\ \text{response}}} + \underbrace{f(t)e^{at}}_{\substack{\text{steady-state}\\ \text{response}}} \\ &= [x(0) - f(0)]e^{at} + f(t)e^{at} \\ &= \underbrace{\gamma e^{-t/\tau}}_{\substack{\text{transient}\\ \text{response}}} + \underbrace{x_{ss}(t)}_{\substack{\text{steady-}\\ \text{state}\\ \text{response}}}, \end{aligned} \tag{10-52}$$

We conclude that the total response for any network to any forcing function can be expressed as a transient term of the form $\gamma e^{-t/\tau}$ plus the steady-state term $x_{ss}(t)$.

In many cases the steady-state response x_{ss} is easy to obtain. For example, for direct-current forcing functions [$u(t)$ = constant], we can assume all inductances to be short circuits and all capacitances to be open circuits, so that only algebraic equations are encountered. For sinusoidal forcing functions, the steady-state response is a sinusoid of the same frequency as the forcing function, and its amplitude and phase can be computed by the method of Chapter 4. For more general forcing functions, other methods that are beyond the scope of this text are available for computing the steady-state response.

Next note that the transient response is an exponential that decays with time constant τ, which is easily computed by writing the network equation to obtain the natural frequency a, using $\tau = -a^{-1}$.

The remaining constant γ can be evaluated from knowledge of the energy state of the network at time $t = 0^-$. To see this, we evaluate Equation (10-52) at time $t = 0^+$; that is,

$$x(0^+) = \gamma + x_{ss}(0^+) \tag{10-53}$$

or

$$\gamma = x(0^+) - x_{ss}(0^+) \tag{10-54}$$

Since we have already evaluated the steady-state response $x_{ss}(t)$ for $t \geq 0$, $x_{ss}(0^+)$ is a known number. Furthermore, since $x(t)$ is either a capacitance voltage or an inductance current, $x(0^+) = x(0^-)$. (This is a result of the fact that the stored energy cannot change instantaneously; the stored energy at $t = 0$ is $\frac{1}{2}Lx^2(0)$ if x is an inductance current, or $\frac{1}{2}Cx^2(0)$ if x is a capacitance voltage. Hence, $x(0^-) = x(0) = x(0^+)$ in all cases.) Consequently,

$$\gamma = x(0^-) - x_{ss}(0^+) \tag{10-55}$$

The initial condition $x(0^-)$ can be evaluated from knowledge of the amount of energy stored in the network at $t = 0^-$.

In summary, the time-constant concept can be used to obtain the solution to any first-order network by following the following four steps:

1. Write the response in the form

$$x(t) = \gamma e^{-t/\tau} + x_{ss}(t)$$

2. Compute $x_{ss}(t)$ by any convenient method.
3. Evaluate the time constant τ.
4. Evaluate $\gamma = x(0^-) - x_{ss}(0^+)$ by determining the amount of energy stored at $t = 0$.

Example 10-6

In this example we use the time-constant concept to compute the voltage across the capacitance of Figure 10-8 for all $t \geq 0$.

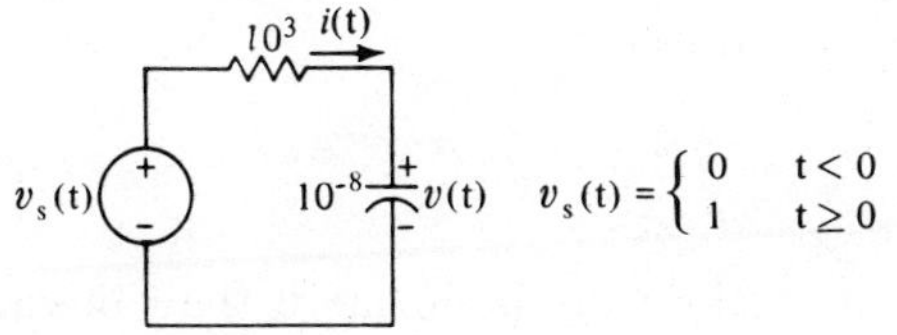

Figure 10-8 Network for Example 10-6.

Since the forcing function is a constant, the steady-state response $x_{ss}(t)$ is a constant. By inspection of the network, we note that the capacitance voltage will eventually reach 1 volt; hence, $v_{ss}(t) = 1$. (Another way to evaluate v_{ss} is to note that the capacitance has infinite impedance for direct-current (zero-frequency) sources. With the capacitance an open circuit, no current flows, and $v_{ss}(t) = v_s(t)$.)

Next, we note that the network time constant is $\tau = RC = 10^{-5}$ seconds. Finally, we note that because the voltage across the capacitance cannot change instantaneously, $v(0^+) = v(0^-) = 0$. Therefore, $\gamma = v(0^-) - v_{ss}(0^+) = 0 - 1 = -1$. In short, the capacitance voltage starts at zero volts and rises exponentially with time constant 10^{-5} seconds to its final value of one volt, as shown in Figure 10-9(a).

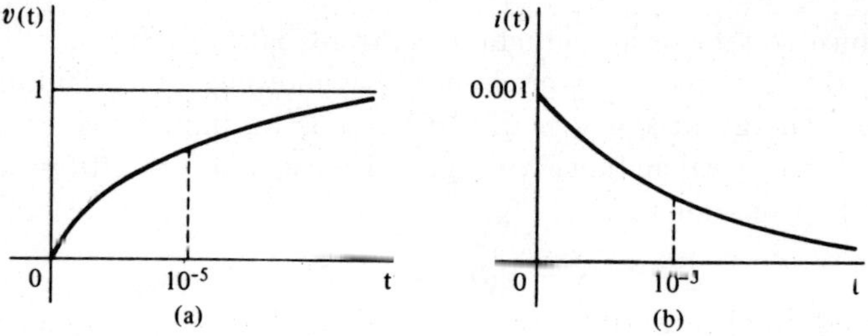

Figure 10-9 (a) Graphs of the capacitance voltage determined in Example 10-6. (b) The capacitance current determined in Example 10-7.

Example 10-7

In this example we compute the current $i(t)$ for the network of Figure 10-8 for $t \geq 0$.

Because the capacitance voltage approaches one volt for large t, the resistance voltage approaches zero volts, and the resistance current approaches zero. Hence, the steady-state current is zero amperes; that is, $i_{ss}(t) = 0$. At $t = 0^-$, the current is zero, since both the input voltage and the capacitance voltage are zero; hence, $i(0^-) = 0$. However, the capacitance voltage cannot change instantaneously, so that at $t = 0^+$ the input voltage is one volt, while the capacitance voltage remains zero volts. Therefore, the resistance voltage at $t = 0^+$ is one volt, so that $i(0^+) = 1/R = 10^{-3}$. We conclude that when the input is applied, the current jumps to 1 MA, and then decays exponentially to zero, as illustrated in Figure 10-9(b).

Example 10-8

In some cases, networks containing more than one energy-storage element are first-order systems. This is true, for example, when all the energy-storage elements can be combined into one equivalent energy-storage element. In this example, we compute a number of responses associated with the network of Figure 10-10.

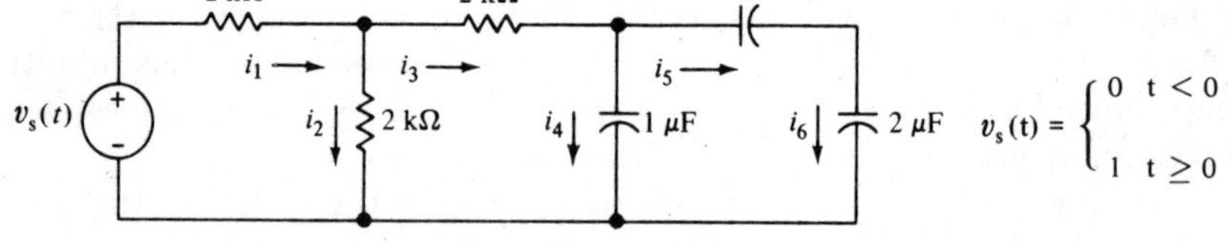

Figure 10-10 Network for Example 10-8.

We first note that the two 2-μF capacitances in series are equivalent to a single 1-μF capacitance. But this 1-μF capacitance, in parallel with the 1-μF capacitance in the network, is equivalent to a 2-μF capacitance.

Hence, the three capacitances in the network of Figure 10-10 can be replaced by their (Thevenin) equivalent capacitance $C_{eq} = 2\ \mu F$, as shown in Figure 10-11(a). Next, we break the network of Figure 10-11(a) at the capacitance terminals and replace the network to the left with its Thevenin's equivalent, as shown in Figure 10-11(b). Note

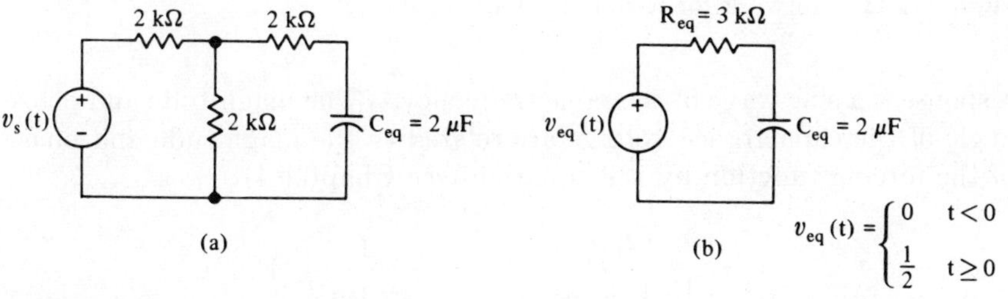

Figure 10-11 Equivalent networks for Example 10-8.

that $v_{eq}(t) = v_s(t)/2$. We conclude that the network of Figure 10-10 is a first-order network with time constant $\tau = R_{eq}C_{eq} = 3 \cdot 10^3 \cdot 2 \cdot 10^{-6} = 6 \cdot 10^{-3} = 6$ msec. Hence, the combined energy storage and transient responses for all the network voltages and currents are exponentials with time constants $\tau = 6 \cdot 10^{-3}$ seconds. The steady-state values of the network voltages and currents are easily computed after we note that, in the steady state, the capacitances offer infinite impedance to constant (zero-frequency) forcing functions. The responses at $t = 0^+$ are easily computed from the responses at $t = 0^-$ by using the fact that the capacitance voltages cannot change instantaneously. All the network voltages and currents grow or decay exponentially with time constant $\tau = 6 \cdot 10^{-3}$ from their values at $t = 0^+$ to their final values at $t = \infty$, as shown in Figure 10-12.

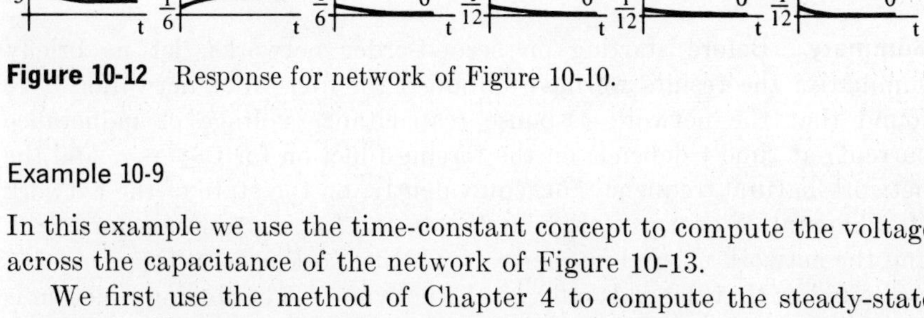

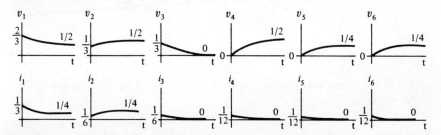

Figure 10-12 Response for network of Figure 10-10.

Example 10-9

In this example we use the time-constant concept to compute the voltage across the capacitance of the network of Figure 10-13.

We first use the method of Chapter 4 to compute the steady-state response. Since the forcing function is a sine wave, the steady-state

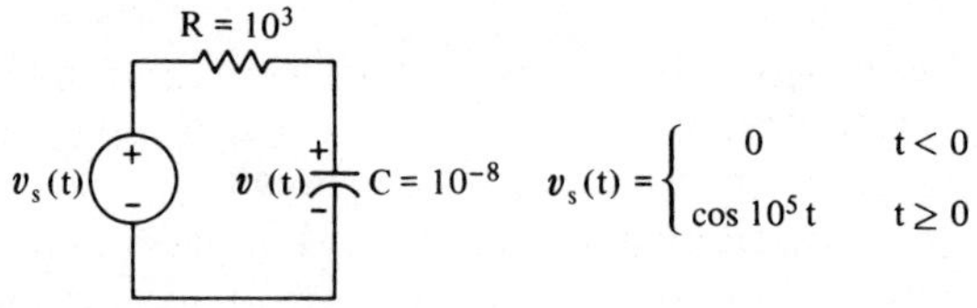

Figure 10-13 Network for Example 10-8.

response is a sine wave of the same frequency. The magnitude and phase angle of the capacitance voltage are related to the magnitude and phase of the forcing function by the equation (see Chapter 4)

$$\begin{aligned} V &= \frac{1/j\omega C}{R + 1/j\omega C} V_s = \frac{1}{1 + j\omega RC} V_s \\ &= \frac{1}{1 + j1} \cdot 1\underline{/0} = 0.707\underline{/-\pi/4} \end{aligned} \tag{10-56}$$

Therefore, the steady-state capacitance voltage is given by

$$v_{ss}(t) = 0.707 \cos (10^5 t - \pi/4) \tag{10-57}$$

The network time constant is given by $\tau = RC = 10^{-5}$. Substituting $v_{ss}(t)$ and τ into Equation (10-52) gives

$$v(t) = \gamma e^{-t/10^{-5}} + 0.707 \cos (10^5 t - \pi/4) \tag{10-58}$$

To find γ we note that at $v(0^-) = 0$ and $v_{ss}(0^+) = 0.707 \cos (-\pi/4)$, so that Equation (10-55) gives

$$\gamma = -0.707 \cos (-\pi/4) = -0.500 \tag{10-59}$$

Therefore

$$v(t) = -0.500 e^{-t/10^{-5}} + 0.707 \cos (10^5 t - \pi/4) \tag{10-60}$$

which is the same result obtained in Example 10-5 by solving the differential equation.

Summary. Before starting on second-order networks, let us briefly summarize the results we have obtained for first-order networks. We found that the network response (capacitance voltage or inductance current) at time t depends on the forcing function for $t \geq -\infty$ and the network natural frequency; or, equivalently, on the state of the network (the amount of energy stored) at $t = 0$, the forcing function for $t \geq 0$, and the network natural frequency. The network natural frequency is a real number that depends only on the network. The forcing function is some constant multiplied by the input time function (the constant depends on the network).

When the forcing function is constant or periodic for $t \geq 0$, the network response can be separated into a transient term plus a steady-state term. The transient response for any linear first-order network is of the form γe^{at}, where a is the network natural frequency and the constant γ depends on both the forcing function and the network. If the natural frequency is negative ($a < 0$), then the transient decays exponentially with time constant $\tau = -a^{-1}$; if the network natural frequency is positive ($a > 0$), then the transient response grows exponentially. Networks containing only +RLC always have negative natural frequencies. With controlled sources positive, negative or zero natural frequencies can be achieved. Clearly, most first-order electronic networks designed to do signal processing have negative natural frequencies.

The steady-state response depends on both the forcing function and the network natural frequency. If the forcing function is a constant (dc level), then the steady-state response is a constant; if the forcing function is periodic with period T, then the steady-state response is periodic with period T; if the forcing function is a sinusoid, then the steady-state response is a sinusoid of the same frequency; if the forcing function is a ramp [$u(t) = t$], then the response is a constant plus a ramp. If the input is a single square pulse, then the output may or may not resemble a square pulse, depending on relative magnitudes of the pulse duration and the network natural frequency.

In a first-order network, once the basic differential equation is solved, all network currents and voltages are related to the solution of the differential equation by linear algebraic equations. Thus the functional form of the solution of the differential equation gives the functional form of all network voltages and currents. In general, each voltage and each current can have a transient and a steady-state part. All transients have the same time constant. All steady-state terms have the same form.

10-4 Second-Order Systems

Linear networks containing two independent energy storage elements can be described by a set of two first-order differential equations in two unknowns by applying the method of Section 10-2. This set of equations is of the form

$$\begin{aligned} \dot{x}_1 &= a_{11}x_1 + a_{12}x_2 + u_1 \\ \dot{x}_2 &= a_{21}x_1 + a_{22}x_2 + u_2 \end{aligned} \tag{10-61}$$

where $x_1 = x_1(t)$ and $x_2 = x_2(t)$ are the two unknowns (capacitance voltages and/or inductance currents) for which we must solve; the a_{ij}

(i, j = 1, 2) are constants that depend only on the network parameters; and $u_1 = u_1(t)$ and $u_2 = u_2(t)$ are forcing functions. The procedure for solving this set of two equations for the two unknowns is identical to the procedure used in Section 10-3 to solve the first-order system. We first solve the set of homogeneous equations by assuming a functional form for the set of solutions and substituting it into the set of homogeneous equations. Then we obtain the solution to the set of complete equations by repeating the same procedure with the set of complete equations.

Solution to the homogeneous equations. The set of homogeneous equations associated with the set of complete equations given by Equation (10-61) is obtained by setting both forcing functions equal to zero; that is,

$$\begin{aligned} \dot{y}_1 &= a_{11}y_1 + a_{12}y_2 \\ \dot{y}_2 &= a_{21}y_1 + a_{22}y_2 \end{aligned} \qquad \textbf{(10-62)}$$

We start by assuming a set of solutions of the form

$$\begin{aligned} y_1 &= c_1e^{pt} \\ y_2 &= c_2e^{qt} \end{aligned} \qquad \textbf{(10-63)}$$

where p, q, c_1, c_2 are constants to be determined. The values of these constants, if any, for which Equations (10-63) satisfy Equations (10-62), can be obtained by substituting Equations (10-63) into Equations (10-62). Proceeding toward this end, we note from Equations (10-63) that

$$\begin{aligned} \dot{y}_1 &= pc_1e^{pt} \\ \dot{y}_2 &= qc_2e^{qt} \end{aligned} \qquad \textbf{(10-64)}$$

Substituting Equations (10-63) and Equations (10-64) into Equations (10-62), we obtain

$$\begin{aligned} pc_1e^{pt} &= a_{11}c_1e^{pt} + a_{12}c_2e^{qt} \\ qc_2e^{qt} &= a_{21}c_1e^{pt} + a_{22}c_2e^{qt} \end{aligned} \qquad \textbf{(10-65)}$$

Moving the e^{pt} terms to one side and the e^{qt} terms to the other side results in

$$\begin{aligned} (p - a_{11})c_1e^{pt} &= a_{12}c_2e^{qt} \\ (q - a_{22})c_2e^{qt} &= a_{21}c_1e^{pt} \end{aligned} \qquad \textbf{(10-66)}$$

Two exponential functions are equal for all t if and only if they have the same amplitudes and the same time constant; that is, $\alpha_1e^{\beta_1 t} = \alpha_2e^{\beta_2 t}$ for all t if and only if $\alpha_1 = \alpha_2$, and $\beta_1 = \beta_2$. Therefore, Equations (10-66) require that

$$p = q \tag{10-67}$$

and also that

$$\begin{aligned} (p - a_{11})c_1 &= a_{12}c_2 \text{ or } c_2 = \frac{p - a_{11}}{a_{12}} c_1 \\ (p - a_{22})c_2 &= a_{21}c_1 \text{ or } c_2 = \frac{a_{21}}{p - a_{22}} c_1 \end{aligned} \tag{10-68}$$

Equation (10-67) states that y_1 and y_2 must have identical exponents, and the first equation of Equations (10-68) states that the two constants c_1 and c_2 are related in a special way. Using these results in Equations (10-63), we have

$$\begin{aligned} y_1 &= c_1e^{pt} \\ y_2 &= \frac{p - a_{11}}{a_{12}} c_1e^{pt} \end{aligned} \tag{10-69}$$

We appear to be making some progress. We are attempting to find the constants p, q, c_1, and c_2, for which Equations (10-63) satisfy Equations (10-62). We have determined that $q = p$ and $c_2 = [(p - a_{11})/a_{12}]c_1$, so that we have reduced the number of constants to be determined from four to two, p and c_1. We next seek the value(s) of p for which Equations (10-63) are solutions to Equations (10-62). Equations (10-68) state that

$$\frac{p - a_{11}}{a_{12}} = \frac{a_{21}}{p - a_{22}} \tag{10-70}$$

which can also be written in the form

$$p^2 - (a_{11} + a_{22})p + a_{11}a_{22} - a_{12}a_{21} = 0 \tag{10-71}$$

which implies

$$p = \frac{a_{11} + a_{22}}{2} \pm \sqrt{\left(\frac{a_{11} + a_{22}}{2}\right)^2 - a_{11}a_{22} + a_{12}a_{21}} \tag{10-72}$$

Hence, there are two values of p for which the set of equations given by Equation (10-69) satisfies the set of homogeneous equations. We label these two values p_1 and p_2; that is,

$$\begin{Bmatrix} p_1 \\ p_2 \end{Bmatrix} = \frac{a_{11} + a_{22}}{2} \pm \sqrt{\left(\frac{a_{11} + a_{22}}{2}\right)^2 - a_{11}a_{22} + a_{12}a_{21}} \tag{10-73}$$

and note that they depend only on the known network parameters $a_{ij}(i, j = 1, 2)$.

We have now determined that

$$\begin{aligned} y_1 &= c_1 e^{p_1 t} \\ y_2 &= \frac{p_1 - a_{11}}{a_{12}} c_1 e^{p_1 t} \end{aligned} \tag{10-74}$$

satisfy Equations (10-62) for any constant c_1. We have also determined that

$$\begin{aligned} y_1 &= c_1 e^{p_2 t} \\ y_2 &= \frac{p_2 - a_{11}}{a_{12}} c_1 e^{p_2 t} \end{aligned} \tag{10-75}$$

satisfy Equations (10-62) for any c_1. Therefore, all linear combinations of these sets of equations are also solutions to the set given by Equations (10-62); that is,

$$\begin{aligned} y_1 &= B_1 e^{p_1 t} + B_2 e^{p_2 t} \\ y_2 &= \frac{p_1 - a_{11}}{a_{12}} B_1 e^{p_1 t} + \frac{p_2 - a_{11}}{a_{12}} B_2 e^{p_2 t} \end{aligned} \tag{10-76}$$

satisfy Equations (10-62) for any B_1 and any B_2.

For convenience, we now define

$$k_1 = \frac{p_1 - a_{11}}{a_{12}}, \qquad k_2 = \frac{p_2 - a_{11}}{a_{12}} \tag{10-77}$$

Since p_1 and p_2 are known constants that depend only on the a_{ij}'s [see Equations (10-73)], k_1 and k_2 are also known constants. Hence, Equations (10-76) can be written

$$\begin{aligned} y_1 &= B_1 e^{p_1 t} + B_2 e^{p_2 t} \\ y_2 &= k_1 B_1 e^{p_1 t} + k_2 B_2 e^{p_2 t} \end{aligned} \tag{10-78}$$

This is as far as we can go with the set of homogeneous equations. In summary, we determined that the set of equations given by Equations (10-78) is a solution to the set of homogeneous equations given by Equations (10-62). p_1, p_2, k_1, and k_2 are known constants that depend only on the network parameters a_{ij} $(i, j = 1, 2)$. B_1 and B_2 are arbitrary numbers. [When we substitute Equations (10-78) into Equations (10-62), B_1 and B_2 cancel.]

Solution to the complete equation. Our method for solving the set of complete equations is the same as that used to solve the set of homogeneous equations: We assume functional forms for the set of solutions and then we substitute the set of assumed solutions into the set of complete equations and solve for the unknown parameters.

The form we assume for a set of solutions to the set of complete equations is the same as the form we assumed for a set of solutions for the set of homogeneous equations, except that we now let the two constants B_1 and B_2 be time functions. That is, we set

$$\begin{aligned} x_1 &= b_1 e^{p_1 t} + b_2 e^{p_2 t} \\ x_2 &= k_1 b_1 e^{p_1 t} + k_2 b_2 e^{p_2 t} \end{aligned} \tag{10-79}$$

where p_1 and p_2 are known constants given by Equations (10-73), k_1 and k_2 are known constants given by Equations (10-77), and $b_1 = b_1(t)$ and $b_2 = b_2(t)$ are time functions to be determined. If Equations (10-79) satisfy Equations (10-61) for some b_1 and b_2, the only remaining problem is to find the appropriate b_1 and b_2. This can be accomplished by substituting Equations (10-79) into Equations (10-61), which gives two equations in the two unknowns b_1 and b_2. These two equations can then be solved for the required b_1 and b_2.

Since b_1 and b_2 are time functions, the time derivatives of Equations (10-79) are given by

$$\begin{aligned} \dot{x}_1 &= \dot{b}_1 e^{p_1 t} + p_1 b_1 e^{p_1 t} + \dot{b}_2 e^{p_2 t} + p_2 b_2 e^{p_2 t} \\ \dot{x}_2 &= k_1 \dot{b}_1 e^{p_1 t} + k_1 p_1 b_1 e^{p_1 t} + k_2 \dot{b}_2 e^{p_2 t} + k_2 p_2 b_2 e^{p_2 t} \end{aligned} \tag{10-80}$$

Substituting Equations (10-79) and (10-80) into Equations (10-61), we obtain

$$\begin{aligned} &\dot{b}_1 e^{p_1 t} + \dot{b}_2 e^{p_2 t} + [p_1 b_1 e^{p_1 t}] + \{p_2 b_2 e^{p_2 t}\} \\ &\quad = [a_{11} b_1 e^{p_1 t}] + \{a_{11} b_2 e^{p_2 t}\} + [a_{12} k_1 b_1 e^{p_1 t}] + \{a_{12} k_2 b_2 e^{p_2 t}\} + u_1 \\ &k_1 \dot{b}_1 e^{p_1 t} + k_2 \dot{b}_2 e^{p_2 t} + [p_1 k_1 b_1 e^{p_1 t}] + \{p_2 k_2 b_2 e^{p_2 t}\} \\ &\quad = [a_{21} b_1 e^{p_1 t}] + \{a_{21} b_2 e^{p_2 t}\} + [a_{22} k_1 b_1 e^{p_1 t}] + \{a_{22} k_2 b_2 e^{p_2 t}\} + u_2 \end{aligned} \tag{10-81}$$

Equations (10-81) have a very large number of terms, but many of them cancel because of the chosen functional forms. The terms in brackets in both equations contain the factor $b_1 e^{p_1 t}$. The terms in the braces contain the factor $b_2 e^{p_2 t}$. Grouping terms with like factors shows that in the first equation, $b_1 e^{p_1 t}$ multiplies $[p_1 - a_{11} - a_{12} k_1]$. The definition of k_1 in Equation (10-77) shows that this term is zero. Regrouping the other bracketed and braced terms in Equation (10-81) gives zero for the other three groups as well. That is,

$$\begin{aligned} p_1 - a_{11} - a_{12} k_1 &= 0 \\ p_2 - a_{11} - a_{12} k_2 &= 0 \\ p_1 - a_{21} - a_{22} k_1 &= 0 \\ p_2 - a_{21} - a_{22} k_2 &= 0 \end{aligned} \tag{10-82}$$

Hence, as a result of Equations (10-62), Equations (10-81) reduce to

$$\begin{aligned} e^{p_1 t} \dot{b}_1 + e^{p_2 t} \dot{b}_2 &= u_1 \\ k_1 e^{p_1 t} \dot{b}_1 + k_2 e^{p_2 t} \dot{b}_2 &= u_2 \end{aligned} \tag{10-83}$$

The final step is to solve Equations (10-83) for $\dot{b}_1$ and $\dot{b}_2$. Equations

(10-83) are easily solved for $\dot{b}_1$ and $\dot{b}_2$ by Cramer's rule:

$$\dot{b}_1 = \frac{\begin{vmatrix} u_1 & e^{p_2t} \\ u_2 & k_2e^{p_2t} \end{vmatrix}}{\begin{vmatrix} e^{p_1t} & e^{p_2t} \\ k_1e^{p_1t} & k_2e^{p_2t} \end{vmatrix}} = \frac{k_2u_1e^{p_2t} - u_2e^{p_2t}}{k_2e^{p_1t}e^{p_2t} - k_1e^{p_1t}e^{p_2t}}$$

$$= \frac{(k_2u_1 - u_2)e^{p_2t}}{(k_2 - k_1)e^{p_1t}e^{p_2t}} = \frac{k_2u_1 - u_2}{k_2 - k_1} e^{-p_1t} \tag{10-84}$$

$$\dot{b}_2 = \frac{\begin{vmatrix} e^{p_1t} & u_1 \\ k_1e^{p_1t} & u_2 \end{vmatrix}}{\begin{vmatrix} e^{p_1t} & e^{p_2t} \\ k_1e^{p_1t} & k_2e^{p_2t} \end{vmatrix}} = \frac{u_2e^{p_1t} - k_1u_1e^{p_1t}}{k_2e^{p_1t}e^{p_2t} - k_1e^{p_1t}e^{p_2t}}$$

$$= \frac{(u_2 - k_1u_1)e^{p_1t}}{(k_2 - k_1)e^{p_1t}e^{p_2t}} = \frac{u_2 - k_1u_1}{k_2 - k_1} e^{-p_2t}$$

Therefore, b_1 and b_2 are given by

$$b_1 = b_1(t) = \int_{-\infty}^{t} \dot{b}_1(\lambda)\, d\lambda = \frac{1}{k_2 - k_1} \int_{-\infty}^{t} [k_2u_1(\lambda) - u_2(\lambda)]e^{-p_1\lambda}\, d\lambda$$

$$\tag{10-85}$$

$$b_2 = b_2(t) = \int_{-\infty}^{t} \dot{b}_2(\lambda)\, d\lambda = \frac{1}{k_2 - k_1} \int_{-\infty}^{t} [u_2(\lambda) - k_1u_1(\lambda)]e^{-p_2\lambda}\, d\lambda$$

And, finally, inserting these values for b_1 and b_2 into Equations (10-79) gives the set of solutions to the set of complete equations.

$$x_1(t) = \frac{e^{p_1t}}{k_2 - k_1} \int_{-\infty}^{t} [k_2u_1(\lambda) - u_2(\lambda)]\, e^{-p_1\lambda}\, d\lambda + \frac{e^{p_2t}}{k_2 - k_1} \int_{-\infty}^{t} [u_2(\lambda) - k_1u_1(\lambda)]e^{-p_2\lambda}\, d\lambda$$

$$\tag{10-86}$$

$$x_2(t) = \frac{k_1e^{p_1t}}{k_2 - k_1} \int_{-\infty}^{t} [k_2u_1(\lambda) - u_2(\lambda)]\, e^{-p_1\lambda}\, d\lambda + \frac{k_2e^{p_2t}}{k_2 - k_1} \int_{-\infty}^{t} [u_2(\lambda) - k_1u_1(\lambda)]e^{-p_2\lambda}\, d\lambda$$

Both x_1 and x_2 depend on the constants p_1, p_2, k_1, and k_2, which are related to the network parameters by Equations (10-73) and (10-77). Both responses x_1 and x_2 also depend on the total past history of both forcing functions; that is, both depend on $u_1(\lambda)$ and $u_2(\lambda)$ for all $-\infty < \lambda \le t$.

In many applications the total past history of the inputs may not be known. If the values of the two variables are known at a particular time t_0, then these two values $x_1(t_0)$ and $x_2(t_0)$ constitute the state of the system at t_0. The forcing functions $u_1(\lambda)$ and $u_2(\lambda)$ for $t_0 \leq \lambda \leq t$ and the state of the system at $t = t_0$ suffice to calculate the values of the variables at any time t. For convenience, we usually set $t_0 = 0$. The solution method above led to Equations (10-84) for $\dot{b}_1(t)$ and $\dot{b}_2(t)$. Since $b_i(t) = x_i(t)e^{-p_i t}$, $b_1(0) = x_1(0)$ and $b_2(0) = x_2(0)$. Thus the constants of integration for determining $b_1(t)$ from $\dot{b}_1(t)$ and $b_2(t)$ from $\dot{b}_2(t)$ are set. For known $x_1(0)$ and $x_2(0)$, the equations for $x_1(t)$ and $x_2(t)$ become

$$\begin{aligned} x_1(t) = {} & \left[\frac{k_2}{k_2 - k_1} x_1(0) + \frac{-1}{k_2 - k_1} x_2(0)\right] e^{p_1 t} \\ & + \left[\frac{-k_1}{k_2 - k_1} x_1(0) + \frac{1}{k_2 - k_1} x_2(0)\right] e^{p_2 t} \\ & + \frac{e^{p_1 t}}{k_2 - k_1} \int_0^t [k_2 u_1(\lambda) - u_2(\lambda)] e^{-p_1 \lambda}\, d\lambda \\ & + \frac{e^{p_2 t}}{k_2 - k_1} \int_0^t [u_2(\lambda) - k_1 u_1(\lambda)] e^{-p_2 \lambda}\, d\lambda \end{aligned} \tag{10-87}$$

$$\begin{aligned} x_2(t) = {} & \left[\frac{k_1 k_2}{k_2 - k_1} x_1(0) + \frac{-k_1}{k_2 - k_1} x_2(0)\right] e^{p_1 t} \\ & + \left[\frac{-k_1 k_2}{k_2 - k_1} x_1(0) + \frac{k_2}{k_2 - k_1} x_2(0)\right] e^{p_2 t} \\ & + \frac{k_1 e^{p_1 t}}{k_2 - k_1} \int_0^t [k_2 u_1(\lambda) - u_2(\lambda)] e^{-p_1 \lambda}\, d\lambda \\ & + \frac{k_2 e^{p_2 t}}{k_2 - k_1} \int_0^t [u_2(\lambda) - k_1 u_1(\lambda)] e^{-p_2 \lambda}\, d\lambda \end{aligned}$$

These formulas give the responses at time t, $x_1(t)$, and $x_2(t)$, in terms of the state of the system at $t = 0$, $x_1(0)$, and $x_2(0)$ (or initial conditions), and the forcing functions $u_1(\lambda)$ and $u_2(\lambda)$ for $0 \leq \lambda \leq t$. By using the same arguments we used in Section 10-3 following Equation (10-32) we can interpret the terms that depend on $x_1(0)$ and $x_2(0)$ as energy response terms, and the terms that depend on the forcing functions as the forced response. The forced response, in turn, consists of a transient term and a steady-state term. Finally, by moving the $e^{p_i t}$ terms under the integrals we can interpret each integral as a convolution integral.

Example 10-10

In this example we compute the voltage $v(t)$ for the network of Figure 10-14(a).

The first step is to write the normal-form equations for the network;

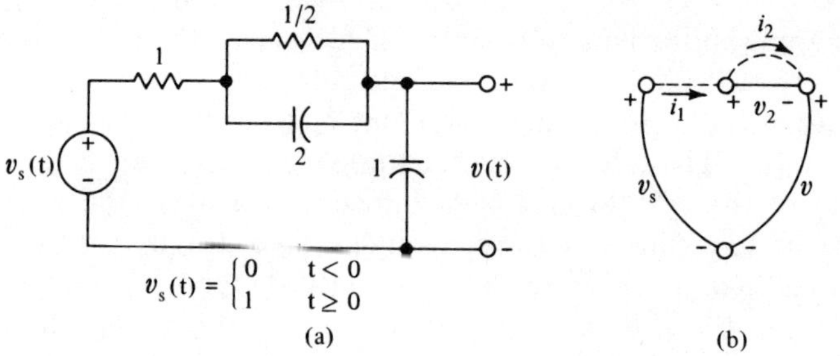

Figure 10-14 Network for Example 10-10.

then solving for $v(t)$ is simply a matter of applying Equations (10-86) or (10-87). Following the normal-form algorithm of Section 10-2 we first construct a normal tree and assign a symbol to each branch, as shown in Figure 10-14(b). For the 1-farad capacitance, we write

$$1\dot{v} = i_1 \tag{10-88}$$

For the 2-farad capacitance, we write

$$2\dot{v}_2 = i_1 - i_2 \tag{10-89}$$

For the two resistances, we use KVL to write

$$\begin{aligned} 1i_1 &= v_s - v_2 - v \\ \frac{i_2}{2} &= v_2 \end{aligned} \tag{10-90}$$

We next solve Equations (10-90) for i_1 and i_2 and substitute these quantities into Equations (10-88) and (10-89) and, finally, arrange the resulting equations into the normal form

$$\begin{aligned} \dot{v} &= v - v_2 + v_s \\ \dot{v}_2 &= -\tfrac{1}{2}v - \tfrac{3}{2}v_2 + \tfrac{1}{2}v_s \end{aligned} \tag{10-91}$$

These equations are identical to Equations (10-61) if we set

$$\begin{aligned} &x_1 = v \quad a_{11} = -1 \quad a_{12} = -1 \quad u_1 = v_s = \begin{cases} 0 & t < 0 \\ 1 & t \geq 0 \end{cases} \\ &x_2 = v_2 \quad a_{21} = -\tfrac{1}{2} \quad a_{22} = -\tfrac{3}{2} \quad u_2 = \frac{v_s}{2} = \begin{cases} 0 & t < 0 \\ \tfrac{1}{2} & t \geq 0 \end{cases} \end{aligned} \tag{10-92}$$

Inserting these quantities into Equations (10-73) and (10-77), we find that the natural frequencies for the network are given by

$$\begin{Bmatrix} p_1 \\ p_2 \end{Bmatrix} = -\tfrac{5}{4} \pm \tfrac{3}{4} = \begin{Bmatrix} -\tfrac{1}{2} \\ -2 \end{Bmatrix} \tag{10-93}$$

and that the constants k_1, k_2, and $k_2 - k_1$ are given by

$$\begin{aligned} k_1 &= -\tfrac{1}{2} \\ k_2 &= 1 \end{aligned} \qquad k_2 - k_1 = \tfrac{3}{2} \tag{10-94}$$

Finally, substituting these values for p_1, p_2, k_1, k_2, u_1 and u_2 into the first equation of Equations (10-86), we obtain

$$v(t) = \begin{cases} 0 & t < 0 \\ \frac{2}{3}e^{-t/2}\int_0^t (1 - \frac{1}{2})e^{\lambda/2}\, d\lambda + \frac{2}{3}e^{-2t}\int_0^t (\frac{1}{2} + \frac{1}{2})e^{2\lambda}\, d\lambda & t \geq 0 \end{cases} \tag{10-95}$$

Simplifying the equation for $t \geq 0$, we find

$$\begin{aligned} v(t) &= \tfrac{1}{3}e^{-t/2}\int_0^t e^{\lambda/2}\, d\lambda + \tfrac{2}{3}e^{-2t}\int_0^t e^{2\lambda}\, d\lambda \\ &= \tfrac{1}{3}e^{-t/2}\left[\frac{e^{\lambda/2}}{\frac{1}{2}}\right]_{\lambda=0}^{\lambda=t} + \tfrac{2}{3}e^{-2t}\left[\frac{e^{2\lambda}}{2}\right]_{\lambda=0}^{\lambda=t} \\ &= \tfrac{1}{3}e^{-t/2}[2e^{t/2} - 2] + \tfrac{2}{3}e^{-2t}[\tfrac{1}{2}e^{2t} - \tfrac{1}{2}] \qquad t \geq 0 \end{aligned} \tag{10-96}$$

so that we have, finally,

$$v(t) = \begin{cases} 0 & t < 0 \\ 1 - \frac{2}{3}e^{-t/2} - \frac{1}{3}e^{-2t} & t \geq 0 \end{cases} \tag{10-97}$$

The voltage $v(t) = 0$ for $t < 0$ because $v_s(t) = 0$ for $t < 0$. For $t \geq 0$, the response consists of two terms, the steady-state response, 1, and the transient response $-\frac{2}{3}e^{-t/2} - \frac{1}{3}e^{-2t}$.

The steady-state response could have been obtained by inspection of the network: In the steady state, both capacitances are open circuits, so that all currents are zero, which implies that the voltages across both resistances are zero. Hence, the output voltage $v(t)$ is the same as the source voltage.

The transient response is a sum of two exponentials, with the constants in the exponents the natural frequencies. These terms are sketched in Figure 10-15. Note that the duration of the transient depends essentially on the magnitude of the smaller natural frequency. The transient response can also be written in the form

$$-\tfrac{2}{3}e^{-t/2} - \tfrac{1}{3}e^{-t/(1/2)} = -\tfrac{2}{3}e^{-t/\tau_1} - \tfrac{1}{3}e^{-t/\tau_2} \tag{10-98}$$

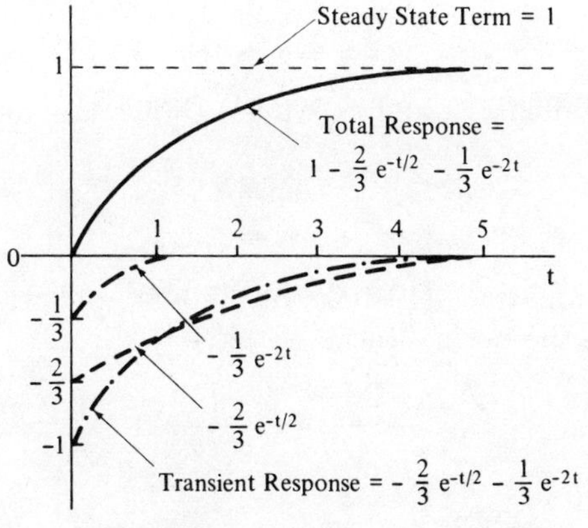

Figure 10-15 Total response for Example 10-10.

where $\tau_1 = 2 = -1/p_1$ and $\tau_2 = \frac{1}{2} = -1/p_2$ are the two network time constants. Hence, the transient duration depends essentially on the longest time constant.

Example 10-11

In this example we compute the response $v(t)$ of the second-order Butterworth insertion-power-gain prototype to a step input, as shown in Figure 10-16(a). We use the same procedure followed in Example 10-10.

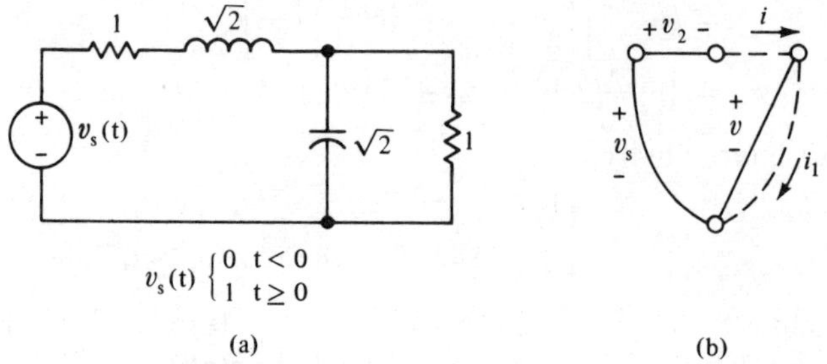

Figure 10-16 Network for Example 10-11.

Before starting, we note that the steady-state solution can be obtained from Figure 10-16(a) by letting the impedances of the inductance and capacitance be zero and infinite, respectively. Hence, the voltage source sees only the two resistances in series, and the steady-state output voltage is $\frac{1}{2}$ volt. We can use this as a check on our later work.

We start by constructing a normal tree, as shown in Figure 10-16(b). Rather than specifically follow the algorithm of Section 10-2 for writing normal-form equations, we shall write a set of normal-form equations by inspection of Figure 10-16(b). Since we want to end up with two equations in the inductance current and the capacitance voltage (i and v), we write

$$\begin{aligned} \sqrt{2}\,\dot{i} &= -v_2 + v_s - v \\ \sqrt{2}\,\dot{v} &= i - i_1 \end{aligned} \tag{10-99}$$

In order to eliminate i_1 and v_2, we use Ohm's law for each resistance to write

$$\begin{aligned} v_2 &= i \\ i_1 &= v \end{aligned} \tag{10-100}$$

Substituting Equations (10-100) into (10-99) and arranging the resulting equations into the normal form, we have

$$\begin{aligned} \dot{i} &= -\frac{1}{\sqrt{2}}i - \frac{1}{\sqrt{2}}v + \frac{1}{\sqrt{2}}v_s \\ \dot{v} &= \frac{1}{\sqrt{2}}i - \frac{1}{\sqrt{2}}v \end{aligned} \tag{10-101}$$

The natural frequencies for the network are given by

$$\begin{Bmatrix} p_1 \\ p_2 \end{Bmatrix} = -\frac{1}{\sqrt{2}} \pm \sqrt{-\frac{1}{2}} = -\frac{1}{\sqrt{2}} \pm j\frac{1}{\sqrt{2}} \tag{10-102}$$

and the constants k_1 and k_2 are given by

$$k_1 = -j \qquad k_2 = +j \tag{10-103}$$

It follows that

$$\frac{k_1}{k_2 - k_1} = -\frac{1}{2} \qquad \frac{k_2}{k_2 - k_1} = +\frac{1}{2} \tag{10-104}$$

From Equations (10-101) we also note that

$$u_1(\lambda) = \frac{v_s(\lambda)}{\sqrt{2}} = \begin{cases} 0 & \lambda < 0 \\ \dfrac{1}{\sqrt{2}} & \lambda \geq 0 \end{cases} \qquad u_2(\lambda) = 0 \tag{10-105}$$

To compute the output voltage $v(t)$, we substitute Equations (10-102) to (10-105) into the second of Equations (10-86):

$$v(t) = \begin{cases} 0 & t < 0 \\ -\frac{1}{2}e^{p_1 t}\displaystyle\int_0^t \frac{j}{\sqrt{2}} e^{-p_1\lambda}\, d\lambda + \frac{1}{2}e^{p_2 t}\int_0^t \frac{j}{\sqrt{2}} e^{-p_2\lambda}\, d\lambda & t \geq 0 \end{cases} \tag{10-106}$$

The integrals are easily evaluated:

$$\int_0^t \frac{j}{\sqrt{2}} e^{-p_\ell \lambda}\, d\lambda = \frac{-j}{\sqrt{2}\, p_\ell} [e^{-p_\ell t} - 1] \qquad \ell = 1, 2 \tag{10-107}$$

Now substituting Equations (10-107) into Equation (10-106), we obtain for $t \geq 0$

$$v(t) = \frac{j}{2\sqrt{2}\, p_1} - \frac{j}{2\sqrt{2}\, p_1} e^{p_1 t} - \frac{j}{2\sqrt{2}\, p_2} + \frac{j}{2\sqrt{2}\, p_2} e^{p_2 t} \tag{10-108}$$

Collecting the constant and exponential terms yields

$$v(t) = \underbrace{\frac{j}{2\sqrt{2}}\left(\frac{1}{p_1} - \frac{1}{p_2}\right)}_{v_{ss}} + \underbrace{\frac{-j}{2\sqrt{2}\, p_1} e^{p_1 t} + \frac{j}{2\sqrt{2}\, p_2} e^{p_2 t}}_{v_{tr}} \tag{10-109}$$

$$= v_{ss} + v_{tr}$$

v_{ss} and v_{tr} are the steady-state and transient responses, respectively. We shall evaluate v_{tr} first and then v_{ss}.

Note that v_{tr} is of the form

$$v_{tr} = \gamma_1 e^{p_1 t} + \gamma_2 e^{p_2 t} \tag{10-110}$$

that is, a sum of two exponentials with the constants in the exponents the natural frequencies. We first evaluate the complex constants $\gamma_1 =$

$-j/(2\sqrt{2}\,p_1)$ and $\gamma_2 = j/(2\sqrt{2}\,p_2)$ by substituting Equations (10-102):

$$\gamma_1 = \frac{-j}{2\sqrt{2}\left(-\frac{1}{\sqrt{2}} + \frac{j}{\sqrt{2}}\right)} = \frac{1}{2}(-j)\left(\frac{1}{-1+j}\right)$$
$$= \frac{1}{2}(1\underline{/-\pi/2})\left(\frac{1}{\sqrt{2}\,\underline{/3\pi/4}}\right)$$
$$= \frac{1}{2\sqrt{2}}\underline{/3\pi/4} = \frac{1}{2\sqrt{2}}e^{j3\pi/4}$$
$$\gamma_2 = \frac{j}{2\sqrt{2}\left(-\frac{1}{\sqrt{2}} - \frac{j}{\sqrt{2}}\right)} = \frac{1}{2}(j)\left(\frac{1}{-1-j}\right)$$
$$= \frac{1}{2}(1\underline{/\pi/2})\left(\frac{1}{\sqrt{2}\,\underline{/-3\pi/4}}\right)$$
$$= \frac{1}{2\sqrt{2}}\underline{/-3\pi/4} = \frac{1}{2\sqrt{2}}e^{-j3\pi/4} \qquad \textbf{(10-111)}$$

We have written the complex numbers γ_1 and γ_2 in polar coordinates. Note that γ_1 and γ_2 are complex conjugates. As we show in the next step, this is necessary for the two complex quantities $\gamma_1 e^{p_1 t}$ and $\gamma_2 e^{p_2 t}$ to add to a real quantity via Euler's equation. We now substitute Equations (10-102) and Equations (10-111) into Equation (10-110) to get

$$v_{tr} = \frac{1}{2\sqrt{2}}e^{+j3\pi/4}e^{(-1/\sqrt{2}+j/\sqrt{2})t} + \frac{1}{2\sqrt{2}}e^{-j3\pi/4}e^{(-1/\sqrt{2}-j/\sqrt{2})t}$$
$$= \frac{1}{2\sqrt{2}}e^{-t/\sqrt{2}+j(t/\sqrt{2}+3\pi/4)} + \frac{1}{2\sqrt{2}}e^{-t/\sqrt{2}-j(t/\sqrt{2}+3\pi/4)}$$
$$= \frac{e^{-t/\sqrt{2}}}{\sqrt{2}}\left(\frac{e^{j(t/\sqrt{2}+3\pi/4)} + e^{-j(t/\sqrt{2}+3\pi/4)}}{2}\right)$$
$$= \frac{1}{\sqrt{2}}e^{-t/\sqrt{2}}\cos\,(t/\sqrt{2} + 3\pi/4) \qquad \textbf{(10-112)}$$

Here we have rearranged the exponential terms in order to factor the common term $e^{-t/\sqrt{2}}$ and to group the remaining term into the form $(e^{jx} - e^{-jx})/2$, which we recognize as $\cos x$.

Note that the transient response is a damped sinusoid. The constant in the exponential damping term $e^{-t/\sqrt{2}}$ is the real part of the natural frequency, and the radian frequency of oscillation of the sinusoid is the imaginary part of the natural frequency. That is, if $p_1 = \sigma + j\omega$ and $p_2 = \sigma - j\omega$, then the transient response is of the form $\nu e^{\sigma t}\cos\,(\omega t + \phi)$.

To evaluate the steady-state response, we start with Equation

(10-109) and substitute Equations (10-102) to obtain

$$v_{ss} = \frac{j}{2\sqrt{2}}\left(\frac{1}{p_1} - \frac{1}{p_2}\right) = \frac{j}{2\sqrt{2}}\left(\frac{p_2 - p_1}{p_1 p_2}\right)$$

$$= \frac{j[(-1/\sqrt{2} + j/\sqrt{2}) - (-1/\sqrt{2} - j/\sqrt{2})]}{2\sqrt{2}\,(-1/\sqrt{2} + j/\sqrt{2})(-1/\sqrt{2} - j/\sqrt{2})}$$

$$= \tfrac{1}{2} \tag{10-113}$$

Hence, the complex quantities combine, as they must, to yield the (real) steady-state voltage.

The final step is to sum the steady-state and transient responses to obtain the total response:

$$v(t) = 0.500 + 0.707e^{-t/\sqrt{2}} \cos (t/\sqrt{2} + 3\pi/4) \tag{10-114}$$

To sketch $v(t)$ vs. t, we first note that at $t = 0$, we have

$$v(0) = 0.500 + 0.707 \cos (3\pi/4) = 0.500 - 0.707 \times 0.707 = 0 \tag{10-115}$$

We also note that, due to the inductance, the capacitance current at $t = 0$ is $0 = C\dot{v}(0)$. Hence, $\dot{v}(0) = 0$.

The transient term is a damped sine wave of frequency $\omega = 1/\sqrt{2} \approx 0.707$ rad/sec. The time for one complete cycle is given by the solution to $\omega t = 0.707t = 2\pi$ or $t = 2\sqrt{2}\,\pi \approx 8.9$ seconds. On the other hand, the sine wave is damped out to 37 percent of its initial value in $t = \sqrt{2} \approx 1.414$ seconds or in $1.414/8.9 \approx 16$ percent of one cycle. Therefore, even though the transient is oscillatory, it is essentially damped out before the first cycle is completed. The result is sketched in Figure 10-17.

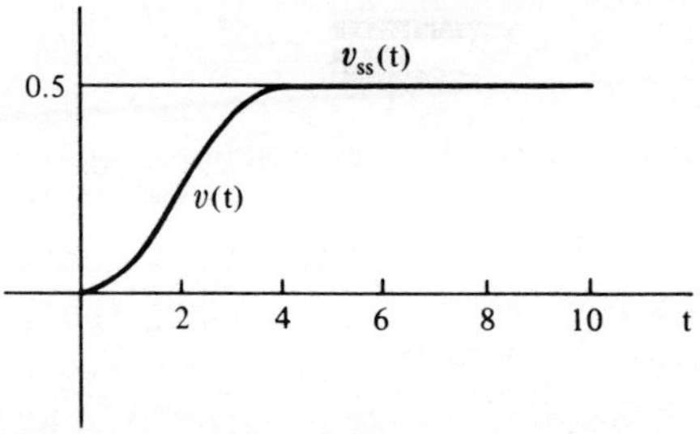

Figure 10-17 Total response for Example 10-11.

In Examples 10-10 and 10-11, we first evaluated the natural frequencies p_1 and p_2, the constants k_1 and k_2, and the forcing functions u_1 and u_2. We then substituted these quantities into Equation (10-86) and evaluated the integrals and grouped the resulting terms into the transient and steady-state responses. When the natural frequencies are real, as in Example 10-10, no particular difficulties are encountered in carrying out these computations. When the natural frequencies are complex, as in Example 10-11, the steps are the same, but complex quantities must be manipulated. To avoid becoming entangled in a jungle of complex quantities, some systematic procedure should be followed. Let us summarize the systematic procedure followed in Example 10-11.

We first note that when a real network is excited (driven) by a real voltage or current source, the network responses (all the network voltages and currents) must be real. Therefore, even when the natural frequencies p_1 and p_2 and the constants k_1 and k_2 are complex, the resulting complex quantities must somehow combine to form real quantities. Hence, the trick is to pair the complex quantities such that they add up to real quantities. The natural frequencies p_1 and p_2 and the constants k_1 and k_2 are always complex conjugate pairs, and Euler's equations [see Chapter 4, Equations (4-11)] provide the link between the real and complex quantities.

We now summarize a systematic procedure for analyzing second-order networks with complex natural frequencies:

1. Write the network equations in normal form.
2. Compute
a. the natural frequencies p_1 and p_2
b. the constants k_1, k_2, $k_1/(k_2 - k_1)$, and $k_2/(k_2 - k_1)$
c. the forcing functions $u_1(\lambda)$ and $u_2(\lambda)$
d. verify that p_1 and p_2 are complex conjugates and that k_1 and k_2 are complex conjugates.
3. Substitute u_1 and u_2 into Equation (10-86) and evaluate the integrals.
4. Substitute the values for k_1, k_2, $k_1/(k_2 - k_1)$, and $k_2/(k_2 - k_1)$.
5. Group the terms into the transient response v_{tr} plus the steady-state response v_{ss}. (v_{tr} is of the form $\gamma_1 e^{p_1 t} + \gamma_2 e^{p_2 t}$, where γ_1 and γ_2 are complex constants. The remaining terms constitute v_{ss}.)
6. Compute the transient response:
a. Evaluate γ_1 and γ_2 (substitute the values for p_1 and p_2) and write them in polar coordinates. Verify that γ_1 and γ_2 are complex conjugates.
b. Substitute the values of γ_1 and γ_2 and p_1 and p_2 into v_{tr}.
c. Factor the common (real) exponential term.
d. Arrange the remaining terms into the form $(e^{jx} + e^{-jx})/2$ and use Euler's formula [see Equation (4-11)].
7. Compute the steady-state response by substituting the values for p_1 and p_2.

The time-constant concept. In Section 10-3 we introduced the time-constant method for analyzing first-order networks. We now extend this concept to the case of second-order networks. The generalization is straightforward.

A general form for the two responses to any second-order linear network for any forcing functions is presented in Equations (10-87). Let us work with the equation for $x_1(t)$. The equation for $x_2(t)$ can be handled by the same procedure. Proceeding as we did for the first-order system, we note that the first equation of Equation (10-87) consists of an energy-storage term of the form $c_1e^{p_1t} + c_2e^{p_2t}$ plus a forced-response term involving two integrals over the interval (0, t). Suppose that we carry out the two integrations and evaluate the two integrals at $\lambda = 0$. This results in two terms (one for each integral), the first of which is of the form $c_1'e^{p_1t}$ and the second of which is of the form $c_2'e^{p_2t}$, where c_1' and c_2' are constants. These two terms comprise the part of the forced response that contributes to the transient response. Therefore, they can be combined with the energy-storage term to yield a transient response of the form $\gamma_1e^{p_1t} + \gamma_2e^{p_2t}$, where $\gamma_1 = c_1 + c_1'$, $\gamma_2 = c_2 + c_2'$. Finally, we evaluate the two integrals at $\lambda = t$ to obtain the steady-state response. Hence, the first of Equations (10-87) can always be written in the form

$$x(t) = \gamma_1e^{p_1t} + \gamma_2e^{p_2t} + x_{ss}(t) \tag{10-116}$$

Therefore, if the steady-state response x_{ss} can be computed by any means whatsoever, we can proceed to compute the total response by next computing the natural frequencies, and, finally, evaluating the constants γ_1 and γ_2.

The natural frequencies can, of course, always be found by writing the normal-form equations to obtain the a_{ij}'s (i, j = 1, 2) and then using Equations (10-73).

The constants γ_1 and γ_2 can be evaluated from initial conditions. That is, the second-order network contains two energy-storage elements. Knowledge of the energy stored in each of these two elements can be translated into two initial conditions on $x(t)$. The exact procedure varies in each case and is best described by examples.

When the natural frequencies are real, γ_1 and γ_2 are real constants, and no particular problems are encountered. When the natural frequencies are complex, then γ_1 and γ_2 are complex constants. In this case, after evaluating γ_1 and γ_2 (which constitute a complex conjugate pair) the (complex) exponential terms $\gamma_1e^{p_1t} + \gamma_2e^{p_2t}$ can be combined to form a single real term (via Euler's formula) by the procedure of step 6 listed at the end of the preceding section. An alternative procedure is to realize at the outset that the transient response is a damped sinusoid, with the damping constant given by the real part of the natural frequencies, and with the frequency given by the imaginary part of p_1 (or, the negative of the imaginary part of p_2). In this case, the two constants to be determined are the amplitude and phase of the sinusoid. That is, for complex natural frequencies, the transient response can always be expressed in the form

$$x_{tr} = \nu e^{\sigma t} \cos(\omega + \phi), \qquad p_1 = \sigma + j\omega \tag{10-117}$$

where $\sigma = \text{Re}\,\{p_1\}$ and $\omega = \text{Im}\,\{p_1\}$, and ν and ϕ are constants to be determined from initial conditions.

In summary, any voltage or current in any second-order linear network with unique natural frequencies can be computed by the following procedure:

1. Compute the steady-state response $x_{ss}(t)$ by any convenient method.[1]

2. Compute the natural frequencies p_1 and p_2.

3a. Real natural frequencies. Write the total response in the form

$$x(t) = \gamma_1 e^{p_1 t} + \gamma_2 e^{p_2 t} + x_{ss}(t) \qquad t \geq 0$$

and evaluate γ_1 and γ_2 from initial conditions determined from the amounts of energies stored in the two energy-storage elements at $t = 0$.

3b. Complex natural frequencies. Write the total response in the form

$$x(t) = \nu e^{\sigma t} \cos(\omega t + \phi) + x_{ss}(t) \qquad t \geq 0$$
$$p_1 = \sigma + j\omega$$

and evaluate ν and ϕ from initial conditions determined from the amounts of energy stored in the two energy-storage elements at $t = 0$.

Example 10-12

In this example we use the time-constant method to solve the same problem considered in Example 10-10. The network is presented again in Figure 10-18.

The steady-state response is easily determined by inspection of the network. Both capacitances offer infinite impedance to the dc source. Using the fact that no current (dc) flows through the capacitances and using KCL, we conclude that all the steady-state network currents are

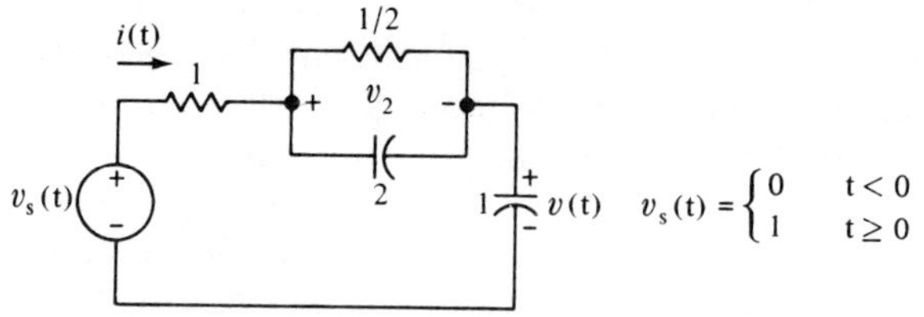

Figure 10-18 Network for Example 10-12.

[1] When the forcing function is a sine wave, step, or set of widely spaced pulses, there is a well-defined steady state. In many signal processing problems, the signal is a shaped pulse that does not have a well-defined steady-state form. In such cases the integrals of Equations (10-87) must be evaluated.

zero. Hence, the voltage across both resistances and the 2-farad capacitance are zero, and the steady-state voltage across the 1-farad capacitance is the same as the source voltage; that is,

$$v_{ss} = 1 \qquad t \geq 0 \tag{10-118}$$

The natural frequencies were computed in Example 10-10, where we found

$$p_1 = -\tfrac{1}{2}, \qquad p_2 = -2 \tag{10-119}$$

Hence, the transient response is of the form

$$v_{tr} = \gamma_1 e^{-t/2} + \gamma_2 e^{-2t} \qquad t \geq 0 \tag{10-120}$$

and the total response is of the form

$$\begin{aligned} v(t) &= v_{tr} + v_{ss}(t) \\ &= \gamma_1 e^{-t/2} + \gamma_2 e^{-2t} + 1 \qquad t \geq 0 \end{aligned} \tag{10-121}$$

The remaining problem is to find values for γ_1 and γ_2 such that $v(t)$ satisfies two initial conditions, one of which is associated with each of the two energy-storage elements (capacitances).

We first consider the 1-farad capacitance. The energy stored at $t = 0^-$ is zero. Since the amount of energy stored cannot change instantaneously, the amount of energy stored at $t = 0^+$ is also zero. The energy stored in the 1-farad capacitance at time t is given by $\frac{1}{2}Cv^2(t)$ with $C = 1$. We conclude that $v(0^-) = v(0^+) = 0$. Therefore, γ_1 and γ_2 must satisfy Equation (10-121) at $t = 0^+$; that is,

$$v(0^+) = 0 = \gamma_1 + \gamma_2 + 1 \tag{10-122}$$

We now consider the 2-farad capacitance. Let $v_2(t)$ represent the voltage across the 2-farad capacitance with the polarity shown in Figure 10-18. The energy stored at $t = 0^-$ is zero. Hence, by the same argument used for the 1-farad capacitance, we conclude that $v_2(0^-) = v_2(0^+) = 0$. In order to find the initial condition on $v(t)$ imposed by this constraint on $v_2(t)$, we refer to Figure 10-18 and note that with $v(0^+) = v_2(0^+) = 0$ and $v_s(0^+) = 1$, KVL requires that the voltage across the 1-ohm resistance at $t = 0^+$ is 1 volt. That is, since both capacitances are short circuits (zero voltage) at $t = 0^+$, the voltage across the 1-ohm resistance is the source voltage $v_s(0^+) = 1$. Hence, the current through the 1-ohm resistance at $t = 0^+$ is $i(0^+) = 1$. Since the 1-ohm resistance and the 1-farad capacitance are in series, $i(t)$ is also the current down through the 1-farad capacitance. The 1-farad capacitance voltage and current are related by $i = C\dot{v}$ so that at $t = 0^+$, $\dot{v}(0^+) = i(0^+)/C = 1$. Hence, we have translated the initial condition on $v_2(t)$ into an initial condition on $v(t)$. γ_1 and γ_2 must satisfy this condition. Taking the derivative of Equation (10-121), we obtain

$$\dot{v}(t) = -\frac{\gamma_1}{2} e^{-t/2} - 2\gamma_2 e^{-2t} \qquad t \geq 0 \tag{10-123}$$

At $t = 0^+$ we have

$$\dot{v}(0^+) = 1 = -\frac{\gamma_1}{2} - 2\gamma_2 \tag{10-124}$$

γ_1 and γ_2 must satisfy Equations (10-122) and (10-124). Solving these equations for γ_1 and γ_2, we obtain

$$\gamma_1 = -\tfrac{2}{3}, \qquad \gamma_2 = -\tfrac{1}{3} \tag{10-125}$$

Finally, we substitute these values in Equation (10-121) to obtain

$$v(t) = \begin{cases} 0 & t < 0 \\ -\frac{2}{3}e^{-t/2} - \frac{1}{3}e^{-2t} + 1 & t \geq 0 \end{cases} \tag{10-126}$$

which checks with Equation (10-97). Equation (10-126) is plotted in Figure 10-15.

Example 10-13

In this example we use the time-constant method to solve the same problem considered in Example 10-11. The network is presented in Figure 10-19.

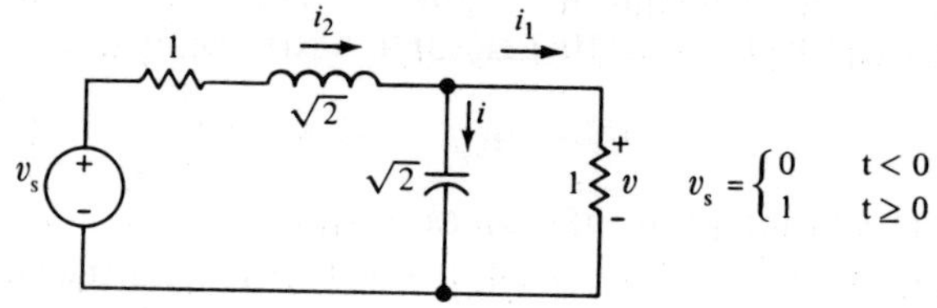

Figure 10-19 Network for Example 10-13.

In the steady state, the capacitance is an open circuit, and the inductance a short circuit. Hence, the steady-state value of $v(t)$ is given by

$$v_{ss} = \tfrac{1}{2}v_s = \begin{cases} 0 & t < 0 \\ \frac{1}{2} & t \geq 0 \end{cases} \tag{10-127}$$

The natural frequencies were computed in Example 10-11, where we found

$$p_1 = -1/\sqrt{2} + j/\sqrt{2}, \qquad p_2 = -1/\sqrt{2} - j/\sqrt{2} \tag{10-128}$$

Therefore, the transient response is a damped sinusoid, with damping time constant of $\sqrt{2}$ seconds and a radian frequency of oscillation of

$1/\sqrt{2}$; that is,

$$v_{tr} = \nu e^{-t/\sqrt{2}} \cos (t/\sqrt{2} + \phi) \tag{10-129}$$

Consequently, the total response can be expressed in the form

$$\begin{aligned} v &= v_{tr} + v_{ss} \\ &= \nu e^{-t/\sqrt{2}} \cos (t/\sqrt{2} + \phi) + \tfrac{1}{2} \end{aligned} \tag{10-130}$$

Equation (10-130) gives the solution to within two constants (ν and ϕ).

The energy stored in the inductance and the energy stored in the capacitance cannot change in zero time. Since $v_s(t) = 0$ for $t < 0$, the inductance energy at $t = 0^+$ is $\frac{1}{2}Li_L{}^2(0^+) = 0$, and the capacitance energy at $t = 0^+$ is $\frac{1}{2}Cv^2(0^+) = 0$. Hence, the two initial conditions imposed by energy considerations are $i_L(0^+) = 0$ and $v(0^+) = 0$. Using the latter constraint and Equation (10-130), we have

$$v(0^+) = 0 = \nu \cos \phi + \tfrac{1}{2} \tag{10-131}$$

In order to determine the effect of $i_L(0^+) = 0$, we refer to Figure 10-19 and note that since $v(0^+) = 0$, the current through the resistance on the right is zero; $i_1(0^+) = 0$. Since $i_L(0^+) = 0$ and $i_1(0^+) = 0$, KCL requires that the current down through the capacitance at $t = 0^+$ is also zero; that is, $i(0^+) = 0$. Furthermore, the capacitance equation requires $i = C\dot{v}$ or $\dot{v} = i/C$. Evaluated at $t = 0^+$, this is $\dot{v}(0^+) = i(0^+)/C = 0/\sqrt{2} = 0$. Therefore, we take the derivative of Equation (10-130):

$$\dot{v} = -(\nu/\sqrt{2})e^{-t/\sqrt{2}} \sin (t/\sqrt{2} + \phi) - (\nu/\sqrt{2})e^{-t/\sqrt{2}} \cos (t/\sqrt{2} + \phi) \tag{10-132}$$

and evaluate it at $t = 0^+$ to obtain

$$\dot{v}(0^+) = 0 = -(\nu/\sqrt{2}) \sin \phi - (\nu/\sqrt{2}) \cos \phi \tag{10-133}$$

or

$$0 = \sin \phi + \cos \phi \tag{10-134}$$

Solving Equations (10-131) and (10-134) for ν and ϕ, we obtain

$$\nu = -1/\sqrt{2}, \qquad \phi = -\pi/4 \tag{10-135}$$

Finally, using these values in Equation (10-130), we have

$$v(t) = \begin{cases} 0 & t < 0 \\ -1/\sqrt{2}\ e^{-t/\sqrt{2}} \cos (1/\sqrt{2} - \pi/4) + 1/2 & t \geq 0 \end{cases} \tag{10-136}$$

A sketch of Equation (10-136) is presented in Figure 10-17.

Summary. Second-order networks have two natural frequencies. The natural frequencies are either real or occur as a complex conjugate pair. Networks for which $p_1 = p_2$ also exist. This case, which represents the point of transition between real and complex natural frequencies, is discussed in Chapter 11 (see Example 11-2).

Second-order networks containing only R's and C's or only R's and L's always have real natural frequencies. Second-order networks containing only R's, L's, and C's can have either real or complex natural frequencies. Second-order networks containing controlled sources can have either real or complex natural frequencies.

When the input is periodic or constant, the network response consists of a transient response plus a steady-state response. The functional form of the steady-state response depends on the functional form of the forcing function. (The comments relative to the steady-state response contained in the summary at the end of Section 10-3 also apply here.) The transient response can be expressed as the sum of two exponentials, with the natural frequencies the constants in the exponents. If the natural frequencies are real, both terms decay exponentially with time constants $\tau_1 = -p_1^{-1}$ and $\tau_2 = -p_2^{-1}$, respectively. One might be tempted to say that the duration of the transient depends essentially on the longest time constant (smallest p_i), but this depends on the definition of the transient duration, since the two exponential terms have different amplitudes as well as different time constants. If the natural frequencies are complex, the transient is an exponentially damped sinusoid. The oscillation decays exponentially with the time constant given by the negative inverse of the real part of the natural frequencies (the larger the real part, the shorter the duration), and the frequency is given by the magnitude of the imaginary part of the natural frequency (the larger the imaginary part, the higher the frequency). Although the decay time constant and the frequency depend only on the network (through the natural frequencies), the amplitude and phase of the transient depend on the state of the network (initial energy storage) and the forcing function as well as on the network.

PROBLEMS

10-1 Write a set of normal-form equations for the network shown in Figure P10-1. Specify the $\boldsymbol{x}$, $\mathbf{a}$, and $\boldsymbol{u}$ matrices.

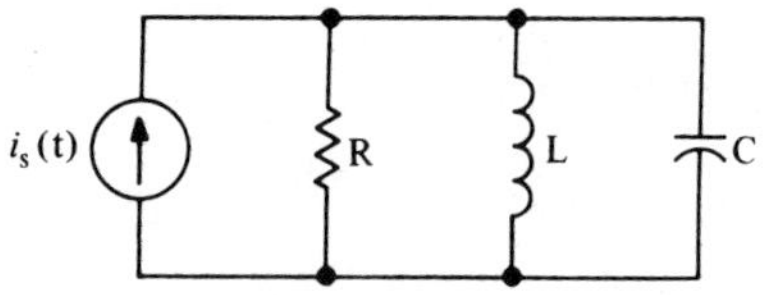

Figure P10-1

10-2 Write a set of normal-form equations for the network shown in Figure P10-2. Specify the $\boldsymbol{x}$, $\mathbf{a}$, and $\boldsymbol{u}$ matrices.

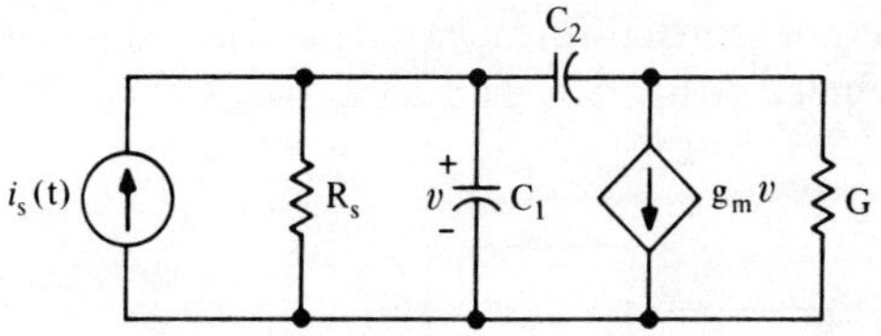

Figure P10-2

10-3 Write a set of normal-form equations for the network shown in Figure P10-3. Show the $\boldsymbol{x}$, **a**, and $\boldsymbol{u}$ matrices.

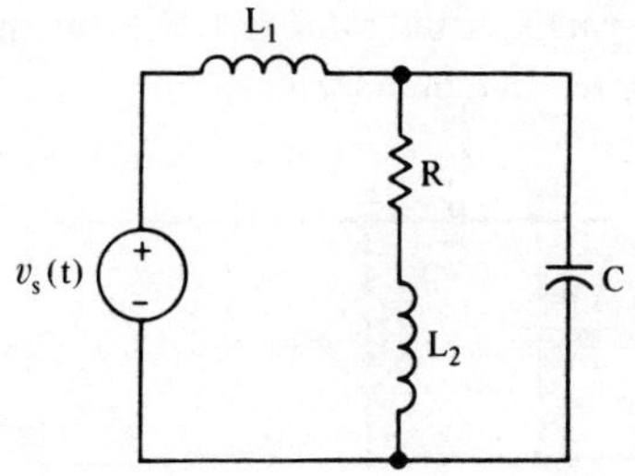

Figure P10-3

10-4 Write a set of normal-form equations for the network shown in Figure P10-4. The transistor model is shown at the right. Show the $\boldsymbol{x}$, **a**, and $\boldsymbol{u}$ matrices.

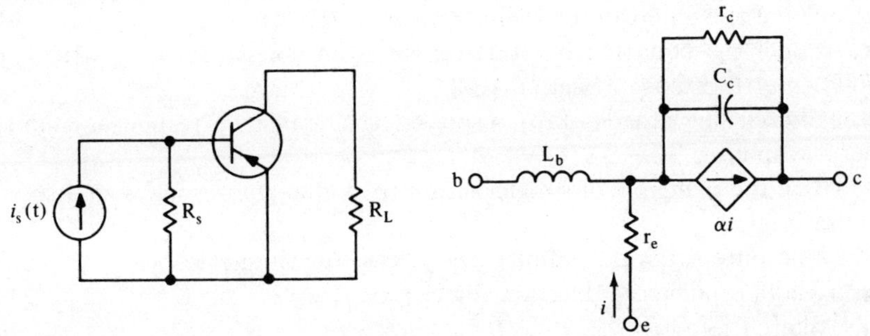

Figure P10-4

10-5 Write a set of normal-form equations for the network shown in Figure P10-5. Specify the $\boldsymbol{x}$, **a**, and $\boldsymbol{u}$ matrices.

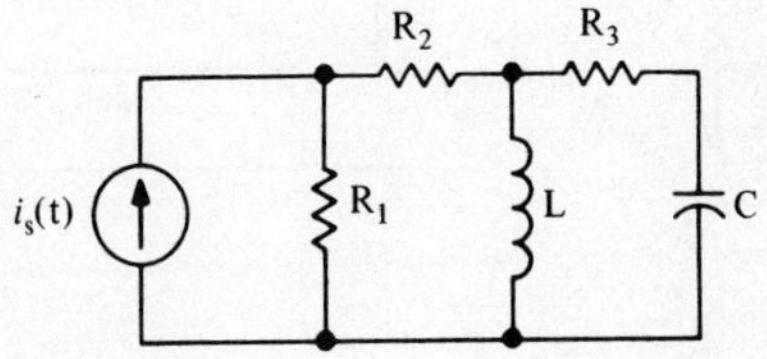

Figure P10-5

10-6 Write a set of normal-form equations for the network shown in Figure P10-6. Specify the $\boldsymbol{x}$, $\mathbf{a}$, and $\boldsymbol{u}$ matrices.

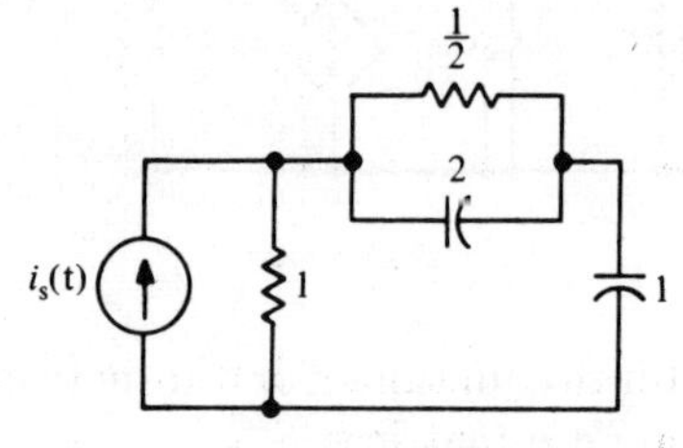

Figure P10-6

10-7 Write a set of normal-form equations for the network shown in Figure P10-7. Specify the $\boldsymbol{x}$, $\mathbf{a}$, and $\boldsymbol{u}$ matrices.

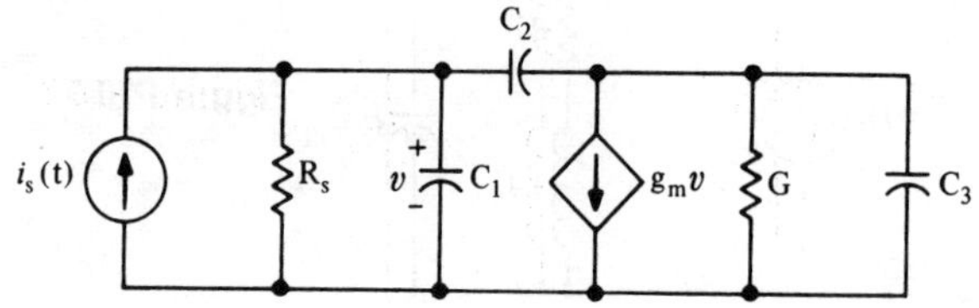

Figure P10-7

10-8 **a.** Write a normal-form equation for the network shown in Figure P10-8.

b. Solve the equation by using Equation (10-32).

c. Solve the equation by starting with the normal-form equation and following the steps of Section 10-3.

d. Sketch the steady-state response, the transient response, and the total response.

e. Use the principle of superposition to sketch the response to a square pulse input.

f. Compute the input admittance $y_{in}(j\omega)$ for the network.

g. Make a pole-zero diagram for $y_{in}(j\omega)$.

h. Sketch $|y_{in}(j\omega)|$ vs. ω.

i. What is the relationship between the network pole and the network natural frequency?

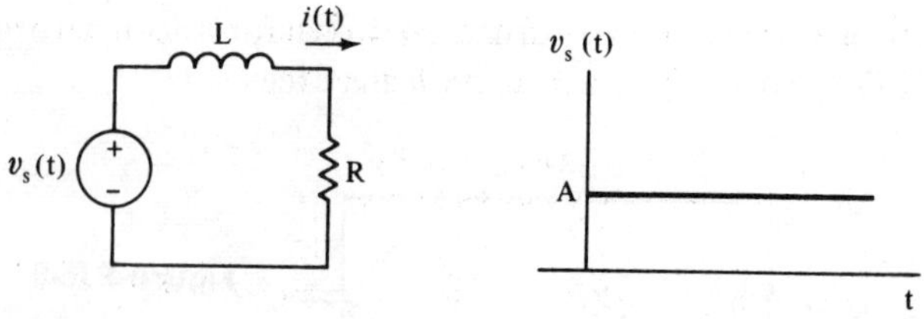

Figure P10-8

10-9 Compute and sketch $v_0(t)$ for all t (see Figure P10-9).

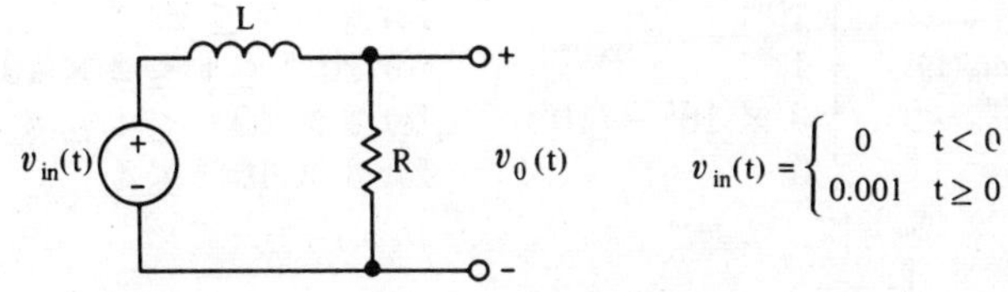

Figure P10-9

10-10 Compute $v_0(t)$ for the network of Problem 10-9 for

$$v_{in}(t) = \begin{cases} 0 & t < 0 \\ 0.001 & 0 \leq t \leq T \\ 0 & T < t \end{cases}$$

Set R = 1000, L = 0.001, and sketch $v_0(t)$ for $T = 10^{-5}$, 10^{-6}, and $T = 10^{-7}$.

10-11 **a.** Write a normal-form equation for the network shown in Figure P10-11.
b. Solve the equation by using Equation (10-32).
c. Sketch the steady-state response, the transient response, and the total response.
d. Use the sinusoidal steady-state analysis of Chapter 4 to compute the steady-state response. Compare to the result obtained in c.
e. Use the principle of superposition to sketch the response to a pulse of sinusoid; that is,

$$v(t) = \begin{cases} 0 & t < 0 \\ \cos \omega t & 0 < t < T, \qquad \omega = 20\pi/T \\ 0 & t > T \end{cases}$$

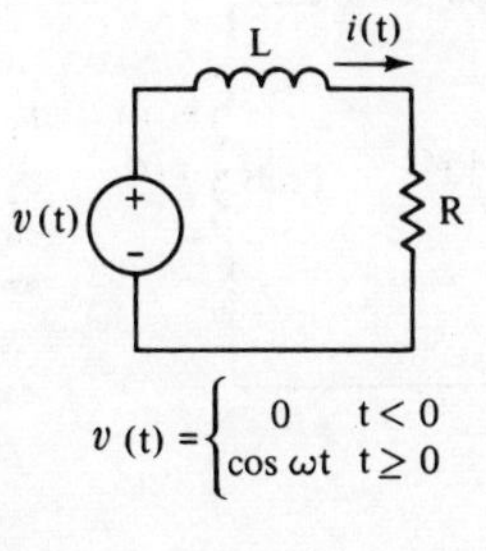

Figure P10-11

10-12 Repeat Problem 10-10 for (see Figure P10-12)

$$v_{in}(t) = \begin{cases} 0 & \text{for } t \le 0 \\ 10^6 t & \text{for } 0 \le t \le 10^{-6} \\ 1 & \text{for } 10^{-6} \le t \le 2 \times 10^{-6} \\ 3 \times 10^6 - 10^6 t & \text{for } 2 \times 10^{-6} \le t \le 3 \times 10^{-6} \\ 0 & \text{for } 3 \times 10^{-6} \le t \end{cases}$$

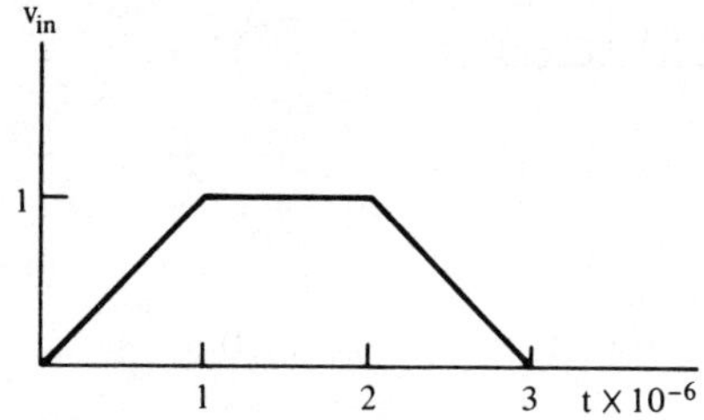

Figure P10-12

10-13 For the network shown in Figure P10-13

a. Show that the network is first-order.
b. Compute the network time constant.
c. What is the natural frequency of the network?
d. Compute the input admittance $y_{in}(j\omega)$ and compute the pole of this transfer function.
e. Sketch all of the branch voltages and currents.

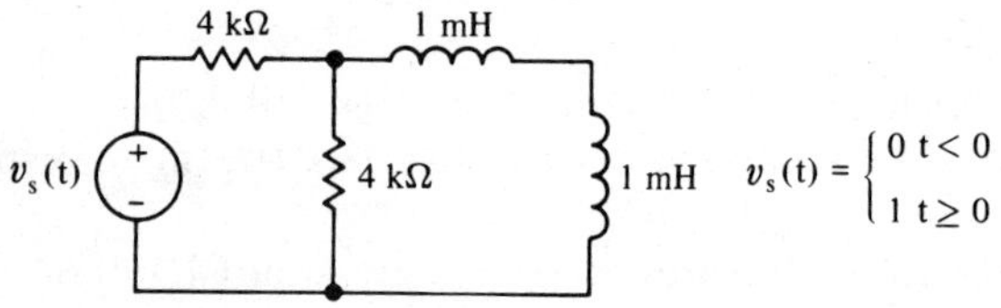

Figure P10-13

10-14 Use the time-constant concept (see, in particular, Example 10-9) to compute $v(t)$ and $i(t)$ for the network shown in Figure P10-14. Sketch $v_s(t)$, $v(t)$, and $i(t)$ for all t. The switch is thrown from position 1 to position 2 at t = 0.

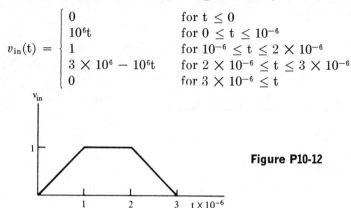

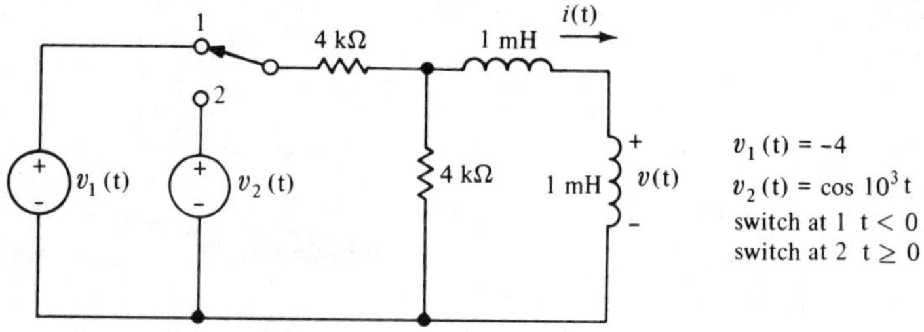

Figure P10-14

10-15 Use the time-constant concept to solve Problem 10-11 for both

$$v(t) = \begin{cases} 0 & t < 0 \\ \cos \omega t & t \geq 0 \end{cases}$$

and

$$v(t) = \begin{cases} 0 & t < 0 \\ \cos \omega t & 0 < t < T, \qquad \omega \gg T^{-1} \\ 0 & t > T \end{cases}$$

10-16 Repeat Example 10-15, but let

$$v(t) = \begin{cases} 0 & t < 0 \\ \cos \omega t & t \geq 0 \end{cases}$$

for $\omega = 2\pi$ and $\omega = 20\pi$.

10-17 Sketch to scale all the voltages and currents in the network shown in Figure P10-17.

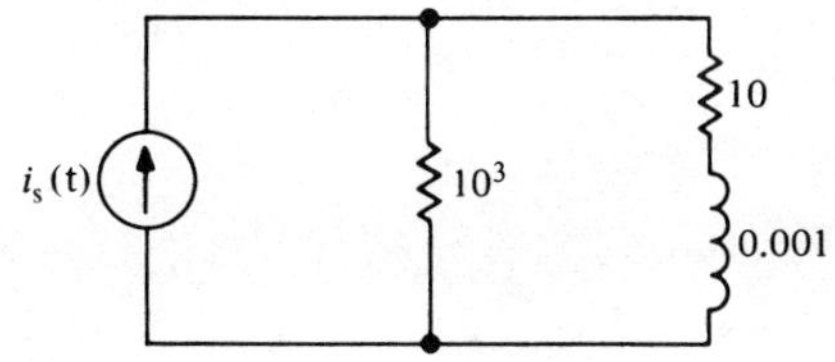

$$i_s(t) = \begin{cases} 1 & t < 0 \\ 0 & t \geq 0 \end{cases}$$

Figure P10-17

10-18 For the network shown in Figure P10-18

a. Write the normal-form equations.

b. Compute p_1, p_2, k_1, k_2 from Equations (10-73) and (10-77).

c. Solve for $v_0(t)$ by using Equation (10-86).

d. Sketch the transient, steady-state, and total responses.

e. Compute the transfer function V_0/I_s and compare its poles to the natural frequencies.

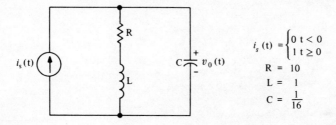

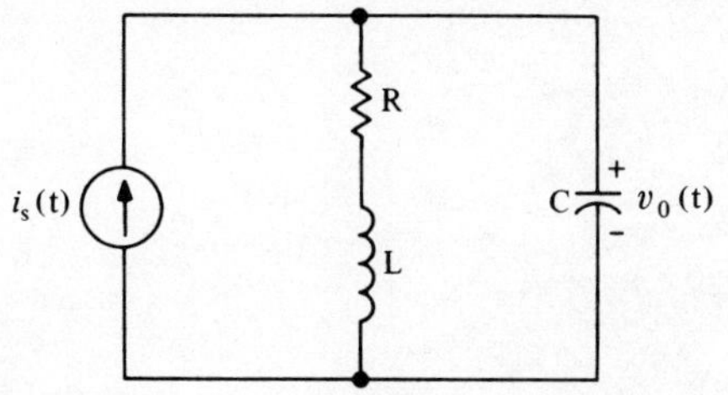

$$i_s(t) = \begin{cases} 0 & t < 0 \\ 1 & t \geq 0 \end{cases}$$

$R = 10$

$L = 1$

$C = \frac{1}{16}$

Figure P10-18

10-19 Repeat Problem 10-18 for $R = 8$, $L = 1$, $C = 1/25$.

10-20 Repeat Problem 10-18 for $R = 10^2$, $L = 10^{-2}$, $C = 10^{-10}$.

10-21 Compute the voltage across the output resistance of the second-order insertion-power-gain prototype filter when the input is a unit voltage step function. Carefully sketch the solution. Some check points are

$$\mathbf{a} = \begin{bmatrix} -1/\sqrt{2} & -1/\sqrt{2} \\ 1/\sqrt{2} & -1/\sqrt{2} \end{bmatrix} \qquad \boldsymbol{u} = \begin{bmatrix} v_s/\sqrt{2} \\ 0 \end{bmatrix}$$

$$\begin{Bmatrix} p_1 \\ p_2 \end{Bmatrix} = -1/\sqrt{2} \pm j/2 \qquad \begin{Bmatrix} k_1 \\ k_2 \end{Bmatrix} = \mp j$$

$$v(t) = \begin{cases} 0, & t < 0 \\ -\dfrac{e^{p_1 t}}{2} \displaystyle\int_0^t \dfrac{j}{\sqrt{2}} e^{-p_1 \lambda}\, d\lambda + \dfrac{e^{p_2 t}}{2} \displaystyle\int_0^t \dfrac{j}{\sqrt{2}} e^{-p_2 \lambda}\, d\lambda, & t \geq 0 \end{cases}$$

$$= \begin{cases} 0, & t < 0 \\ \dfrac{j}{2\sqrt{2}} \left(\dfrac{p_2 - p_1}{p_1 p_2} \right) + \dfrac{j}{2\sqrt{2}} \left(\dfrac{-p_2}{p_1 p_2} e^{p_1 t} + \dfrac{p_1}{p_1 p_2} e^{p_2 t} \right), & t \geq 0 \end{cases}$$

Answer:

$$v(t) = \begin{cases} 0 & t < 0 \\ 1/2 - 1/\sqrt{2}\, e^{-t/\sqrt{2}} \cos\,(t/\sqrt{2} - \pi/4), & t \geq 0 \end{cases}$$

CHAPTER

11

General Solution of Linear Network Equations—nth-Order Systems

11-1 Introduction

In Chapter 10, Section 10-2, we derived an algorithm for writing a set of normal-form differential equations for any linear network. By utilizing this algorithm we can construct sets of n linearly independent first-order differential equations for linear networks containing n independent energy storage elements. In Section 10-3 we solved this set of normal-form equations for the first-order case (n = 1). In Section 10-4 we solved this set of normal-form equations for the second-order case (n = 2). In this chapter we solve this set of normal-form equations for the general nth-order case.

The procedure for solving the normal-form equations for the nth-order case is identical to the procedure used to solve the 1st-order case and the 2nd-order case. The procedure consists of two steps. First, we solve the homogeneous equation $\dot{\boldsymbol{y}} = \boldsymbol{a}\boldsymbol{y}$ for $\boldsymbol{y}$ by assuming an exponential form for the solution and substituting this assumed solution into the homogeneous equation. This gives the solution to within n unknown constants. Second, we solve the complete equation $\dot{\boldsymbol{x}} = \boldsymbol{a}\boldsymbol{x} + \boldsymbol{u}$ by assuming the solution to have the same functional form as the solution to the homogeneous equation but with the unknown constants replaced by unknown time functions. This allows us to solve for the remaining n parameters associated with the assumed solution.

Before proceeding, we make some preliminary observations based on what we found for 1st- and 2nd-order networks. First-order networks had one natural frequency; second-order networks had two natural frequencies. Consequently, we conjecture that nth-order networks have n natural frequencies. Furthermore, we suspect that these n natural frequencies will depend only on the network through the network matrix a and not on the n forcing functions. We also project that the total response will consist of an energy storage term (due to energy already stored in the network) plus a forced term (due to the forcing functions). For periodic forcing functions, the forced response will consist of a transient term plus a steady-state response. As for 1st- and 2nd-order systems the transient part of the forced response can be combined with the energy storage term to form the transient response. Hence, the total response can be expressed as a transient response plus a steady-state response. The transient response for first-order networks was an exponential with the constant in the exponent the network natural frequency; for second-order networks the transient response was a sum of two exponentials with the constants in the exponents the natural frequencies. Based on this and the conjecture that an nth-order network has n natural frequencies, we project that the transient response for an nth-order network will consist of a sum of n exponential functions with the constants in the exponents the n natural frequencies. Since the natural frequencies may be either real or complex whereas the transient response must be real, we suspect that any complex natural frequency must be accompanied by another natural frequency that is its complex conjugate. Thus, a 3rd-order network can have either three real natural frequencies or one real and two complex (a complex conjugate pair) natural frequencies; a 4th-order network allows three possibilities—four real natural frequencies, four complex natural frequencies (two complex conjugate pairs) or two real and two complex (a complex conjugate pair). For each complex conjugate pair of natural frequencies, the two corresponding (complex) exponential terms in the transient response combine to form (via Euler's equation) an exponentially damped sinusoid. Hence, the transient responses for nth-order networks consist of sums of exponentially decaying terms (one for each real natural frequency), sums of exponentially decaying sinusoids (one for each complex conjugate pair), or mixtures of these terms. For example, the transient response of a 3rd-order network can take one of two forms: A sum of three exponentially decaying terms (for the case of three real natural frequencies) or one exponentially decaying term plus an exponentially decaying sinusoid (for the case of one real and two complex natural frequencies). Finally, we suspect that the functional form of the steady-state response will depend on the functional form of the forcing functions as was the case for both 1st- and 2nd-order networks.

11-2 Solution to the Homogeneous Equation

In this section we solve the homogeneous equation associated with the complete equation $\dot{x} = \mathbf{a}x + u$. The homogeneous equation is given by

$$\dot{y} = \mathbf{a}y \tag{11-1}$$

where $\mathbf{a}$ is an n × n matrix of known constants and y is a 1 × n matrix of unknown time functions. The problem is to solve Equation (11-1) for y.

We start by assuming a solution of the form

$$y = \mathbf{C}e^{pt} \tag{11-2}$$

where p is an unknown constant (to be determined) and $\mathbf{C}$ is a 1 × n matrix of unknown constants (to be determined).[1] The time derivative of y is given by

$$\dot{y} = \mathbf{C}pe^{pt} \tag{11-3}$$

Substituting the assumed solution into the homogeneous equation, that is, substituting Equations (11-2) and (11-3) into Equation (11-1), we obtain

$$\mathbf{C}pe^{pt} = \mathbf{aC}e^{pt}$$

or

$$\mathbf{C}p = \mathbf{aC}$$

or

$$\mathbf{0} = \mathbf{aC} - p\mathbf{C} \tag{11-4}$$

where $\mathbf{0}$ is a 1 × n matrix with all entries zero. Since $\mathbf{C} = \mathbf{1C}$ where $\mathbf{1}$ is the unit (or identity) matrix defined by

$$\mathbf{1} = \begin{bmatrix} 1 & 0 & \cdots & 0 \\ 0 & 1 & \cdots & 0 \\ \cdot & \cdot & \cdot & \cdot \\ \cdot & \cdot & \cdot & \cdot \\ \cdot & \cdot & \cdot & \cdot \\ 0 & 0 & \cdots & 1 \end{bmatrix} \tag{11-5}$$

we can write Equation (11-4) in the form

$$\begin{aligned} \mathbf{0} &= \mathbf{aC} - p\mathbf{1C} \\ &= (\mathbf{a} - p\mathbf{1})\mathbf{C} \end{aligned} \tag{11-6}$$

[1] A more general form for the assumed solution would result if we assumed a different natural frequency for each y_j, that is, if we assumed $y_j = c_j e^{p_j t}$ (j = 1, 2, · · · , n) rather than $y_j = c_j e^{pt}$. However, if we were to start with this more general form we would soon discover that Equation (11-2) satisfies Equation (11-1) if, and only if, all the natural frequencies are identical. The argument is identical to the one used in Section 10-4 for the 2nd-order case (see pages 344, 345).

Equation (11-6) is a shorthand notation for

$$\begin{bmatrix} 0 \\ 0 \\ \cdot \\ \cdot \\ \cdot \\ 0 \end{bmatrix} = \begin{bmatrix} a_{11} - p & a_{12} & \cdots & a_{1n} \\ a_{21} & a_{22} - p & \cdots & a_{2n} \\ \cdot & \cdot & \cdot & \cdot \\ \cdot & \cdot & \cdot & \cdot \\ \cdot & \cdot & \cdot & \cdot \\ a_{n1} & a_{n2} & \cdots & a_{nn} - p \end{bmatrix} \begin{bmatrix} c_1 \\ c_2 \\ \cdot \\ \cdot \\ \cdot \\ c_n \end{bmatrix} \tag{11-7}$$

Equation (11-7) is satisfied if $c_1 = c_2 = \cdots = c_n = 0$, but this is a rather uninteresting solution. A more interesting solution requires only that the determinant $|\mathbf{a} - p\mathbf{1}| = 0$ (see Appendix I), that is,

$$\begin{vmatrix} a_{11} - p & a_{12} & \cdots & a_{1n} \\ a_{21} & a_{22} - p & \cdots & a_{2n} \\ \cdot & \cdot & \cdot & \cdot \\ \cdot & \cdot & \cdot & \cdot \\ \cdot & \cdot & \cdot & \cdot \\ a_{n1} & a_{n2} & \cdots & a_{nn} - p \end{vmatrix} = 0 \tag{11-8}$$

The determinant in Equation (11-8) is an nth-order polynomial in p and the equation $|\mathbf{a} - p\mathbf{1}| = 0$ has n solutions which are the n roots of this polynomial. Let us label these n roots $p_1, p_2, \cdots, p_n$. Then we conclude that a necessary condition for $\boldsymbol{y} = \mathbf{C}e^{pt}$ to satisfy $\dot{\boldsymbol{y}} = \boldsymbol{a}\boldsymbol{y}$ is that $p = p_1$, or $p_2, \cdots$, or p_n. These n values of p are called the natural frequencies of the network. Hence, an nth-order network has n natural frequencies which depend only on the elements of the network matrix **a**. For the moment we assume that the n natural frequencies are distinct. The special case of one or more identical natural frequencies is considered in Section 11-5.

Equations of the form $|\mathbf{a} - p\mathbf{1}| = 0$ occur so often in mathematical physics that they have been given a name. The equation $|\mathbf{a} - p\mathbf{1}| = 0$ is called the characteristic equation of the matrix **a**. The solutions to the characteristic equation are called the eigenvalues (or characteristic values) of the matrix **a**. Therefore, in this terminology we would state a necessary condition for $\boldsymbol{y} = \mathbf{C}e^{pt}$ to satisfy $\dot{\boldsymbol{y}} = \boldsymbol{a}\boldsymbol{y}$ that the natural frequency p be an eigenvalue of the matrix **a**. Since an nth-order network has an $n \times n$ network matrix, and since an $n \times n$ matrix has n eigenvalues, we conclude that an nth-order network has n natural frequencies $p_1, p_2, \cdots, p_n$ and that $\boldsymbol{y} = \boldsymbol{a}\boldsymbol{y}$ has n solutions $\boldsymbol{y} = \mathbf{C}e^{p_1 t}$, $\boldsymbol{y} = \mathbf{C}e^{p_2 t}$, $\cdots$, $\boldsymbol{y} = \mathbf{C}e^{p_n t}$. We now return to Equation (11-7) and write it in the form

$$\begin{aligned} 0 &= (a_{11} - p)c_1 + (a_{12} \quad)c_2 \cdots + (a_{1n} \quad)c_n \\ 0 &= (a_{21} \quad)c_1 + (a_{22} - p)c_2 \cdots + (a_{2n} \quad)c_n \\ &\cdot \qquad \cdot \qquad \cdot \qquad \cdot \\ &\cdot \qquad \cdot \qquad \cdot \qquad \cdot \\ &\cdot \qquad \cdot \qquad \cdot \qquad \cdot \\ 0 &= (a_{n1} \quad)c_1 + (a_{n2} \quad)c_2 \cdots + (a_{nn} - p)c_n \end{aligned} \tag{11-9}$$

The n equations given by Equations (11-9) are not linearly independent. However, any $n-1$ of these equations are linearly independent. For the next step we can use any $n-1$ of these n equations, that is, we could discard any one of these equations. Let us (arbitrarily) choose to discard the last equation. Discarding the last equation and transferring the terms containing c_1 to the left-hand sides in the first $n-1$ equations we obtain:

$$\begin{aligned}
-(a_{11}-p)c_1 &= (a_{12})c_2 + \cdots + (a_{1n})c_n \\
-(a_{21})c_1 &= (a_{22}-p)c_2 + \cdots + (a_{2n})c_n \\
&\;\;\vdots \\
-(a_{n-1,1})c_1 &= (a_{n-1,2})c_2 + \cdots + (a_{n-1,n})c_n
\end{aligned} \tag{11-10}$$

Recall that $c_1, c_2, \cdots, c_n$ are unknown constants. Nonetheless, let us assume that c_1 is known (we will show how to find it later) and use Cramer's rule to solve for $c_2, c_3, \cdots, c_n$ in terms of c_1.

$$c_2 = \frac{\begin{vmatrix} -(a_{11}-p)c_1 & \cdots & a_{1n} \\ \vdots & & \vdots \\ -(a_{n-1,1})c_1 & \cdots & a_{n-1,n} \end{vmatrix}}{\begin{vmatrix} a_{12} & \cdots & a_{1n} \\ \vdots & & \vdots \\ a_{n-1,2} & \cdots & a_{n-1,n} \end{vmatrix}} = \frac{\begin{vmatrix} -(a_{11}-p) & \cdots & a_{1n} \\ \vdots & & \vdots \\ -(a_{n-1,1}) & \cdots & a_{n-1,n} \end{vmatrix}}{\begin{vmatrix} a_{12} & \cdots & a_{1n} \\ \vdots & & \vdots \\ a_{n-1,2} & \cdots & a_{n-1,n} \end{vmatrix}} c_1 = k_2(p)c_1$$

$$\vdots \tag{11-11}$$

$$c_n = \frac{\begin{vmatrix} a_{12} & \cdots & -(a_{11}-p)c_1 \\ \vdots & & \vdots \\ a_{n-1,2} & \cdots & -(a_{n-1,1}c_1) \end{vmatrix}}{\begin{vmatrix} a_{12} & \cdots & a_{1n} \\ \vdots & & \vdots \\ a_{n-1,2} & \cdots & a_{n-1,n} \end{vmatrix}} = \frac{\begin{vmatrix} a_{12} & \cdots & -(a_{11}-p) \\ \vdots & & \vdots \\ a_{n-1,2} & \cdots & -(a_{n-1,1}) \end{vmatrix}}{\begin{vmatrix} a_{12} & \cdots & a_{1n} \\ \vdots & & \vdots \\ a_{n-1,2} & \cdots & a_{n-1,n} \end{vmatrix}} c_1 = k_n(p)c_1$$

Here, the $n - 1$ constants $k_2(p), k_3(p), \cdots, k_n(p)$ are uniquely defined as ratios of two determinants that depend only on the a_{ij}'s and p.[2]

Therefore, Equation (11-7) not only specifies the n natural frequencies $p_1, p_2, \cdots, p_n$ for which Equation (11-2) satisfies Equation (11-1), but also specifies $n - 1$ constraints on the n constants $c_1, c_2, \cdots, c_n$. From Equations (11-2) and (11-11), we conclude that

$$y = \begin{bmatrix} c_1 e^{pt} \\ k_2(p) c_1 e^{pt} \\ \cdot \\ \cdot \\ \cdot \\ k_n(p) c_1 e^{pt} \end{bmatrix} \tag{11-12}$$

is a solution to Equation (11-1) for any c_1 (it cancels) provided we set $p = p_1$, or $p_2, \cdots$, or p_n. Hence

$$y = \begin{bmatrix} e^{p_1 t} \\ k_2(p_1) e^{p_1 t} \\ \cdot \\ \cdot \\ \cdot \\ k_n(p_1) e^{p_1 t} \end{bmatrix}, \quad y = \begin{bmatrix} e^{p_2 t} \\ k_2(p_2) e^{p_2 t} \\ \cdot \\ \cdot \\ \cdot \\ k_n(p_2) e^{p_2 t} \end{bmatrix}, \quad \cdots, \quad y = \begin{bmatrix} e^{p_n t} \\ k_2(p_n) e^{p_n t} \\ \cdot \\ \cdot \\ \cdot \\ k_n(p_n) e^{p_n t} \end{bmatrix} \tag{11-13}$$

are n different solutions to the homogeneous equation. For notational convenience, we set

$$k_{ij} = \begin{cases} 1 & i = 1; \quad j = 1, 2, \cdots, n \\ k_i(p_j) & i = 2, 3, \cdots, n; \quad j = 1, 2, \cdots, n \end{cases} \tag{11-14}$$

Then Equations (11-13) become

$$y = \begin{bmatrix} k_{11} e^{p_1 t} \\ k_{21} e^{p_1 t} \\ \cdot \\ \cdot \\ \cdot \\ k_{n1} e^{p_1 t} \end{bmatrix}, \quad y = \begin{bmatrix} k_{12} e^{p_2 t} \\ k_{22} e^{p_2 t} \\ \cdot \\ \cdot \\ \cdot \\ k_{n2} e^{p_2 t} \end{bmatrix}, \quad \cdots, \quad y = \begin{bmatrix} k_{1n} e^{p_n t} \\ k_{2n} e^{p_n t} \\ \cdot \\ \cdot \\ \cdot \\ k_{nn} e^{p_n t} \end{bmatrix} \tag{11-15}$$

Since each of Equations (11-15) is a solution to Equation (11-1), any linear combination of them is also a solution. It follows that the most

[2] In special cases we may find that the denominator determinant is zero and the method seems to fail. Often in such cases the difficulty can be surmounted by taking some other c_i rather than c_1 as the left-hand side of Equation (11-10). If this fails, the methods of Reference [35] can be used. Such special cases are beyond the scope of this text.

general solution to the homogeneous equation is given by

$$y = \begin{bmatrix} k_{11}B_1e^{p_1t} + k_{12}B_2e^{p_2t} + \cdots + k_{1n}B_ne^{p_nt} \\ k_{21}B_1e^{p_1t} + k_{22}B_2e^{p_2t} + \cdots + k_{2n}B_ne^{p_nt} \\ \cdot \\ \cdot \\ \cdot \\ k_{n1}B_1e^{p_1t} + k_{n2}B_2e^{p_2t} + \cdots + k_{nn}B_ne^{p_nt} \end{bmatrix} \tag{11-16}$$

where $B_1, B_2, \cdots, B_n$ are arbitrary constants. Equation (11-16) can also be written in the form

$$\begin{aligned} y &= \begin{bmatrix} k_{11}e^{p_1t} & k_{12}e^{p_2t} & \cdots & k_{1n}e^{p_nt} \\ k_{21}e^{p_1t} & k_{22}e^{p_2t} & \cdots & k_{2n}e^{p_nt} \\ \cdot & \cdot & \cdot & \cdot \\ \cdot & \cdot & \cdot & \cdot \\ \cdot & \cdot & \cdot & \cdot \\ k_{n1}e^{p_1t} & k_{n2}e^{p_2t} & \cdots & k_{nn}e^{p_nt} \end{bmatrix} \begin{bmatrix} B_1 \\ B_2 \\ \cdot \\ \cdot \\ \cdot \\ B_n \end{bmatrix} \\ &= \varphi \mathbf{B} \end{aligned} \tag{11-17}$$

which is more convenient for our purposes.

Note that since φ is a matrix of time functions while **B** is a matrix of constants, the time derivative of Equation (11-17) is given by

$$\dot{y} = \dot{\varphi}\mathbf{B} \tag{11-18}$$

Substituting Equations (11-17) and (11-18) into Equation (11-1), we obtain the relationship

$$\dot{\varphi}\mathbf{B} = \mathbf{a}\varphi\mathbf{B} \tag{11-19}$$

We will need this equation in the next section.

This is as far as we can go with the homogeneous equation. In summary, we have determined that Equation (11-17) is a solution to the homogeneous equation. In Equation (11-17), **B** is a $1 \times n$ matrix whose n elements are n arbitrary constants; that is, Equation (11-17) satisfies Equation (11-1) for any **B**. φ is an $n \times n$ matrix of known quantities; the n natural frequencies $p_1, p_2, \cdots, p_n$ are the n eigenvalues of the network matrix; the n^2 numbers $k_{11}, \cdots, k_{nn}$ also depend only on the elements of the network matrix and are given by Equation (11-14).

11-3 Solution to the Complete Equation

To obtain the solution to the general nth-order complete equation

$$\dot{x} = \mathbf{a}x + u \tag{11-20}$$

we assume a solution of the form

$$x = \varphi b \tag{11-21}$$

where $\boldsymbol{\varphi}$ is a known n × n matrix defined by Equation (11-17), and $\boldsymbol{b}$ is a 1 × n matrix of unknown time functions

$$\boldsymbol{b} = \begin{bmatrix} b_1(t) \\ b_2(t) \\ \cdot \\ \cdot \\ \cdot \\ b_n(t) \end{bmatrix} \tag{11-22}$$

to be determined. Note that the assumed form for $\boldsymbol{x}$ is identical to $\boldsymbol{y}$ (the solution to the homogeneous equation), except that we have replaced the matrix of constants **B** with a matrix of time functions $\boldsymbol{b}$. Note also that we have specified the solution to the n equations of Equation (11-20) to within n unknowns. Hence, the remaining step is to substitute Equation (11-21) into Equation (11-20) and solve for the n unknowns $\boldsymbol{b}$.

Since both $\boldsymbol{\varphi}$ and $\boldsymbol{b}$ are matrices whose elements are time functions, the time derivative of $\boldsymbol{x}$ is given by

$$\dot{\boldsymbol{x}} = \boldsymbol{\varphi}\dot{\boldsymbol{b}} + \dot{\boldsymbol{\varphi}}\boldsymbol{b} \tag{11-23}$$

where $\dot{\boldsymbol{b}}$ and $\dot{\boldsymbol{\varphi}}$ are matrices whose elements are the time derivatives of the corresponding elements of $\boldsymbol{b}$ and $\boldsymbol{\varphi}$.[3]

Substituting Equations (11-21) and (11-23) into Equation (11-20), we obtain

$$\boldsymbol{\varphi}\dot{\boldsymbol{b}} + \dot{\boldsymbol{\varphi}}\boldsymbol{b} = \mathbf{a}\boldsymbol{\varphi}\boldsymbol{b} + \boldsymbol{u} \tag{11-24}$$

But $\dot{\boldsymbol{\varphi}}\boldsymbol{b} = \mathbf{a}\boldsymbol{\varphi}\boldsymbol{b}$ for any $\boldsymbol{b}$ [this is the homogeneous equation as given by Equation (11-19)] so that Equation (11-24) reduces to

$$\boldsymbol{\varphi}\dot{\boldsymbol{b}} = \boldsymbol{u} \tag{11-25}$$

Premultiplying both sides of this by $\boldsymbol{\varphi}^{-1}$ (the inverse of $\boldsymbol{\varphi}$), we obtain[4]

$$\dot{\boldsymbol{b}} = \boldsymbol{\varphi}^{-1}\boldsymbol{u} \tag{11-26}$$

Taking the time integral of Equation (11-26), we obtain

$$\boldsymbol{b} = \int_{-\infty}^{t} \boldsymbol{\varphi}^{-1}(\lambda)\boldsymbol{u}(\lambda)\, d\lambda \tag{11-27}$$

Hence, the solution to the complete equation $\dot{\boldsymbol{x}} = \mathbf{a}\boldsymbol{x} + \boldsymbol{u}$ is given by

$$\boldsymbol{x}(t) = \boldsymbol{\varphi}(t) \int_{-\infty}^{t} \boldsymbol{\varphi}^{-1}(\lambda)\boldsymbol{u}(\lambda)\, d\lambda \tag{11-28}$$

[3] The validity of Equation (11-23), that is, the product rule for the differentiation of matrices, can be established by writing out the equations and differentiating.

[4] To be able to write Equation (11-26), we must be sure that $\boldsymbol{\varphi}^{-1}$ exists. Since the columns of $\boldsymbol{\varphi}$ are solutions to the homogeneous equation, the inverse will exist if these solutions are linearly independent. When the eigenvalues, p_i, are distinct, the $\boldsymbol{\varphi}$ matrix defined by Equation (11-17) has an inverse. If two or more of the p_i are identical, the special techniques discussed in Section 11-5 are required.

Equation (11-28) gives the n time functions $x_1(t)$, $x_2(t)$, $\cdots$, $x_n(t)$ in terms of known constants, known exponential functions, and the n forcing functions $u_1(\lambda)$, $u_2(\lambda)$, $\cdots$, $u_n(\lambda)$ on $-\infty \leq \lambda \leq t$.

By writing the integral over $(-\infty, t)$ as the integral over $(-\infty, 0)$ plus the integral over $(0, t)$, Equation (11-28) can also be written in the form

$$\begin{aligned} \boldsymbol{x}(t) &= \boldsymbol{\varphi}(t) \int_{-\infty}^{0} \boldsymbol{\varphi}^{-1}(\lambda)\boldsymbol{u}(\lambda)\, d\lambda + \boldsymbol{\varphi}(t) \int_{0}^{t} \boldsymbol{\varphi}^{-1}(\lambda)\boldsymbol{u}(\lambda)\, d\lambda \\ &= \boldsymbol{\varphi}(t)\boldsymbol{x}(0) + \boldsymbol{\varphi}(t) \int_{0}^{t} \boldsymbol{\varphi}^{-1}(\lambda)\boldsymbol{u}(\lambda)\, d\lambda \end{aligned} \tag{11-29}$$

where $\boldsymbol{x}(0)$ can be interpreted as a matrix of the n initial conditions $x_1(0)$, $x_2(0)$, $\cdots$, $x_n(0)$. Hence, Equation (11-29) gives the n time functions $x_1(t)$, $x_2(t)$, $\cdots$, $x_n(t)$ in terms of $\boldsymbol{\varphi}$, the n initial conditions $x_1(0)$, $x_2(0)$, $\cdots$, $x_n(0)$, and the n forcing functions $u_1(\lambda)$, $u_2(\lambda)$, $\cdots$, $u_n(\lambda)$ for $0 \leq \lambda \leq t$. The first term in Equation (11-29) is called the energy-storage term because $\boldsymbol{x}(0)$ depends on the energy stored in the network at $t = 0$. If the network contains no stored energy at $t = 0$, then $\boldsymbol{x}(0) = \mathbf{0}$. The second term depends on the forcing functions and is called the forced response. If the network contains no forcing functions, that is, $\boldsymbol{u}(\lambda) = \mathbf{0}$, then the forced response is zero.

11-4 A Solution Algorithm

In this section we summarize the procedure of the preceding sections in the form of an algorithm for solving any set of normal-form differential equations.

An algorithm for solving any set of n normal-form differential equation is as follows:

1. Compute the eigenvalues of the network matrix. This gives the n natural frequencies p_i $(i = 1, 2, \cdots, n)$.
2. Use Equations (11-11) and (11-14) to compute the n^2 constants k_{ij} $(i, j = 1, 2, \cdots, n)$.
3. Construct the $\boldsymbol{\varphi}$ matrix according to Equation (11-17).
4. Compute $\boldsymbol{\varphi}^{-1}$.
5. Substitute $\boldsymbol{\varphi}$, $\boldsymbol{\varphi}^{-1}$ and the matrix of forcing functions $\boldsymbol{u}$ into Equation (11-28):
a. Compute $\boldsymbol{\varphi}^{-1}\boldsymbol{u}$.
b. Integrate each element of the column matrix $\boldsymbol{\varphi}^{-1}\boldsymbol{u}$.
c. Premultiply the matrix obtained in part b by $\boldsymbol{\varphi}$.

This algorithm assumes all p_i computed in step 1 are distinct. If they are not, modifications as given in Section 11-5 are required.

Example 11-1

In this example we use the nth-order solution algorithm of Section 11-4 to compute the two capacitance voltages for the network presented in Figure 11-1. This is the same problem considered in Examples 10-10 and 10-12.

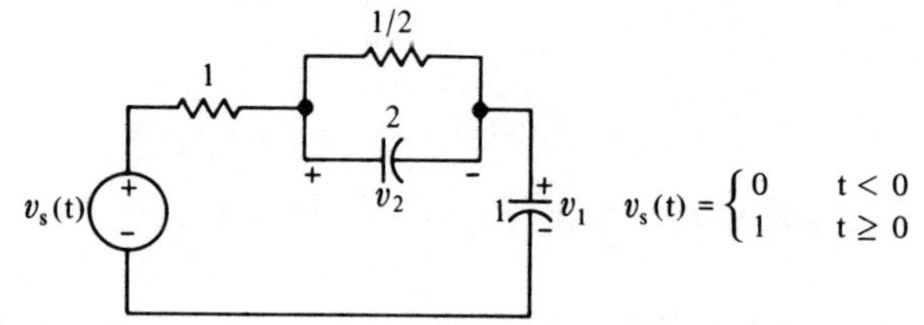

Figure 11-1 Network for Example 11-1.

Step 1 of the solution algorithm dictates that we compute the eigenvalues of the network matrix. The **a** matrix for this network was computed in Example 10-10, where we found

$$\mathbf{a} = \begin{bmatrix} -1 & -1 \\ -\frac{1}{2} & -\frac{3}{2} \end{bmatrix} \tag{11-30}$$

The eiginvalues of this matrix are given by the solution to the equation

$$\begin{aligned} 0 &= \begin{vmatrix} -1 - p & -1 \\ -\frac{1}{2} & -\frac{3}{2} - p \end{vmatrix} = (p + 1)(p + \tfrac{3}{2}) - \tfrac{1}{2} \\ &= p^2 + \tfrac{5}{2}p + 1 \end{aligned} \tag{11-31}$$

Hence, we find

$$\begin{Bmatrix} p_1 \\ p_2 \end{Bmatrix} = -\tfrac{5}{4} \pm \sqrt{(\tfrac{5}{4})^2 - 1} = -\tfrac{5}{4} \pm \tfrac{3}{4} = \begin{Bmatrix} -\frac{1}{2} \\ -2 \end{Bmatrix} \tag{11-32}$$

$p_1 = -\frac{1}{2}$ and $p_2 = -2$ are the two natural frequencies of the network. According to step 2, we must use Equations (11-11) and (11-14) to compute k_{11}, k_{12}, k_{21}, and k_{22}. The first two are easy; that is, from Equation (11-14) we write

$$k_{11} = 1, \qquad k_{12} = 1 \tag{11-33}$$

The second two are given by Equation (11-11). For this case the $(n - 1) \times (n - 1)$ determinants are 1×1's, and we have

$$k_{2i} = k_2(p_i) = \frac{|-(a_{11} - p_i)|}{|a_{12}|} = \frac{-a_{11} + p_i}{a_{12}}, \qquad i = 1, 2 \tag{11-34}$$

Hence

$$\begin{aligned} k_{21} &= \frac{-(-1) + (-\frac{1}{2})}{-1} = -\tfrac{1}{2} \\ k_{22} &= \frac{-(-1) + (-2)}{-1} = 1 \end{aligned} \tag{11-35}$$

The third step is to construct the $\boldsymbol{\varphi}$ matrix according to Equation (11-17). Inserting Equations (11-32), (11-33), and (11-35) into Equation (11-17), we obtain

$$\boldsymbol{\varphi} = \begin{bmatrix} e^{-(1/2)t} & e^{-2t} \\ -\frac{1}{2}e^{-(1/2)t} & e^{-2t} \end{bmatrix} \tag{11-36}$$

Step 4 requires the inverse of $\boldsymbol{\varphi}$. The formula for the inverse of a 2 × 2 matrix is derived in Appendix I as Equation (AI-16). Applying this formula to Equation (11-36) gives

$$\begin{aligned} \boldsymbol{\varphi}^{-1} &= \frac{1}{e^{-(1/2)t}e^{-2t} + \frac{1}{2}e^{-(1/2)t}e^{-2t}} \begin{bmatrix} e^{-2t} & -e^{-2t} \\ \frac{1}{2}e^{(1/2)t} & e^{-(1/2)t} \end{bmatrix} \\ &= \tfrac{2}{3}e^{(5/2)t} \begin{bmatrix} e^{-2t} & -e^{-2t} \\ \frac{1}{2}e^{-(1/2)t} & e^{-(1/2)t} \end{bmatrix} \\ &= \begin{bmatrix} \frac{2}{3}e^{(1/2)t} & -\frac{2}{3}e^{(1/2)t} \\ \frac{1}{3}e^{2t} & \frac{2}{3}e^{2t} \end{bmatrix} \end{aligned} \tag{11-37}$$

The final step is to substitute the matrices $\boldsymbol{\varphi}$, $\boldsymbol{\varphi}^{-1}$, and $\boldsymbol{u}$ into the solution equation given by Equation (11-28). In Example 10-10 we found that $\boldsymbol{u}$ is given by

$$\begin{aligned} \boldsymbol{u} &= \begin{bmatrix} v_s \\ \frac{1}{2}v_s \end{bmatrix} \\ &= \begin{cases} \begin{bmatrix} 0 \\ 0 \end{bmatrix} & \lambda < 0 \\ \begin{bmatrix} 1 \\ \frac{1}{2} \end{bmatrix} & \lambda \geq 0 \end{cases} \end{aligned} \tag{11-38}$$

Therefore,

$$\begin{aligned} \boldsymbol{\varphi}^{-1}(\lambda)\boldsymbol{u}(\lambda) &= \begin{bmatrix} \frac{2}{3}e^{(1/2)\lambda} & -\frac{2}{3}e^{(1/2)\lambda} \\ \frac{1}{3}e^{2\lambda} & \frac{2}{3}e^{2\lambda} \end{bmatrix} \begin{bmatrix} 0 \\ 0 \end{bmatrix} = \begin{bmatrix} 0 \\ 0 \end{bmatrix} & \lambda < 0 \\ \boldsymbol{\varphi}^{-1}(\lambda)\boldsymbol{u}(\lambda) &= \begin{bmatrix} \frac{2}{3}e^{(1/2)\lambda} & -\frac{2}{3}e^{(1/2)\lambda} \\ \frac{1}{3}e^{2\lambda} & \frac{2}{3}e^{2\lambda} \end{bmatrix} \begin{bmatrix} 1 \\ \frac{1}{2} \end{bmatrix} = \begin{bmatrix} \frac{1}{3}e^{(1/2)\lambda} \\ \frac{2}{3}e^{2\lambda} \end{bmatrix} & \lambda \geq 0 \end{aligned} \tag{11-39}$$

We can now evaluate the integrals:

$$\int_{-\infty}^{t} \boldsymbol{\varphi}^{-1}(\lambda)\boldsymbol{u}(\lambda)\, d\lambda = \int_{-\infty}^{0} \boldsymbol{\varphi}^{-1}(\lambda)\boldsymbol{u}(\lambda)\, d\lambda + \int_{0}^{t} \boldsymbol{\varphi}^{-1}(\lambda)\boldsymbol{u}(\lambda)\, d\lambda \tag{11-40}$$

But

$$\begin{aligned} \int_{-\infty}^{0} \boldsymbol{\varphi}^{-1}(\lambda)\boldsymbol{u}(\lambda)\, d\lambda &= \begin{bmatrix} \int_{-\infty}^{0} 0\, d\lambda \\ \int_{-\infty}^{0} 0\, d\lambda \end{bmatrix} = \begin{bmatrix} 0 \\ 0 \end{bmatrix} \\ \int_{0}^{t} \boldsymbol{\varphi}^{-1}(\lambda)\boldsymbol{u}(\lambda)\, d\lambda &= \begin{bmatrix} \int_{0}^{t} \frac{1}{3}e^{(1/2)\lambda}\, d\lambda \\ \int_{0}^{t} \frac{2}{3}e^{2\lambda}\, d\lambda \end{bmatrix} = \begin{bmatrix} (\frac{2}{3}e^{(1/2)t} - \frac{2}{3}) \\ (\frac{1}{3}e^{2t} - \frac{1}{3}) \end{bmatrix} \end{aligned} \tag{11-41}$$

Finally, we premultiply Equation (11-41) by $\boldsymbol{\varphi}$ to obtain the required solution:

$$
\begin{aligned}
\boldsymbol{x} &= \boldsymbol{\varphi} \int_0^t \boldsymbol{\varphi}^{-1}(\lambda)\boldsymbol{u}(\lambda)\, d\lambda \\
&= \begin{bmatrix} e^{-(1/2)t} & e^{-2t} \\ -\frac{1}{2}e^{-(1/2)t} & e^{-2t} \end{bmatrix} \begin{bmatrix} (\frac{2}{3}e^{(1/2)t} - \frac{2}{3}) \\ (\frac{1}{3}e^{2t} - \frac{1}{3}) \end{bmatrix} \\
&= \begin{bmatrix} (1 - \frac{2}{3}e^{-(1/2)t} - \frac{1}{3}e^{-2t}) \\ (\frac{1}{3}e^{-(1/2)t} - \frac{1}{3}e^{-2t}) \end{bmatrix}
\end{aligned}
\qquad \textbf{(11-42)}
$$

The result for $v_1(t)$ is the same as the result obtained in Example 10-10.

11-5 Multiple Natural Frequencies

In Chapter 10 and in the preceding sections of Chapter 11, we mentioned the possibility of multiple natural frequencies (two or more natural frequencies having the same value), but we excluded these cases from our analysis. In this section we consider the case of multiple natural frequencies.

The solution algorithm of Section 11-4 is not valid for networks with multiple natural frequencies. However, the principles used to formulate the algorithm do apply. To illustrate the method for multiple natural frequencies, we first consider a second-order network whose two natural frequencies are identical. Recall that the network natural frequencies are the eigenvalues of the network matrix. Hence, we first consider a second-order network for which **a** has two identical eigenvalues. We then generalize the method for the case of m identical natural frequencies.

Example 11-2

Consider the parallel RLC network of Figure 11-2. The numbers are considerably simplified if we set $L = C = 1$ and select R to give equal natural frequencies.

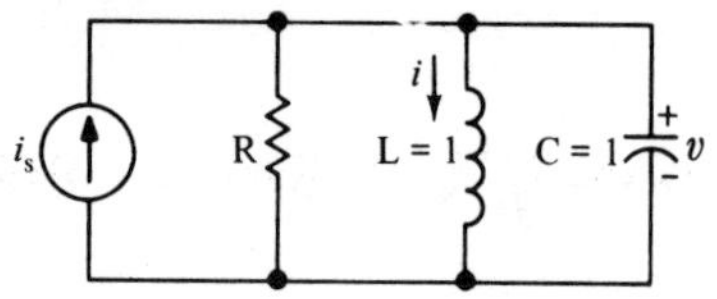

Figure 11-2 Network for Example 11-2.

To obtain the normal-form equations for this network, we could resort to the normal-form equation algorithm of Section 10-2. However, for this simple network it is easier to formulate the two equations directly from the network. We know that the two unknowns are the inductance current i and the capacitance voltage v. Furthermore, the derivative terms are due to the inductance voltage $L\dot{i}$ and the capacitance current

$C\dot{v}$. Therefore, we use KVL to write $L\dot{i}$ in terms of i and v:

$$L\dot{i} = \dot{i} = v \tag{11-43}$$

and we use KCL to write $C\dot{v}$ in terms of i and v:

$$C\dot{v} = \dot{v} = \left(i_s - \frac{1}{R}v\right) - i \tag{11-44}$$

Rearranging these equations into the normal form, we have for $L = C = 1$

$$\begin{aligned} \dot{v} &= -\frac{1}{R}v - i + i_s \\ \dot{i} &= v \end{aligned} \tag{11-45}$$

Therefore, the network matrix is given by

$$a = \begin{bmatrix} -1/R & -1 \\ 1 & 0 \end{bmatrix} \tag{11-46}$$

and the eigenvalues of the network matrix are given by the solutions to the equation

$$\begin{aligned} 0 &= |\mathbf{a} - p\mathbf{1}| \\ &= \begin{vmatrix} -\frac{1}{R} - p & -1 \\ 1 & -p \end{vmatrix} \\ &= \left(-\frac{1}{R} - p\right)(-p) - (-1)(1) = p^2 + \frac{1}{R}p + 1 \end{aligned} \tag{11-47}$$

Hence, the eigenvalues are given by

$$\left\{\begin{matrix} p_1 \\ p_2 \end{matrix}\right\} = -\frac{1}{2R} \pm \sqrt{\frac{1}{4R^2} - 1} \tag{11-48}$$

If $R = \frac{1}{2}$, then

$$p_1 = p_2 = -1; \qquad R = \tfrac{1}{2} \tag{11-49}$$

Note that for $R > \frac{1}{2}$ then both natural frequencies (eigenvalues) are real; for $R < \frac{1}{2}$ both are complex.

For $R = \frac{1}{2}$ the homogeneous equation is given by

$$\begin{aligned} \dot{v} &= -2v - i \\ \dot{i} &= v \end{aligned} \tag{11-50}$$

One solution to the homogeneous equation is given by

$$\begin{aligned} v &= B_1e^{p_1t} = B_1e^{-t} \\ i &= p_1B_1e^{p_1t} = -B_1e^{-t} \end{aligned} \tag{11-51}$$

as described in Section 10-3. For $p_1 \neq p_2$, a second solution is given by

$$\begin{aligned} v &= B_2 e^{p_2 t} \\ i &= p_2 B_2 e^{p_2 t} \end{aligned} \tag{11-52}$$

as also determined in Section 10-3. For $p_1 = p_2$, Equations (11-52) and Equations (11-51) are identical, so that Equations (11-52) do not provide a second solution to the homogeneous equation.

It is known that for $p_2 = p_1$, a second solution to the homogeneous equation can be obtained by choosing v to be of the form

$$\begin{aligned} v &= B_2 t e^{p_2 t} \\ &= B_2 t e^{-t} \end{aligned} \tag{11-53}$$

With this form for v, the corresponding form for i can be found by solving the first of Equations (11-50) for i; that is,

$$i = -2v - \dot{v} \tag{11-54}$$

and substituting v from Equation (11-53):

$$\begin{aligned} i &= -2B_2 t e^{-t} - (B_2 e^{-t} - B_2 t e^{-t}) \\ &= -B_2(t + 1)e^{-t} \end{aligned} \tag{11-55}$$

Note that for the second solution, v [Equation (11-53)] and i [Equation (11-55)] are not related by a simple constant, as they are in the first solution [Equations (11-51)].

The general solution to the homogeneous equation is the sum of these two solutions; that is,

$$\begin{aligned} v &= B_1 e^{-t} + B_2 t e^{-t} \\ i &= -B_1 e^{-t} - B_2(t + 1)e^{-t} \end{aligned} \tag{11-56}$$

As with the case of distinct roots, this is as far as we can go with the homogeneous equation. In summary, we have found the solution to the homogeneous equation given by Equations (11-50) to within the two arbitrary constants B_1 and B_2.

The procedure for finding the solution to the complete equation is the same as the procedure followed for the case of distinct natural frequencies. That is, we assume a solution to the complete equations

$$\begin{aligned} \dot{v} &= -2v - i + i_s \\ \dot{i} &= v \end{aligned} \tag{11-57}$$

to be of a form similar to the solution to the homogeneous equation, but with the constants B_1 and B_2 replaced by time functions b_1 and b_2; that is,

$$\begin{aligned} v &= b_1 e^{-t} + b_2 t e^{-t} \\ i &= -b_1 e^{-t} - b_2(t + 1)e^{-t} \end{aligned} \tag{11-58}$$

The remaining step is to substitute Equations (11-58) into Equations

(11-57) and solve for the required values of b_1 and b_2. Proceeding toward this end, we compute the derivatives of Equations (11-58):

$$\begin{aligned} \dot{v} &= \dot{b}_1 e^{-t} - b_1 e^{-t} + \dot{b}_2 te^{-t} + b_2[e^{-t} - te^{-t}] \\ \dot{i} &= -\dot{b}_1 e^{-t} + b_1 e^{-t} - \dot{b}_2(t+1)e^{-t} - b_2[e^{-t} - (t+1)e^{-t}] \end{aligned} \tag{11-59}$$

Now substituting Equations (11-58) and (11-59) into Equations (11-57) and grouping terms, we obtain

$$\begin{aligned} \dot{b}_1 e^{-t} + \dot{b}_2 te^{-t} &= i_s \\ -\dot{b}_1 e^{-t} - \dot{b}_2(t+1)e^{-t} &= 0 \end{aligned} \tag{11-60}$$

Note that the terms containing b_1 and b_2 dropped out because their coefficients summed to zero. This great simplification, as in the case of distinct natural frequencies [see Equations (10-81) and (11-24)], occurs because of the assumed functional form. It is a consequence of the method (and the reason we use this method). Equations (11-60) can also be written in matrix notation:

$$\begin{bmatrix} e^{-t} & te^{-t} \\ -e^{-t} & -(t+1)e^{-t} \end{bmatrix} \begin{bmatrix} \dot{b}_1 \\ \dot{b}_2 \end{bmatrix} = \begin{bmatrix} i_s \\ 0 \end{bmatrix} \tag{11-61}$$

The coefficient matrix on the left of Equation (11-61) is a matrix whose columns are solutions to the homogeneous equation. This is exactly the property of the φ matrix defined in Equation (11-17). We can solve for $\dot{b}_1$ and $\dot{b}_2$, provided Equations (11-60) are linearly independent. This requires that the coefficient matrix on the left of Equations (11-61) (the φ matrix) be nonsingular. The φ matrix is nonsingular if its columns are linearly independent. The columns of the φ matrix in Equation (11-61) are obviously linearly independent. (This was guaranteed from the start, since the columns of the φ matrix are the solutions to the homogeneous equation, and the two solutions to the homogeneous equation were chosen to be linearly independent at the outset.)

Solving Equations (11-60) for $\dot{b}_1$ and $\dot{b}_2$, we obtain

$$\begin{aligned} \dot{b}_1 &= (t+1)e^{t} i_s \\ \dot{b}_2 &= -e^{t} i_s \end{aligned} \tag{11-62}$$

Integrating Equations (11-62) and substituting these values for b_1 and b_2 into Equations (11-58) we obtain the final result:

$$\begin{aligned} v(t) &= e^{-t} \int_{-\infty}^{t} (\lambda+1)e^{\lambda} i_s(\lambda)\, d\lambda - te^{-t} \int_{-\infty}^{t} e^{\lambda} i_s(\lambda)\, d\lambda \\ i(t) &= -e^{-t} \int_{-\infty}^{t} (\lambda+1)e^{\lambda} i_s(\lambda)\, d\lambda - (t+1)e^{-t} \int e^{\lambda} i_s(\lambda)\, d\lambda \end{aligned} \tag{11-63}$$

Equations (11-63) can also be arranged into the form

$$\begin{aligned} v(t) &= \int_{-\infty}^{t} [1 - (t-\lambda)]e^{-(t-\lambda)} i_s(\lambda)\, d\lambda \\ i(t) &= \int_{-\infty}^{t} (t-\lambda)e^{-(t-\lambda)} i_s(\lambda)\, d\lambda \end{aligned} \tag{11-64}$$

Equations (11-64) express the result in the form of a convolution integral like Equation (10-34). In both of Equations (11-64) i_s corresponds to the function g in Equation (10-34); in the first equation, $[1 - (t - \lambda)] e^{-(t-\lambda)}$ corresponds to the function $h(t - \lambda)$ and in the second equation, $(t - \lambda)e^{-(t-\lambda)}$ corresponds to $h(t - \lambda)$.

In the preceding example the network matrix had two identical eigenvalues. For this case two linearly independent solutions to the homogeneous equation were derived by assigning y_1 the forms $y_1 = B_1e^{pt}$, and $y_1 = B_1te^{pt}$, where p is the eigenvalue. The required forms for y_2 were then computed from the homogeneous equation. Except for special cases, such as the one illustrated in Example 11-3, if an eigenvalue is repeated k times, then k linearly independent solutions to the homogeneous equation can be generated by setting

$$\begin{aligned} y_1 &= B_1e^{pt} \\ y_1 &= B_1te^{pt} \\ y_1 &= B_1t^2e^{pt} \\ &\;\;\vdots \\ y_1 &= B_1t^{k-1}e^{pt} \end{aligned} \tag{11-65}$$

The required forms for $y_2, y_3, \cdots, y_k$ can then be obtained from the homogeneous equation. Once the k linearly independent solutions to the homogeneous equation are generated, the solution to the complete equation is the same as for the case of distinct eigenvalues. That is, Equation (11-28) is valid for all cases. Only the method for generating the $\boldsymbol{\varphi}$ matrix depends on whether or not the network matrix has multiple eigenvalues.

Suppose we have an nth-order network with n natural frequencies and that $k \leq n$ of the natural frequencies are identical and the remaining n-k natural frequencies are distinct. Then n linearly independent solutions to the homogeneous equation can be generated by setting

$$\begin{aligned} y_1 &= B_1e^{pt} \\ y_1 &= B_1te^{pt} \\ &\;\;\vdots \\ y_1 &= B_1t^{k-1}e^{pt} \\ y_1 &= B_1e^{p_{k+1}t} \\ y_1 &= B_1e^{p_{k+2}t} \\ &\;\;\vdots \\ y_1 &= B_1e^{p_nt} \end{aligned} \tag{11-66}$$

where p is the natural frequency that is repeated k times, and p_{k+1},

$p_{k+2}, \cdots, p_n$ are the remaining n-k distinct natural frequencies. For example, if the network matrix has the seven eigenvalues $[-1, -1, -2, -2, -2, -3, -4]$, then seven linearly independent solutions to the homogeneous equation can be generated by setting

$$\begin{aligned} y_1 &= B_1e^{-t} \\ y_1 &= B_1te^{-t} \\ y_1 &= B_1e^{-2t} \\ y_1 &= B_1te^{-2t} \\ y_1 &= B_1t^2e^{-2t} \\ y_1 &= B_1e^{-3t} \\ y_1 &= B_1e^{-4t} \end{aligned} \tag{11-67}$$

Example 11-3

Sometimes a network will have two or more simultaneous roots to its characteristic polynomial and not require the special form of solution given by Equation (11-66). Such networks can be separated into separate pieces, as this example shows. Sometimes the separation is not obvious from the network diagram or from the equations.

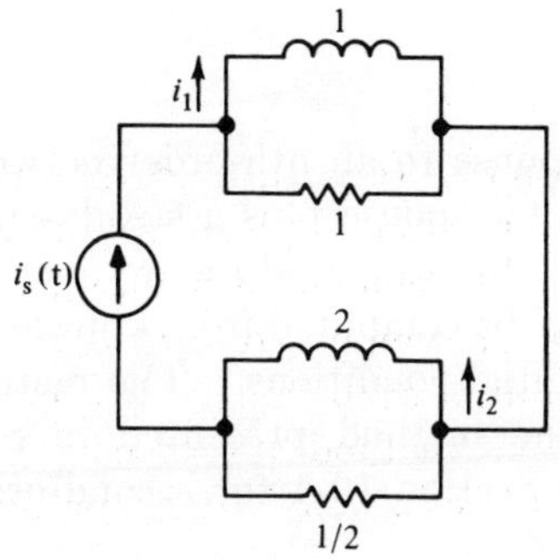

Figure 11-3 Network for Example 11-3.

Consider the network presented in Figure 11-3. The equations for this network are

$$\begin{aligned} \dot{i}_1 &= -i_1 + i_s \\ \dot{i}_2 &= -i_2 + i_s \end{aligned} \tag{11-68}$$

The eigenvalues of the **a** matrix are given by

$$\begin{vmatrix} p+1 & 0 \\ 0 & p+1 \end{vmatrix} = (p+1)^2 \quad \text{or} \quad p_1 = p_2 = -1 \tag{11-69}$$

For the solution to the homogeneous equation we let

$$y_1 = C_1e^{-t}; \qquad y_2 = C_2e^{-t} \tag{11-70}$$

Substituting into the homogeneous part of Equation (11-68) gives

$$\begin{aligned} -C_1e^{-t} &= -C_1e^{-t} \\ -C_2e^{-t} &= -C_2e^{-t} \end{aligned} \tag{11-71}$$

There is no relation between C_1 and C_2. If we had used the formula (11-11) above, we would have gotten a zero denominator either way. Thus two linearly independent solutions are

$$\boldsymbol{y}_1 = \begin{bmatrix} C_1 \\ 0 \end{bmatrix} e^{pt} \quad \text{and} \quad \boldsymbol{y}_2 = \begin{bmatrix} 0 \\ C_2 \end{bmatrix} e^{pt} \tag{11-72}$$

The two vectors in Equation (11-72) are linearly independent. Had we tried a solution in the form te^{-t}, we would not have been able to satisfy the equation. From Equation (11-72) the remainder of the solution is straightforward.

The problem with the example is obvious from Figure 11-3 and Equations (11-68). This is two separate first-order networks connected at only one common node. They operate independently as the equations indicate. There are more complicated examples where the degeneracy is not so obvious. If these cases are encountered, the methods discussed above will fail because of zeros in the denominators of Equations (11-11). In those cases, one can check for such degenerate situations by the methods of Reference [35].

11-6 The Time-Constant Concept

When the input is periodic, the total response to an nth-order network can be expressed as the sum of a transient response plus a steady-state response. If the steady-state response can be computed by any method whatsoever, then the transient response can be computed from knowledge of the n natural frequencies and the n initial conditions. The method is a straightforward generalization of the method presented in Section 10-3 for first-order networks and in Section 10-4 for second-order networks.

We start with Equation (11-29) and assume that the integral can be evaluated to give

$$\int \boldsymbol{\varphi}^{-1}(\lambda)\boldsymbol{u}(\lambda) = \boldsymbol{h}(\lambda) \tag{11-73}$$

Then

$$\int_0^t \boldsymbol{\varphi}^{-1}(\lambda)\boldsymbol{u}(\lambda)\, d\lambda = \boldsymbol{h}(t) - \boldsymbol{h}(0) \tag{11-74}$$

and Equation (11-29) can be written in the form

$$\begin{aligned} \boldsymbol{x}(t) &= \boldsymbol{\varphi x}(0) + \boldsymbol{\varphi h}(t) - \boldsymbol{\varphi h}(0) \\ &= \boldsymbol{\varphi}[\boldsymbol{x}(0) - \boldsymbol{h}(0)] + \boldsymbol{\varphi h}(t) \\ &= \boldsymbol{x}_{tr} + \boldsymbol{x}_{ss} \end{aligned} \tag{11-75}$$

Each element of the matrix $\boldsymbol{x}(0) - \boldsymbol{h}(0)$ is a constant. Hence, the jth response (the jth element of $\boldsymbol{x}$) has a transient response (the jth element of $\boldsymbol{x}_{tr}$) that is a linear combination of the elements in the jth row of $\boldsymbol{\varphi}$.

Therefore, for an nth-order network with n distinct roots, each total response is of the form

$$x(t) = \gamma_1 e^{p_1 t} + \gamma_2 e^{p_2 t} + \cdots + \gamma_n e^{p_n t} + x_{ss}(t) \qquad \textbf{(11-76)}$$

where $p_1, p_2, \cdots, p_n$ are the n natural frequencies. Furthermore, an nth-order network has n (independent) energy-storage elements and the n amounts of energy stored at $t = 0$ can be translated into n initial conditions on $x(t)$. These n equations can be used to evaluate the n constants $\gamma_1, \gamma_2, \cdots, \gamma_n$.

For one or more multiple natural frequencies, the procedure is the same except that the elements of the matrix $\boldsymbol{\varphi}$ have a different form, as discussed in Section 11-5. For example, for a seventh-order network with natural frequencies $[-1, -1, -2, -2, -2, -3, -4]$, the response has the form

$$x(t) = \gamma_1 e^{-t} + \gamma_2 t e^{-t} + \gamma_3 e^{-2t} + \gamma_4 t e^{-2t} + \gamma_5 t^2 e^{-2t} + \gamma_6 e^{-3t} + \gamma_7 e^{-4t} + x_{ss}(t) \qquad \textbf{(11-77)}$$

After computing the steady-state response x_{ss} and the eigenvalues of the network matrix, the seven constants $\gamma_1, \gamma_2, \cdots, \gamma_7$ can be evaluated from using Equation (11-77) and the seven initial conditions.

Finally, we point out that for either case (distinct or multiple natural frequencies), terms corresponding to complex natural frequencies are best handled by expressing the pairs of exponential terms as damped sinusoids via Euler's equation. (See Section 10-4.)

11-7 Summary

In this chapter we have constructed the general solution for the equations of networks of the elements defined in Chapter 1. The analysis process begins with a network model from Chapter 1. Next we apply the network laws of Chapter 2 and the network topology concepts of Chapter 3 and formulate a set of differential equations as discussed in Section 10-2. The basic ideas behind developing a solution to these equations were presented for simple networks in Chapter 10. With Chapters 1, 2, 3 and 10 as background, we were able to develop the solution to the network equations in this chapter. In developing this solution the concepts of matrix theory are helpful in simplifying the notation and making the steps understandable.

The solution starts with a set of normal-form equations:

$$\dot{\mathbf{x}} = \mathbf{a}\mathbf{x} + \mathbf{u}$$

The first concept from matrix theory that was utilized was the construction of the characteristic polynomial of the matrix **a**. The roots of this polynomial, called the eigenvalues of the matrix **a**, were also the natural frequencies of the network. With these natural frequencies computed, we could construct the basic functional form of the solution of the equations.

The constants of the solution are constructed by the formulas (11-11) and displayed in the fundamental solution matrix $\boldsymbol{\varphi}$ defined by Equation (11-17). The formulas (11-11) provide a column of $\boldsymbol{\varphi}$ for each eigenvalue of $\mathbf{a}$. The column vectors of constants in each column of $\boldsymbol{\varphi}$ are often called eigenvectors in the matrix theory literature. The final form of the solution to the network equations is given in Equation (11-29) in terms of the $\boldsymbol{\varphi}$ matrix and its inverse, the vector of initial conditions $\boldsymbol{x}(0)$, and the vector of forcing functions $\boldsymbol{u}(t)$. Each element of the solution vector $\boldsymbol{x}(t)$ has the form of a convolution integral plus a solution to the homogeneous equation. When $\mathbf{a}$ is nth-order, there are n convolution terms and n homogeneous solution terms. Each convolution is that of an exponentional function and the forcing function. Each homogeneous solution term is an exponentional function. When the eigenvalues are distinct, which is the normal situation in engineering problems, the exponentials are all different. Should two or more eigenvalues be identical, then polynomials multiplied by exponentials may occur in the $\boldsymbol{\varphi}$ matrix and consequently in the convolutions and the natural response terms.

In the case of forcing functions with a well-defined steady state, the time-constant concept introduced in the previous chapter applies. The forced response will contain a transient term plus a steady-state term. The steady-state term has the same form as that of the forcing function. The transient term has the form of the natural (or energy-storage) response.

The development in this chapter emphasizes the form of the network response. With a knowledge of this form we can use the analysis methods of Chapter 4 to develop simpler computational techniques. These are discussed in the next chapter.

PROBLEMS

11-1 Write an equation (a third-order polynomial in p) whose solution yields the three natural frequencies of the network shown in Figure P11-1.

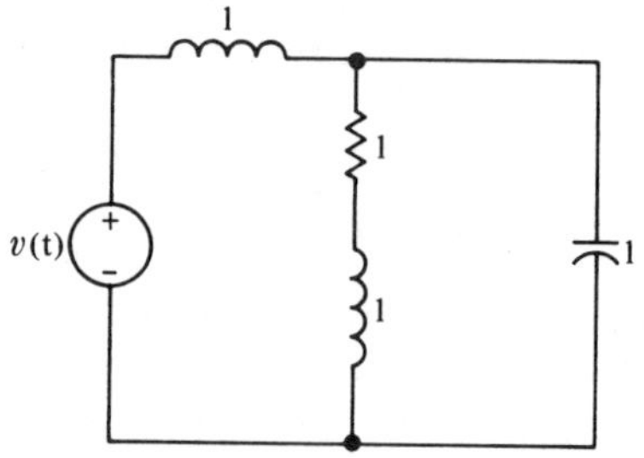

Figure P11-1

11-2 The network shown is a third-order transfer impedance Butterworth prototype filter. Thus you know the location of the poles.

a. Specify $\boldsymbol{x}$ and find the $\mathbf{a}$ and $\boldsymbol{u}$ matrices.

b. Find the solution matrix $\varphi(t)$.

c. Find the output voltage when the input is a step. (See Figure P11-2.)

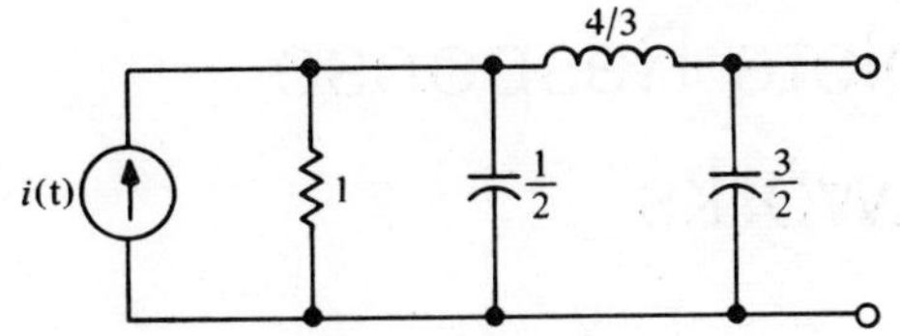

Figure P11-2

11-3 A network has the following $\mathbf{a}$ matrix:

$$\mathbf{a} = \begin{bmatrix} 1 & 1 & 0 \\ 0 & 1 & 0 \\ 0 & 0 & 1 \end{bmatrix}$$

Furthermore, the $\boldsymbol{u}$ vector can be written as a vector of constants time a scalar voltage.

$$\boldsymbol{u} = \begin{bmatrix} 1 \\ 1 \\ 1 \end{bmatrix} v(t)$$

a. Find the solution matrix $\varphi(t)$ and its inverse $\varphi^{-1}(t)$.

b. Compute the matrix product $\varphi(t)\varphi^{-1}(\tau)$.

CHAPTER

12

Computational Methods for the Complete Response of Linear Networks

12-1 Introduction

In this chapter we first discuss the relation of the port and terminal descriptions of networks (Chapters 5 and 6) to the normal-form equations (Chapters 10 and 11). Next, we relate the constants of the complete response to the steady-state analysis method of Chapter 4. The third section deals with a very useful class of forcing functions called the singularity functions. The final section relates these analysis techniques to the Laplace-Fourier transform methods that are common in the network literature.

12-2 Input–Output Descriptions of Networks

In Chapter 10 we developed network equations in the normal form

$$\dot{x} = \mathbf{a}x + u \tag{12-1}$$

where x is an n vector (n $\times$ 1 column matrix) whose elements are the capacitance voltages and inductance currents; **a** is an n × n matrix of constants, each element of which depends on the network parameters; ***u*** is an n vector (n × 1 matrix), called the forcing function, whose elements are constants times the independent source quantities. The constants depend on the network parameters.

Often a network has only certain voltages and currents that are of interest called *outputs*. The outputs may not be the capacitance voltages and the inductance currents defined by ***x***. Therefore, after Equation (12-1) has been solved for the capacitance voltages and inductance currents, the outputs can be computed by solving the algebraic equations of a network of resistances, controlled sources, and independent sources. This network is derived from the original network by replacing all capacitances by independent voltage sources and inductances with independent current sources, with the source quantities the capacitance voltages and inductance currents that are the solution to Equation (12-1). We first consider only networks containing no inductance cut sets or capacitance loops. Later, we show how to handle these cases. (See Example 12-3.)

Example 12-1

As an example, let us again consider the network considered in Examples 10-10, 10-12, and 11-1. The network is presented in Figure 12-1(a).

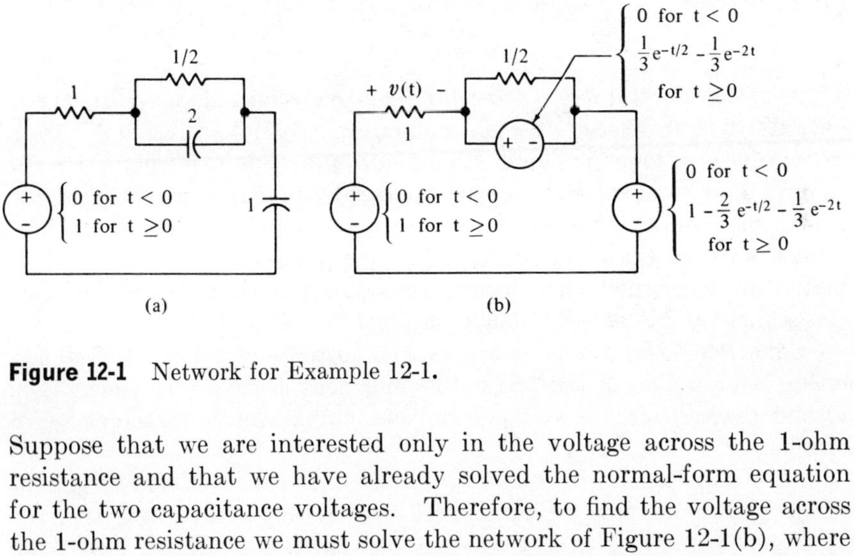

Figure 12-1 Network for Example 12-1.

Suppose that we are interested only in the voltage across the 1-ohm resistance and that we have already solved the normal-form equation for the two capacitance voltages. Therefore, to find the voltage across the 1-ohm resistance we must solve the network of Figure 12-1(b), where the capacitances have been replaced by independent voltage sources whose values are given by the known capacitance voltages. [These were computed in Example 11-1 by solving the normal-form equation and are given by Equation (11-42).] The voltage v across the 1-ohm resistance with the polarity shown in Figure 12-1(b) is easily found by KVL:

$$v(t) = \begin{cases} 0 & t < 0 \\ 1 - (1 - \frac{2}{3}e^{-t/2} - \frac{1}{3}e^{-2t}) - (\frac{1}{3}e^{-1/2t} - \frac{1}{3}e^{-2t}) & t \geq 0 \end{cases}$$
$$= \begin{cases} 0 & t < 0 \\ \frac{1}{3}e^{-t/2} + \frac{2}{3}e^{-2t} & t \geq 0 \end{cases} \quad \textbf{(12-2)}$$

In general, after the capacitance voltages and inductance currents have been computed by solving the normal-form (differential) equations, the remaining voltages and currents can be computed by solving a set of algebraic equations. The unknowns include the outputs (desired quantities) and sometimes some other voltages and currents that are not required and need not be computed. The knowns are independent source quantities that include the original independent sources, plus the capacitance voltages and inductance currents obtained by solving the normal-form equation. Since this network is linear, superposition applies and the effects of the original sources can be separated from the effects of the capacitances and inductances. Furthermore, since the network is resistive, each output is a linear combination of the independent source quantities.

Suppose that there are m voltages and currents considered as outputs, q original independent sources, and a total of n capacitances and inductances. Then each of the m outputs is a linear combination of the q (original) independent sources and the n capacitance voltages and inductance currents. Hence, we can write

$$\boldsymbol{w} = \mathbf{c}\boldsymbol{x} + \mathbf{d}\boldsymbol{v} \quad \textbf{(12-3)}$$

where $\boldsymbol{w}$ is an m vector (m $\times$ 1 column matrix) whose elements are the m outputs; $\boldsymbol{x}$ is an n vector (n $\times$ 1 column matrix) whose elements are n capacitance voltages and inductance currents; $\boldsymbol{v}$ is a q vector (q $\times$ 1 column matrix) whose elements are the q independent source parameters; $\mathbf{c}$ is an m $\times$ n matrix of constants that depend on the network resistive elements (resistances and controlled source parameters); $\mathbf{d}$ is an m $\times$ q matrix of constants that depend on the network resistive elements (resistances and controlled source parameters).

Equation (12-3) simply expresses each output (each element of $\boldsymbol{w}$) as a linear combination of the original independent sources (the elements of $\boldsymbol{v}$) and the capacitance voltages and inductance currents (the elements of $\boldsymbol{x}$).

Returning to Example 12-1, we note that Equation (12-2) can be arranged into the form of Equation (12-3) by noting that (for t $\geq$ 0)

$$\boldsymbol{w} = [v]$$
$$\boldsymbol{x} = \begin{bmatrix} (1 - \frac{2}{3}e^{-t/2} - \frac{1}{3}e^{-2t}) \\ (\frac{1}{3}e^{-t/2} - \frac{1}{3}e^{-2t}) \end{bmatrix} \quad \textbf{(12-4)}$$
$$\boldsymbol{v} = [1]$$

Hence for this case we have m = 1, q = 1, and n = 2. Writing Equation (12-2) in the form of Equation (12-3), we have

$$v = [c_{11} \quad c_{12}] \begin{bmatrix} (1 - \frac{2}{3}e^{-t/2} - \frac{1}{3}e^{-2t}) \\ (\frac{1}{3}e^{-t/2} - \frac{1}{3}e^{-2t}) \end{bmatrix} + [d][1] \tag{12-5}$$

Therefore, by comparing Equations (12-3) and (12-5), we recognize that **c** is the 2 × 1 matrix $[-1 \quad -1]$, and **d** is the 1 × 1 matrix [1].

Example 12-2

As a second illustration, we consider the network of Figure 12-2. This network is the same as that of Example 10-1, except for the addition of the current source i_0 at the right-hand terminal pair.

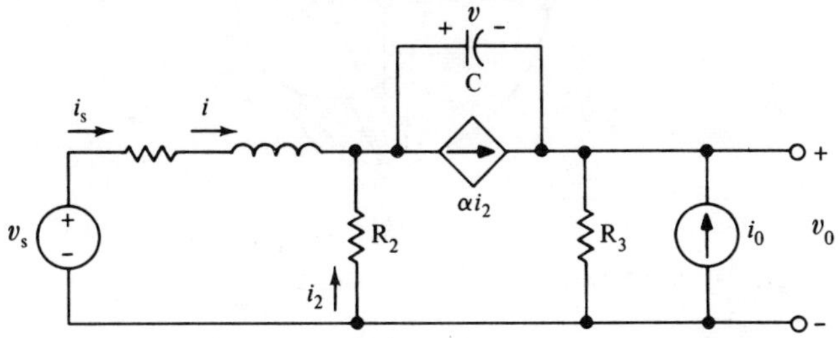

Figure 12-2 Network for Example 12-2.

The input vector $\boldsymbol{v}$ is a 2 × 1 matrix whose elements are the independent sources quantities v_s and i_0, and the vector $\boldsymbol{x}$ consists of the capacitance voltage and the inductance current (v and i). Let us take the output vector to be the current i_s and the voltage v_0; that is, we set

$$\boldsymbol{v} = \begin{bmatrix} v_s \\ i_0 \end{bmatrix} \qquad \boldsymbol{x} = \begin{bmatrix} v \\ i \end{bmatrix} \qquad \boldsymbol{w} = \begin{bmatrix} i_s \\ v_0 \end{bmatrix} \tag{12-6}$$

Therefore, for this example, m = 2, q = 2, n = 2, and Equation (12-3) is of the form

$$\begin{bmatrix} i_s \\ v_0 \end{bmatrix} = \begin{bmatrix} c_{11} & c_{12} \\ c_{21} & c_{22} \end{bmatrix} \begin{bmatrix} v \\ i \end{bmatrix} + \begin{bmatrix} d_{11} & d_{12} \\ d_{21} & d_{22} \end{bmatrix} \begin{bmatrix} v_s \\ i_0 \end{bmatrix} \tag{12-7}$$

In order to compute the matrices **c** and **d**, we refer to Figure 12-3(a). This network contains only resistances, controlled sources, and independent sources. We call such a network a resistive network, since it has no energy-storage elements and, therefore, no differentials or integrals appear in the equations. Furthermore, all Thevenin impedances are real (resistances). For this network, we see immediately that i_s equals i. To find v_0, we select a tree consisting of the R_3 branch, the R_1 branch, and the two voltage sources, as illustrated in Figure 11-3(b). Then the cut set of the R_3 branch is the R_2 link and the current sources

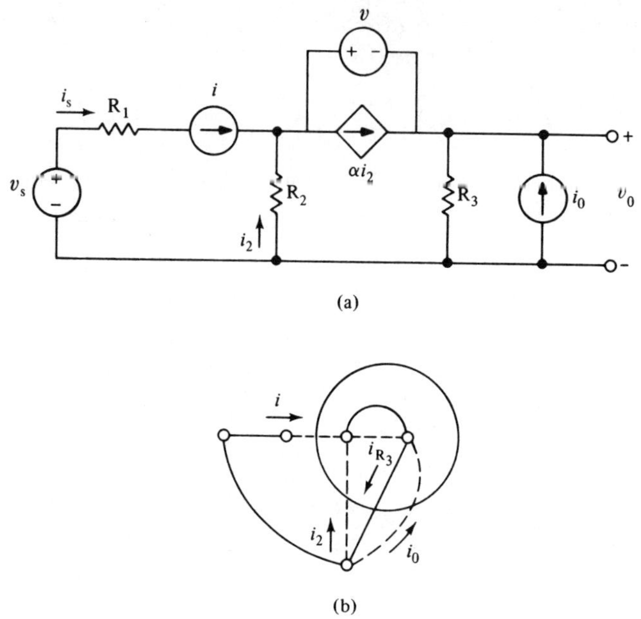

Figure 12-3 Resistive network for Example 12-2.

i_0 and i. Thus the current down through R_3 is given by

$$i_{R_3} = \frac{v_0}{R_3} = i + i_2 + i_0 \tag{12-8}$$

But the voltage across R_2 is $v + v_0$ so that $i_2 = -(v + v_0)/R_2$. Substituting this quantity for i_2 in Equation (12-8) gives

$$v_0 = R_3\left(i - \frac{v}{R_2} - \frac{v_0}{R_2} + i_0\right) \tag{12-9}$$

Solving for v_0 gives

$$v_0 = \frac{R_2R_3}{R_2 + R_3}\,i - \frac{R_3}{R_2 + R_3}\,v + \frac{R_2R_3}{R_2 + R_3}\,i_0 \tag{12-10}$$

In matrix notation the output is

$$\begin{bmatrix} i_s \\ v_0 \end{bmatrix} = \begin{bmatrix} 0 & 1 \\ -\dfrac{R_3}{R_2 + R_3} & \dfrac{R_2R_3}{R_2 + R_3} \end{bmatrix}\begin{bmatrix} v \\ i \end{bmatrix} + \begin{bmatrix} 0 & 0 \\ 0 & \dfrac{R_2R_3}{R_2 + R_3} \end{bmatrix}\begin{bmatrix} v_s \\ i_0 \end{bmatrix} \tag{12-11}$$

Hence, we have evaluated the matrices **c** and **d** of Equation (12-7).

When there are inductance and current source cut sets or capacitance and voltage source loops, then it is not easy to list all contingencies for getting the network equations in a standard form. Kirchhoff's laws are always correct, as are the rules of linear algebra. These and a little ingenuity are all that is needed. We illustrate this point by an example.

Example 12-3

As an example, let us reconsider the network of Example 10-2. This network is presented in Figure 12-4. For this example we have a single

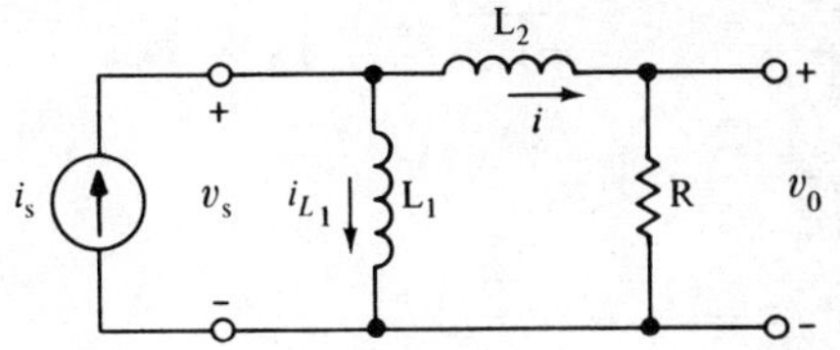

Figure 12-4

input, i_s (q = 1), and two outputs, v_s and v_0 (m = 2). As shown in Example 10-2, this network is characterized by a first-order differential equation given by

$$\dot{i} = -\frac{R}{L_1 + L_2} i + \frac{L_1}{L_1 + L_2} \dot{i}_s \tag{12-12}$$

The outputs in this case are readily computed by inspection of the network

$$\begin{aligned} v_s &= L_1 \dot{i}_{L_1} = -L_1 \dot{i} + L_1 \dot{i}_s \\ v_0 &= Ri \end{aligned} \tag{12-13}$$

Equations (12-12) and (12-13) give a complete description of the network in terms of the input i_s, the outputs v_1 and v_0, and i. If i is first computed by solving the normal-form equation [Equation (12-12)], then v_s and v_0 are easily computed by using Equations (12-13). If we attempt to write Equations (12-13) in the form of Equation (12-3), we encounter a difficulty—v_s depends on $\dot{i}$ and $\dot{i}_s$ as well as on i and i_s. Hence, a more general form given by

$$\mathbf{w} = \mathbf{cx} + \mathbf{dv} + \mathbf{e\dot{v}} \tag{12-14}$$

is required. Equations (12-13) can be put into this form by making the change of variable

$$x = i - \frac{L_1}{L_1 + L_2} i_s \tag{12-15}$$

or

$$i = x + \frac{L_1}{L_1 + L_2} i_s \tag{12-16}$$

Substituting Equation (12-16) into Equation (12-12) and rearranging terms, we find that the normal-form equation is now given by

$$\dot{x} = -\frac{R}{L_1 + L_2} x - \frac{L_1 R}{(L_1 + L_2)^2} i_s \tag{12-17}$$

Note that with x as the unknown, the normal-form equation involves only i_s and not $\dot{i}_s$. Substituting Equation (12-16) into Equations (12-13) and rearranging terms, we obtain

$$\begin{aligned} v_s &= \frac{L_1 R}{L_1 + L_2} x + \frac{L_1^2 R}{(L_1 + L_2)^2} i_s + \frac{L_1 L_2}{L_1 + L_2} \dot{i}_s \\ v_0 &= Rx + \frac{L_1 R}{L_1 + L_2} i_s \end{aligned} \tag{12-18}$$

Or, in matrix notation, we have

$$\begin{bmatrix} v_s \\ v_0 \end{bmatrix} = \begin{bmatrix} \dfrac{L_1 R}{L_1 + L_2} \\ R \end{bmatrix} [x] + \begin{bmatrix} \dfrac{L_1^2 R}{(L_1 + L_2)^2} \\ \dfrac{L_1 R}{L_1 + L_2} \end{bmatrix} [i_s] + \begin{bmatrix} \dfrac{L_1 L_2}{L_1 + L_2} \\ 0 \end{bmatrix} [\dot{i}_s] \tag{12-19}$$

Therefore, we have succeeded in getting Equation (12-12) into the normal form given by Equation (12-1), and we have succeeded in getting Equations (12-13) into the form of Equation (12-14).

12-3 The State-Variable Description of Networks

In the network literature the standard form for describing an nth-order network with q independent sources and m outputs is given by the pair of equations (called the state-variable equations)

$$\begin{aligned} \dot{\boldsymbol{x}} &= \mathbf{a}\boldsymbol{x} + \mathbf{b}\boldsymbol{v} \\ \boldsymbol{w} &= \mathbf{c}\boldsymbol{x} + \mathbf{d}\boldsymbol{v} + \mathbf{e}\dot{\boldsymbol{v}} \end{aligned} \tag{12-20}$$

where $\boldsymbol{v}$ is a q × 1 (column) matrix whose q elements are the q independent source quantities; $\dot{\boldsymbol{v}}$ is a q × 1 (column) matrix whose q elements are the derivatives of the corresponding elements of $\boldsymbol{v}$; $\boldsymbol{x}$ is a n × 1 (column) matrix whose n elements are time functions related to certain network voltages and currents; $\boldsymbol{w}$ is a m × 1 (column) matrix whose m elements are the m voltages and currents chosen as outputs; **a** is the n × n network matrix described in Chapter 10; **b** is a m × q matrix of constants defined by the equation $\boldsymbol{u} = \mathbf{b}\boldsymbol{v}$, where $\boldsymbol{u}$ is the matrix of forcing functions defined in Chapter 10; **c** is a n × m matrix of constants discussed in Section 12-2; **d** is a q × m matrix of constants discussed in Section 12-2; **e** is a q × m matrix of constants discussed in Section 12-2.

Equations (12-20) represent a total of n + m equations (n differential equations and m algebraic equations). Any linear nth-order network (consisting of the network elements defined in Chapter 1) with m outputs can be described in state-variable form. The state-variable form separates the network equations into the smallest possible number of first-order linear differential equations (n, where n is the number of independent energy-storage elements) and the smallest number of algebraic equations required to compute the m outputs (m).

The first equation of Equations (12-20) is the normal-form equation discussed at length in Chapters 10 and 11, except that we have separated the forcing function $\boldsymbol{u}$ into two components $\mathbf{b}$ and $\boldsymbol{v}$. The second equation of Equations (12-20) is the output description discussed at length in Section 12-2. Note that $\mathbf{a}$, $\mathbf{b}$, $\mathbf{c}$, $\mathbf{d}$, and $\mathbf{e}$ are matrices of constants whose elements depend on the network parameters (R's, L's, C's, M's, μ's, g_m's, r_m's, and α's). $\boldsymbol{v}$, $\dot{\boldsymbol{v}}$, $\boldsymbol{x}$, and $\boldsymbol{w}$ are column matrices of time functions. $\boldsymbol{v}$ is sometimes called the system *input* and $\boldsymbol{w}$ is called the system *output*. $\boldsymbol{x}$ is called the system *internal dynamic variable* or *state variable*.

We are usually given a network with one or more inputs and asked to compute one or more outputs. The network defines the matrices $\mathbf{a}$, $\mathbf{b}$, $\mathbf{c}$, $\mathbf{d}$, and $\mathbf{e}$, and the inputs define $\boldsymbol{v}$. From these we are to compute the outputs $\boldsymbol{w}$. The usual procedure is first to solve the first of Equations (12-20) for $\boldsymbol{x}$, and then to substitute $\boldsymbol{x}$ into the second of Equations (12-20) and solve for $\boldsymbol{w}$. Hence, the solution to the normal-form differential equation $\boldsymbol{x}$ may not be of interest in itself, but represents an intermediate step in the analysis. For networks containing no loops of capacitances and voltage sources or cut sets of inductances and current sources, the state variables $\boldsymbol{x}$ are the capacitance voltages and inductance currents. For networks containing loops of capacitances and/or cut sets of inductances and current sources, the network equations can still be put in the form of Equations (12-20), but the elements of $\boldsymbol{x}$ (the state variables) are linear combinations of the loop voltages or the cut set currents, as illustrated in Example 12-3.

From the results of Chapters 10 and 11 we know that if $\boldsymbol{x}(t_0)$ is known and the input $\boldsymbol{v}(t)$ is known between t_0 and t, then the state variables and the outputs at time t can be found. Furthermore, the form of the solution of the first of Equations (12-20) for the state variables is that of a convolution integral. These facts will be used in the next section to relate the state-variable description to the steady-state analysis methods of Chapter 4. From this relationship, we can obtain simpler methods of finding the constants in the various convolution expressions.

12-4 Relationship between Steady-State Analysis and the Complete Solution

The method of variation of parameters presented in Chapter 11 is a straightforward way to derive the n complete responses for any nth-order lumped, linear, time-invariant network to excitation by one or more independent sources of arbitrary form. For large n, say $n > 2$, the method is quite tedious. If all n responses for the nth-order system must be evaluated, this method is as good as any. On the other hand, when only one or a few of the n responses is required, the method still requires that we compute the matrix φ and its inverse φ^{-1} so that almost as much work is required to compute only one of the n responses as is required to compute all n responses. When only one (or a few) of the n responses is required, much of this tedium can be circumvented by making use of our knowledge of the form of the response. The idea is to relate the constants of the known form of the complete solution to constants of the steady-state response derived by the methods of Chapters 4, 5 and 6. These methods, along with the pole-zero concept of Chapter 8, allow us to construct the complete solution for any forcing functions with a minimum of computations.

Rather than going directly into the nth-order case, we first consider a first-order example and then a second-order example. Finally, we present an algorithm for the nth-order case and a partial proof that the method is valid.

Example 12-4

Consider the network presented in Figure 12-5(a). The independent voltage source can be any arbitrary function of time.

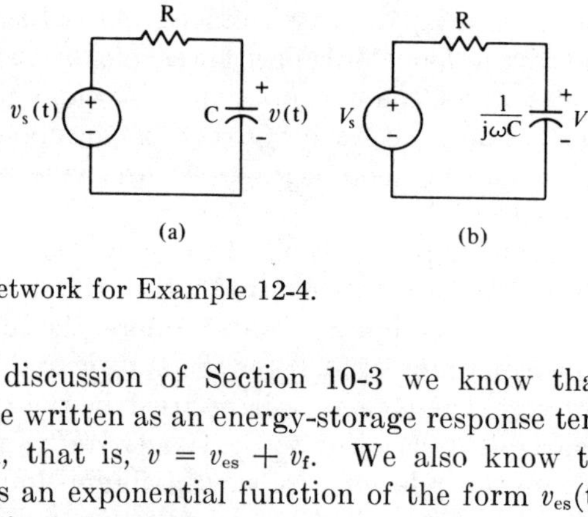

Figure 12-5 Network for Example 12-4.

From the discussion of Section 10-3 we know that the complete response can be written as an energy-storage response term plus a forced-response term, that is, $v = v_{es} + v_f$. We also know that the energy-storage term is an exponential function of the form $v_{es}(t) = \gamma e^{pt}$, where the constant p is the network natural frequency. We further know from Equation (10-32) that the forced term is given by the convolution integral

$$v_f(t) = \int_0^t u(\lambda)e^{p(t-\lambda)}\,d\lambda$$

where the constant p is again the natural frequency and the forcing function $u(\lambda)$ is some constant multiplied by the independent source;

that is, $u(\lambda) = \beta v_s(\lambda)$. Hence, we can write the complete solution in the form

$$v(t) = \gamma e^{pt} + \beta e^{pt} \int_0^t v_s(\lambda) e^{-p\lambda} \, d\lambda \tag{12-21}$$

The response for any first-order network can be written in this form. Hence, to solve any particular example, we can immediately write Equation (12-21); the only remaining problem is to evaluate the three constants γ, p, and β. In this example, we present one method for evaluating these constants.

Before we allow $v_s(t)$ to be an arbitrary time function, let us, for the moment, assume that $v_s(t)$ is a cosine wave of frequency ω. In this case the total response was found in Example 10-5 as Equation (10-50). In accordance with the technique of Chapter 4, we can construct the complex equivalent network illustrated in Figure 12-5(b). Next, we use the transfer-function concept of Chapter 8 to construct the transfer function from V_s to V; that is,

$$\frac{V}{V_s} = \frac{\dfrac{1}{RC}}{p + \dfrac{1}{RC}} = \frac{1}{RCp + 1} \tag{12-22}$$

The transfer function has a pole at $p = -1/RC$. Comparing this with Equation (10-50), we see that the pole is identical to the natural frequency. Hence, we have computed the constant p by computing the pole of the transfer function from the independent source to the response of interest.

When we expand the transfer function in its partial function expansion with respect to the variable p, we get

$$\frac{V}{V_s} = \frac{\dfrac{1}{RC}}{p + \dfrac{1}{RC}} \tag{12-23}$$

Comparing this with Equation (10-50) we see that the coefficient (numerator) of the partial fraction expansion is precisely the constant β; that is, $\beta = 1/RC$. Hence, we have computed the constant β by expanding the transfer function in a partial fraction expansion.

The remaining constant γ can be evaluated from the initial conditions by following a procedure similar to the one used in the time-constant method of Chapters 10 and 11.

Setting $t = 0+$ in Equation (12-21), we obtain

$$v(0+) = \gamma \tag{12-24}$$

The capacitance voltage cannot change instantaneously, so that $v(0-) = v(0+)$. Hence, if we are given the state of the network at $t = 0-$, for example, the energy stored in the capacitance at $t = 0-$, the constant γ is easily evaluated.

We conclude that the response $v(t)$, for any arbitrary independent source $v_s(t)$, is given by

$$v(t) = v(0)e^{-t/RC} + e^{-t/RC}\int_0^t \frac{1}{RC} v_s(\lambda)e^{\lambda/RC}\, d\lambda \tag{12-25}$$

This result can be verified by comparing it to the first equation of Equations (10-48) of Example 10-3, where we solved the same problem [with $v(0) = 0$, $v_s(\lambda) = \cos \omega\lambda$] by the method of variation of parameters.

In summary, we have computed the response to the network of Figure 12-5(a) by the following:

1. We wrote the known form of the response to within three constants.
2. We evaluated one constant (the network natural frequency) by writing the transfer function from the independent source to the response and finding the pole of the transfer function.
3. We evaluated the constant associated with the forced response (convolutional integral) by expanding the transfer function from the independent source to the response in its partial fraction expansion and noting the coefficient.
4. The constant associated with the energy-storage response was related to the network initial conditions.

Example 12-5

Consider the network presented in Figure 12-6(a). The independent voltage source can be any arbitrary function of time.

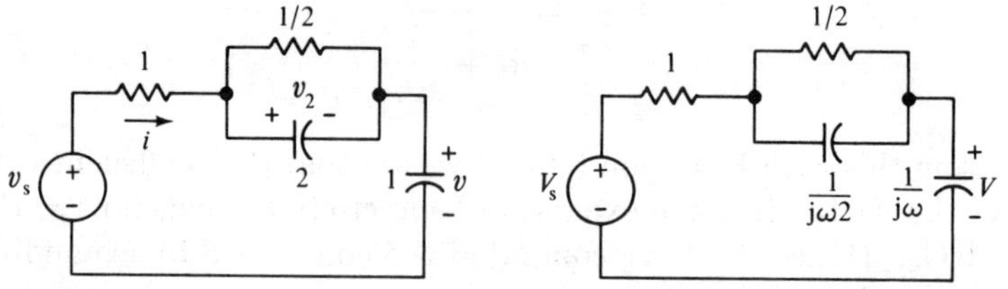

Figure 12-6 Network for Example 12-5.

The response v is known to consist of an energy-storage term plus a forced term; that is, $v = v_{es} + v_f$. The energy-storage term is known to be a sum of two exponentials [see, for example, Equation (10-87)]: $v_{es}(t) = \gamma_1 e^{p_1 t} + \gamma_2 e^{p_2 t}$. The forced response is known to be a sum of two convolution integrals [see, for example, Equation (10-87)]

$$v_f(t) = \beta_1 \int_0^t v_s(\lambda)e^{p_1(t-\lambda)}\,d\lambda + \beta_2 \int_0^t v_s(\lambda)e^{p_2(t-\lambda)}\,d\lambda$$

Therefore, the total response for any linear second-order network (with distinct natural frequencies) can be expressed in the form

$$v(t) = \gamma_1 e^{p_1 t} + \gamma_2 e^{p_2 t} + \beta_1 \int_0^t v_s(\lambda)e^{p_1(t-\lambda)}\,d\lambda + \beta_2 \int_0^t v_s(\lambda)e^{p_2(t-\lambda)}\,d\lambda \tag{12-26}$$

Therefore, to solve any particular second-order network we can start by writing the total response in the form given by Equation (12-26). The remaining problem is to evaluate the six constants p_1, p_2, γ_1, γ_2, β_1, β_2. In this example, we present a method for evaluating p_1, p_2, β_1, and β_2 from the sinusoidal steady-state transfer function and γ_1 and γ_2 from initial conditions.

Following the same procedure used in Example 12-4, let us assume a sinusoidal source with frequency ω and construct the complex equivalent network, as illustrated in Figure 12-6(b). The transfer function from the independent source to the quantity of interest is given by

$$\frac{V}{V_s} = \frac{p + 1}{p^2 + \frac{5}{2}p + 1} \tag{12-27}$$

The poles of the transfer function are the roots of the denominator polynomial $p^2 + \frac{5}{2}p + 1$. The solutions to $p^2 + \frac{5}{2}p + 1 = 0$ are given by

$$p_1 = -\tfrac{1}{2}, \qquad p_2 = -2 \tag{12-28}$$

The poles are also the network natural frequencies. (See Examples 10-10, 12.) Hence, we have evaluated the two constants p_1 and p_2 of Equation (12-26).

The next step is to expand the transfer function in its partial fraction expansion

$$\frac{V}{V_s} = \frac{p + 1}{p^2 + \frac{5}{2}p + 1} = \frac{\frac{1}{3}}{p + \frac{1}{2}} + \frac{\frac{2}{3}}{p + 2} = \frac{\beta_1}{p - p_1} + \frac{\beta_2}{p - p_2} \tag{12-29}$$

The coefficient of the p_1 term is β_1, and the coefficient of the p_2 term is β_2. Hence, the constants β_1 and β_2 are given by

$$\beta_1 = \tfrac{1}{3}, \qquad \beta_2 = \tfrac{2}{3} \tag{12-30}$$

Hence, the forced response is given by

$$v_f(t) = \tfrac{1}{3}e^{-(1/2)t} \int_0^t v_s(\lambda)e^{(1/2)\lambda}\,d\lambda + \tfrac{2}{3}e^{-2t} \int_0^t v_s(\lambda)e^{2\lambda}\,d\lambda \tag{12-31}$$

and the complete response is given by

$$v(t) = \gamma_1 e^{-(1/2)t} + \gamma_2 e^{-2t} + \tfrac{1}{3}e^{-(1/2)t}\int_0^t v_s(\lambda)e^{(1/2)\lambda}\,d\lambda + \tfrac{2}{3}e^{-2t}\int_0^t v_s(\lambda)e^{2\lambda}\,d\lambda \quad \text{(12-32)}$$

The remaining two constants, γ_1 and γ_2, can be evaluated from knowledge of the initial conditions, that is, the state of the system at $t = 0-$. For example, knowledge of the energy stored in each capacitance at $t = 0-$ is equivalent to knowledge of $v(0-)$ and $v_2(0-)$, which, in turn, are equal to $v(0+)$ and $v_2(0+)$ because the capacitance voltages cannot change instantaneously. Hence, $v(0+)$ is known so that we can set $t = 0+$ in Equation (12-32) to obtain $v(0+) = \gamma_1 + \gamma_2$. To obtain a second equation, we note from Figure 12-6(a) that at $t = 0+$ the current through the 1-ohm resistance is given by $i(0+) = v_s(0+) - v_2(0+) - v(0+)$. Since $v_s(0+)$, $v_2(0+)$, and $v(0+)$ are known, so is $i(0+)$. Note further that the current through the 1-farad capacitance is also $i(t)$, so that the element equation for the 1-farad capacitance $[i(t) = C(dv/dt)]$ evaluated at $t = 0+$ gives $i(0+) = \dot{v}(0+)$. We conclude that knowledge of the state of the system at $t = 0-$ implies knowledge of $\dot{v}(0+)$. Hence, we differentiate Equation (12-32) and set $t = 0+$ to obtain $\dot{v}(0+) = -\frac{1}{2}\gamma_1 - 2\gamma_2 + \frac{1}{3}v_s(0+) + \frac{2}{3}v_s(0+)$. We now have two equations in which the only two unknowns are γ_1 and γ_2. Hence, they can be solved for these two constants.

As a check on this result, we note that if the energy-storage term is zero $[v_{es} = 0]$, then the total response is the forced response $[v = v_f]$. Setting $v_s(t) = 1$ in Equation (12-31), we note that we obtain the same result obtained in Example 10-10. [See the first of Equations (10-96).]

At this point, let us pause and summarize the procedure used in Example 12-5. We first wrote the form of the response to within six constants (p_1, p_2, β_1, β_2, γ_1, γ_2). These six unknown constants were evaluated by the following:

1. The two network natural frequencies (p_1, p_2) were evaluated by writing the transfer function from the independent source to the response of interest and finding the poles of the transfer function.
2. The two constants associated with the forced response (β_1, β_2) were evaluated by expanding the transfer function from the independent source to the response of interest in its partial fraction expansion and noting the coefficients.
3. The two constants associated with the energy-storage response (γ_1, γ_2) were evaluated from known initial conditions.

The extension to the general nth-order case is now obvious. All n responses are given by Equation (11-29). Each response (each element of the matrix $\boldsymbol{x}$) can be written as an energy-storage term plus a forced term; that is,

$$x = x_{es} + x_f \tag{12-33}$$

where x_{es} is a sum of n exponentials, and x_f is a sum of n convolutions.

$$x_{es}(t) = \sum_{i=1}^{n} \gamma_i e^{p_i t}$$

$$x_f(t) = \sum_{i=1}^{n} \beta_i \int_0^t v_s(\lambda) e^{p_i(t-\lambda)}\, d\lambda \tag{12-34}$$

These equations are valid for any linear nth-order network having n distinct natural frequencies. For networks with multiple natural frequencies, these equations must be modified as discussed in Section 11-5. Equations (12-34) give the total response to within three n constants ($p_1, p_2, \cdots, p_n, \beta_1, \beta_2, \cdots, \beta_n, \gamma_1, \gamma_2, \cdots, \gamma_n$). These three n constants can be evaluated by the same procedure used in Examples 12-4 and 12-5.

An algorithm for computing any one of the n responses. The following algorithm can be used to compute any one of the n responses for any nth-order network for any one independent source. If the network contains more than one independent source, the procedure can be repeated for each independent source and the results summed in accordance with the principle of superposition.

1. Write the form of the response to within three n constants:

$$x(t) = \sum_{i=1}^{n} \gamma_i e^{p_i t} + \sum_{i=1}^{n} \beta_i \int_0^t v_s(\lambda) e^{p_i(t-\lambda)}\, d\lambda \tag{12-35}$$

2. Write the transfer function from the independent source to the response of interest and compute the poles of the transfer function to obtain the n constants $p_1, p_2, \cdots, p_n$.
3. Make a partial fraction expansion of the transfer function to obtain the n constants $\beta_1, \beta_2, \cdots, \beta_n$.
4. An nth-order network has n (equivalent) energy storage elements. The final (and hardest) step is to write n equations relating $x(0)$ to these n energy-storage constants and $v_s(0)$ and solve for the n constants $\gamma_1, \gamma_2, \cdots, \gamma_n$. However, if there is no energy stored in the network at $t = 0$, then the energy-storage term is zero; that is, $\gamma_1 = \gamma_2 = \cdots = \gamma_n = 0$.

Justification for the algorithm. So far we have given no justification for the algorithm of the preceding section. Indeed, our only justification for the method of Examples 12-4 and 12-5 is that we ended up with the correct answers. The method for relating the γ_i to the initial conditions is straightforward and needs no justification, and a proof that the β_i are the coefficients of the partial fraction expansion of the transfer function is beyond the scope of this text. In this section we verify that the poles of the transfer function are the network natural frequencies.

Consider an nth-order network characterized by the state-variable description

$$\begin{aligned} \dot{x} &= \mathbf{a}x + \mathbf{b}v \\ w &= \mathbf{c}x + \mathbf{d}v + \mathbf{e}\dot{v} \end{aligned} \tag{12-36}$$

These equations are a set of equilibrium equations formulated from a network graph. For the steady-state analysis of Chapter 4 these equations are as good as any other. From that chapter we recall that so far as steady-state amplitudes are concerned, a set of network integro-differential equations can be converted to a set of algebraic equations in complex voltage and current amplitudes as follows. Each time function in the integro-differential equation is replaced by its complex amplitude. The derivative of a time function is replaced by $j\omega$ multiplied by the complex amplitude of that function. The integral of a time function is replaced by $1/j\omega$ multiplied by the corresponding complex amplitudes. Since we are considering only steady-state quantities, limits of integration must drop out.

Using capital letters for complex amplitudes in Equations (12-36) gives

$$\begin{aligned} j\omega X &= \mathbf{a}X + \mathbf{b}V \\ W &= \mathbf{c}X + \mathbf{d}V + j\omega\mathbf{e}V \end{aligned} \tag{12-37}$$

To find the relation between the complex amplitudes of the outputs, W, and those of the input, V, we solve the first set of Equations (12-37) for X and substitute it into the second set. Symbolically, by solving the first of Equations (12-37), we obtain

$$X = [j\omega\mathbf{1} - \mathbf{a}]^{-1}\mathbf{b}V \tag{12-38}$$

and substituting this into the second of Equations (12-37), we obtain

$$W = \{\mathbf{c}[j\omega\mathbf{1} - \mathbf{a}]^{-1}\mathbf{b} + \mathbf{d} + j\omega\mathbf{e}\}V \tag{12-39}$$

The braces in Equation (12-39) show the transfer-function matrix.

The transfer-function matrix, as made evident by Equation (12-39), is the sum of three matrices. The first of the three is the most complicated. It is discussed in more detail below. The second is a matrix of constants. The third is a matrix of terms linear in $j\omega$. Since the elements of the transfer-function matrix are respective sums of elements of the three matrices, the occurrence of poles in individual transfer functions must be due to the first term. The poles in the transfer functions must appear in the matrix $[j\omega\mathbf{1} - \mathbf{a}]^{-1}$. Premultiplication by $\mathbf{c}$ and postmultiplication by $\mathbf{b}$ gives a resultant matrix whose elements are linear combinations of the elements of $[j\omega\mathbf{1} - \mathbf{a}]^{-1}$. The elements of an inverse matrix (see Appendix I) can be expressed symbolically as the ratio of two determinants. All elements have the same denominator determinant, the determinant of the original matrix. The numerator determinants are cofactors of the original matrix. Thus all transfer functions have the same poles except where a particular transfer function has fewer poles due to cancellation of a denominator and numerator factor.

The poles of all transfer functions are those values of p that satisfy the equation

$$|p\mathbf{1} - \mathbf{a}| = 0 \tag{12-40}$$

But Equation (12-40) is the same as Equation (11-8), the equation for determining the eigenvalues of the **a** matrix that are the natural frequencies of the network. Thus, the poles of all transfer functions are natural frequencies of the network.

12-5 Singularity Functions and the Complete Response

In several examples in Chapter 10, we choose inputs that are zero for t less than zero and one for t greater than zero. At t equals zero, there is a discontinuity. Such a function, called the unit step function, is one member of an important class of functions called singularity functions.[1] In this section the singularity functions are presented as they apply to network analysis. Such functions provide a physical intuition about the relations between the forced and energy-storage responses of a network. From this knowledge we shall see how to obtain the complete response by the methods of the previous section.

Properties of convolution integrals. Before discussing the singularity functions as they apply to networks, we need to derive two properties of convolutions. We need to know the derivative and the integral of a convolution integral in terms of the derivatives and integrals of the two functions in the integrand. Specifically, we consider the convolution

$$\int_{t_0}^{t} f(t - \lambda)g(\lambda)\, d\lambda \tag{12-41}$$

The expression (12-41) is a function of t. It can be differentiated and integrated with respect to t over any interval.

In order to differentiate the convolution, we can either go to the definition of the derivative, or we can apply the Leibniz rule for differentiating integrals, with the variable both as a limit and in the integrand.

[1] As pointed out below, half of these singularity functions are not functions in the mathematical sense. Putting them in a correct mathematical framework requires more advanced mathematics. However, the engineering usefulness of these functions does not require this high degree of sophistication.

Let us apply the definition

$$\frac{d}{dt}\int_{t_0}^{t} f(t-\lambda)g(\lambda)\,d\lambda$$
$$= \lim_{\Delta\to 0}\left[\frac{\int_{t_0}^{t+\Delta} f(t+\Delta-\lambda)g(\lambda)\,d\lambda - \int_{t_0}^{t} f(t-\lambda)g(\lambda)\,d\lambda}{\Delta}\right] \tag{12-42}$$

In order to see what the limit is, we need to manipulate the integrals.

$$\int_{t_0}^{t+\Delta} f(t+\Delta-\lambda)g(\lambda)\,d\lambda = \int_{t_0}^{t} f(t+\Delta-\lambda)g(\lambda)\,d\lambda + \int_{t}^{t+\Delta} f(t+\Delta-\lambda)g(\lambda)\,d\lambda \tag{12-43}$$

Thus the bracket on the right of Equation (11-42) can be written as

$$\int_{t_0}^{t} \frac{f(t+\Delta-\lambda)-f(t-\lambda)}{\Delta}\,g(\lambda)\,d\lambda + \frac{1}{\Delta}\int_{t}^{t+\Delta} f(t+\Delta-\lambda)g(\lambda)\,d\lambda \tag{12-44}$$

The limit as Δ approaches zero of the expression (12-44) is the desired result.

If the functions f and g are uniformly bounded over the domain required for the integrals, then the limit can be brought inside the integral of the first term. The result is

$$\lim_{\Delta\to 0}\int_{t_0}^{t} \frac{f(t+\Delta-\lambda)-f(t-\lambda)}{\Delta}\,g(\lambda)\,d\lambda = \int_{t_0}^{t} \dot{f}(t-\lambda)g(\lambda)\,d\lambda \tag{12-45}$$

where $\dot{f}(t-\lambda)$ means the derivative of the function f evaluated at $t-\lambda$; for example, if $f(t)$ is sin t, $\dot{f}(t-\lambda)$ is cos $(t-\lambda)$. The second term in the expression (12-44) is easily evaluated if f is continuous at zero and g is continuous at t. In this case, when Δ is very small, $f(t+\Delta-\lambda)$ and $g(\lambda)$ are essentially constant as λ takes on values between t and $t+\Delta$. Thus an approximation that becomes better as $\Delta\to 0$ is

$$\int_{t}^{t+\Delta} f(t+\Delta-\lambda)g(\lambda)\,d\lambda \approx f(0)g(t)\Delta \tag{12-46}$$

Using the approximation (12-46) in the second part of the expression (12-44) and taking the limit gives

$$\lim_{\Delta\to 0}\frac{1}{\Delta}\int_{t}^{t+\Delta} f(t+\Delta-\lambda)g(\lambda)\,d\lambda = f(0)g(t) \tag{12-47}$$

Combining Equations (12-45) and (12-47) gives the final result

$$\frac{d}{dt}\int_{t_0}^{t} f(t-\lambda)g(\lambda)\,d\lambda = \int_{t_0}^{t} \dot{f}(t-\lambda)g(\lambda)\,d\lambda + f(0)g(t) \tag{12-48}$$

The derivation of Equation (12-48) required certain restrictions on the functions f and g. First, $\dot{f}(\tau)$ must exist for $0 < \tau < t - t_0$. Next, f must be continuous at zero, at least for non-negative values of the argument, and g must be continuous at t, at least for values of the argument less than or equal to t.[2] Finally, f and g must be well-behaved enough over the entire interval so that the interchange of limit and integration in Equation (12-45) is valid.

Equation (12-48) gives a formula for the derivative of a convolution in terms of the derivative of the function f. An alternative formula involving the derivative of g can be derived if we first make a change of variables in the convolution integral. If we let

$$\zeta = t - \lambda \tag{12-49}$$

then the convolution becomes

$$\int_{t_0}^{t} f(t - \lambda)g(\lambda)\, d\lambda = \int_{0}^{t-t_0} f(\zeta)g(t - \zeta)\, d\zeta \tag{12-50}$$

Applying the definition of the derivative to the right side of Equation (12-50) and proceeding as above gives

$$\frac{d}{dt}\int_{t_0}^{t} f(t - \lambda)g(\lambda)\, d\lambda = \int_{0}^{t-t_0} f(\zeta)\dot{g}(t - \zeta)\, d\zeta + f(t - t_0)g(t_0) \tag{12-51}$$

Note that if $t_0 = 0$, Equation (12-51) is of the same form as Equation (12-48), with f and g interchanged.

For the integral of a convolution, we use the same limits of integration as on the convolution itself. Thus we compute

$$\int_{t_0}^{t}\int_{t_0}^{\tau} f(\tau - \lambda)g(\lambda)\, d\lambda\, d\tau \tag{12-52}$$

To convert the integral (12-52) to a simple convolution, we integrate by parts. The result is

$$\int_{t_0}^{\tau} f(\tau - \lambda)g(\lambda)\, d\lambda = \left[f(\tau - \lambda)\int_{t_0}^{\lambda} g(\zeta)\, d\zeta\right]_{\lambda = t_0}^{\lambda=\tau} - \int_{t_0}^{\tau} -\dot{f}(\tau - \lambda)\int_{t_0}^{\lambda} g(\zeta)\, d\zeta\, d\lambda \tag{12-53}$$

The first term on the right evaluated at $\lambda = t_0$ is zero because of the integral in the bracket. Thus we have

$$\int_{t_0}^{\tau} f(\tau - \lambda)g(\lambda)\, d\lambda = f(0)\int_{t_0}^{\tau} f(\zeta)\, d\zeta + \int_{t_0}^{\tau} \dot{f}(\tau - \lambda)\int_{t_0}^{\lambda} g(\zeta)\, d\zeta\, d\lambda \tag{12-54}$$

The right side of Equation (12-54) is exactly the formula for the derivative of the convolution of $f(t)$ with the integral of $g(t)$. When Equation (12-54) is substituted into the expression (12-52), the first integration

[2] These requirements of continuity on one side of the end points assumes that $t > t_0$ in the original convolution.

cancels the differentiation. Thus

$$\int_{t_0}^{t}\int_{t_0}^{\tau} f(\tau-\lambda)g(\lambda)\,d\lambda\,d\tau = \int_{t_0}^{t} f(t-\lambda)\int_{t_0}^{\lambda} g(\zeta)\,d\zeta\,d\lambda \tag{12-55}$$

If we had started with the change of variables (12-49) before integrating by parts, the result would be

$$\int_{t_0}^{t}\int_{t_0}^{\tau} f(\tau-\lambda)g(\lambda)\,d\lambda\,d\tau = \int_{0}^{t-t_0} g(t-\lambda)\int_{t_0}^{\lambda} f(\zeta)\,d\zeta\,d\lambda \tag{12-56}$$

The desired formula for the integral of a convolution is either Equation (12-55) or Equation (12-56). The choice depends on the ease of integrating f compared to the ease of integrating g. As with the derivative formulas when $t_0 = 0$, Equations (12-55) and (12-56) have identical form, with f and g interchanged.

Singularity functions defined. The singularity functions are a set of functions related to the unit step function. Their usefulness in engineering occurs because of the results of convolutions, where one function, say g in the formulas above, is a singularity function. To keep from obscuring the singularity functions with notation from the convolution integrals, we shall only consider convolutions where the lower limit is zero and the upper limit t is greater than zero. Thus we consider

$$\int_{0}^{t} f(t-\lambda)g(\lambda)\,d\lambda = \int_{0}^{t} g(t-\lambda)f(\lambda)\,d\lambda \tag{12-57}$$

$$\begin{aligned}\frac{d}{dt}\left[\int_{0}^{t} f(t-\lambda)g(\lambda)\,d\lambda\right] &= \int_{0}^{t} \dot{f}(t-\lambda)g(\lambda)\,d\lambda + f(0)g(t)\\ &= \int_{0}^{t} \dot{g}(t-\lambda)f(\lambda)\,d\lambda + f(t)g(0)\end{aligned} \tag{12-58}$$

$$\begin{aligned}\int_{0}^{t}\left[\int_{0}^{\tau} f(\tau-\lambda)g(\lambda)\,d\lambda\right]d\tau &= \int_{0}^{t} f(t-\lambda)\left[\int_{t_0}^{\lambda} g(\zeta)\,d\zeta\right]d\lambda\\ &= \int_{0}^{t} g(t-\lambda)\left[\int_{t_0}^{\lambda} f(\zeta)\,d\zeta\right]d\lambda\end{aligned} \tag{12-59}$$

The unit step function is defined as

$$U(t) = \begin{cases} 0 \text{ for } t < 0\\ 1 \text{ for } t \geq 0\end{cases} \tag{12-60}$$

In the convolutions above, there are situations when we should like to consider steps that start at some other time than $t = 0$, the start of the convolution. Thus we shall often use the displaced unit step

$$U(t-t_1) = \begin{cases} 0 \text{ for } t < t_1\\ 1 \text{ for } t \geq t_1\end{cases} \tag{12-61}$$

Since our convolutions all start at t = 0, Equation (12-61) will be used with $t_1 \geq 0$.

When g(t) in the formulas (12-57) through (12-59) is the displaced unit step, the important properties of singularity functions appear. First, in Equation (12-57)

$$\int_0^t f(t - \lambda)U(\lambda - t_1)\, d\lambda = \int_{t_1}^t f(t - \lambda)\, d\lambda \tag{12-62}$$

$$\int_0^t U(t - \lambda - t_1)f(\lambda)\, d\lambda = \int_0^{t-t_1} f(\lambda)\, d\lambda \tag{12-63}$$

A change of variables in the right side of Equation (12-62) also yields the right side of Equation (12-63). The more interesting results occur when the displaced step is used in the derivative formula (12-58). The first form presents no problem. The second form must be handled with care, since the derivative of $U(t - t_1)$ does not exist at t_1. This special handling will lead to the definition of the second singularity function, the unit impulse.

In order to apply the second form of Equation (12-58) to the case where $g(t)$ is a displaced step, we must first break the integration into two pieces—one from zero to t_1 and the other from t_1 to t. That is,

$$\int_0^t f(t - \lambda)U(\lambda - t_1)\, d\lambda = \int_0^{t_1} f(t - \lambda)U(\lambda - t_1)\, d\lambda + \int_{t_1}^t f(t - \lambda)U(\lambda - t_1)\, d\lambda \tag{12-64}$$

The first term on the right of Equation (12-64) is identically zero, since $U(\lambda - t_1)$ is zero over the entire interval of integration. The second term is a convolution with lower limit t_1. It can be differentiated by the formula (12-51). Thus

$$\frac{d}{dt}\left[\int_0^t f(t - \lambda)U(\lambda - t_1)\, d\lambda\right] = \int_0^{t-t_1} f(\zeta)\dot{U}(\zeta)\, d\zeta + f(t - t_1)U(t_1) = f(t - t_1) \tag{12-65}$$

The integral on the right of Equation (12-64) is zero, since U is constant over the interval $[t_1, t]$. Furthermore, $U(t_1)$ is one, so the second term reduces to the simple expression. There is a minor problem in applying the proof above, which required continuity at the end points. If the limits are taken using only values of the domain of U that appear in the integration, the results are seen to be correct. The definition of U outside the interval is irrelevant. Of course, the result of Equation (12-64) could also have been obtained from the first derivative formula in Equation (12-58) or by differentiating Equation (12-62).

Often we wish to proceed symbolically with operations on convolutions without performing the actual operations until later in a derivation. In such situations it is convenient to have a symbolism for the derivative of the unit step. To this end, we define an entity $\delta(t)$, called the unit impulse. The basic property of $\delta(t)$ is its behavior as one of the functions in a convolution. Specifically,

$$\int_0^t f(t - \lambda)\delta(\lambda)\, d\lambda = f(t) \qquad \text{(12-66)}$$

When δ is convolved with any function, the result is that function evaluated where the argument of δ is zero. Thus, for $\delta(t - t_1)$, $0 \leq t_1 \leq t$

$$\int_0^t f(t - \lambda)\delta(\lambda - t_1)\, d\lambda = f(t - t_1) \qquad \text{(12-67)}$$

If t_1 is not in the interval of integration, the convolution is zero for all t. To show that $\delta(t)$ has the property of $\dot{U}(t)$, we examine the two forms of Equation (12-58). Specifically, when $g(t) = \dot{U}(t - t_1)$, Equation (12-58) states the following for $0 \leq t_1 \leq t$

$$\begin{aligned} \int_0^t \dot{U}(t - t_1 - \lambda)f(\lambda)\, d\lambda &= \int_0^t \dot{f}(t - \lambda)U(\lambda - t_1)\, d\lambda \\ &\quad + f(0)U(t - t_1) - f(t)U(-t_1) \\ &= \int_{t_1}^t \dot{f}(t - \lambda)\, d\lambda + f(0) \\ &= \int_0^{t-t_1} \dot{f}(\tau)\, d\tau + f(0) \\ &= f(t - t_1) - f(0) + f(0) = f(t - t_1) \end{aligned} \qquad \text{(12-68)}$$

Symbolically the impulse behaves like the derivative of a step function under a convolution integral. Proving that both symbolisms are valid requires the theory of distributions (see Reference [39]). This mathematical theory, which was developed long after engineers and physicists had made good use of the impulse, does justify the symbols and show that the results are correct. We shall use the symbolism without further proof.

The unit impulse is often called the delta function, or the Dirac delta function, after P.A.M. Dirac, a physicist who was one of the first to use the notation extensively. The impulse is not a function. Its value is zero everywhere except at the origin, where it is not well defined. Consequently, an impulse $\delta(t)$ standing alone has no real meaning in the sense of elementary mathematics. Nevertheless, there are some pseudomathematical arguments about the impulse as a limit that are very useful for a physical intuition of the significance of $\delta(t)$. The derivative of the unit step does not exist at $t = 0$ because the function is discontinuous there. If the unit step is approximated by a continuous, piecewise continuously differentiable function, then the derivative does exist. A very common approximation to the unit step is the function

$$U_\epsilon(t) = \begin{cases} 0 & \text{for } t < -\dfrac{\epsilon}{2} \\ \dfrac{1}{\epsilon}\left(t + \dfrac{\epsilon}{2}\right) & \text{for } -\dfrac{\epsilon}{2} \le t \le \dfrac{\epsilon}{2} \\ 1 & \text{for } t > \dfrac{\epsilon}{2} \end{cases} \tag{12-69}$$

Then the derivative is

$$\delta_\epsilon(t) = \dot{U}_\epsilon(t) = \begin{cases} 0 & \text{for } t < -\dfrac{\epsilon}{2} \\ \dfrac{1}{\epsilon} & \text{for } -\dfrac{\epsilon}{2} \le t \le \dfrac{\epsilon}{2} \\ 0 & \text{for } t > \dfrac{\epsilon}{2} \end{cases} \tag{12-70}$$

If $\delta_\epsilon(t)$ is used in a convolution integral and the limit as $\epsilon \to 0$ is taken after integration, the result is the same as the definition of $\delta(t)$ when the origin of the impulse is interior to the convolution integral interval. That is, if $\delta(t - t_1)$ is used in the formula (12-66), t_1 must be greater than zero and less than t. It cannot be equal to 0 or t. When the origin of the impulse is an end point of the integral, then only half the area of $\delta_\epsilon(t)$ is in the integration. There are situations where the $\frac{1}{2}$ is correct that are beyond the scope of this text. Of course the limit of Equation (12-70) as $\epsilon \to 0$ does not exist.

The approximating function $\delta_\epsilon(t)$ all are zero away from the origin and have fixed area one in the small region about the origin. These are a sequence of constant area pulses. In the limit (nonexistent) there is a pulse that delivers all its area at once. Hence the name "impulse." Often it is convenient to talk about a voltage source or current source that is an impulse. Since source quantities are convolved with various exponentials to give the response, such an undefined source gives a well-defined response. In the next subsection we discuss such applications of the impulse.

When the unit step is used in the integral of the convolution formula Equation (12-59), the first form gives the convolution of *f* with the integral of the unit step. This integral, called the unit ramp and denoted by $U_1(t)$, is well defined. It is

$$U_1(t) = \begin{cases} 0 & \text{for } t < 0 \\ t & \text{for } t \ge 0 \end{cases} \tag{12-71}$$

If the unit ramp is convolved with a function *f* and the result integrated, then this operation is equivalent to convolution of f with the unit quadratic defined by

$$U_2(t) = \begin{cases} 0 & \text{for } t < 0 \\ \dfrac{t^2}{2} & \text{for } t \geq 0 \end{cases} \tag{12-72}$$

This process can be continued to generate a whole family of unit functions

$$U_k(t) = \begin{cases} 0 & \text{for } t < 0 \\ \dfrac{t^k}{k!} & \text{for } t \geq 0 \end{cases} \tag{12-73}$$

This family, with $k \geq 0$, is the well-defined half of the set of singularity functions.

The remainder of the set of singularity functions can be generated by using $\delta(t)$ or $\delta(t - t_1)$ as $g(t)$ in the derivative formula Equation (12-58). The integrals can be performed without differentiating δ to obtain the correct symbolism. The result is a set of symbols $\delta_k(t)$ (not functions) defined by

$$\int_0^t f(t - \lambda)\delta_k(\lambda - t_1)\, d\lambda = \frac{d^k}{dt^k} f(t - t_1) \tag{12-74}$$

The successive $\delta_k(t)$ can be "approximated" by approximating first δ_ϵ of Equation (12-70) and then successive $\delta_{k\epsilon}$ by piecewise continuously differentiable functions. For $\delta_{1\epsilon}$, the result is

$$\delta_{1\epsilon}(t) = \begin{cases} \dfrac{1}{\epsilon^2} & \text{for } -\epsilon \leq t < 0 \\ -\dfrac{1}{\epsilon^2} & \text{for } 0 \leq t \leq \epsilon \\ 0 & \text{otherwise} \end{cases} \tag{12-75}$$

Because of the shape of $\delta_{1\epsilon}(t)$, the symbol $\delta_1(t)$ is often called the unit doublet. Successive derivatives are called the unit triplet, quadruplet, and so on.

The use of the impulse in understanding the complete response. The easiest way to introduce the use of the impulse in network analysis is through an example.

Example 12-6

Let us consider the simple parallel RC network excited by an impulse current source, as shown in Figure 12-7. The output is taken as the voltage across the capacitance. Thus the state-variable description of the network is

$$\dot{v} = -\frac{1}{RC}v + \frac{1}{C}K\delta(t - t_1)$$
$$v = v \tag{12-76}$$

When the voltage across the capacitance is known at $t_0 < 0$, then the voltage for $t \geq t_0$ is the solution to Equations (12-76). From Chapter 10 we know this to be

$$\begin{aligned} v(t) &= v(t_0)e^{-(1/RC)(t-t_0)} - \frac{1}{C}\int_{t_0}^{t} K\delta(\lambda - t_1)e^{-(1/RC)(t-\lambda)}\,d\lambda \\ &= v(t_0)e^{-(1/RC)(t-t_0)} + \frac{1}{C}Ke^{-(1/RC)(t-t_1)} \end{aligned} \tag{12-77}$$

Note that in Equation (12-77) the forced response and the energy-storage response have exactly the same form. Thus, if the current source in Figure 12-7 had value $Cv(t_0)\delta(t - t_0) + K\delta(t - t_1)$, and there were

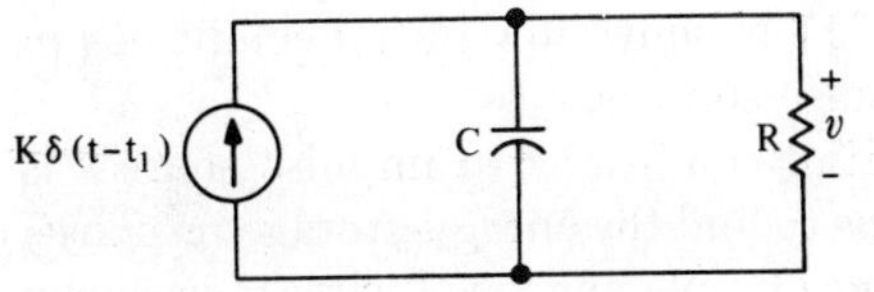

Figure 12-7 Network for Example 12-6.

no additional energy storage, the total response would be exactly the same as that of Equation (12-77). In the discussions of complete response and the time-constant concept in Chapters 10 and 11, it is stated that the voltage across a capacitance cannot change instantaneously. Now we find that an impulse current source makes the capacitance voltage change instantaneously by $1/C$ times the constant multiplying the impulse. If we modify the statement of Chapters 10 and 11 to be that the voltage across a capacitance cannot change instantaneously (be discontinuous) as long as excitations to the network containing the capacitance are well defined (finite), then impulse sources are excluded. The statement is now correct for networks with impulses.

If we think of the impulse as a pulse of zero width, infinite height, and fixed area, then we can discuss the situation of the circuit of Figure 12-7 physically. Since the integral of the current is the charge, the area under an impulse of current represents a fixed charge. The effect of the impulse current source is to place a fixed charge on the capacitance at a specific time. By Kirchhoff's laws we can see that all of the charge must go to the capacitance. If part of the impulse of current flowed through the resistance, the resulting voltage would contain an impulse. This voltage also appears across the capacitance. Since the current through a capacitance is proportional to the derivative of the voltage, such an impulse of voltage would produce a doublet of current. Since the current source does not contain a doublet, this doublet of current through the capacitance must flow through the resistance and produce a doublet of voltage. Such an argument goes on indefinitely. The only solution is that all the impulse of current from the current source must flow through the capacitance.

An extension of the ideas of the above example shows that the energy-storage response of a network can be considered as a forced response due to impulse sources. A capacitance with voltage v_0 at t_0 can be replaced by a capacitance with zero volts prior to t_0 excited by a parallel current source whose value is an impulse at $t = t_0$ multiplied by v_0C. Similarly, an inductance with current i_0 at t_0 can be replaced by an inductance with no current prior to t_0 excited by a series voltage source whose value is an impulse at $t = t_0$ multiplied by i_0L. Both these statements apply only to the case where there are no loops of capacitances and voltage sources or cut sets of inductances and current sources. In these two special cases, impulse sources can still be used to replace initial conditions, but the substitution is not quite so simple. Kirchhoff's law arguments, such as those at the end of Example 11-6, show how initial conditions in such cases can be considered as impulse sources.

The advantage of converting initial conditions to impulse sources is that we can now use transfer functions to find the energy-storage response. We form a port across each capacitance by placing two "wires," as shown in Figure 12-8(a). We form a port in series with each inductance by

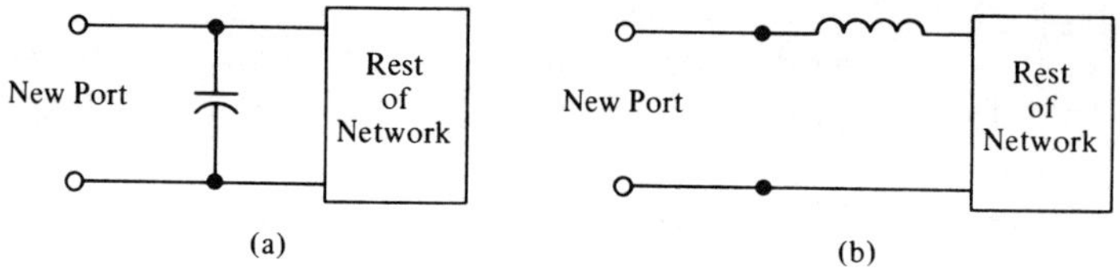

Figure 12-8 Placing ports for the energy-storage response.

cutting the inductance "wire," as shown in Figure 12-8(b). The partial fraction expansion of the transfer function from these ports to the output gives the constants of the energy storage response.

Example 12-7

As an example of the use of transfer-function methods to compute the energy-storage response of a network, let us consider the third-order Butterworth insertion power-gain filter of Figure 12-9. At $t = 0$, the current in the inductance is $i(0)$ and the capacitance voltages are $v_1(0)$

and $v_2(0)$, respectively. We wish to compute the energy-storage part of $v_2(t)$. As pointed out above, the energy-storage response of the network of Figure 12-9(a) is the same as the forced response of the network

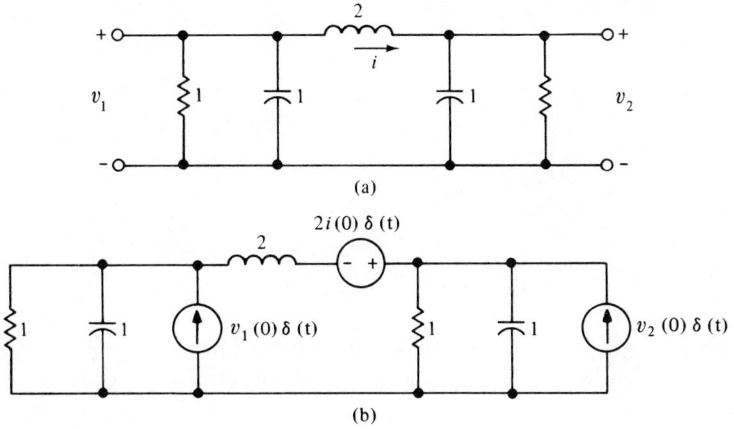

Figure 12-9 Networks for Example 12-7.

of Figure 12-9(b). We shall consider the second network by the method of Section 12-2.

The first step in finding the response is to find the steady-state transfer functions. That is, we must analyze the circuit of Figure 12-10 and

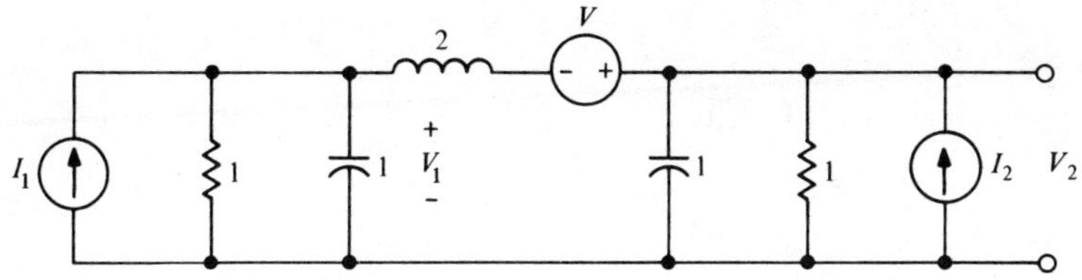

Figure 12-10 Steady-state network for Example 12-7.

find the transfer functions from the three sources to the voltage V_2. The easiest approach is to write two node-to-datum equations. They are

$$\begin{aligned}\left(1 + j\omega + \frac{1}{2j\omega}\right) V_1 - \frac{1}{2j\omega}(V_2 - V) &= I_1 \\ -\frac{1}{2j\omega}(V_1 + V) + \left(1 + j\omega + \frac{1}{2j\omega}\right) V_2 &= I_2\end{aligned} \tag{12-78}$$

Solving gives

$$V_2 = \frac{\begin{vmatrix} 1 + j\omega + \dfrac{1}{2j\omega} & I_1 - \dfrac{1}{2j\omega} V \\ -\dfrac{1}{2j\omega} & I_2 + \dfrac{1}{2j\omega} V \end{vmatrix}}{\begin{vmatrix} 1 + j\omega + \dfrac{1}{2j\omega} & -\dfrac{1}{2j\omega} \\ -\dfrac{1}{2j\omega} & 1 + j\omega + \dfrac{1}{2j\omega} \end{vmatrix}} \tag{12-79}$$

$$= \frac{\frac{1}{2}I_1 + [(j\omega)^2 + j\omega + \frac{1}{2}]I_2 + \frac{1}{2}(1 + j\omega)V}{(j\omega)^3 + 2(j\omega)^2 + 2j\omega + 1}$$

The three roots of the denominator polynomial are -1, $-\frac{1}{2} + j(\sqrt{3}/2)$, and $-\frac{1}{2} - j(\sqrt{3}/2)$. Thus the transfer functions in partial fraction form are

$$\frac{V_2}{I_1} = \frac{\frac{1}{2}}{(p + 1)} + \frac{\dfrac{1}{-3 + j\sqrt{3}}}{\left(p + \dfrac{1}{2} + j\dfrac{\sqrt{3}}{2}\right)} + \frac{\dfrac{1}{-3 - j\sqrt{3}}}{\left(p + \dfrac{1}{2} - j\dfrac{\sqrt{3}}{2}\right)} \tag{12-80}$$

$$\frac{V_2}{I_2} = \frac{\frac{1}{2}}{p + 1} + \frac{\dfrac{2 + \sqrt{3} - j\sqrt{3}}{-3 + j\sqrt{3}}}{p + \dfrac{1}{2} + j\dfrac{\sqrt{3}}{2}} + \frac{\dfrac{2 + \sqrt{3} + j\sqrt{3}}{-3 - j\sqrt{3}}}{p + \dfrac{1}{2} - j\dfrac{\sqrt{3}}{2}} \tag{12-81}$$

$$\frac{V_2}{V} = \frac{\dfrac{1 - j\sqrt{3}}{2(-3 + j\sqrt{3})}}{p + \dfrac{1}{2} + j\dfrac{\sqrt{3}}{2}} + \frac{\dfrac{1 + j\sqrt{3}}{2(-3 - j\sqrt{3})}}{p + \dfrac{1}{2} - j\dfrac{\sqrt{3}}{2}} \tag{12-82}$$

From these three transfer functions and the coefficients of the three impulse sources in Figure 12-9, we can write down the energy-storage response by inspection. It is for $t \geq 0$.

$$\begin{aligned} v_2(t) &= [\tfrac{1}{2}v_1(0) + \tfrac{1}{2}v_2(0)]e^{-t} \\ &+ \left[\frac{v_1(0) + (2 + \sqrt{3} - j\sqrt{3})v_2(0) + (1 - j\sqrt{3})i(0)}{-3 + j\sqrt{3}}\right] e^{[-(1/2) - j(\sqrt{3}/2)]t} \\ &+ \left[\frac{v_1(0) + (2 + \sqrt{3} + j\sqrt{3})v_2(0) + (1 - j\sqrt{3})i(0)}{-3 - j\sqrt{3}}\right] e^{[-(1/2) + j(\sqrt{3}/2)]t} \end{aligned} \tag{12-83}$$

In Chapter 8 we defined the transfer function as the ratio of a response voltage (or current) amplitude to a source voltage (or current) amplitude in the sinusoidal steady state. In this chapter we have found a time function, derivable from the transfer function that determines the response of any network to arbitrary inputs. When the transfer function has a denominator of higher degree than the numerator, the associated time function is a sum of exponentials. The exponents are the roots of the denominator polynomial from the transfer function, and the coefficients of the exponentials are the corresponding coefficients of the partial fraction expansion of the transfer function. This time function is called the *impulse response function* of the network. The reason for this name should now be obvious, because when the input is an impulse, the output is exactly the impulse response function.

When the transfer function has a numerator whose degree is equal to or greater than that of the denominator, the impulse response function is still defined. For a constant in the partial fraction expansion, the impulse response is that constant multiplied by an impulse. For a linear term in the partial fraction expansion, the impulse response is the coefficient of the linear term multiplied by a doublet. In such cases, the impulse response function is really not a function. Thus the transfer function and the corresponding impulse response function for the voltage across the one ohm resistor in Figure 12-6(b) are

$$\frac{V_R}{V_s} = 1 + \frac{-\frac{1}{6}}{j\omega + \frac{1}{2}} + \frac{-\frac{4}{3}}{j\omega + 2} \tag{12-84}$$

$$v_R(t)\Big|_{v_s(t)=\delta(t)} = \delta(t) - \tfrac{1}{6}e^{-(1/2)t} - \tfrac{4}{3}e^{-2t} \tag{12-85}$$

The forced response, $v_R(t)$, due to any input $v_s(t)$ is the convolution of the impulse response function of Equation (12-85) with the input $v_s(t)$.

12-6 The Laplace–Fourier Transform as an Alternative Method

The computational method for obtaining the complete response, energy storage plus forced, can also be derived by a different method — the Laplace-Fourier transform. In signal theory, transform techniques are extremely powerful (see Cooper and McGillem, Reference [14], or Papoulis, Reference [35]). For analysis of networks of elements defined in Chapter 1, the transform methods give no more than that which has already been derived. Furthermore, there are many mathematical subtleties in transform theory that tend to obscure the basic properties of these networks. Nevertheless, many elementary texts in network analysis introduce the other topics of this chapter via the Laplace-Fourier transform. Thus for completeness and for helping the engineer communicate with his colleagues who had the other approach, a brief discussion of the method is in order.

The Laplace–Fourier transform starts with a function $f(t)$ and converts it to a function $F(s)$ by the following integral:

$$F(s) = \int_{-\infty}^{\infty} f(t)e^{-st}\,dt \tag{12-86}$$

The integral in Equation (12-86) is improper because of the infinite limits. It is well-defined for only certain values of s. These values depend on the nature of the function $f(t)$. In many discussions of Laplace–Fourier transforms, only functions $f(t)$ that are zero for $t < 0$ are considered. Then the lower limit in Equation (12-86) becomes zero, and the convergence difficulties are, so to speak, cut in half. For our discussions, we shall stay with Equation (12-86) because this is required to handle all the analysis discussed in Chapter 11. Furthermore, there are nonlinear network and systems problems where the transform starting at $t = 0$ is inadequate.

In order to apply the Laplace–Fourier transform to integro-differential equations such as the equations of linear, time-invariant networks, we must have formulas for the transforms of the derivative and integral of a function $f(t)$. For the derivative, we examine

$$\int_{-\infty}^{\infty} \dot{f}(t)e^{-st}\,dt$$

Integrating by parts gives

$$\int_{-\infty}^{\infty} \dot{f}(t)e^{-st}\,dt = [f(t)e^{-st}]_{-\infty}^{\infty} - \int_{-\infty}^{\infty} f(t)(-s)e^{-st}\,dt \tag{12-87}$$

When the function $f(t)$ is well-behaved enough at $\pm\infty$ so that

$$\lim_{t\to\pm\infty} [f(t)e^{-st}] = 0$$

for a range of values of s, then Equation (11-87) becomes

$$\int_{-\infty}^{\infty} \dot{f}(t)e^{-st}\,dt = sF(s) \tag{12-88}$$

For the integral of $f(t)$, since we are considering the domain of f to be $(-\infty, \infty)$, we must use the integral from $-\infty$ to t. Thus the required transform integral is

$$\int_{-\infty}^{\infty} \left[\int_{-\infty}^{t} f(\tau)\,d\tau\right] e^{-st}\,dt$$

Again, we integrate by parts to get

$$\int_{-\infty}^{\infty} \left[\int_{-\infty}^{t} f(\tau)\,d\tau\right] e^{-st}\,dt = \left[f(t)\,\frac{e^{-st}}{-s}\right]_{-\infty}^{\infty} + \frac{1}{s}\int_{-\infty}^{\infty} f(t)e^{-st}\,dt \tag{12-89}$$

Again, we require f to be sufficiently well-behaved at $\pm\infty$ so that

$$\lim_{t\to\pm\infty}\left[\frac{f(t)e^{-st}}{s}\right] = 0$$

for a range of values of s. Then

$$\int_{-\infty}^{\infty}\left[\int_{-\infty}^{t} f(\tau)\,d\tau\right] e^{-st}\,dt = \frac{F(s)}{s} \tag{12-90}$$

Of course the validity of the formula requires that

$$\int_{-\infty}^{t} f(\tau)\,d\tau$$

be well-defined for all t, just as the derivative formula requires the derivative to exist for all t.

Example 12-8

Before discussing additional properties of Laplace–Fourier transforms, let us examine a simple example to show how the properties derived so far apply to network analysis. We consider the series RLC circuit of

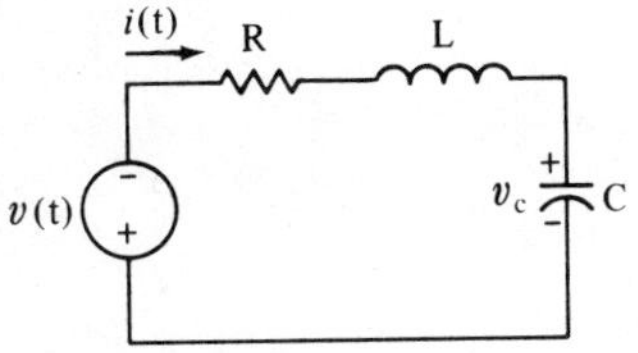

Figure 12-11 Network for Example 12-8.

Figure 12-11. The integro-differential equation for this circuit is

$$v(t) = Ri(t) + L\dot{i} + \frac{1}{C}\int_{-\infty}^{t} i(\tau)\,d\tau \tag{12-91}$$

If $v(t)$, $\dot{i}$, and

$$\int_{-\infty}^{t} i(\tau)\,d\tau$$

all have well-defined Laplace–Fourier transforms for an overlapping range of values for s, then Equation (12-91) can be transformed to an algebraic equation in s. The result is

$$V(s) = \left(R + Ls + \frac{1}{Cs}\right) I(s) \tag{12-92}$$

If $v(t)$ is known, then $V(s)$ can be computed. Equation (11-92) can then be solved for $I(s)$.

The problem now is, given $I(s)$, how does one find the associated $i(t)$. From the previous sections we know that the forced response of the network of Figure 12-11 is the convolution of $v(t)$ with two exponentials. The exponents are the roots of a polynomial — exactly the coefficient polynomial of $I(s)$ in Equation (12-92). Thus the example directs us to investigate the transform of a convolution.

The next property of Laplace–Fourier transforms that we wish to find is the transform of a convolution. Suppose that we have two functions $f(t)$ and $g(t)$. Furthermore, for overlapping ranges of s, they have Laplace–Fourier transforms $F(s)$ and $G(s)$, respectively. Since the transform integral requires values of t from $-\infty$ to $+\infty$, we must examine a very special convolution integral — a convolution that we shall see is actually more general than the one previously used. It is

$$\int_{-\infty}^{\infty} f(t-\tau)g(\tau)\,d\tau \tag{12-93}$$

The Laplace–Fourier transform is

$$\int_{-\infty}^{\infty}\left[\int_{-\infty}^{\infty} f(t-\tau)g(\tau)\,d\tau\right]e^{-st}\,dt \tag{12-94}$$

If both functions, f and g are sufficiently well-behaved at $\pm\infty$ for the transforms to exist and the convolution to exist, then we can interchange orders of integration with impunity.

Now

$$\begin{aligned}\int_{-\infty}^{\infty} f(t-\tau)e^{-st}\,dt &= \int_{-\infty}^{\infty} f(\zeta)e^{-s(\zeta+\tau)}\,d\zeta\\ &= e^{-s\tau}\int_{-\infty}^{\infty} f(\zeta)e^{-s\zeta}\,d\zeta\\ &= e^{-s\tau}F(s)\end{aligned} \tag{12-95}$$

Substituting Equation (12-95) into the formula of Equation (12-94) gives

$$\begin{aligned}\int_{-\infty}^{\infty}\left[\int_{-\infty}^{\infty} f(t-\tau)g(\tau)\,d\tau\right]e^{-st}\,dt &= F(s)\int_{-\infty}^{\infty} g(\tau)e^{-s\tau}\,d\tau\\ &= F(s)G(s)\end{aligned} \tag{12-96}$$

This last expression shows that the Laplace–Fourier transform of the convolution Equation (12-93) is the product of the transforms of the separate functions, provided the convolution is well-defined and the transforms of both functions are well-defined for the same value of s. From the form of the right side of Equation (12-96), it appears that one can interchange the roles of f and g in the convolution Equation (12-93). A change of variables $\zeta = t - \tau$ shows this to be true.

The convolutions that arose in Chapter 10 are similar to the form of Equation (12-93), except that the upper limit was t instead of ∞. If the function $f(t)$ used in the formula of Equation (12-93) is zero for $t < 0$, then the result is

$$\int_{-\infty}^{\infty} f(t-\tau)g(\tau)\,d\tau = \int_{-\infty}^{t} f(t-\tau)g(\tau)\,d\tau \tag{12-97}$$

where $f(t) = 0$ for $t < 0$. Convolutions such as Equation (12-97) in Chapters 10 and 11 always had $f(t)$ as an exponential or polynomial multiplied by an exponential. Thus, we should compute the transforms of these functions in order to relate Laplace–Fourier transforms to Chapter 11. Let

$$f(t) = \begin{cases} 0 & \text{for } t < 0 \\ e^{at} & \text{for } t \geq 0 \end{cases} \tag{12-98}$$

where a is, in general, complex. Then the limits on the transform integral become 0 and ∞, so that

$$F(s) = \int_0^\infty e^{at}e^{-st}\,dt = \frac{1}{s-a} \text{ for Re [s]} > \text{Re [a]} \tag{12-99}$$

When

$$f(t) = \begin{cases} 0 & \text{for } t < 0 \\ te^{at} & \text{for } t \geq 0 \end{cases}$$

then integrating by parts gives

$$F(s) = \int_0^\infty te^{at}e^{-st}\,dt = \left[\frac{te^{(a-s)t}}{a-s}\right]_0^\infty - \int_0^\infty \frac{e^{(a-s)t}}{a-s}\,dt \tag{12-100}$$

$$= \frac{1}{(s+a)^2} \qquad \text{for Re [s]} > \text{Re [a]} \tag{12-101}$$

Example 12-8 *(continued)*

With the specific transforms derived above, we are in a position to complete Example 12-8. First, we solve Equation (12-92) for I(s) as follows:

$$\begin{aligned} I(s) &= \frac{Cs}{CLs^2 + CRs + 1} V(s) \\ &= \frac{\frac{1}{L}s}{(s-p_1)(s-p_2)} V(s) \\ &= \frac{\frac{p_1}{L(p_1-p_2)}}{s-p_1} V(s) + \frac{\frac{p_2}{L(p_2-p_1)}}{s-p_2} V(s) \end{aligned} \tag{12-102}$$

where

$$p_1 = -\frac{R}{2L} + \sqrt{\left(\frac{R}{2L}\right)^2 - \frac{1}{LC}}$$

$$p_2 = -\frac{R}{2L} - \sqrt{\left(\frac{R}{2L}\right)^2 - \frac{1}{LC}}$$

Each term on the right side of Equation (12-102) is in the form of a constant over a linear function of s multiplied by V(s). Such a function of s could arise from a convolution of a function such as that of Equation (12-98) with the voltage source function $v(t)$. Thus, we conclude that a possible solution is

$$i(t) = \frac{p_1}{L(p_1 - p_2)} \int_{-\infty}^{t} e^{p_1(t-\tau)} v(\tau)\, d\tau + \frac{p_2}{L(p_2 - p_1)} \int_{-\infty}^{t} e^{p_2(t-\tau)} v(\tau)\, d\tau \tag{12-103}$$

This is exactly the same solution that we would derive for the network by the methods of Chapter 11.

Initial conditions via Laplace–Fourier transforms. There are two ways to bring initial conditions into Laplace–Fourier transform analysis of linear networks. One is to use impulse sources such as those of the previous section to establish initial conditions. The other is to pick a particular time, usually $t = 0$, and ascertain the transforms of functions that are zero prior to this time but have jump discontinuities there. These jump discontinuities put special constraints on the derivative and integral formulas. The constraints take care of the initial conditions. The transform of an impulse at $t = 0$ is

$$\int_{-\infty}^{\infty} \delta(t) e^{-st}\, dt = 1 \qquad \text{for all s} \tag{12-104}$$

For an impulse at t_0, the transform is

$$\int_{-\infty}^{\infty} \delta(t - t_0) e^{-st}\, dt = e^{-st_0} \qquad \text{for all s} \tag{12-105}$$

To introduce initial conditions via functions that are zero for $t < 0$, we first note that there is no problem taking the Laplace–Fourier transform of such a function. For the derivative formula we have

$$\begin{aligned} \int_0^{\infty} \dot{f}(t) e^{-st}\, dt &= [f(t)e^{-st}]_0^{\infty} + s \int_0^{\infty} f(t) e^{-st}\, dt \\ &= f(0) + sF(s) \end{aligned} \tag{12-106}$$

The integral formula of Equation (12-90) required only the existence of the integral. Thus it is valid for functions that are zero for $t < 0$ as is. In the case of an integral in a network equation, such as the situation for the capacitance in the network of Figure 12-11 as manifested in Equation (12-91), assuming $i(t)$ is zero for $t < 0$ sets the constant of integration. In order to handle initial conditions we must allow for a different constant. Thus, instead of only using the indefinite integral in Equation (12-91), we must use

$$\frac{1}{C} \int_0^{t} i(\tau)\, d\tau + v_c(0) \tag{12-107}$$

where $v_c(0)$, the required constant of integration, is the voltage across the capacitance at t = 0. With Equation (12-106) for derivatives and the constant of integrations included with integrals, the Laplace–Fourier transform can be used, if we assume all functions are zero for t < 0. In this case, the transform is called only the Laplace transform.

Example 12-8 *(continued)*

Let us suppose that $v(t)$ in Equation (12-10) is known only for $t \geq 0$ and that at t = 0, both $i(0)$ and $v_c(0)$ are known. We can now analyze the network, using Laplace–Fourier transforms with impulse sources in the network, as shown in Figure 12-12. We can also use the strict Laplace transform as just described. First let us use the impulses.

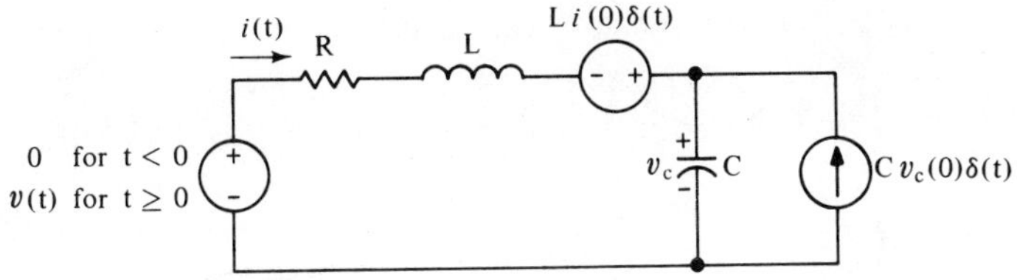

Figure 12-12 Network for Example 12-8 with initial conditions.

The equation for the circuit of Figure 12-12 is

$$v(t) = Ri(t) + L\dot{i} + Li(0)\,\delta(t) + \frac{1}{C}\int_0^t [i(\tau) + Cv_c(0)\,\delta(\tau)]\,d\tau \qquad \textbf{(12-108)}$$

The Laplace–Fourier transform of Equation (12-108) is

$$V(s) = \left(R + Ls + \frac{1}{Cs}\right) I(s) + Li(0) + \frac{1}{s}\,v_c(0) \qquad \textbf{(12-109)}$$

Solving gives

$$I(s) = \frac{V(s)Cs}{LCs^2 + RCs + 1} + \frac{Li(0)Cs}{LCs^2 + RCs + 1} + \frac{Cv_c(0)}{LCs^2 + RCs + 1} \qquad \textbf{(12-110)}$$

The first term in Equation (12-110) could be from the same convolution as found before. Since $v(t)$ is zero for t < 0, the lower limit of integration is zero instead of $-\infty$. The other two terms, when expanded in partial fractions, are seen to be the transforms of exponentials that are zero for t < 0. Thus, one possible time function that leads to the transform Equation (12-110) is

$$i(t) = \int_0^t \left[\frac{p_1 e^{p_2(t-\tau)}}{L(p_1 - p_2)} + \frac{p_2 e^{p_2(t-\tau)}}{L(p_2 - p_1)} \right] v(\tau)\, d\tau + \frac{p_1 i(0)}{p_1 - p_2} e^{p_1 t}$$
$$+ \frac{p_2 i(0)}{p_2 - p_1} e^{p_2 t} + \frac{v_c(0)}{L(p_1 - p_2)} e^{p_1 t} + \frac{v_c(0)}{L(p_2 - p_1)} e^{p_2 t} \qquad \textbf{(12-111)}$$

This is the same result that would be obtained by the methods of Chapter 10.

If we use the Laplace transform, starting everything at $t = 0$, the differential equation for the network of Figure 12-11 is

$$v(t) = Ri(t) + L\dot{i} + \frac{1}{C}\int_0^t i(\tau)\, d\tau + v_c(0) \qquad \textbf{(12-112)}$$

Transforming with the derivative formula Equation (12-106) gives

$$V(s) = \left(R + Ls + \frac{1}{Cs}\right) I(s) + Li(0) + \int_0^\infty v_c(0) e^{-st}\, dt$$
$$= \left(R + Ls + \frac{1}{Cs}\right) I(s) + Li(0) + \frac{v_c(0)}{s} \qquad \textbf{(12-113)}$$

Since Equation (12-113) is the same as Equation (12-109), the resulting time-function answer is the same.

The Laplace–Fourier transform converts network integro-differential equations to algebraic equations. For both the forced and the energy-storage response, these algebraic equations are the same as those derived by steady-state analysis methods. In both cases, we were able to relate the solution of the algebraic equations to the time response because we knew the form of the time response from Chapter 11. By a more complete study of transforms, the same results can be derived directly, without relying on the known form of the complete network response. Many other network-analysis texts use this approach. Some of the questions that must be answered are as follows. What is the meaning of the new variables? For what classes of functions do the various transform integrals have overlapping ranges of s? Once the algebraic equations are solved, how can the associated time function be found without having to know the result in advance.

Answering the above questions with complete rigor requires a level of mathematical maturity well beyond that of the normal undergraduate engineering student. The questions can be answered in a manner plausible to engineering students, as is done in undergraduate texts that treat transforms. Examples are Cooper and McGillem, Reference [14]; Van Valkenberg, Reference [13]; Kuo, Reference [11]; and many others. More advanced answers can be found in books such as Papoulis, Reference

[35]. Handling all signals of engineering interest requires the theory of distributions. Zemanian, Reference [39], gives such a treatment. For the networks of this text, Laplace–Fourier transforms give no additional information over those methods already presented. Furthermore, the improper integrals can lead to confusion if not well understood.

12-7 Signal Theory: The Next Subject for Study

The primary purpose of electronic circuits is the processing of electrical signals (voltages and currents). This text has discussed networks, the model for circuits. We have seen that networks of certain linear elements have their signal-processing function in the form of a convolution integral. Furthermore, we have seen the results of this operation on steady-state sinusoidal signals and on some simple rectangular pulses. From here, there are two directions for further study. One is the investigation of more complicated networks. The other is the investigation of convolutions of more complicated signals.

In this chapter we have introduced two concepts that are extremely useful in signal theory. One is the singularity function. The other is the Laplace–Fourier transform. These concepts had only minor usefulness in understanding linear networks. For an understanding of signals, they, plus the concept of orthogonal expansions, are essential. We leave this material for a subsequent course.

PROBLEMS

12-1 Set up state-variable equations showing the input $i_1(t)$ and output $v_2(t)$ as separate variables for the second-order Butterworth filters shown in Figure P12-1.

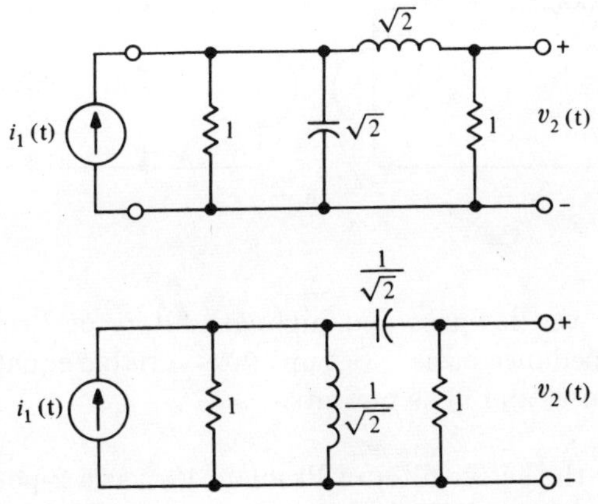

Figure P12-1

12-2 Set up state-variable equations showing the input $v_1(t)$ and output $v_{out}(t)$ for the second-order active filter shown in Figure P12-2.

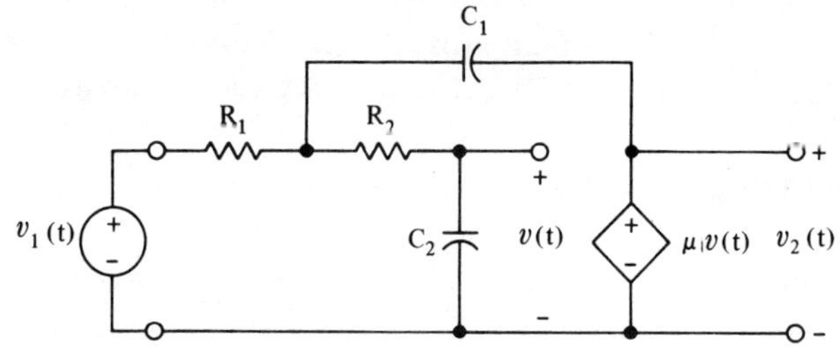

Figure P12-2

12-3 A gyrator is a nonreciprocal 2-port device (it does not satisfy the reciprocity theorem). Its admittance matrix is

$$\mathbf{Y} = \begin{bmatrix} 0 & 1 \\ -1 & 0 \end{bmatrix}$$

The symbol for a gyrator is as shown in Figure P12-3(a). It can be constructed from two controlled sources as shown in Figure P12-3(b). Set up state-variable equations for the circuit in Figure P12-13(c).

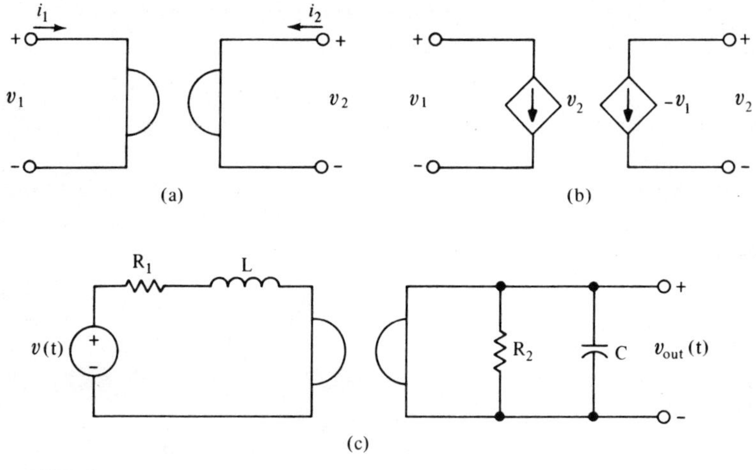

Figure P12-3

12-4 Consider the lowpass and highpass filters of Problem 12-1 as 2-ports on an impedance basis. Set up state-variable equations with i_1 and i_2 as inputs and v_1 and v_2 as outputs.

12-5 Consider the active filter of Problem 12-2 as a 2-port on a g parameter basis. Set up state-variable equations with v_1 and i_2 as inputs and i_1 and v_2 as outputs.

Problems 12-6 through 12-15 deal with networks that were introduced in problems of Chapters 4 through 9. The steady-state computations required in those problems can be used to obtain the constants of the solutions required here. The number in parentheses is the steady-state problem with the same network.

12-6 (4-6) For the network shown in Figure P12-6 the input is

$$v_1(t) = \begin{cases} 0 & \text{for } t < 0 \\ 1 & \text{for } t \geq 0 \end{cases}$$

Find $v_2(t)$.

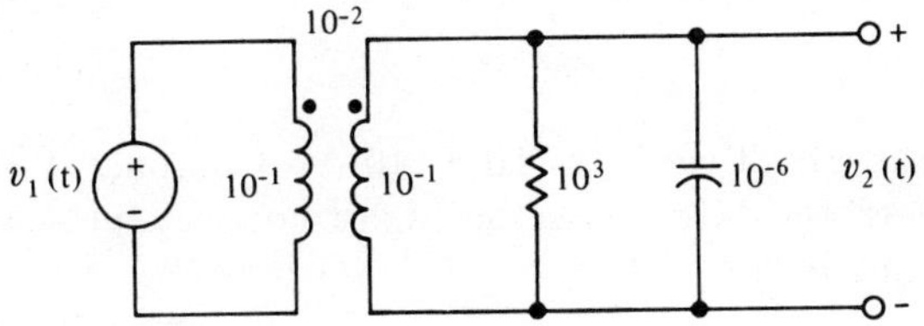

Figure P12-6

12-7 (4-5) For the network shown in Figure P12-17

$$v_1(t) = \begin{cases} 0 & \text{for } t \leq 0 \\ t & \text{for } 0 \leq t \leq 1 \\ 1 & \text{for } 1 \leq t \leq 2 \\ 3 - t & \text{for } 2 \leq t \leq 3 \\ 0 & \text{for } t \geq 3 \end{cases}$$

Find $v_2(t)$.

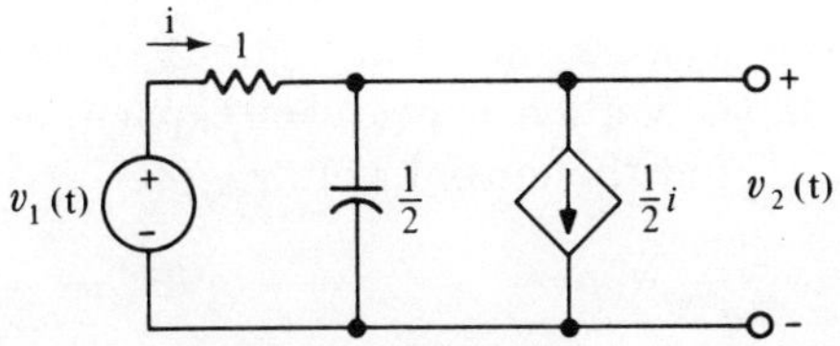

Figure P12-7

12-8 (4-6) Repeat Problem 12-6 with

$$v_1(t) = \begin{cases} 0 & \text{for } t < 0 \\ \cos 10^3 t & \text{for } 0 \leq t \leq 0.01 \\ 0 & \text{for } t > 0.01 \end{cases}$$

12-9 (9-11) Consider a third-order Butterworth highpass filter as an amplifier for short pulses. To be specific, let the break point be at 10^5 Hz. Consider three pairs of pulses: one at frequency below the break frequency; one at the break frequency; one well above the break frequency. Specifically, let the input be

$$v(t) = \begin{cases} 0 & \text{for } t < 0 \text{ and } t > \dfrac{3T}{2} \\ 1 & \text{for } 0 \le t \le \dfrac{T}{2} \text{ and } T \le t \le \dfrac{3T}{2} \\ 0 & \text{for } \dfrac{T}{2} < t < T \end{cases}$$

The three cases are $T = 10^{-4}, 10^{-5}, 10^{-6}$. Compute the output for each case. Does the filter pass the high-frequency pulses better than the low-frequency ones?

12-10 (9-17) Consider the response of a second-order Butterworth bandpass filter to a pulse cosine wave at the center frequency. Consider three cases: (a) a pulse of only one cycle; (b) a pulse of 10 cycles; (c) a pulse of 100 cycles. To be specific, use a filter with 50.3-kHz center frequency and 17.5 kHz bandwidth. This is a fourth-order system, but you know the location of the poles of the transfer function.

12-11 (9-15) Consider the response of a second-order Butterworth bandreject filter to a pulse cosine wave at the center frequency. Use the same three inputs as in Problem 12-10. Also make the bandwidth and center frequency the same.

12-12 (4-9) For the network shown in Figure P12-12, there is no input for $t > 0$. At $t = 0$, the voltage across each capacitance is one volt with the polarity shown. Find the output voltage $v_2(t)$.

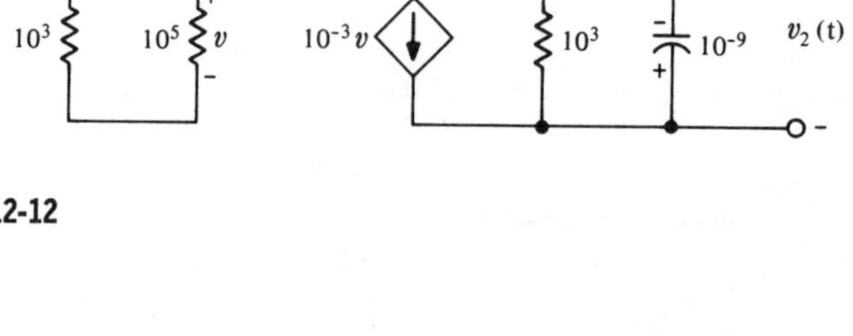

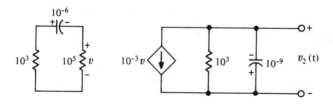

Figure P12-12

12-13 (8-4) For the network shown in Figure P12-13 the input $i_s(t)$ is a 1 mA step applied at t = 0. At t = 0 the voltage v is one volt, and the voltage across the other capacitance with the polarity shown is 5 volts. Find the voltage $v_2(t)$ for t ≥ 0.

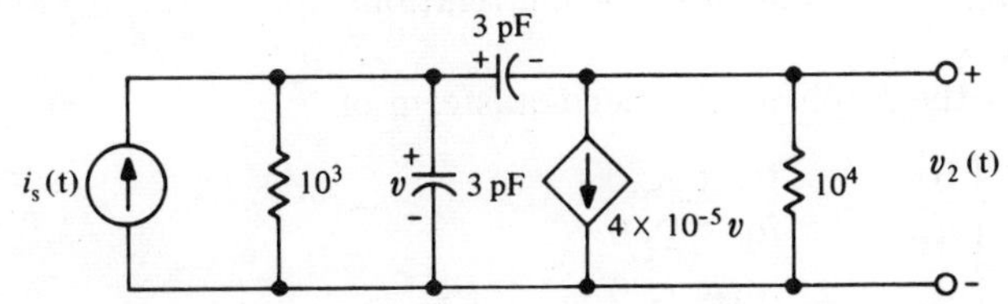

Figure P12-13

12-14 (6-17) For the twin T network shown in Figure P12-14, each capacitance has one volt in the polarity shown at t = 0. Compute the short-circuit currents at the two ports for t ≥ 0.

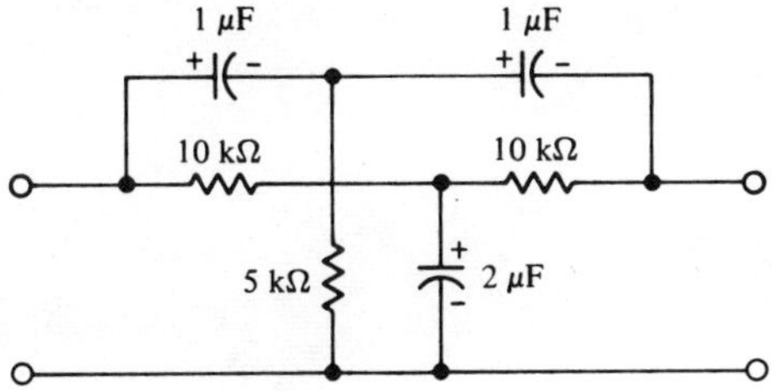

Figure P12-14

12-15 (5-7) For the transformer, there is one ampere flowing down in each winding at t = 0. Compute the voltages across the resistances for t ≥ 0. (See Figure P12-15.)

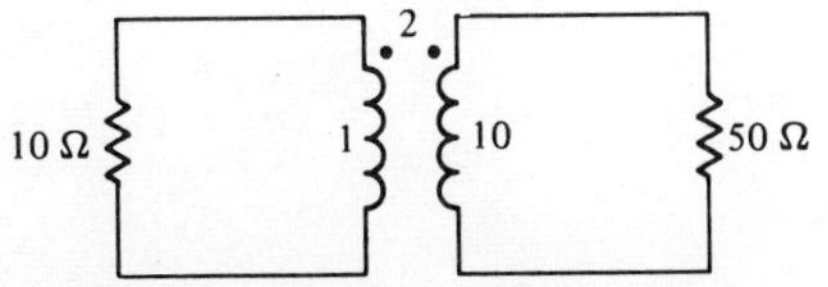

Figure P12-15

12-16 Show that the step response of a network is the integral of the impulse response.

12-17 Consider the network of Problem 12-7.

a. Let the input be the derivative of the input from Problem 12-7. Show that the output is the derivative of the answer to Problem 12-7.

b. Let the input be

$$v_1(t) = \delta(t) - \delta(t-1) - \delta(t-2) + \delta(t-3)$$

This is the second derivative of the input of Problem 12-7. Show that the resulting outputs have the same relationship.

12-18 Find the Laplace–Fourier transform of

a. $f(t) = \begin{cases} 0 & \text{for } t < 0 \\ \cos t & \text{for } t \geq 0 \end{cases}$

b. $f(t) = \begin{cases} 0 & \text{for } t < 0 \\ \sin t & \text{for } t \geq 0 \end{cases}$

12-19 Find the Laplace–Fourier transform of the input voltage of Problem 12-7.

Bibliography

In engineering, the discipline called circuits, or networks with modifiers electrical, electric, or electronic, is an extensive one. A listing of all pertinent published material relative to this discipline or even to this text would require more space than the presentation of the material. Thus, we present a few selected works that compete with and complement the material presented here. The list is confined to texts grouped by their relation to the present text. In the listing of books on basic and intermediate circuits, only a small fraction of the possible candidates are listed.

Introductory Background

The texts listed here present an elementary view of electrical engineering. A course from one of these texts may precede the course for which the present text was written. Reference [2] was used in a preceding course at Purdue University.

[1] del Toro, V., *Principles of Electrical Engineering* (Englewood Cliffs, N.J.: Prentice-Hall, 1965).

[2] Hayt, W. J., Jr., and Hughes, G. W., *Introduction to Electrical Engineering* (New York: McGraw-Hill, 1968).

[3] Pederson, D. O., Whinnery, J. R., and Studer, J. J., *Introduction to Electronic Systems, Circuits, and Devices* (New York: McGraw-Hill, 1966).

[4] Smith, R. J., *Circuits, Devices, and Systems: A First Course in Electrical Engineering* (New York: John Wiley and Sons, 1966).

Elementary and Intermediate Circuits and Networks

The texts listed here are, in part, competitors to the present text. Most have different approaches, and some, particularly References [5] and [7], present material not included in this text. They may be used in subsequent courses with the overlapping material deleted from the course. The list includes only those texts that have sufficient material over that presented by their competition at the time of publication. Texts whose only advantage is a "better" presentation of the same material are not included, as there are hundreds. We begin the chronology with Reference [9].

[5] Carlin, H. J. and Giordaro, A. B., *Network Theory* (Englewood Cliffs, N.J.: Prentice-Hall, 1964).

[6] Cruz, J. B., Jr., and Van Valkenburg, M. E., *Introductory Signals and Circuits* (Waltham, Mass.: Blaisdell, 1967).

[7] Desoer, C. A. and Kuh, E. S., *Basic Circuit Theory* (New York: McGraw-Hill, 1969).

[8] Friedland, B. J., Wing, O., and Ash, R., *Principles of Linear Networks* (New York: McGraw-Hill, 1961).

[9] Guillemin, E. A., *Introductory Circuit Theory* (New York: John Wiley and Sons, 1953).

[10] Huelsman, L. P., *Circuits, Matrices, and Linear Vector Spaces* (New York: McGraw-Hill, 1963).

[11] Kuo, F. F., *Network Analysis and Synthesis* (New York: John Wiley and Sons, 1962).

[12] Merriam, C. W., III, *Analysis of Lumped Electrical Systems* (New York: John Wiley and Sons, 1969).

[13] Van Valkenburg, M. E., *Network Analysis* (Englewood Cliffs, N.J.: Prentice-Hall, second ed., 1964).

Systems

In recent years the course that follows a first course on circuits often generalizes the circuits concepts to other engineering systems. Some texts begin with the systems concept, as Reference [16]. Most build on a circuits background, such as that of the present text. The texts listed here are suitable for a subsequent course, perhaps with additional mathematical background.

[14] Cooper, G. R. and McGillen, C. D., *Methods of Signal and System Analysis* (New York: Holt, Rinehart and Winston, 1967).

[15] DeRusso, P. M., Roy, R. J., and Close, C. M., *State Variables for Engineers* (New York: John Wiley and Sons, 1965).

[16] Koenig, H. E., Tokad, Y., and Kesavan, H. K., *Analysis of Discrete Physical Systems* (New York: McGraw-Hill, 1967).

[17] Leon, B. J., *Lumped Systems* (New York: Holt, Rinehart and Winston, 1968).

[18] Schwarz, R. J. and Friedland, B., *Linear Systems* (New York: McGraw-Hill, 1965).

[19] Zadeh, L. A. and Desoer, C. A., *Linear System Theory* (New York: McGraw-Hill, 1963).

Advanced Circuits and Networks

The lore of circuits and networks is sufficiently extensive so that the student can spend three full years in courses, all containing different topics. Topics include graph theory, References [20] and [29]; passive, reciprocal networks [23], [24], [32], and [33]; active networks, with [26], [27], and [28] or without [25] inductors; nonlinear networks, [21] and [30]; and distributed parameter networks [22]. Some of these references are elementary, such as [21], which is taught in parallel with this text at Purdue University, while others, particularly [28], require a considerable degree of mathematical sophistication in addition to the network theory of this text.

[20] Chan, S. P., *Introductory Topological Analysis of Electrical Networks* (New York: Holt, Rinehart and Winston, 1969).

[21] Chua, L. O., *Introduction to Nonlinear Network Theory* (New York: McGraw-Hill, 1969).

[22] Ghausi, M. S. and Kelly, J. J., *Introduction to Distributed Parameter Networks* (New York: Holt, Rinehart and Winston, 1968).

[23] Guillemin, E. A., *Synthesis of Passive Networks* (New York: John Wiley and Sons, 1956).

[24] Hazony, D., *Elements of Network Synthesis* (New York: Rinehold Publishing Corporation, 1963).

[25] Huelsman, L. P., *Theory and Design of Active RC Circuits* (New York: McGraw-Hill, 1968).

[26] Kuh, E. S. and Rohrer, R. A., *Theory of Linear Active Networks* (San Francisco: Holden-Day, 1967).

[27] Mitra, S. K., *Analysis and Synthesis of Linear Active Networks* (New York: John Wiley and Sons, 1969).

[28] Newcomb, R. W., *Linear Multiport Synthesis* (New York: McGraw-Hill, 1966).

[29] Seshu, S. and Reed, M. B., *Linear Graphs and Electrical Networks* (Reading, Mass.: Addison-Wesley, 1961).

[30] Stern, T. E., *Theory of Nonlinear Networks and Systems* (Reading, Mass.: Addison-Wesley, 1965).

[31] Su, K. L., *Active Network Synthesis* (New York: McGraw-Hill, 1965).

[32] Van Valkenberg, M. E., *Introduction to Modern Network Synthesis* (New York: John Wiley and Sons, 1960).

[33] Weinberg, L., *Network Analysis and Synthesis* (New York: McGraw-Hill, 1962).

Mathematics

The mathematical background required for this text includes calculus, linear algebra, and differential equations—all at an elementary level. As originally taught, the course of this text paralleled the differential equations course in the sophomore year. The available texts for these math courses are numerous: [34], [36], [37] and [38] are typical. References [35] and [39] deal in advanced topics in applied mathematics.

[34] Golomb, M. and Shanks, M., *Elements of Ordinary Differential Equations* (New York: McGraw-Hill, 2nd ed., 1965).

[35] Papoulis, A., *The Fourier Integral* (New York: McGraw-Hill, 1962).

[36] Schneider, H. and Barker, G. P., *Matrices and Linear Algebra* (New York: Holt, Rinehart and Winston, 1968).

[37] Sokolnikoff, I. S. and Redheffer, R. M., *Mathematics of Physics and Modern Engineering* (New York: McGraw-Hill, 2nd ed., 1966).

[38] Tropper, A. M., *Matrix Theory for Electrical Engineers* (Reading, Mass.: Addison-Wesley, 1962).

[39] Zemanian, A., *Distribution Theory and Transform Analysis* (New York: McGraw-Hill, 1965).

APPENDIX

I

Linear Algebraic Equations

Virtually all the mathematical techniques of this text are derived from the problem of solving a set of linear algebraic equations with constant coefficients. In this appendix the basic steps of the problem are detailed as they are needed for the book. The student who has further interest in linear algebra is referred to any of the many excellent texts on the subject, such as References [36] and [38].

AI-1 A Simple Example

The simplest case is illustrated by the following set of two equations in two unknowns. The problem is to find x_1 and x_2 when

$$\begin{aligned} x_1 + x_2 &= 1 \\ 3x_1 + 4x_2 &= 2 \end{aligned} \tag{AI-1}$$

We shall consider two methods for solving this problem—systematic elimination and Cramer's rule. Systematic elimination is the most efficient method for solving the equations numerically. Cramer's rule is a convenient symbolic method for discussing properties of the solution without carrying out all the computations.

For the method of systematic elimination, the second of Equations (AI-1) is multiplied by the appropriate constant so that when the first equation is substracted from the modified second equation, the resultant does not contain the variable x_1. In this case the constant is $\frac{1}{3}$. The modified second equation is

$$x_1 + \tfrac{4}{3}x_2 = \tfrac{2}{3} \tag{AI-2}$$

When the first of Equations (AI-1) is subtracted from (AI-2), the result is

$$\tfrac{1}{3}x_2 = -\tfrac{1}{3}$$

or

$$x_2 = -1 \tag{AI-3}$$

Substituting Equation (AI-3) into either of the Equations (AI-1) shows that

$$x_1 = 2 \tag{AI-4}$$

Thus the solution to Equations (AI-1) is $x_1 = 2$, $x_2 = -1$.

For Cramer's rule, we must first define a quantity called a determinant. A *determinant* is a number computed from an array of n-squared numbers as follows:

$$\text{for } n = 2;\ \begin{vmatrix} a_{11} & a_{12} \\ a_{21} & a_{22} \end{vmatrix} = a_{11}a_{22} - a_{21}a_{12}$$

$$\text{for } n = 3;\ \begin{vmatrix} a_{11} & a_{12} & a_{13} \\ a_{21} & a_{22} & a_{23} \\ a_{31} & a_{32} & a_{33} \end{vmatrix}$$

$$= a_{11}\begin{vmatrix} a_{22} & a_{23} \\ a_{32} & a_{33} \end{vmatrix} - a_{21}\begin{vmatrix} a_{12} & a_{13} \\ a_{32} & a_{33} \end{vmatrix} + a_{31}\begin{vmatrix} a_{12} & a_{13} \\ a_{22} & a_{23} \end{vmatrix}$$

$$\text{for } n > 3;\ \begin{vmatrix} a_{11} & a_{12} & \cdots & a_{1n} \\ a_{21} & a_{22} & \cdots & a_{2n} \\ \cdot & \cdot & \cdot & \cdot \\ \cdot & \cdot & \cdot & \cdot \\ \cdot & \cdot & \cdot & \cdot \\ a_{n1} & a_{n2} & \cdots & a_{nn} \end{vmatrix} = \sum_{i=1}^{n} (-1)^{i+1}a_{i1} \left[\begin{array}{l}\text{The determinant of} \\ \text{the } (n-1) \times \\ (n-1) \text{ array} \\ \text{formed by crossing} \\ \text{out the first column} \\ \text{and ith row of the} \\ \text{given } n \times n \text{ array.}\end{array}\right]$$

For Equations (AI-1), Cramer's rule states that

$$x_1 = \frac{\begin{vmatrix} 1 & 1 \\ 2 & 4 \end{vmatrix}}{\begin{vmatrix} 1 & 2 \\ 3 & 4 \end{vmatrix}} = \frac{2}{1} = 2$$

$$x_2 = \frac{\begin{vmatrix} 1 & 1 \\ 3 & 2 \end{vmatrix}}{\begin{vmatrix} 1 & 1 \\ 3 & 4 \end{vmatrix}} = \frac{-1}{1} = -1 \qquad \text{(AI-5)}$$

AI-2 The General Solution of Algebraic Equations

The general problem of solving a set of n linear algebraic equations in n unknowns can be stated symbolically as follows: Given n^2 numbers a_{ij}, $i = 1, 2, \cdots n$; $j = 1, 2, \cdots n$ and n other numbers y_i, $i = 1, 2, \cdots n$, find the n numbers x_j satisfying the equations

$$\begin{array}{ccccccc} a_{11}x_1 & + & a_{12}x_2 & + \cdots + & a_{1n}x_n & = & y_1 \\ a_{21}x_1 & + & a_{22}x_2 & + \cdots + & a_{2n}x_n & = & y_2 \\ \cdot & & \cdot & \cdot & \cdot & & \cdot \\ \cdot & & \cdot & \cdot & \cdot & & \cdot \\ \cdot & & \cdot & \cdot & \cdot & & \cdot \\ a_{n1}x_1 & + & a_{n2}x_2 & + \cdots + & a_{nn}x_n & = & y_n \end{array} \qquad \text{(AI-6)}$$

The systematic elimination procedure is as follows:

1. Multiply the second equation by a_{11}/a_{21}, then multiply the third equation by a_{11}/a_{31}, and so on, until the nth equation has been multiplied by a_{11}/a_{n1}. The result is $(n - 1)$ modified equations.

2. Subtract the first of Equations (AI-6) from the $(n - 1)$ modified equations of step (1). The result is $(n - 1)$ equations with unknowns $x_2, x_3, \cdots, x_n$.

3. Relabel the coefficients of the $(n - 1)$ equations in $(n - 1)$ unknowns that resulted from step (2) so that they have the basic form of Equation (AI-6). Then repeat steps (1) and (2). The result is $(n - 2)$ equations in $(n - 2)$ unknowns $x_3, x_4, \cdots, x_n$. The procedure is continued until there remains only one equation in the one unknown x_n.

4. Solve the final equation of step (3) for x_n and substitute this value into the previous result of step (3). That previous result was two equations in two unknowns x_{n-1} and x_n. With x_n known, x_{n-1} is readily computed from either of the two equations. Now with x_n and x_{n-1} known, the stage of step (3) that resulted in three equations in three unknowns can be used to compute x_{n-2}. The procedure is continued until all the x_1, i = 1, 2, · · · , n have been computed.

For the set of Equations (AI-6), Cramer's rule states that

$$
\begin{aligned}
x_1 &= \frac{\begin{vmatrix} y_1 & a_{12} & \cdots & a_{1n} \\ y_2 & a_{22} & \cdots & a_{2n} \\ \vdots & \vdots & & \vdots \\ y_n & a_{2n} & \cdots & a_{nn} \end{vmatrix}}{\begin{vmatrix} a_{11} & a_{12} & \cdots & a_{1n} \\ a_{21} & a_{22} & \cdots & a_{2n} \\ \vdots & \vdots & & \vdots \\ a_{n1} & a_{n2} & \cdots & a_{nn} \end{vmatrix}} \\
x_2 &= \frac{\begin{vmatrix} a_{11} & y_1 & \cdots & a_{1n} \\ a_{12} & y_2 & \cdots & a_{2n} \\ \vdots & \vdots & & \vdots \\ a_{1n} & y_n & \cdots & a_{nn} \end{vmatrix}}{\begin{vmatrix} a_{11} & a_{12} & \cdots & a_{1n} \\ a_{21} & a_{22} & \cdots & a_{2n} \\ \vdots & \vdots & & \vdots \\ a_{n1} & a_{n2} & \cdots & a_{nn} \end{vmatrix}} \\
&\ \ \vdots \\
x_n &= \frac{\begin{vmatrix} a_{11} & a_{12} & \cdots & y_1 \\ a_{21} & a_{22} & \cdots & y_2 \\ \vdots & \vdots & & \vdots \\ a_{1n} & a_{2n} & \cdots & y_n \end{vmatrix}}{\begin{vmatrix} a_{11} & a_{12} & \cdots & a_{1n} \\ a_{21} & a_{22} & \cdots & a_{2n} \\ \vdots & \vdots & & \vdots \\ a_{n1} & a_{2n} & \cdots & a_{nn} \end{vmatrix}}
\end{aligned}
\tag{AI-7}
$$

As a symbolic solution for each x_i, Equation (AI-7) is much simpler than anything we could get from the method of systematic elimination. If one figures the number of computational steps required to evaluate the determinants of Equation (AI-7), he would find this procedure much more involved than the method of systematic elimination.

AI-3 Linear Independence

In stating the method of systematic elimination and Cramer's rule above, no mention was made of cases wherein the methods may fail to yield a solution. From Equation (AI-7) it is clear that if the denominator determinant is zero, Cramer's rule does not yield any results, because all the x_i will have a zero in the denominator.

The systematic elimination procedure may fail for two reasons, one correctable and one fundamental. In the first step, the construction of the modified equations, the procedure fails if one of the equations does not contain x_1 as a variable. This failure is easily corrected by applying steps 1 and 2 to only those equations containing x_1 as a variable. The equations that do not contain x_1 are in the same form as the results of step 2, because the equations that result from the subtractions of step 2 do not contain x_1 either. Thus at the end of step 2, there are still $(n - 1)$ equations in the $(n - 1)$ unknowns $x_2, x_3, \cdots, x_n$. The problem of dividing by zero in step 1 can be corrected by appropriate grouping of equations before starting the procedure.

The fundamental problem that can occur will appear in the last execution of step 2. When there remain only two equations in two unknowns, there is the possibility that step 1, the modification step, makes the two equations identical. Then the subtraction step, step 2, does not yield an equation that can be solved for x_n. It yields an equation with the left side zero. If the right side is not also zero, there is an inconsistency. Even if the right side is zero, the equations cannot be solved for all the x_n.

When the above problem occurs, the equations are not independent. Let us state the definition of linear independence and show that the problem is a consequence of not satisfying this definition.

Definition: linear independence. A set of n objects $u_1, u_2, \cdots, u_n$ is said to be linearly independent if there exist *no* constants $k_1, k_2, \cdots, k_n$, not all zero, such that

$$k_1 u_1 + k_2 u_2 + \cdots + k_n u_n = 0 \tag{AI-8}$$

If such constants can be found, then the objects are said to be dependent.

The application of the definition of linear independence to the simultaneous equations requires the identification of the left side of each

Equation (AI-6) as one of the u_1 in the definition. The execution of steps 2 and 3 is equivalent to multiplication of the equations by constants and adding. If at the final step the result is zero, then we have displayed a set of k_i such that Equation (AI-8) is satisfied, and the equations are dependent. It can be shown that if the procedure does succeed, then there are no k_i satisfying Equation (AI-8), and the equations are linearly independent.

It can also be shown that a set of equations in the form of Equation (AI-8) is linearly independent if and only if the determinant of the coefficients is not zero. Thus, when the equations are independent, both systematic elimination and Cramer's rule give a solution. When they are not independent, there is, in general, no solution.

AI-4 Matrix Notation

The notation of linear algebraic equations can be simplified by using matrices. A matrix is a group of numbers written in a special format. This format is an array consisting of m rows with n numbers in each row. The rows are written one over the other so that the resulting array consists of m $\times$ n numbers arranged in n columns.

As an example, let m = 2 and n = 3. Then, an example of a 2 by 3 matrix is the numbers

$$\begin{bmatrix} 10 & 3 & 5 \\ 2 & 7 & 14 \end{bmatrix}$$

We designate the fact that the array is a matrix by enclosing the array in brackets.

There are rules for adding and multiplying matrices. To state these rules, we consider three matrices. Let the element in the ith row and the jth column of the first matrix be a_{ij}. For the second matrix, the corresponding element is b_{ij}, and for the third, it is c_{ij}. For the operation of addition, let us add the first matrix to the second to get the third. The rule is

$$c_{ij} = a_{ij} + b_{ij} \tag{AI-9}$$

When we add two matrices, each element of the sum is the sum of the corresponding elements of the two original matrices. For addition, both matrices must have the same number of rows and the same number of columns.

For matrix multiplication, the number of columns of the first matrix must equal the number of rows of the second matrix. Let this number be n. Then the element c_{ij} of the resultant matrix is

$$c_{ij} = \sum_{k=1}^{n} a_{ik}b_{kj} \tag{AI-10}$$

The number of rows of the resultant is the number of rows of the first matrix, and the number of columns of the resultant is the number of columns of the second matrix.

In matrix notation, the set of Equations (AI-1) becomes

$$\begin{bmatrix} 1 & 1 \\ 3 & 4 \end{bmatrix} \begin{bmatrix} x_1 \\ x_2 \end{bmatrix} = \begin{bmatrix} 1 \\ 2 \end{bmatrix} \tag{AI-11}$$

Two types of matrices are particularly important in discussing linear algebraic equations. One is the *square matrix*—a matrix with the same number of rows and columns. The other is the *column matrix* (or vector)—a matrix with any number of rows, but only one column. Note that Equation (AI-11) involves only these two types of matrices. In the text we use bold-faced letters to represent matrices.

AI-5 Relations between Sets of Variables

In Chapter 5, linear algebraic equations are used to relate certain circuit variables to other circuit variables. In matrix notation, such a relationship takes the form

$$\mathbf{Z}\mathbf{I} = \mathbf{V} \tag{AI-12}$$

When $\mathbf{Z}$ is an impedance matrix, $\mathbf{I}$ is a vector of current amplitudes, and $\mathbf{V}$ is a vector of the corresponding voltage amplitudes. The Equation (AI-12) gives each voltage amplitude as a sum of impedances multiplied by current amplitudes; that is,

$$V_i = \sum_{j=1}^{n} z_{ij} I_j \tag{AI-13}$$

A set of equations such as Equation (AI-12) can be solved by systematic elimination or Cramer's rule to obtain a set of equations that gives each I_i in terms of the V_j. To see the form of such a solution, let us examine a set of two simultaneous equations in the form of Equation (AI-12). Thus,

$$\begin{aligned} z_{11} I_1 + z_{12} I_2 &= V_1 \\ z_{21} I_1 + z_{22} I_2 &= V_2 \end{aligned} \tag{AI-14}$$

By Cramer's rule,

$$\begin{aligned} I_1 &= \frac{\begin{vmatrix} V_1 & z_{12} \\ V_2 & z_{22} \end{vmatrix}}{\begin{vmatrix} z_{11} & z_{12} \\ z_{21} & z_{22} \end{vmatrix}} = \frac{z_{22}}{z_{11}z_{22} - z_{12}z_{21}} V_1 - \frac{z_{12}}{z_{11}z_{22} - z_{12}z_{21}} V_2 \\ I_2 &= \frac{\begin{vmatrix} z_{11} & V_1 \\ z_{21} & V_2 \end{vmatrix}}{\begin{vmatrix} z_{11} & z_{22} \\ z_{12} & z_{21} \end{vmatrix}} = \frac{-z_{21}}{z_{11}z_{22} - z_{12}z_{21}} V_1 + \frac{z_{11}}{z_{11}z_{22} - z_{12}z_{21}} V_2 \end{aligned} \tag{AI-15}$$

In matrix notation Equation (AI-15) can be written

$$\begin{bmatrix} \dfrac{z_{22}}{z_{11}z_{22} - z_{12}z_{21}} & \dfrac{-z_{12}}{z_{11}z_{22} - z_{12}z_{21}} \\ \dfrac{-z_{21}}{z_{11}z_{22} - z_{12}z_{21}} & \dfrac{z_{11}}{z_{11}z_{22} - z_{12}z_{21}} \end{bmatrix} \begin{bmatrix} V_1 \\ V_2 \end{bmatrix} = \begin{bmatrix} I_1 \\ I_2 \end{bmatrix} \qquad \text{(AI-16)}$$

The square matrix in Equation (AI-16) is called the *inverse* to the original **Z** matrix. It is often designated by $\mathbf{Z}^{-1}$. The reader can verify that for the 2 × 2 matrix, the product of a matrix and its inverse is the identity matrix—a matrix with one in the ii positions and zero in the ij positions i ≠ j. Symbolically for a 2 × 2 matrix **Z**

$$\mathbf{ZZ}^{-1} = \mathbf{Z}^{-1}\mathbf{Z} = \mathbf{1} = \begin{bmatrix} 1 & 0 \\ 0 & 1 \end{bmatrix} \qquad \text{(AI-17)}$$

For larger than 2 × 2 square matrices, Cramer's rule or systematic elimination can be used to construct the inverse matrix, provided the rows are linearly independent, and thus the determinant constructed from the matrix is nonzero.

A second situation where matrix notation is convenient in relating sets of variables is the situation where the variables x_i are related to variables y_i by a set of equations, and the y_i are in turn related to variables z_i by a second set of equations. In matrix notation, we have square matrices **A** and **B** that relate the vectors $\boldsymbol{x}$, $\boldsymbol{y}$, and $\boldsymbol{z}$ by

$$\mathbf{A}\boldsymbol{x} = \boldsymbol{y}; \qquad \mathbf{B}\boldsymbol{y} = \boldsymbol{z} \qquad \text{(AI-18)}$$

If we write out the equations and make the substitutions that eliminate the variables y_i, the result is a new square matrix **C** relating $\boldsymbol{x}$ and $\boldsymbol{z}$. Symbolically

$$\mathbf{C}\boldsymbol{x} = \boldsymbol{z} \qquad \text{(AI-19)}$$

If we examine the matrix **C** in the light of the rule for matrix multiplication, Equation (AI-10), we see that **C** is the matrix product **BA**. This result is exactly what we get if we substitute the first matrix equation of Equation (AI-18) into the second. This is

$$\mathbf{B}[\mathbf{A}\boldsymbol{x}] = \boldsymbol{z} \qquad \text{(AI-20)}$$

AI-6 Homogeneous Linear Algebraic Equations

In Chapter 11 there is a set of homogeneous linear algebraic equations, that is, a set of equations in the form

$$\mathbf{A}\boldsymbol{x} = \mathbf{0} \qquad \text{(AI-21)}$$

where **0** is the vector whose components are all zero.

When the solution to such a set of equations is written down via Cramer's rule, the numerator determinant for each component of $\boldsymbol{x}$ contains a column of zeros. The value of this determinant is zero. Thus, each x_i is zero unless the determinant of $\mathbf{A}$ is also zero. If this latter situation is the case, then each x_i is of the indeterminant form zero over zero. In other words, if the equations in (AI-21) are linearly independent, then all x_i are zero. If the equations are not independent, then there may be another solution.

If the determinant of the coefficient matrix of a set of linear algebraic equations such as (AI-21) is zero, then the equations are dependent. Then any one of the equations can be written as a linear combination of the others. To be specific, let us consider the last (nth) equation to be superfluous. The remaining (n − 1) equations can be written in the form

$$\mathbf{A}'\boldsymbol{x}' = \mathbf{a}x_n \tag{AI-22}$$

where $\mathbf{A}'$ is the square matrix formed by deleting the last row and last column of $\mathbf{A}$. $\boldsymbol{x}'$ is the (n − 1) vector formed from $\boldsymbol{x}$ by deleting x_n. $\mathbf{a}$ is the vector consisting of the last column of $\mathbf{A}$ with a_{nn} deleted. The product $\mathbf{a}x_n$ means a vector formed from $\mathbf{a}$ by multiplying each component by x_n.

If the determinant of $\mathbf{A}'$ is not zero, then the Equations (AI-22) can be solved for each x_i, $i = 1, 2, \cdots, (n - 1)$ in terms of x_n. Thus not only are the equations dependent, but also the components of the vector $\boldsymbol{x}$ are dependent. If one component is known through some other condition than Equation (AI-21), all the others can be found by solving Equation (AI-22).

APPENDIX

II

Proofs of Stated, Well-Known Theorems

In the text, several well-known theorems of network theory were stated and used without proof. These proofs were not inserted in the text because the significant point of development was the application of the theorem, not the methods of proof; the authors believe that covering the proofs would detract from the main theme of the text. Proofs of network theorems are important to all electrical engineers. The network theorist is primarily interested in ascertaining new network theorems. The circuit engineer, who is interested in using the theorems, must be able to check a proof to be sure it is correct. Thus, those theorems that can be proved within the level of mathematical maturity required for this text are proved here. The theorems are the following:

1. The linear independence of the equilibrium equations derived from a tree (Chapter 3).
2. Superposition (Chapter 2).
3. Thevenin's and Norton's theorems (Chapter 2).
4. The maximum power-transfer theorem (Chapter 2).
5. Tellegin's theorem (Chapters 2 and 5).
6. The reciprocity theorem (Chapter 5).

AII-1 Linear Independence of Equilibrium Equations

In Chapter II we saw that a network with b branches had 2b electrical quantities—b voltages and b currents. In that chapter an algorithm for setting up 2b equations with these 2b quantities as unknowns was derived. It can be shown that the equations are linearly independent except in certain degenerate cases where the network is inconsistent or indeterminant. The complete proof is beyond the scope of this text. Those steps that are within the level of the text are given here.

For a network with b branches, n nodes, s independent sources, c controlled sources, and e other elements, the 2b equations were set up in the following groups:

(n − 1)	Kirchhoff current law equations in the form one tree branch current equals a sum of link currents
(b − n + 1)	Kirchhoff voltage law equations in the form one link voltage equals a sum of tree branch voltages.
s	source equations
e	element equations
c	control equations

In the discussion that follows, we shall divide the source, element, and control equations, again depending on whether or not the element is in the tree or cotree.

In order to see the independence of the various equations, let us write them so that the column vector of variables has the tree branch currents in the first (n − 1) places, the link voltages in the next (b − n + 1) places, the tree branch voltages in the next (n − 1) places, and the link currents in the last (b − n + 1) places. Furthermore, let us write the KCL equations first and the KVL equations next, with the first KCL equation being that giving the first tree branch current in terms of link currents. Then the second equation is for the second tree branch current, and so on. When we get to the KVL equations, the first should give the first link voltage in terms of tree branch voltages, and so on. With this ordering the first b equations (the KCL and KVL equations taken together) have a coefficient matrix as follows:

	(n − 1)	(b − n + 1)	(n − 1)	(b − n + 1)
(n − 1)	ones on diagonal; all others zero	all zeros	all zeros	+1, −1, or zero
(b − n + 1)	all zeros	ones on diagonal; all others zero	+1, −1, or zero	all zeros

By considering the left half of this matrix, we see that the equations are linearly independent. Each equation contains one variable that is not contained in any other equation. This is manifested in the fact that the left half of the coefficient matrix is a diagonal matrix.

The next block of equations in this systematic listing contains the source, element, and control equations for tree branches. For definiteness, let the first variables be tree branch currents through the independent voltage sources, the next be the RLC elements, and then the controlled voltage sources. The matrix for this block of $(n - 1)$ equations is as follows:

source equations	zeros	zeros	ones on diagonal; all others zero	zeros
element equations	R or $j\omega L$ or $1/j\omega C$ on diagonal	zeros	ones on diagonal; all others zero	zeros
control equations	?	?	ones on diagonal; all others zero	?

The statements about diagonal elements refer to the diagonal of the submatrix of all rows and columns between partitioning lines. The final $(b - n + 1)$ rows for the source, element, and control equations of the links will have essentially the same form when the variables are chosen in the corresponding order. That is, the last $(b - n + 1)$ rows of the coefficient matrix will be as follows:

source equations	zeros	zeros	zeros	ones on diagonal
element equations	zeros	$1/R$ or $1/j\omega L$ or $j\omega C$ on diagonal; otherwise zeros	zeros	ones on diagonal
control equations	?	?	?	ones on diagonal

Example AII-1

Before proceeding, let us set up an example to illustrate the equation grouping. Consider the network of Figure AII-1. The tree is shown by the heavy lines. The assumed polarity for the voltages and currents is indicated by arrows on all branches. The numbers on the branches will be used as subscripts on the voltage and current variables. The

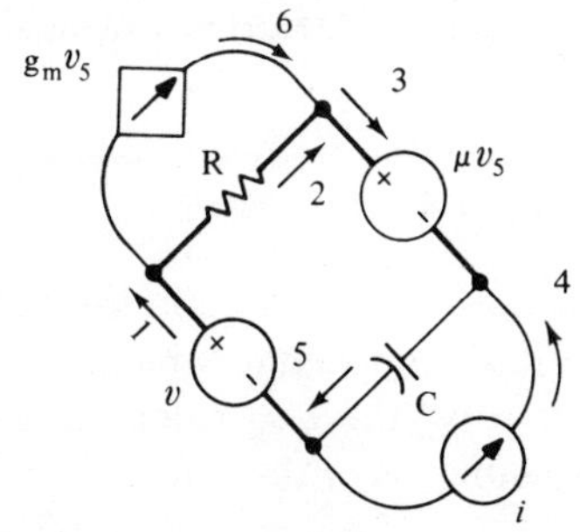

Figure AII-1 An example for equation formulation.

equations in the format outlined above are as follows:

$$
\begin{array}{l}
\text{KCL} \\ \\ \\ \\ \text{KVL} \\ \\ \\ \\ \text{tree branch} \\ \\ \\ \\ \text{link} \\ \\
\end{array}
\left[\begin{array}{ccc|ccc|ccc|ccc}
1 & 0 & 0 & 0 & 0 & 0 & 0 & 0 & 0 & 1 & -1 & 0 \\
0 & 1 & 0 & 0 & 0 & 0 & 0 & 0 & 0 & 1 & -1 & 1 \\
0 & 0 & 1 & 0 & 0 & 0 & 0 & 0 & 0 & 1 & -1 & 0 \\
 & & & & & & & & & & & \\
0 & 0 & 0 & 1 & 0 & 0 & -1 & -1 & -1 & 0 & 0 & 0 \\
0 & 0 & 0 & 0 & 1 & 0 & 1 & 1 & 1 & 0 & 0 & 0 \\
0 & 0 & 0 & 0 & 0 & 1 & 0 & -1 & 0 & 0 & 0 & 0 \\
 & & & & & & & & & & & \\
0 & 0 & 0 & 0 & 0 & 0 & 1 & 0 & 0 & 0 & 0 & 0 \\
0 & -R & 0 & 0 & 0 & 0 & 0 & 1 & 0 & 0 & 0 & 0 \\
0 & 0 & 0 & 0 & -\mu & 0 & 0 & 0 & 1 & 0 & 0 & 0 \\
 & & & & & & & & & & & \\
0 & 0 & 0 & 0 & 0 & 0 & 0 & 0 & 0 & 1 & 0 & 0 \\
0 & 0 & 0 & 0 & -j\omega C & 0 & 0 & 0 & 0 & 0 & 1 & 0 \\
0 & 0 & 0 & 0 & -g_m & 0 & 0 & 0 & 0 & 0 & 0 & 1
\end{array}\right]
\left[\begin{array}{c}
I_1 \\ I_2 \\ I_3 \\ \\ V_4 \\ V_5 \\ V_6 \\ \\ V_1 \\ V_2 \\ V_3 \\ \\ I_4 \\ I_5 \\ I_6
\end{array}\right]
=
\left[\begin{array}{c}
0 \\ 0 \\ 0 \\ \\ 0 \\ 0 \\ 0 \\ \\ -V \\ 0 \\ 0 \\ \\ I \\ 0 \\ 0
\end{array}\right]
\tag{AII-1}
$$

With the above system for listing the various equations, the coefficient matrix always has ones on the main diagonal. In all cases, these ones are the only entries in the b $\times$ b submatrix in the upper left-hand corner. For this reason, the first b equation can always be used to eliminate the first b variables—the tree branch currents and link voltages. The result is b equations with tree branch voltages and link currents as variables. If the original 2b equations are independent, the remaining b equations are still independent. This is all that can be said about the linear independence of the network equations without considering the specific types of elements and element values. It can be shown (see Weinberg, Reference [33]) that if all elements are positive resistances and independent sources, then the equations are independent. This proof requires knowledge of the subtle details of the matrices of ± 1's and zeros. We shall not discuss these here. All we can say here is that, if the equations formulated are not independent, then there are no independent equations. Either the network is indeterminant or inconsistent.

To show that there are such inconsistent and indeterminant networks, let us consider two examples.

Example AII-2

As an example of an inconsistent network, let us consider Figure AII-2. At the frequency where $1/j\omega C = j\omega L$, the inductance and capacitance in series constitute a short circuit. Consequently, the two elements have zero voltage across them, and the voltage source says they have V volts.

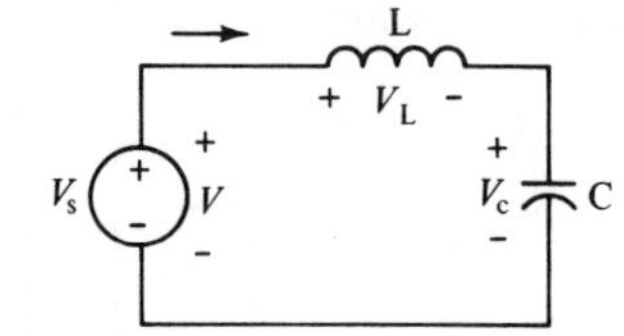

Figure AII-2 Example AII-2.

Thus, there is an inconsistency. The equations, placing the voltage source and the inductance in the tree and the capacitance in the cotree, are

$$\begin{bmatrix} 1 & 0 & 0 & 0 & 0 & -1 \\ 0 & 1 & 0 & 0 & 0 & -1 \\ 0 & 0 & 1 & 1 & -1 & 0 \\ 0 & 0 & 0 & 1 & 0 & 0 \\ 0 & -j\omega L & 0 & 0 & 1 & 0 \\ 0 & 0 & -j\omega C & 0 & 0 & 1 \end{bmatrix} \begin{bmatrix} I \\ I_L \\ V_C \\ V \\ V_L \\ I_L \end{bmatrix} = \begin{bmatrix} 0 \\ 0 \\ 0 \\ V_s \\ 0 \\ 0 \end{bmatrix} \tag{AII-2}$$

The determinant of the coefficient matrix is $(1 + \omega^2 LC)$. If $\omega^2 = 1/LC$, then the determinant is zero, and the equations are not linearly independent.

Example AII-3

As a second example, let us consider the circuit of Figure AII-3. Since the voltage across the resistance is $I_R R$, and the controlled source voltage is $I_R r_m$, we expect a problem if $r_m \neq R$. In setting up the equa-

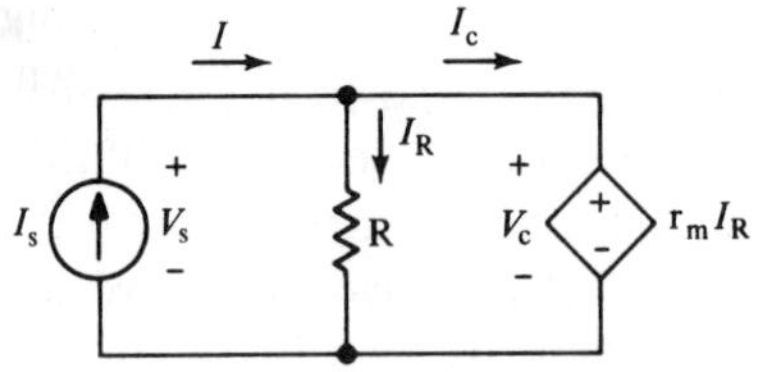

Figure AII-3 Example AII-3.

tions, we place the controlled source in the tree and the independent source and the resistance in the cotree. The equations are

$$\begin{bmatrix} 1 & 0 & 0 & 0 & -1 & 1 \\ 0 & 1 & 0 & -1 & 0 & 0 \\ 0 & 0 & 1 & -1 & 0 & 0 \\ 0 & 0 & 0 & 1 & 0 & -\mathrm{r_m} \\ 0 & 0 & 0 & 0 & 1 & 0 \\ 0 & 0 & -\dfrac{1}{\mathrm{R}} & 0 & 0 & 1 \end{bmatrix} \begin{bmatrix} I_C \\ V_s \\ V_R \\ V_c \\ I \\ I_R \end{bmatrix} = \begin{bmatrix} 0 \\ 0 \\ 0 \\ 0 \\ I_s \\ 0 \end{bmatrix} \tag{AI-13}$$

The determinant of the coefficient matrix is $1 - \mathrm{r_m}/\mathrm{R}$. This indicates trouble only if $\mathrm{r_m} = \mathrm{R}$. Thus, we must examine the numerator to see the problem that we predicted from our first look at the circuit.

The numerator determinant is $I_s(1 - (\mathrm{r_m}/\mathrm{R}))$ when we solve for I and I_c, and it is zero for the other four variables. Thus, when $\mathrm{r_m} \neq \mathrm{R}$, the independent source current all flows through the controlled source. The voltage across all three elements and the current through the resistance is zero in this case. This is the only solution. When $\mathrm{r_m} = \mathrm{R}$, Cramer's rule gives zero over zero for all quantities. Thus, the circuit seems indeterminant. In fact, a closer examination of the circuit shows that any voltage is possible. Then the other variables take appropriate values via the definition of the resistance and Kirchhoff's current law. In this last case the circuit is indeterminant.

AII-2 Superposition

The superposition theorem states that, given a network of elements from Chapter 1 with several independent sources, any variable can be found as the sum of corresponding variables in a set of simpler networks. The simpler networks are formed from the original network by setting all independent sources except one equal to zero.

The proof follows from the rules for evaluating a determinant by expanding on one column. In the previous section (see Section AII-1), we listed the network equilibrium equations in such a way that each independent source quantity was listed as a separate entry (element) of the right-hand side (forcing-function) vector. When Cramer's rule is used for solving for any variable, then this forcing function vector is one column of the numerator determinant. When the determinant is expanded on that column, the result is a sum of terms, each having a particular independent source quantity as a multiplicative factor. Each such term is exactly the response of the network when its source is present and all other sources are set to zero.

AII-3 Thevenin's and Norton's Theorems

In order to prove Thevenin's theorem, we use the network shown in Fig. AII-4. We include an impedance $\mathfrak{z}$ across the accessible terminals rather than specifying V and I separately, as we did in Chapter 2. We write the equilibrium equations for the network of Figure AII-4(a) in

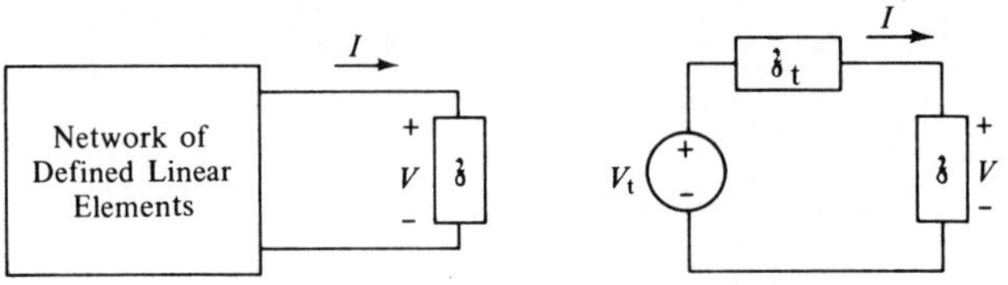

Figure AII-4 Thevenin's theorem network with steady-state voltage and current amplitudes.

the format of Section AII-1. To be specific, we choose the impedance $\mathfrak{z}$ as a link and place the voltage V in the nth place in the vector of voltages and currents. Then the current I will be in the (2b − n)th place.

If the voltage V is not the control voltage for a voltage-controlled source in the network, then in the coefficient matrix of the equilibrium equations, the nth column contains $+1$ in the nth place and $-1/\mathfrak{z}$ in the (2b − n)th place. All other elements are zero. Thus, the determinant of the coefficient matrix can be written symbolically as

$$\Delta = \Delta_{n,n} - \frac{1}{\mathfrak{z}} \Delta_{(2b-n),n} \tag{AII-4}$$

where Δ_{ij} is the cofactor of the i, j element of the matrix. A cofactor is $(-1)^{i+j}$ times the determinant formed by crossing out the ith row and jth column of the original determinant. With the equations written in the standard format, the voltage V can be written symbolically via Cramer's rule. It is

$$V = \frac{N}{\Delta} = \frac{N}{\Delta_{n,n} - \dfrac{1}{\mathfrak{z}} \Delta_{(2b-n),n}} \tag{AII-5}$$

where N is the numerator determinant formed by replacing the nth column of the coefficient determinant by the vector of independent source quantities from the right-hand side of the equilibrium equations. To prove Thevenin's theorem, we must show that the voltage V in the circuit of Figure AII-4(b) is the same.

Both the Thevenin voltage V_t and the Thevenin impedance $\mathfrak{z}_t$ can be written symbolically in terms of determinants. To see this, we must establish the appropriate networks for computing the two quantities.

The Thevenin voltage V_t is the voltage that the network in the box of Figure AII-4(a) develops when the impedance $\mathfrak{z}$ is replaced by an open circuit. When this is done, the equilibrium equations for the resulting network are the same as before, except that the term $-1/\mathfrak{z}$ in the (2b − n, n) place of the coefficient matrix is replaced by zero. Thus, the Thevenin voltage can be written symbolically as

$$V_t = \frac{N}{\Delta_{n,n}} \tag{AII-6}$$

where N and Δ are the same as before. To compute the Thevenin impedance, we set all independent sources in the box to zero and compute the input impedance of the box. The input impedance computation is equivalent to replacing $\mathfrak{z}$ by a 1-ampere current source (directed so that I is minus one) and computing the voltage. When this change is made, the following changes take place in the equilibrium equations. In the coefficient matrix the (2b − n, n)th element is zero instead of $1/\mathfrak{z}$. In the forcing vector, all elements are zero except the (2b − n)th, which is one. Thus, the Thevenin impedance is

$$\mathfrak{z}_t = -\frac{\Delta_{(2b-n),n}}{\Delta_{n,n}} \tag{AII-7}$$

where the minus occurs because the current source is opposite I in the figure.

The final step in the proof of Thevenin's theorem is to write down the voltage V for the network of Figure AII-2(b) and then show that this is the same as that of Figure AII-2(a). Specifically,

$$\begin{aligned} V &= \frac{V_t\,\mathfrak{z}}{\mathfrak{z}_t + \mathfrak{z}} = \frac{\dfrac{N}{\Delta_{n,n}}\,\mathfrak{z}}{\dfrac{\Delta_{(2b-n),n}}{\Delta_{n,n}} + \mathfrak{z}} \\ &= \frac{N}{\Delta_{n,n} - \dfrac{1}{\mathfrak{z}}\,\Delta_{(2b-n),n}} \end{aligned} \tag{AII-8}$$

Thus, Thevenin's theorem is proved.

The proof of Norton's theorem is exactly the same as the proof of Thevenin's theorem. All that is required are the following interchanges of wording:

for	current	use	voltage
	voltage		current
	tree branch		link
	impedance		admittance

Two facts that can be related by such a word interchange are said to be dual. Thus, Norton's theorem is the dual of Thevenin's theorem.

In the above proof we neglected the possibility that the voltage V might be the control voltage for one or more controlled sources. Had this not been true, there would have been more cofactors multiplied by μ's or g_m's to carry along and complicate the expressions. The truth of the theorem will still be apparent.

AII-4 The Maximum-Power Transfer Theorem

In the sinusoidal steady state, the maximum-power transfer theorem states: Given a voltage source V in series with a fixed impedance $\mathfrak{z}$, $\text{Re}\,[\mathfrak{z}] \geq 0$, and a variable impedance $\mathfrak{z}_L$, $\text{Re}\,[\mathfrak{z}_L] \geq 0$, the maximum power is delivered to $\mathfrak{z}_L$ when $\mathfrak{z}_L = \mathfrak{z}^*$.

To prove the theorem, we write down the expression for the power in $\mathfrak{z}_L$ and show that a maximum occurs when the theorem is satisfied. The power, to within a factor of two, is

$$\begin{aligned}\langle p \rangle &= |I|^2\, \text{Re}\,[\mathfrak{z}_L] = \left|\frac{V}{\mathfrak{z} + \mathfrak{z}_L}\right|^2 \text{Re}\,[\mathfrak{z}_L] \\ &= |V|^2 \frac{\text{Re}\,[\mathfrak{z}_L]}{\text{Re}^2\,[\mathfrak{z} + \mathfrak{z}_L] + \text{Im}^2\,[\mathfrak{z} + \mathfrak{z}_L]}\end{aligned} \tag{AII-9}$$

The imaginary part of $\mathfrak{z}_L$ appears only in the denominator. As $\text{Im}\,[\mathfrak{z}_L]$ is varied, the denominator is smallest when $\text{Im}\,[\mathfrak{z} + \mathfrak{z}_L]$ is zero. Thus, $\text{Im}\,[\mathfrak{z}_L] = -\text{Im}\,[\mathfrak{z}]$ is one condition for maximum power. With the imaginary part of $\mathfrak{z}_L$ set, the power is proportional to a real-number formula. It is

$$\langle p \rangle \sim \frac{x}{(x + a)^2}$$

where

$x = \text{Re}\,[\mathfrak{z}_L]$ is allowed to take on any non-negative value

$a = \text{Re}\,[\mathfrak{z}]$ is a fixed non-negative number.

To find the maximum of $\langle p \rangle$ for real non-negative x, we differentiate and set the result to zero.

$$\frac{d}{dx}\left[\frac{x}{(x + a)^2}\right] = \frac{(x + a)^2 - 2(x + a)x}{(x + a)^4} = 0 \tag{AII-10}$$

or

$$x + a - 2x = 0$$

or

$$x = a$$

Thus, there is only one extremum. Since $\langle p \rangle = 0$ for $x = 0$ and $x \to \infty$, and $\langle p \rangle > 0$ for $x > 0$, this extremum is the desired maximum, and the theorem is proved.

AII-5 Tellegen's Theorem

Tellegen's theorem states that the net power delivered to all the branches in a network is zero. Thus, some branches deliver power to the rest of the network and others absorb power from the rest of the network, but all power is accounted for. From a physical point of view, we might say that Tellegen's theorem as stated above is obvious because of conservation of energy. We must remember that the network elements were defined independently of power considerations. Thus, the theorem shows that the networks as defined are consistent with conservation of energy. After proving Tellegen's theorem, we shall see that the theorem can be generalized to say more about networks than only conservation of energy. The application of this generalization is beyond the scope of this text.

The proof of Tellegen's theorems depends on the relation between the two submatrices of $\pm$ ones and zeros that appeared in the statements of the KCL and KVL equations in Section AII-1 of this appendix. In that section we saw that Kirchhoff's current law could be stated symbolically as

$$[\mathbf{1}_{(n-1)} \quad \mathbf{X}] \begin{bmatrix} \boldsymbol{I}_{\mathrm{T}} \\ \boldsymbol{I}_{\mathrm{L}} \end{bmatrix} = \mathbf{0}_{(n-1)} \tag{AII-11}$$

where $\mathbf{1}_{n-1}$ is the $(n-1) \times (n-1)$ identity matrix; $\mathbf{X}$ is an $(n-1)$ by $(b-n+1)$ matrix of ± 1's and zeros; $\boldsymbol{I}_{\mathrm{T}}$ is the vector of tree branch currents; $\boldsymbol{I}_{\mathrm{L}}$ is the vector of link currents; $\mathbf{0}_{(n-1)}$ is the $(n-1)$ vector of zeros. Similarly, Kirchhoff's voltage law is

$$[\mathbf{1}_{(b-n+1)} \quad \mathbf{W}] \begin{bmatrix} \boldsymbol{V}_{\mathrm{L}} \\ \boldsymbol{V}_{\mathrm{T}} \end{bmatrix} = \mathbf{0}_{(b-n+1)} \tag{AII-12}$$

where $\mathbf{W}$ is a $(b-n+1)$ by $(n-1)$ vector of ± 1's and zeros. The significant step of the proof of Tellegen's theorem involves the relation between the matrix $\mathbf{X}$ and the matrix $\mathbf{W}$.

Each row of the matrix $\mathbf{X}$ is constructed by considering one tree branch current. The entries in the appropriate columns are nonzero when that column corresponds to a link current that is in the cut set of the particular tree branch. The sign is + if, when that link is connected to the tree, the link current flows through the tree branch in question in the negative direction. Otherwise, the entry in that column of the appropriate row is -1. Thus, as we go down the columns of $\mathbf{X}$, we get a nonzero element when the link current and its associated loop pass through the tree branch corresponding to the row.

Each row of the matrix $\mathbf{W}$ is constructed by considering one link. The entries in the appropriate columns are nonzero when that column corresponds to a tree branch that is in the loop of the particular link. The sign is + if, when the link is connected to the tree, the path around the loop finds the tree branch voltage in the same polarity as the link voltage. Otherwise, the entry in that column of the appropriate row is -1. This process of going across rows in **W** is exactly the same as the process of going down columns in **X**. The only difference is that the sign is changed. Symbolically this means

$$\mathbf{W} = -\mathbf{X}^{\mathrm{T}} \tag{AII-13}$$

where the superscript T on **X** indicates to transpose the rows and columns. Thus, $\mathbf{X}^{\mathrm{T}}$ is $(\mathrm{b} - \mathrm{n} + 1) \times (\mathrm{n} - 1)$ as is **W**.

To prove Tellegen's theorem we first write all currents in terms of link currents. Thus

$$\begin{bmatrix} I_{\mathrm{T}} \\ I_{\mathrm{L}} \end{bmatrix} = \begin{bmatrix} -\mathbf{X} \\ \mathbf{1}_{(\mathrm{b-n+1})} \end{bmatrix} [I_{\mathrm{L}}] \tag{AII-14}$$

Next we multiply each row of Equation (AII-14) by the corresponding voltage and add. This gives the net power. Symbolically we have

$$[V_{\mathrm{T}}{}^{\mathrm{T}} \quad V_{\mathrm{L}}{}^{\mathrm{T}}] \begin{bmatrix} I_{\mathrm{T}} \\ I_{\mathrm{L}} \end{bmatrix} = [V_{\mathrm{T}}{}^{\mathrm{T}} \quad V_{\mathrm{L}}{}^{\mathrm{T}}] \begin{bmatrix} -\mathbf{X} \\ \mathbf{1}_{(\mathrm{b-n+1})} \end{bmatrix} [I_{\mathrm{L}}] \tag{AII-15}$$

A well-known and easily verified property of matrices is, given two commensurate matrices **A** and **B**, then

$$[\mathbf{AB}]^{\mathrm{T}} = \mathbf{B}^{\mathrm{T}}\mathbf{A}^{\mathrm{T}} \tag{AII-16}$$

With this property, the first two matrices on the right side of Equation (AII-15) can be written

$$\begin{aligned} [V_{\mathrm{T}}{}^{\mathrm{T}} \quad V_{\mathrm{L}}{}^{\mathrm{T}}] \begin{bmatrix} -\mathbf{X} \\ \mathbf{1}_{\mathrm{b-n+1}} \end{bmatrix} &= \begin{bmatrix} [-\mathbf{X}^{\mathrm{T}} \quad \mathbf{1}_{(\mathrm{b-n+1})}] \begin{bmatrix} V_{\mathrm{T}} \\ V_{\mathrm{L}} \end{bmatrix} \end{bmatrix}^{\mathrm{T}} \\ &= \begin{bmatrix} [\mathbf{W} \quad \mathbf{1}_{(\mathrm{b-n+1})}] \begin{bmatrix} V_{\mathrm{T}} \\ V_{\mathrm{L}} \end{bmatrix} \end{bmatrix}^{\mathrm{T}} \end{aligned} \tag{AII-17}$$

where the last line uses Equation (AII-13). But the last line of Equation (AII-17) is a reordering of the Equations (AII-12). The result is still the zero vector. Thus Equation (AII-15) becomes

$$[V_{\mathrm{T}}{}^{\mathrm{T}} \quad V_{\mathrm{L}}{}^{\mathrm{T}}] \begin{bmatrix} I_{\mathrm{T}} \\ I_{\mathrm{L}} \end{bmatrix} = \mathbf{0}^{\mathrm{T}}_{(\mathrm{b-n+1})} I_{\mathrm{L}} = 0 \tag{AII-18}$$

Thus Tellegen's theorem is proved.

The proof of Tellegen's theorem is independent of the elements of the network. All that is required is that the currents satisfy KCL and the voltages satisfy KVL. This fact is useful in advanced network theory.

AII-6 The Reciprocity Theorem

The reciprocity theorem states that any network containing only resistances, inductances, capacitance, and mutual inductances has an admittance matrix that is symmetric.

The proof follows by considering the general form of the indefinite admittance matrix for such a network. This IAM is symmetric. The steps required to reduce the IAM to a port-admittance matrix do not disturb the symmetry. For an RLC network with every node considered, the entries in the off-diagonal elements are all minus the admittance connected between the two nodes in question. That is, y_{ij} is minus the admittance of the branch joining nodes i and j. y_{ji} is the same. The IAM for mutual inductance is symmetric. The IAM for the interconnection of two networks is obtaining by adding the two IAM's. Adding symmetric matrices results in a symmetric matrix. Thus, the IAM for an RLC plus mutual inductance network is symmetric. When a symmetric IAM is reduced by property 6, the reduced matrix is still symmetric. This follows immediately from Equation (6-43). If the original matrix is symmetric, so is $\mathbf{I}_*$. Furthermore, the elements of $\mathbf{R}_*$ and $\mathbf{C}_*$ are the same. The product of a column matrix and its transpose yields a symmetric matrix.

When a symmetric IAM is reduced by property 7, this result is symmetric. This is obvious from Equation (6-54).

When a symmetric n × n IAM is converted to an (n − 1) × (n − 1) port-admittance matrix by property 5, the result is symmetric because crossing out the last row and column does not destroy symmetry. Thus, we proved the reciprocity theorem for networks where all ports have a common ground. For a network having a port that does not have a terminal common with other ports, we must show that the additional reductions do not destroy symmetry. To show the final step, let us consider the 2-port of Figure AII-5. Let us suppose that we have

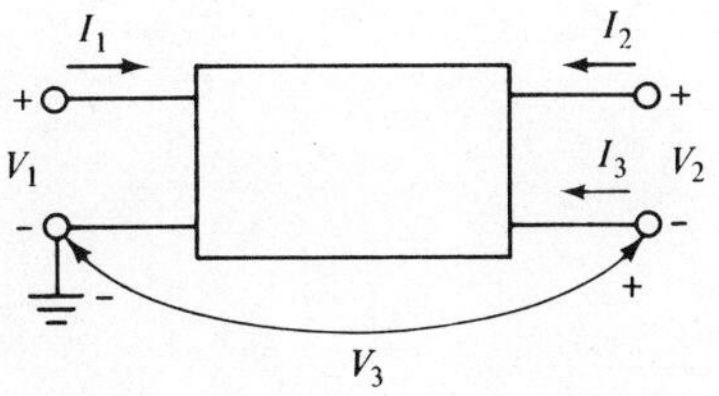

Figure AII-5 A reciprocal 2-port.

analyzed this network via the IAM and reduced the network matrix to a 3-port admittance matrix with common ground, as shown. We know we must show that, if the 3-port admittance matrix is symmetric, so is the 2-port matrix that relates V_1 and I_1 to V_2 and I_2. The 3-port equations are

$$\begin{aligned} I_1 &= {}_3y_{11}V_1 + {}_3y_{12}(V_2 + V_3) + {}_3y_{13}V_3 \\ I_2 &= {}_3y_{12}V_1 + {}_3y_{22}(V_2 + V_3) + {}_3y_{23}V_3 \\ I_3 &= {}_3y_{13}V_1 + {}_3y_{23}(V_2 + V_3) + {}_3y_{33}V_3 \end{aligned}$$

When the network is a 2-port, $I_3 = -I_2$. If we add the last two equations, we can solve for V_3 in terms of V_1 and V_2. Then we can eliminate V_3 from the first two equations. Adding gives

$$0 = ({}_3y_{12} + {}_3y_{13})V_1 + ({}_3y_{22} + {}_3y_{23})V_2 + ({}_3y_{22} + {}_3y_{33} + 2{}_3y_{23})V_3$$

Solving for V_3 and substituting gives

$$I_1 = \left[{}_3y_{11} + \frac{({}_3y_{12} + {}_3y_{13})^2}{{}_3y_{22} + {}_3y_{33} + 2{}_3y_{23}}\right]V_1 + \left[{}_3y_{12} + \frac{({}_3y_{12} + {}_3y_{13})({}_3y_{22} + {}_3y_{23})}{{}_3y_{22} + {}_3y_{33} + 2{}_3y_{23}}\right]V_2$$

$$I_2 = \left[{}_3y_{12} + \frac{({}_3y_{22} + {}_3y_{23})({}_3y_{12} + {}_3y_{13})}{{}_3y_{22} + {}_3y_{33} + 2{}_3y_{23}}\right]V_1 + \left[{}_3y_{22} + \frac{({}_3y_{22} + {}_3y_{33})^2}{{}_3y_{22} + {}_3y_{33} + 2{}_3y_{23}}\right]V_2$$

This matrix is symmetric.

For networks with more ports, more operations like that above are required. The results of each are the same. Thus, the theorem is proved.

APPENDIX

III Complex Number Formulas

In this appendix we review briefly the rectangular and polar forms for complex numbers and the corresponding formula for the products and sums of complex numbers.

COMPLEX NUMBERS	RECTANGULAR FORM	POLAR FORM
A	$a_1 + ja_2$	$\lvert A\rvert\underline{/\alpha}$ } where $\lvert A\rvert = (a_1^2 + a_2^2)^{1/2}$
A*	$a_1 - ja_2$	$\lvert A\rvert\underline{/-\alpha}$ } where $\alpha = \tan^{-1} a_2/a_1$
B	$b_1 + jb_2$	$\lvert B\rvert\underline{/\beta}$
A + B	$a_1 + b_1 + j(a_2 + b_2)$	No simple formula relative to $\lvert A\rvert$, $\lvert B\rvert$, α, β
A − B	$a_1 - b_1 + j(a_2 - b_2)$	No simple formula relative to $\lvert A\rvert$, $\lvert B\rvert$, α, β
AB	$a_1a_2 - b_1b_2 + j(a_1b_2 + b_1a_2)$	$\lvert A\rvert\,\lvert B\rvert\underline{/\alpha + \beta}$
$\frac{A}{B}$	$\frac{a_1a_2 + b_1b_2 + j(a_2b_1 - a_1b_2)}{b_1^2 + b_2^2}$	$\frac{\lvert A\rvert}{\lvert B\rvert}\underline{/\alpha - \beta}$
$e^{j\theta}$	$\cos\theta + j\sin\theta$	$1\underline{/\theta}$

APPENDIX

IV

A Computer Program for Reducing Indefinite Admittance Matrices

With the general availability of large-scale, high-speed digital computers, the engineer can perform detailed analyses of complex networks—networks for which he would have to be satisfied with approximate analyses in years past. Some say that the availability of user-oriented programs for network analysis makes a knowledge of network theory unnecessary. The intelligent use of network models in the design of signal-processing circuits requires a deeper understanding than the ability to analyze a network that someone else has designed. The course for which this text was written was developed to give such understanding. Many insights into the operation of signal processors can be obtained by working with very simple networks such as the ones given as problems at the ends of the chapters here.

For a more thorough knowledge of networks, the student should be exposed to a few situations that are beyond the realm of pencil, paper, and slide-rule analysis. The indefinite admittance matrix, introduced in

Chapter 6, is an excellent vehicle for starting the analysis of complicated linear networks. All that is needed is a computer program to execute the simple but tedious computations of IAM reduction. Such a program,[1] written by S. Bass and R. Bode, is used in the course. The program was placed on a disc file for use by students. The user's manual for an IAM program for the students follows. Slight modifications in control cards may be required for different computer systems.

Purpose

This program was designed to provide a method of examining the characteristics of a linear network by calculating the IAM of the network as a function of frequency, and then displaying any chosen element or elements graphically.

Its second function is to calculate the elements of the IAM at some fixed frequency while varying any chosen network element or source.

Capabilities and Limitations

This program can handle linear networks or circuits that can be linearized. In general, these networks will contain resistances, inductances, capacitances, and controlled sources. No independent sources can be present in the network.

The network may have any number of nodes, ranging from 2 to 20. This does not mean that the network can have only twenty elements, since any number of elements may be placed between two nodes. This will allow the user to employ very complicated models for the circuit elements, which will give good results at high and low frequencies.

The only limitation upon the frequency range is that $f = 0.0$, that is dc, cannot be chosen. It is, however, possible to use $f = 10^{-10}$ Hz. This frequency will provide a close approximation to dc, unless the circuit has an *unusual* physical length.

Preparation of Main Program

The IAM Program is actually a subroutine, which has been placed on the disc of the computer. The user must provide a main program and the control cards necessary to copy the IAM subroutine from disc into the central memory of the computer.

[1] The program is available to instructors adopting this text from Holt, Rinehart and Winston, Inc.

The user's deck must have the following form:

```
ACCOUNT NUMBER, NAME, CM60000, T30
RUN(S)
LIBCOPY (CSCBIN, LGO, IAM, NODCOR, LIMITS, SKIP,
                                        APLOT, INCREM)
LGO
7-8-9
      7
      Program MAIN (Input, Output, Input = Tape 5,
                                        Output = Tape 6)
       Complex     NETWOR (3, 3, 100)
           .
           .
           .
           .     Fortran Program
           .
           .
         Stop
         End
7-8-9
Data deck
6-7-8-9
```

Whenever the user wishes to obtain the IAM of a circuit, he places a card with Call IAM (NETWOR) on it in his main program.

The Array NETWOR

The input to the IAM subroutine is a listing of node numbers and element types and values for a network of up to 20 nodes, as detailed under "Preparation of the Data Deck" below. The output is the elements of an IAM. Because of storage limitations, the final output IAM can be at most 4×4. Since the last row and column of an IAM are readily computed from the other rows and columns (properties 1 and 2 in Chapter 6), only nine complex numbers representing the elements in the first three rows and columns of a 4×4 IAM are stored and then outputed. If a 3×3 IAM is called for, only the first two rows and columns will be outputed. Then only one complex number is outputed for a 2×2 IAM. The first two dimensions of the NETWOR array are $(N - 1)$, $(N - 1)$ for an $N \times N$ IAM.

The third dimension of the array is for variation of frequency or for variation of a network parameter. This dimension is limited to 100 points. Loading of the data for either frequency or parameter variation is given in the next section. As an added convenience in examining network performance as a function of parameter values, the subroutine has provisions for varying as many as five different elements in the network, one at a time. The array NETWOR is not large enough to store all of the variations in five elements. If it is desired to vary more than one element in a single network, the IAM program must be "called", for each element. (This could be done by placing the call statement in a "Do" loop.)

Example: Four elements to be varied.

```
Do  10 T = 1, 4
Call IAM (Networ)
  |
  | <------------calculations
  |
10 continue
```

This would be placed in the main program.

After the control cards have been prepared, the next step is to construct the data deck, which will be placed immediately after the 7–8–9 card.

Preparation of the Data Deck

The data deck consists of five sections, which are as follows:

1. Section one indicates the range of frequencies desired.
2. Section two indicates those elements of the IAM to be plotted, if any.
3. Section three indicates the locations and values of all passive elements and controlled sources, along with the necessary information to vary the elements if this option is desired.
4. Section four indicates any nodes that are to be shorted.
5. Section five indicated any nodes that are to be suppressed.

Example 1—In order to explain how these cards are to be filled out, a simple bandpass amplifier will be used as an example. (See Figure AIV-1.)

The first step is to replace the nonlinear elements, the transistors, with an equivalent linear model. The one indicated in Figure AIV-2 will be used. Notice that there is no longer any reason to choose the simplest model; instead choose one that is applicable over the frequency range desired. The circuit of Figure AIV-1 has the network model of Figure AIV-3.

The next step is to label all of the nodes with a number between 1 and 99. The numbers need not be consecutive.

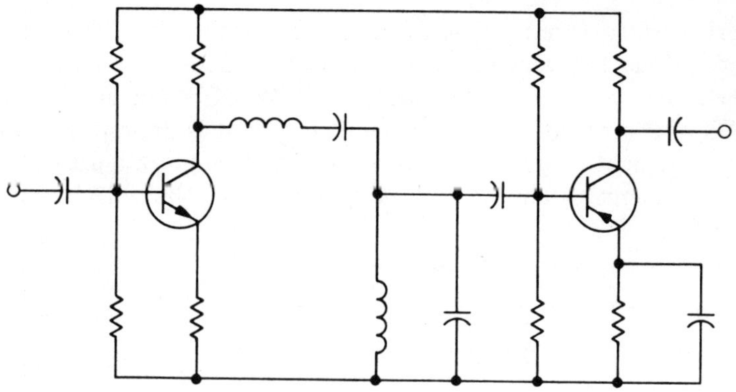

Figure AIV-1

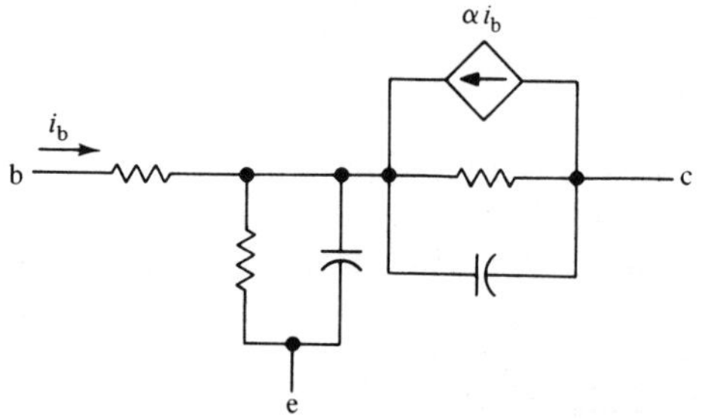

Figure AIV-2

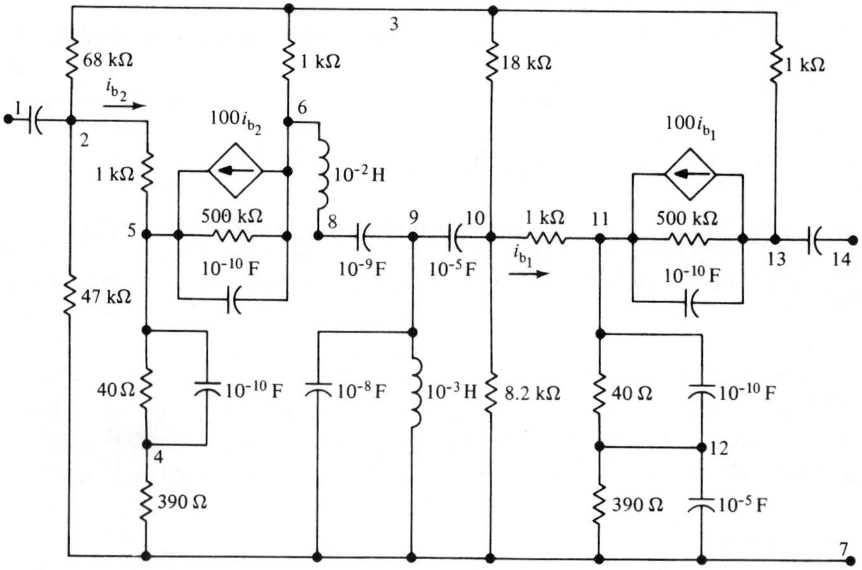

Figure AIV-3

Suppose that the 3×3 IAM as a function of frequency is desired for this network with nodes 1, 7, and 14 accessible.

Section I: Frequency Range

On a single card place the following information in the indicated columns with the indicated format.

COLUMN	
1	R—This will cause the real and imaginary parts of the elements of the IAM to be printed. M—This will cause the magnitude and phase of the elements to be printed. N—None of the elements of the IAM will be printed.
2–16	Indicate the initial frequency at which the IAM is to be calculated—E format.
17–31	Indicate the final frequency at which the IAM is to be calculated—E format.
32–46	Indicate the frequency increment—E format.

NOTE: No more than 100 frequencies can be calculated. All frequencies are in hertz. If the IAM is desired for only one frequency, set the maximum and minimum frequencies equal to each other.

For the network in the example, the resonant frequency is approximately 5×10^4 Hz. Suppose we wish to examine the network performance around this frequency. The most meaningful form of the output would be in terms of phase and magnitude, so an M should be placed in column one. Suppose a range of frequencies from 10 kHz to 0.1 MHz is chosen with an increment of 50 kHz. The first card in the data deck would be the following:

CARD	COLUMN 1	2 – 16	17 – 31	32 – 46	
17	M	1.E + 4	1.E + 5	5.E + 3	

The exponents in an E format are integers and as such must be right-justified. If the five in the above example had been written in column 30 instead of 31, the computer would have assumed that 10^{50} was meant instead of the desired 10^5.
(All of the following are equivalent to 100k: 1.E05, 1.E+05, 1.E+5, and 1. + 5; in all of these cases the exponent must be punched at the extreme right of the last column in the field.)

Section II: Plotting

The next card in the data deck indicates which elements of the IAM will be plotted. No more than five elements can be plotted at one time, and you must choose an increment on the card in the section above such that no more than 100 frequency points will be calculated. (In the example, 20 points will be calculated.) The card must be filled out as follows:

COLUMN	
1	R—Real and imaginary parts of the elements will be plotted. M—Magnitude and phase will be plotted.

The rest of the columns are to be filled out as follows: Y(nn, nn), Y(nn, nn), and so on for as many as five y's. For instance, if y(1, 1) is to be plotted, it must be entered as Y(01, 01). In the example, suppose we were interested in looking at the relationship between the input current and voltage and the output current and voltage [y(1, 1) and y(14, 14)]. Suppose we wish the plots to be the magnitude of each element versus frequency and phase of each element versus frequency. The data card would look like this:

CARD	COLUMN 1	COLUMN 9	COLUMN 17
18	M	Y(01, 01),	Y(14, 14)

If no elements are to be plotted, place a blank card after the card on which the frequency range was indicated. Note that you can obtain a plot of magnitude and phase by selecting M in column 1 above, while obtaining real and imaginary parts data by putting an R in column 1 of the preceding card.

Section III: Passive and Active Elements

There is no specific order required for source and element cards and, if desired, they can be mixed. These cards must contain the following information:

COLUMN	
1–2	PE—This indicates that the card contains data on a passive element (resistance, capacitance, or inductance).
3–4	Select any two nodes having one or more elements between them and place one of the node numbers here—I format. (Right-justified of course.)
5–6	Enter the other node here—I format.
30–39	If there is a capacitance between the two nodes given above, enter its value in farads here—E format.
40–49	If there is a resistance between the two nodes, enter its value in ohms here. If it is desired to create a short circuit between two nodes, enter as small a nonzero value as desired—E format, for example, 1.E-10.
50–59	If there is an inductance between the two nodes given above, enter its value in henries—E format.
60–69	If one of the elements on this card is to be varied (either a resistance, capacitance, or inductance), the final value that this element will have should be placed here—E format.
70–79	The desired increment or decrement should be placed—E format.
80	Place either an R, L, or C in this column to indicate whether a resistance, inductance, or capacitance is to be varied.

In the example, none of the elements is to be varied, so the columns from 60 on should be left blank. (The use of these columns will be illustrated later in another example.) One of these cards must be filled out for every pair of nodes in the circuit that has a passive element between them.

For the example, the following cards would be needed.

CARD	COLUMN 2	4	6		Capacitance 30–39	Resistance 40–49	Inductance 50–59
19	PE	1	2		1. − 5		
20	PE	2	3			6.8 + 4	
21	PE	2	7			4.7 + 4	
22	PE	2	5			1.0 + 3	
23	PE	5	4		1. − 10	4. + 1	
24	PE	4	7			3.9 + 2	
25	PE	5	6		1.0 − 10	5.0 + 5	
26	PE	3	6			1.0 + 3	
27	PE	6	8				1. + 2
28	PE	8	9		1.0 − 9		
29	PE	9	7		1.0 − 8		1. − 3
30	PE	9	10		1. − 5		
31	PE	3	10			1.8 + 4	
32	PE	10	7			8.2 + 3	
33	PE	10	11			1. + 3	
34	PE	11	12		1. − 10	4. + 1	
35	PE	12	7		1. − 5	3.9 + 2	
36	PE	11	13		1. + 10	5.0 + 5	
37	PE	3	13			1. + 3	
38	PE	13	14		1. − 5		

Each one of the above lines is on a separate card. At this point the program has all the information that it needs concerning the passive elements; however, there remain two controlled sources, which must be entered into the network. This will require two more cards, which must be filled out as follows:

COLUMN	
1–2	VV—Use these letters if this card will describe a voltage-controlled voltage source.
1–2	VC—This indicates a voltage-controlled current source will be described.
	CC—This indicates a current-controlled current source will be described.
	CV—This indicates a current-controlled voltage source will be described.
3–4	Place one of the nodes that the source is connected to here, according to the following convention: If a voltage source is being described, place the positive terminal of the source here, and if a current source is being described, enter the number of the node into which sources current is assumed to flow.
5–6	Enter the other node number of the source here.
7–8	Enter the number of the node from which controlling current is assumed to flow or the positive node of a controlling voltage.
9–10	Here indicate the node into which controlling current flows or the negative terminal of the controlling voltage.
11–20	Indicate the parameter of the controlled source; that is, μ, g_m, α, r_m—E format.
21–29	If this card describes a CV or a VV, it is necessary to place a small series resistor in the circuit. It can be made as small as desired, but not zero—E format, for example, 1.E-20.
30–39	If the element through which current or across which voltage is being sensed for the source is a capacitance, enter its value here.
40–49	If the element through which current or across which voltage is being sensed for the source is a resistance, enter its value here.
50–59	If the element through which current or across which voltage is being sensed for the source is an inductance, enter its value here.
60–69	If the parameter of the controlled source is to be varied, then the final value that it would have is to be placed here.
70–79	The increment or decrement should be placed here.
80	If the controlled source is either of type CC or CV and the element through which the current is being sensed is also to be varied, indicate this here by placing a Y in the column. Otherwise leave it blank.

In this example none of the sources is to be varied, so columns 60–80 will be left blank.

Since the example has two controlled sources, there must be two data cards to represent them. For the example, they will be filled out as follows:

CARD		COLUMN 12	4	6	8	10	11 20	2 20	30 39	40 49
39	1	CC	5	6	2	5	1. + 2			1. + 3
40	2	CC	11	13	10	11	1. + 2			1. + 3

50 – 59

At this point the computer has a complete description of the network and it could calculate the IAM. Since the network has 14 nodes, the IAM would be a 14 × 14 matrix. In most cases this is not of very much use or interest.

Before the cards for section four are filled out, it is necessary to place a card that has a 9 punched in column one immediately after Section III.

Section IV: Shorting Nodes

If it is desired to short two nodes together, a card must be filled out for each pair as follows:

COLUMN	
1–2	SH—These letters indicate that two nodes are to be shorted.
3–4	The first node to be shorted—I format.
5–6	The second node to be shorted—I format.

Once these two nodes have been shorted, the second node will be eliminated and must not be mentioned again.

In the example, nodes 3 and 7 must be shorted, since they are connected by a direct-current source. The card would look like this:

CARD	COLUMN		
41	S H	7	3

At the end of this section, it is necessary to place another card with a 9 punched in column one. Even if Section IV is not used, two nine-cards must be placed in the deck before Section V begins.

Section V: Suppressing Nodes

In this section the IAM will be reduced in size to something more workable. Usually the most useful size will be a three by three, so that by striking out one row and one column, it can be reduced to a description of a 2-port device. This card must contain the following information.

COLUMN	
1–2	SU—These letters indicate a node is to be suppressed.
3–4	The number of the node to be suppressed—I format.

At the end of this section, another card with a 9 punched in column one must be added.

For the example, all of the nodes except 1, 7, and 14 will be suppressed. This will require eleven different cards.

CARD	COLUMN 12	COLUMN 14
42	SU	2
43	SU	3
44	SU	4
45	SU	5
46	SU	6
47	SU	8
48	SU	9
49	SU	10
50	SU	11
51	SU	12
52	SU	13

The output from the program for this example will be twenty 3×3 IAM's at the frequencies specified, with each of the elements of the IAM's written in terms of magnitude and phase. Following this will be two plots, one of magnitude versus frequency of the elements $\mathbf{y}(1, 1)$ and $\mathbf{y}(14, 14)$, and another of phase versus frequency for the same two elements.

If it is desired to run another network, make up a data deck for that network in the same manner and place it immediately after this one. Any number of circuits can be analyzed in this manner.

Suppose that, instead of varying the frequency, one of the inductances was to be varied, and the alpha of one of the sources was to be varied. (A maximum of five elements can be varied in any one network.) Each element will be varied separately. Suppose the inductance between nodes 6 and 8 was to be varied from 10 mH to 20 mH, and the alpha of the second transistor was to be varied from 100 to 120.

In the previous example, the following changes would have to be made:

1. Elements can be varied at one fixed frequency only, so for the example let us choose $f = 5 \times 10^5$ Hz. On the first card, the minimum and maximum frequencies would have to be changed so that they were both equal to the above frequency.
2. Two of the cards in Section III will have more information added to them. Card number 27 would have added to it the final value of the inductance, the desired increment, and the information that an inductance was being varied. The new card would look like this:

COLUMN

12	4	6	30 39	40 49	50 59	60 69	70 79	80
PE	6	8			1. − 2	2. − 2	1. − 2	L

Similar changes must be made to card number 40. Assuming an increment of one, the new card would look like this:

COLUMN

12	4	6	8	10 11	20	21 29	30 39	40 49	50 59	60 69	70 79	80
CC	11	13	10	11	1. + 2			1. + 3		1.2 + 2	1. + 0	

Example 2—A FET differential amplifier. (See Figure AIV-4.) The first step is to replace the FET's with an appropriate linear model and label the node numbers. (See Figure AIV-5.)

The FET's were both given exactly the same set of parameters; of course, with real FET's this would be impossible. To examine the effect of unequal characteristics, all that has to be done is to vary the

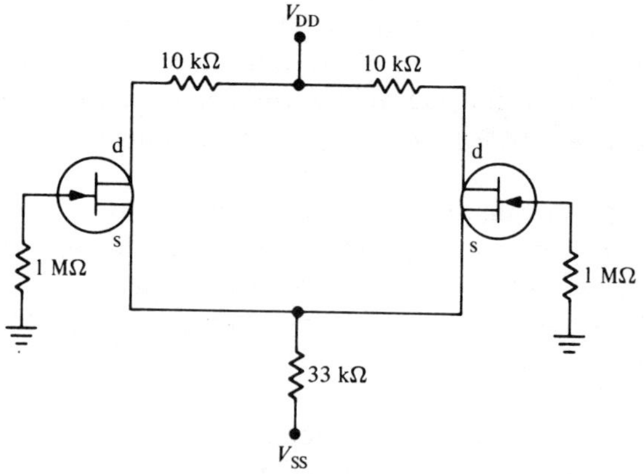

Figure AIV-4

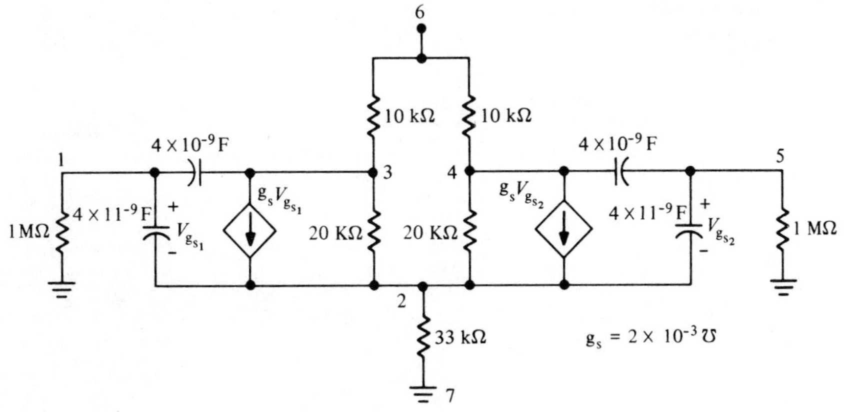

Figure AIV-5

circuit elements. In this example g_{fs} and the drain-to-source resistance will be varied at a fixed frequency; for example, 50 kHz.**y**(1, 1) and **y**(5, 5) will be plotted. The deck required to do this would have to contain the following cards:

COLUMN

123456789 16	20	31	39	49	69	79	80
M 5. + 4		5. + 4					
MY (01,01,Y(05,05))							
PE 1 7				1. + 6			
PE 1 2			4. − 9				
PE 1 3			4. − 9				
PE 3 2				2. + 4			
PE 3 6				1. + 4			
PE 4 6				1. + 4			
PE 4 2				2. + 4	1. + 4	−1. + 3	R
PE 4 5			4. − 9				
PE 5 2			4. − 9				
PE 5 7				1. + 6			
PE 2 7				3.3 + 4			
VC 2 3 1 2	2. − 3						
VC 2 4 5 2	2. − 3				1. − 2	1. − 3	
9							
SH 7 6							
9							
SU 2							
SU 3							
SU 4							
9							

Index

Index

D

E

F

G

H

I

U

V

Y

Z